Developments in Petroleum Science
Volume 59

Basin Evolution and Petroleum Prospectivity of the Continental Margins of India

Developments in Petroleum Science
Volume 59

Basin Evolution and Petroleum Prospectivity of the Continental Margins of India

R. Bastia
Oil and Gas Consultant
Mulund (West)
Mumbai, India

M. Radhakrishna
Department of Earth Sciences
Indian Institute of Technology Bombay
Mumbai, India

ELSEVIER
AMSTERDAM • BOSTON • HEIDELBERG • LONDON • NEW YORK • OXFORD
PARIS • SAN DIEGO • SAN FRANCISCO • SINGAPORE • SYDNEY • TOKYO

Elsevier
Radarweg 29, PO Box 211, 1000 AE Amsterdam, The Netherlands
The Boulevard, Langford Lane, Kidlington, Oxford OX5 1GB, UK

First edition 2012

British Library Cataloguing in Publication Data
A catalogue record for this book is available from the British Library

Library of Congress Cataloging-in-Publication Data
A catalog record for this book is available from the Library of Congress

ISBN: 978-0-444-53604-4
ISSN: 0376-7361

For information on all Elsevier publications
visit our web site at store.elsevier.com

Printed and bound in Great Britain

12 13 14 15 16 10 9 8 7 6 5 4 3 2 1

Contents

Preface

Passive continental margins are the sites of a huge repository of sediments and are known to contain a significant fraction of the hydrocarbon reserves of the world. However, for a detailed evaluation of hydrocarbon resources at these margins, we require knowledge of the regional geological history, factors that contributed to the basin formation in both space and time, and the conditions favorable for the existence of all essential elements of a petroleum system in the basin. The passive continental margin of India, which evolved as a result of rifting and breakup of India from Eastern Antarctica, Madagascar, and the Seychelles in different geological times, is characterized by vast sedimentary cover having a thick pile of sediments both on land as well as in offshore areas. The breakup process and the subsequent geological events at the margins have led to wide variations in the geological setup for different segments of the margin and gave rise to complexities in their geological evolution. The exploration efforts initiated by the National Oil Company ONGC since 1960 in both onshore and the offshore areas of the Indian continental margins resulted in a huge success with the discovery of the giant Mumbai High oil fields in the early 1970s. This gave a major impetus to hydrocarbon exploration activity in the shelfal areas all along the margins. In spite of these intense exploration efforts, most of the information about the hydrocarbon potential of the Indian sedimentary basins (with special reference to coastal and offshore) was known to the geosciences community through only a few publications up until 1990. However, the pioneering efforts by S. K. Biswas and his associates at the KDMIPE in 1991 resulted in the publication of two very significant edited volumes on the petroliferous basins of India in 1993 and these together provided the baseline information for all major sedimentary basins in India in terms of regional tectonic framework, classification, and hydrocarbon occurrence. For those students and teachers of geology and geophysics interested in the petroleum geology of Indian basins, these two volumes remain a major source of information. Although the Association of Petroleum Geologists (AAPG), an extended arm of the AAPG, the Society of Petroleum Geophysicists, and the biyearly PETROTECH volumes have contributed immensely to the generation of the knowledge base and provided suitable platform for showcasing the research work carried out by petroleum geoscientists in India and abroad, no major synthesis of the continental margins connecting the shelf and the deepwater petroleum systems has been taken up in a comprehensive manner. Fortunately, these two above organizations keenly pursued and stressed the need for improving the industry–academia interaction in the country. Further, the stupendous initiatives by the Government of India through the Directorate General of Hydrocarbons (DGH) opened up a vast stretch of the Indian sedimentary basins with a series of Exploration Lease rounds under a New Exploration Licensing Policy (NELP). This is when private players like Reliance, BG, Cairn Energy, BHP Billiton, ENI, etc., from India and abroad started joining the efforts of National Oil Companies to accelerate the exploration, which resulted in huge data generation especially in the deepwater frontiers and led to several new discoveries. While these developments were taking place on the petroleum exploration front in India, following the International Indian Ocean Expedition in 1964, detailed geological and geophysical investigations were undertaken in the Indian Ocean by several Indian as well as international organizations, which brought out a wealth of information on the paleogeography and Gondwana reconstruction, regional tectonics, crustal structure, sedimentation, volcanism, and plume–lithosphere interactions in the oceanic areas surrounding the Indian subcontinent.

Despite all the above developments, there is no basic reference book that deals with all the aspects of basin evolution (from rifting to reservoirs) with special reference to the Indian continental margins. This gap in the knowledge basically stimulated us to undertake the task of writing the present book. In this book, we made a sincere effort to provide a detailed synthesis of the regional tectonic framework and evolution of both eastern and western continental margins and integrate the regional datasets for the development of petroleum systems in the individual basins. This book is intended for geologists, geophysicists, petrophysicists, and engineers interested in building regional geologic models for the individual basins and, at the same time, was prepared keeping in mind the requirement of students and researchers interested in knowing more about the Indian offshore basins by providing several case studies as well as relevant examples.

The book is divided into eight chapters: while Chapters 1 and 8 deal with introduction and summary of the subject matter of the book, respectively, Chapters 2–7 contain the main content of the book. In Chapter 2, we discuss several aspects involved in the reconstruction of Gondwanaland for proper understanding of the breakup history. The tectonic framework and crustal structure along the eastern and western continental margins are discussed in Chapter 3. In Chapter 4, we highlight the importance of the uplift history of the Himalayas, the regional sedimentation pattern, and the development of the Indus and Bengal submarine fans along the Indian continental margins. Chapters 5 and 6 describe the structure, tectonics, stratigraphy, and the development of petroleum systems in various basins along the eastern and western continental margins, respectively. The advancements in imaging and drilling technologies have enhanced our understanding for better characterization of the subsurface as well as help to calibrate the proposed conceptual models. Some of these aspects with reference to the Indian offshore have been included in Chapter 7. While every effort was made to provide as much new information as possible, the material provided in Chapters 5 and 6 is mainly constrained and guided by the availability of information. In Chapter 6, we boldly and intentionally include a discussion on the probable Mesozoic prospectivity of the Konkan–Kerala offshore. In the absence of deep drilled well information at this stage, this is purely conjectural and the information compiled by the authors therefore could be subjected to scrutiny. There is always scope for improvement; therefore, as and when more data, particularly the drilled well information, are available, some of the aspects discussed in the book may undergo further refinement.

The writing of this book would have been impossible without the active support of many individuals. A significant part of this book contains results from the research work carried out by us over the past two decades and we thankfully acknowledge the contributions of all our past research students as well as coauthors in several of our earlier publications. During the initial stages, Octavian Cateneanu provided many valuable suggestions for the framework of the book and Collin Reeves shared his ideas on the tectonic reconstruction. Roberto Fainstein and Rajesh Kalra have provided important input on the challenges and imaging issues discussed in Chapters 1 and 7. Roberto gave us permission to use his figure on comparative stratigraphy of the east and west coasts; Brian Storey gave permission to use his figure on Gondwana reconstruction; Inna Polyakova gave permission to use the compiled map of giant oil fields map of the world; J. N. Sahu gave permission to compile the required writeup on the stratigraphy as well as several figures for the Cauvery and K-G basins discussed in Chapter 5 from two of his earlier publications; S. K. Biswas provided useful comments on the draft version of Chapter 1; Anoop Tripathi, MD Indian Petroleum Publishers, Dehradun, provided permission to use some of the published figures from volumes I&II of Petroliferous Basins of India; Arun Singhal, MD Technology Publications, provided permission to use the figures as well as some portions of the text from the books written by J. N. Sahu on Cauvery and K-G basins; the Geological Society of India gave permission to use figures published in *JGSI*; and the Director, KDMIPE, ONGC, permitted us to use published figures from the Bulletin ONGC. The Society of Petroleum Geophysicists (SPG), GEO India, Indian Academy of Sciences, Indian Association of Sedimentologists, Society of Exploration Geophysicists (SEG) also gave us permission to use some of the figures from their publications. The EAGE Publications permitted us to use part of the text on Mesozoic stratigraphy from the paper by Roberts et al. (2008) in First Break. The authors Hamblin and Christiansen allowed us to use the figure published in their book published by Prentice-Hall Publishers. We are thankful to all the above mentioned people for their kind help. The regional geoscientific information compiled by the DGH through several NELP rounds for the individual basins has been very useful. We used their basin classification scheme, stratigraphic information, and the basin-wise statistical data wherever required in the book and the same is acknowledged. We sincerely acknowledge the scientific discussions and suggestions by the domain experts of our industry colleagues from the Reliance, ONGC, DGH, and other oil companies working in Indian sedimentary basins for presenting various chapters of the book most appropriately. We thank the faculty members in the Department of Earth Sciences, IIT Bombay, particularly D. Chandrasekharam, T. K. Biswal, Kanchan Pande, George Mathew, S. Banerjee, G. Mohan, and E. Chandrasekhar, who have helped in various ways during the preparation of the chapter. The excellent help received from the research scholars D. Twinkle and G. Srinivasa Rao throughout the progress of the book is specially acknowledged. We also thank other research scholars B. Someswara Rao, Himagshu Kaushik, and Manish for their help in the preparation of a few figures and tables. We specially thank Derek, Susan, and Sharmila from Elsevier for their help at different times and the trouble they have taken to ensure timely completion of the book. The painstaking efforts of all those geoscientists in unraveling the subsurface geological information through their contributions in data acquisition, processing, and interpretation became the backbone material for this book and made this gigantic task a reality. We sincerely thank and dedicate this book to all of them. Last but not least, we are deeply indebted to our families for their sacrifices and unstinting support during the entire process of completing the book.

Rabi Bastia
M. Radhakrishna

Acknowledgments

Several publishers listed below are gratefully acknowledged for giving their permission to use the figures / tables published in the journals and books for which they hold the copyright. The original authors of these figures and tables are cited in the figure captions. While every effort has been made to obtain permission to use the copyrighted material, we sincerely apologize for any errors or omissions.

Publisher	Figure/Table Number
American Association of Petroleum Geologists	3.11, 4.5A, 4.8, 4.11, 7.1
American Geophysical Union.	2.2, 2.5, 2.6, 2.10B, 3.13, 3.25e, 3.25f, 3.25 g, 3.25h, 3.28, 3.35
DGH, New Delhi	1.8, 1.9, 1.11, 1.12, 5.62
Elsevier Publishers	1.4, 1.5, 1.6, 2.11, 2.15, 2.17, 3.5, 3.7, 3.8A, 3.8 C, 3.9b, 3.14, 3.17, 3.18, 3.19, 3.20, 3.22, 3.24, 3.25b, 3.25c, 3.25d, 3.26, 3.29, 3.30, 3.31, 3.32, 3.33, 3.34, 3.36, 3.37, 3.38, 4.1, 4.5B, 4.6, 4.7, 4.13, 4.15, 4.16, 4.17, 4.18, 4.19, 4.20, 4.21, 4.22, 4.24, 4.25, 4.26, 4.27, 5.15, 5.23, 5.31, 5.32, 5.34, 5.40, 5.42, 5.43B, 5.44B, 5.45B, 5.46, 6.4A, 6.6B, 6.7A, Table: 6.1, 7.24, 7.25, 8.1
Geological Society London,	2.1, 3.3, 3.13
Geological Society of America	3.6, 3.27, 4.2, 4.3
Himachal Times Group	5.29
Indian Petroleum publishers, Dehra dun	1.8, 3.8D, 5.5, 5.8, 5.11, 5.22, 5.24, 5.27, 5.28, 5.29, 5.30, 5.63, 5.64, 6.2, 6.9B, 6.11, 6.13, 6.14, 6.23, Tables: 5.1, 5.2
John Wiley & Sons, Inc.	2.13, 3.9c
Geological Society of India	3.9a, 3.9c, 3.25a, 5.1, 5.3, 5.25, 6.1, 6.3, 6.4B, 6.7B, 6.10, 6.12, 6.20, Tables: 6.2, 6.3, 6.4
Nature Publishing Group, Macmillan Publishers Limited.	2.10A, 2.14
KDMIPE, ONGC	6.15, 6.16, 6.17, 6.18

Publisher	Figure/Table Number
Petromin, AP Energy Business Publications PTE Ltd	7.7, 7.9, 7.10
Prentice Hall	1.2, 1.3
Society of Petroleum Engineers	7.8, 7.14, 7.15, 7.16, 7.17
Society of Exploration Geophysicist (SEG)	5.18, 5.48, 5.49, 5.50, 5.52, 5.53, 5.54, 5.55, 5.56, 5.57, 5.58, 5.59, 5.66, 7.5
Springer Publishers	1.1, 2.16, 3.8B, 3.10, 3.12, 3.14, 3.23, 4.9A, 4.9B, 4.10, 4.12, 7.19, 7.20, 7.21, 7.22, 7.23, 7.26, 7.27, 7.28
Technology Publishers, Dehradun	5.6, 5.7, 5.9a, 5.9b, 5.10, 5.12, 5.19, 5.21, 5.35, 5.36, 5.37, 5.51, 7.6, Tables: 5.3–5.8
Schlumberger	5.33, 5.65, 6.9A
Society of Petroleum Geophysicists (SPG)	5.39, 5.41, 5.43A, 5.44A, 5.45A, 5.47, 5.60, 6.24, 6.25, 7.2, 7.3, 7.4
GEO India	5.61, 6.26
Indian Academy of Sciences	6.5, 6.6A
Indian Association of Sedimentologists	7.12
Wiley-Blackwell Publishers	6.21, 6.22
SPWLA	7.11, 7.13

Chapter 1

Introduction

1.1. GENERAL INTRODUCTION

Passive continental margins constitute nearly half of the present-day continental margins with an aggregate length of approximately 105,000 km (Levell et al., 2010), and major part of the offshore petroleum-producing basins are situated at these margins. Passive continental margins are partly underlain by continental crust and represent the transition zone between more complex continental regions to simpler tectonic setting of ocean basins. Most conspicuous feature of the passive continental margins formed during the Gondwana disintegration is the presence of thick Mesozoic and Tertiary sedimentary deposits that were collected contemporaneously with steady-state subsidence.

Petroleum exploration of continental margins began in earnest during the 1950s and 1960s when exploratory wells were drilled in the Gulf of Mexico, North Sea, West Africa, and East Brazil. Major discoveries soon followed in all of these regions, encouraging further exploration of the outer shelf and slope areas. Detailed geological and geophysical investigations of the deepwater realm of continental margins carried out during the last two decades further provided significant new information with regard to the crustal architecture, nature of ocean continent transition, and structure as well as stratigraphy of sediments at the margin. Increased understanding of the complex geometry of sedimentary layers at the continental margins subsequently led to the discovery of large petroleum fields located in the offshore Brazil, West Africa, Trinidad, Gulf of Mexico, East Canada, North Sea, Northwest Australia, India, and the Southeast Asia that includes Northwest Java, East Kalimantan, Brunei, Malaysia, and Thailand (Khain and Polyakova, 2008). While the total reserves estimated from the 20 deepwater basins in the year 2001 is around 9 Btpe (Shirley, 2002), several new discoveries since then have substantially increased this estimate. Khain and Polyakova (2008) presented a compilation of the basins lying in the shelf-slope areas of the world ocean margins in which large and giant hydrocarbon accumulations have been encountered and shown in (Fig. 1.1). In view of the promising results obtained from several offshore areas of the margins particularly along the passive continental margins, and the higher potential for future discoveries, a complete

Developments in Petroleum Science, Vol. 59. http://dx.doi.org/10.1016/B978-0-444-53604-4.00001-6

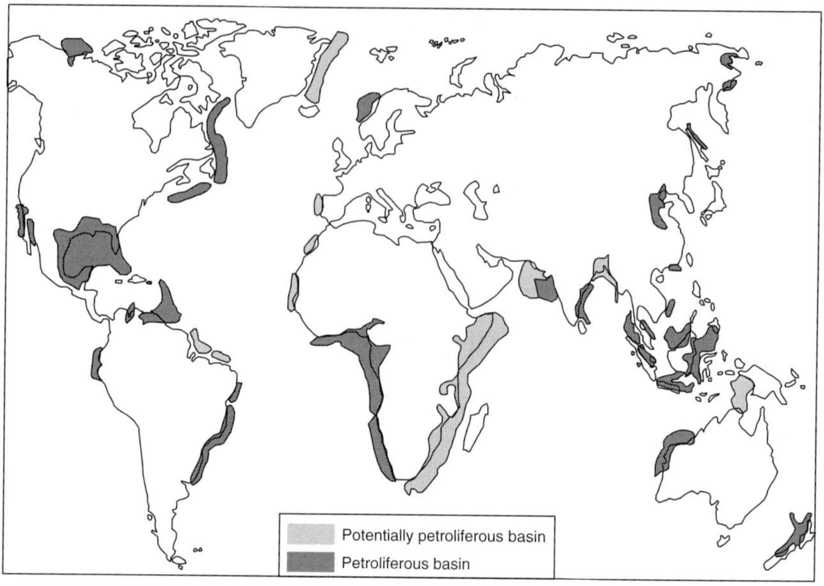

FIGURE 1.1 The shelf-continental slope basins of the World Ocean in which large and giant hydrocarbon accumulations have been encountered *(adapted from Khain and Polyakova, 2008).*

understanding of individual margin segments with regard to its tectonic setup and geological history is essential.

It is imperative that detailed studies be carried out in order to assess further petroleum potential of passive continental margins. Tectonic framework of continental margins and associated basins and their evolution need to be worked out and modeled using geological and geophysical data. This would help better understanding of the genetic link between tectonic, magmatic, geomorphic, and sedimentary processes involved in the basin evolution. Since the sedimentation history is predominantly influenced by major tectonic trends and fluid flow circulation, the sea-level fluctuations, climate, rate of sedimentation, flexure (Watts, 2001) are critical elements to be detailed.

1.2. BREAKUP OF CONTINENTS AND DEVELOPMENT OF OCEAN BASINS AND MARGINS

1.2.1 Framework of Plate Tectonics

The surface of the Earth is covered by approximately 70% of the oceans and 30% of continents. The oceans are relatively much younger when compared to the continents. During the geological history, different cycles of continental breakup events led to the creation of new oceans and closure of older ocean basins. It is well known from the plate tectonic theory that outer shell of the

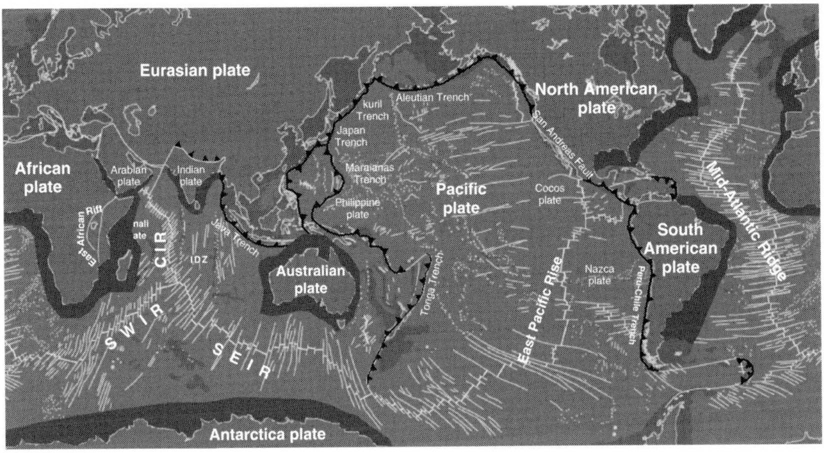

FIGURE 1.2 Map showing the geometry of tectonic plates (modified after Hamblin and Christiansen, 2003). The global distribution of passive continental margins is marked in brownish red *(source: Wikipedia), and active continental margins characterized by subduction zones are marked with thrust symbol. CIR, Central Indian Ridge; SEIR, Southeast Indian Ridge; SWIR, Southwest Indian Ridge; IDZ, Intense Deformation Zone. (For interpretation of the references to color in this figure legend, the reader is referred to the online version of this chapter.)*

Earth is divided into a number of thin, rigid plates of about 100 km thickness (Kearey et al., 2009) and are called as the Lithosphere (Fig. 1.2). The lithospheric plates, comprising of continents, oceans, or both, are in relative motion with respect to one another. While the present-day oceanic lithosphere provides information about various geological events during the past 200 Ma, the continental lithosphere is highly heterogeneous and complex due to different superposed geological events for the past 4 billion years (Fowler, 1990). The lithospheric plates are internally rigid and act as extremely efficient stress guides, which mean that deformation mostly takes place at the plate margins (Bott, 1982). Hence, the boundaries of these lithospheric plates are characterized globally by the occurrence of earthquakes, volcanic eruptions, and mountain building activity. According to Isacks et al. (1968), three different plate boundaries (Fig. 1.3) can be envisaged: (i) the accreting plate boundaries, where plates diverge from each other due to upwelling of magma thereby generating new oceanic lithosphere, (ii) convergent plate boundaries, where the oceanic lithosphere of one of the plates is thrust below the other, and (iii) conservative plate boundaries are marked by strike slip motions in which plates undergo neither destruction nor construction.

1.2.2 Plate-Driving Forces

Due to continuous tectonic activity taking place at the plate boundaries caused by large horizontal movements of the lithosphere associated with creation of

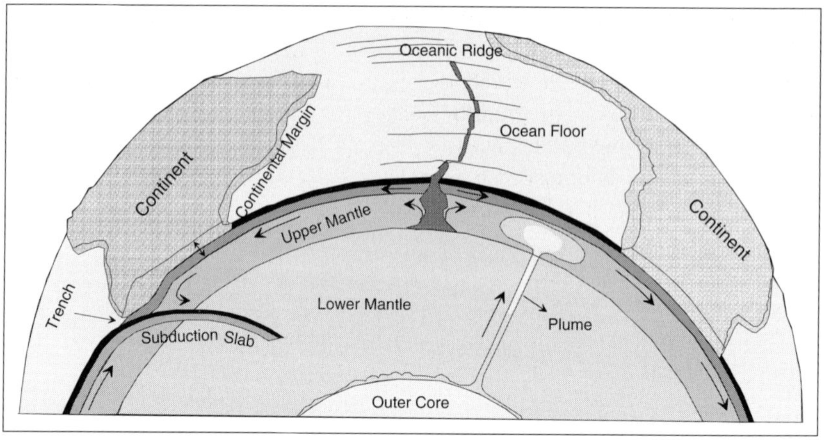

FIGURE 1.3 Schematic model depicting the plate tectonic system operative in the Earth's interior *(adapted with permission from Hamblin and Christiansen, 2003 © Hamblin and Christiansen).*

new seafloor at the ridges as well as continental drift processes, the surface of the Earth undergoes significant changes over geological time periods. It is therefore relevant as well as imperative to understand the causative mechanism underlying such global tectonic activity, most importantly, the mechanism of continental rift initiation and development, and the forces driving the lithospheric plates. As far as the plate-driving force is concerned, of the several mechanisms that were used to explain the motion of lithospheric plates, the convection current hypothesis could be a plausible mechanism, though this classical hypothesis is difficult to reconcile with the observed pattern of present and past plate motions (Bott, 1982). Alternatively, quantitative studies demonstrated that a system of forces acting on plate boundaries at ridges and subduction zones (Forsyth and Uyeda, 1975; Richardson et al., 1979) can satisfactorily account for present-day plate motions. Taking into consideration of the above studies, Bott (1982) in his arguments concluded that plates, if considered to be driven by edge forces, are thermodynamically acceptable and contain adequate power to drive the plates.

1.2.3 Mechanism of Continental Rifting

Understanding the mechanism as well as processes that led to continental rifting is far more important in the sense that they represent the initial stage of continental breakup and the start of the Wilson cycle of plate tectonics, that is, possible sites of incipient lithospheric separation and formation of new ocean basin (Olsen and Morgan, 1995). According to Burke (1977), continental rifts are "zones along which the entire lithosphere has ruptured in extension." Olsen and Morgan (1995), in their classic review, slightly modified this definition and stated that continental rift is an "elongate tectonic depression associated with which the entire lithosphere has been modified in extension." However, in situations, such as, small extensional features or grabens, highly extended terranes, and rifted passive continental margins (Fig. 1.4), the entire lithosphere is not modified or involved in extension and differs from the above restricted definition of continental rifts in their

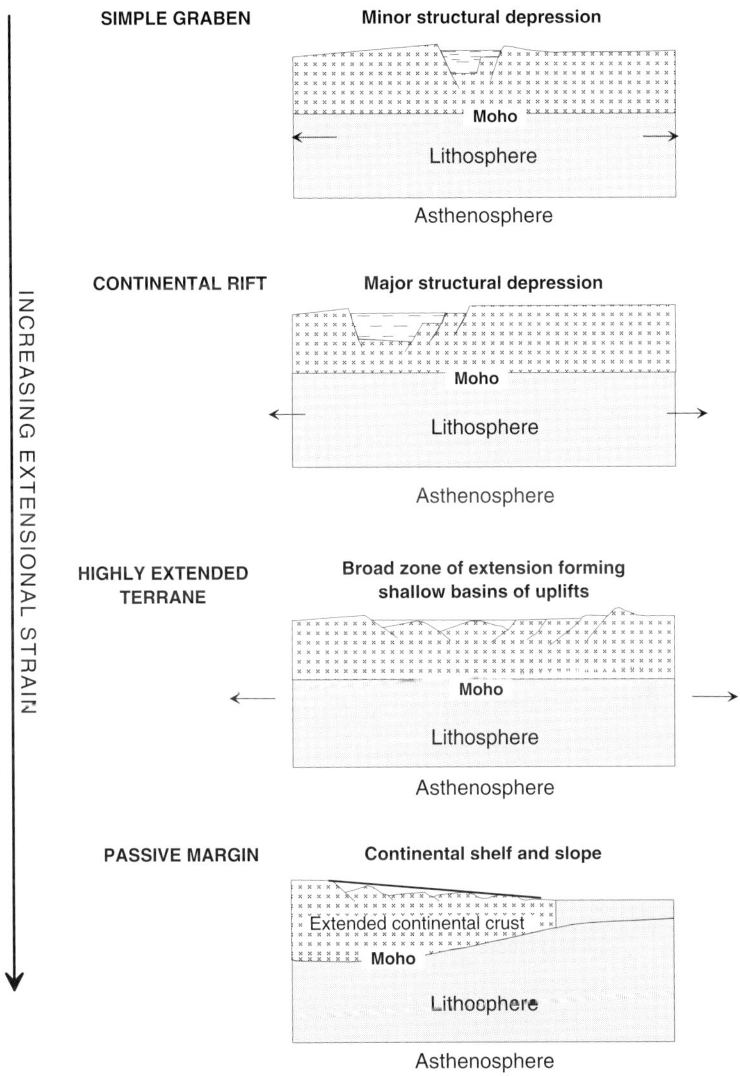

FIGURE 1.4 Various styles associated with the extension of continental lithosphere *(adapted from Olsen and Morgan, 1995).*

magnitude or mode of extension (Olsen and Morgan, 1995). So far as the mechanism of initiation of continental rifting is concerned, they fall under two groups (Sengor and Burke, 1978): (i) passive rifting and (ii) active rifting (Fig. 1.5). In the case of passive rifting, the lithospheric tension of distant origin causes failure of the continental lithosphere and thinning takes place in response to extension. In the active rifting, the anomalous upper mantle develops first by some form of convective upwelling, where causative stresses are associated with lateral thermal density variations in the lithosphere and the underlying asthenosphere. Apart from the above causal mechanisms, several other secondary processes may modify the lithosphere in response to lithospheric thinning (Fig. 1.6A) and give rise to final lithospheric structure. These are (i) secondary convection, (ii) mantle lithosphere delamination,

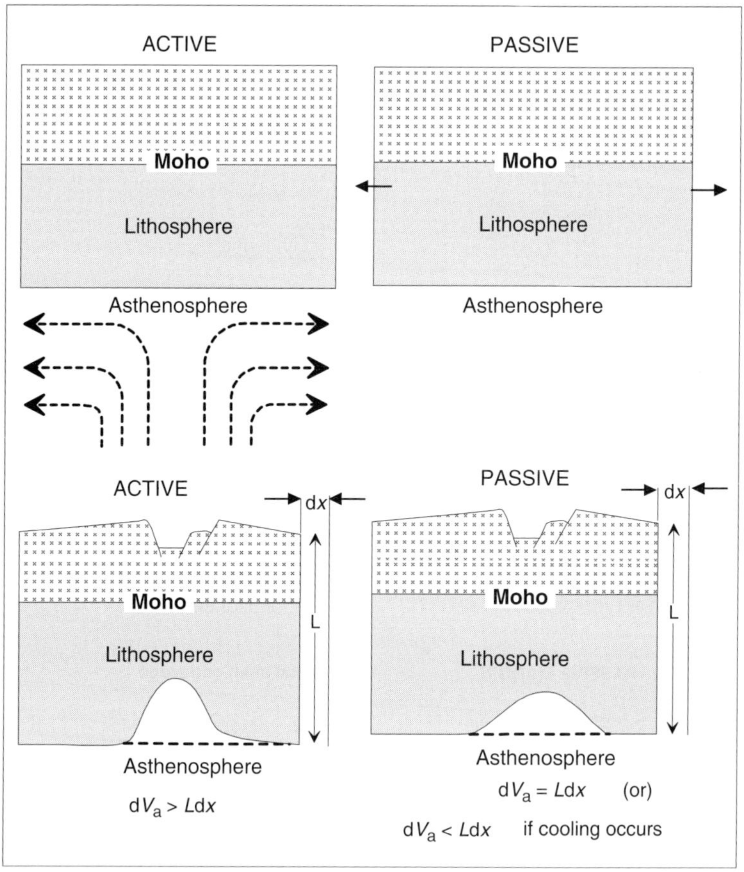

FIGURE 1.5 Simple models depicting causative stresses in the generation of active and passive continental rifting (top panel) and implications for lithospheric thinning (bottom panel) *(adapted from Olsen and Morgan, 1995).*

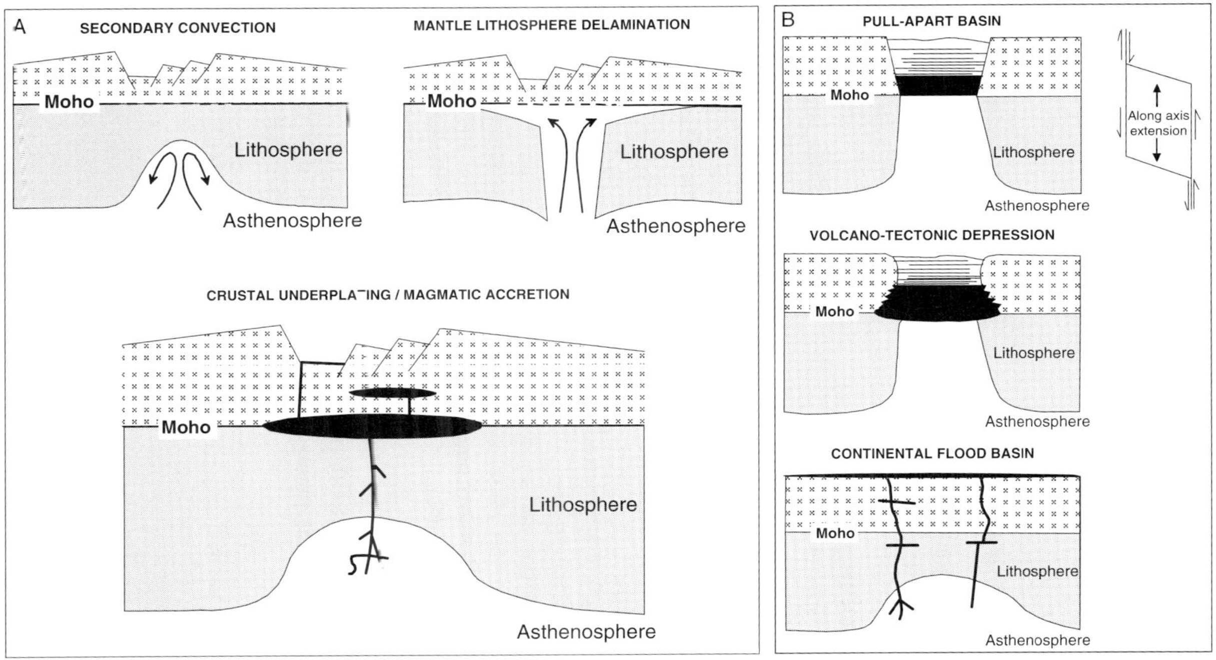

FIGURE 1.6 (A) Various secondary processes that modify the lithosphere in response to thinning. (B) Various tectonic and magmatic structures possessing rift-like characteristics *(adapted from Olsen and Morgan, 1995)*.

and (iii) magmatic accretion/crustal underplating. In addition, other major thermotectonic features, such as pull-apart basins, volcano-tectonic depression, and continental flood basalt regions (Fig. 1.6B), also display rift-like characteristics which sometimes may/may not have genetic association with continental rifts (Olsen and Morgan, 1995). An elaborate description of these processes is out of scope of present objectives of this book, and readers are encouraged to refer Olsen (1995).

1.2.4 Evolution of World Ocean Basins and Margins

The present-day topography of the ocean floor covered below the world's oceans provides a detailed and excellent Mesozoic–Cenozoic geological history of the Earth with regard to the early breakup history of continents, present as well as past plate motions, plume–lithosphere interactions in various times, formation of ocean basins, etc. Until 1970s, such information was available only in certain parts of the oceans due to limited availability of marine geophysical measurements. Subsequently, advances in technology, exploration, and vast developments in space-borne geophysical techniques made possible the high-resolution imaging of the ocean floor in terms of morphology, structure, and tectonics. The most striking features of the ocean floor (Fig. 1.7) include mid-oceanic ridges (MORs), abyssal plains, seamounts, fracture zones, and trenches. A detailed morphotectonic investigation of the ocean basins provides valuable information on the history of creation of seafloor and motions of continents in time. From Fig. 1.2, it can be seen that the MOR system is present in the middle of the world oceans and extends rather continuously from the Arctic and Atlantic Ocean into the Indian Ocean and across the South Pacific Ocean (Hamblin and Christiansen, 2003). Along its length, at many areas, the MOR system is offset by large fracture zones trending perpendicular to the ridge. The vast, deep, and relatively smooth oceanic areas lying on either side of the MOR are known as abyssal plains which generally lie at depths of more than

FIGURE 1.7 Various morphological features of the ocean floor. (For color version of this figure, the reader is referred to the online version of this chapter.)

4000 m up to the continental margins. The continental margin which is the zone of transition between the continent and the ocean basin, where thick continental crust juxtaposes the thin oceanic crust, is also the depocenter of major sedimentation (Gallagher and Brown, 1997). As the continental margins are the main focus area of the present book, we present here a detailed account of their types, characteristics, and mode of their evolution with a view to understanding their petroleum systems.

1.3. CONTINENTAL MARGINS—TYPES, CHARACTERISTICS, AND PROCESSES

The continental margins cover approximately 12% of the global ocean floor and shelf-continental-slope petroleum basins are located at the continental margins of all terrestrial oceans (Khain and Polyakova, 2008). Therefore, a study of the continental margins is as important as the study of continental shield regions. Depending on the tectonic setting, the continental margins are classified into different groups as discussed below.

1.3.1 Types of Continental Margins

Continental margins can be fundamentally divided into three basic types worldwide. These are

- **i.** The passive continental margins are formed under divergent plate boundary setting. These are aseismic (less seismic) and often referred as the Atlantic-type margins having a distinct transition zone between the continental and oceanic crustal regions.
- **ii.** The active continental margins referred as the Pacific-type margins are the zones of seismically active convergent plate boundaries. These are characterized by subduction zones formed under variety of settings such as oceanic–oceanic, oceanic–continental, etc. and are mostly observed along the periphery of the Pacific Ocean.
- **iii.** Transform continental margins (sheared margins) are formed due to lateral plate movements or shearing motions during the continental breakup process in the divergent margin setting. Identifying this category of margins is also in a way important, as they represent an original offset in the line of continental breakup and can normally be traced laterally into an oceanic fracture zone (Bird, 2001). Further, transform margins are also noticed along the Pacific type continental margins.

1.3.2 Crustal Characteristics at the Passive Continental Margins

Passive continental margins are formed by extension and thinning of continental lithosphere (Le Pichon and Sibuet, 1981) and evolve in a wide range of tectonic settings. Once the continental breakup/splitting takes place, the resulting ocean basin might later be affected by tectonic, magmatic, and sedimentation processes of varying magnitude (Watts, 2001). The passive continental margins mostly exhibit a bi-polar gravity anomaly, that is, a positive free-air gravity anomaly at the shelf and negative free-air gravity along the slope (Karner and Watts, 1982), and the amplitude as well as the wavelength of this characteristic anomaly signature is representative of the type of margin and sedimentation pattern (Watts, 1988). The evolutionary process of the passive margins could be divided into several distinctly different stages, but falling broadly under two main phases: an initial "Rift phase" which is prior to the breakup of the continent and a later "drift phase or post rift phase" which occurs subsequent to the onset of seafloor spreading.

The passive margins could be classified into different categories depending on the influence of various tectonic and magmatic processes during their evolution. These are rifted or sheared margins, volcanic or nonvolcanic margins, sediment-rich or sediment-starved margins, and wide or narrow margins (Davison, 1997; White, 1992). The volcanic margins are formed due to active rifting (Bott and Kusznir, 1979; Dewey and Burke, 1974; Spohn and Schubert, 1982), whereas the nonvolcanic margins are formed due to passive rifting (Khain, 1992; McKenzie, 1978; McKenzie and Bickle, 1989). As the magmatic input to the crust during continental rifting is extraordinarily variable, both in time and space, clear distinction between active and passive rifting is not useful and hence is only conditionally justified (Mutter, 1993). In the case of volcanic margin, large melt volumes are emplaced during the continental breakup within a short time period, whereas along the nonvolcanic margin, only a small amount of melt is produced as the continental breakup progresses (Fowler et al., 1989; Horsfield et al., 1993; Morgan et al., 1989; White, 1992). Volcanic passive margins are characterized by igneous intrusions and intense dyke activity and give rise to lava flows on top of the subsided continental crust. This large-scale volcanic activity which obscures the rift-related sequences can be noticed on the seismic sections as the Seaward Dipping Reflector (SDR) sequences (White, 1992). A number of other factors such as segmentation, sediment supply, climate, and sea-level changes affect the development of passive margin in different regions.

1.3.3 Sedimentation

Most conspicuous feature of passive continental margins is that they contain thick sequences of Mesozoic and Tertiary sediments having an average thickness of about 7–8 km sometimes up to 18–20 km. These sediments are divided into two groups: earlier sediment layers deposited during rifting stage and later sequences formed during drifting stage after the breakup. The rift-stage sediments are deposited in fault-bounded troughs along the line of initial continental breakup but now present at the deepest parts of the basin due to burial effect of overlying sediments (Watts, 2001). The drift-stage sediments would be deposited as the basin sags progressively downward without any faulting.

1.3.3.1 *Role of Flexure at the Passive Margins*

The main factors responsible for the subsidence of passive margin basins are crustal thinning, lithospheric cooling, and sediment loading (Watts and Torné, 1992). The geometry of the basin at a passive margin is also controlled partly by the sedimentation and the associated flexural deformation. The accumulation of sediments at the margin represents loads over longer geological periods and contributes significantly to the gravitational edge effect anomaly (Watts and Fairhead, 1999). It is important to understand how the lithosphere responds to the geological loads, as isostasy is operative over timescales ranging from a few thousands to few million years. It is well known that the lithosphere responds to long-term geological loads not locally, as Airy and Pratt models would predict, but regionally by flexure (Watts, 2001). Combined backstripping and gravity modeling of passive margins by several investigators indicated loading of sediments on the stretched continental crust, having wide ranging flexural rigidities, such as low flexural rigidity in the case of East Coast of USA (Fowler and McKenzie, 1989; Watts, 1988), and having high flexural rigidity offshore New Zealand (Holt and Stern, 1991), the Gulf of Lion (Kooi et al., 1992), and offshore East Coast of India (Radhakrishna et al., 2000). Further, Watts and Marr (1995) pointed out some cases where weak segments of margins abutted the strong ones.

1.3.4 Passive Continental Margins and Habitat of Petroleum

Passive continental margins represent a vast and long-lived exploration domain with huge depository of sediments stored during their evolution. The mean preservation lifetime of the Phanerozoic passive margins including the modern margins is about 135 Ma (Bradley, 2008). Mann et al. (2001) estimated that the postrift passive margin sequences host nearly 35% of all giant field discoveries which represent more than half of discovered conventional hydrocarbons. The remaining giant oil fields are found to be occurring in basins associated with collision zones and subduction zones. Passive margins are petroleum storehouses because these are associated with favorable conditions for accumulation and maturation of organic matter. The classical models of passive margin evolution (Bott, 1980; Falvey, 1974) involve mainly three stages of development, such as (i) a rifting phase characterized by complex horst–graben tectonics, (ii) onset of drifting that involves separation of continental lithosphere, often can be dated by the breakup unconformity, and (iii) the main drifting phase involving massive subsidence and accumulation of thick prograding sedimentary sequences separated by unconformities. However, in an excellent review of the passive margins, Levell et al. (2010)

highlighted the diversity in geodynamic process-based models that have emerged from the high-resolution seismic and deep drilling data sets with regard to the structural and tectonic styles and their possible implications on the petroleum occurrence. According to them, some insights to interpret the deeper seismic data include (i) common occurrence of multiple rifting events with an irregular relationship to the line of final breakup, (ii) possibility of polyphase faulting in any given rift event, (iii) better understanding of magmatic addition in the lower crust, (iv) occurrence of wide sag-type basins related to differential lower to mid-crustal stretching instead of thermal sag, and (v) rifting with intracrustal/supracrustal magmatic addition or amagmatic rifting. Improved geodynamic models of passive margins would be possible with better understanding of physical mechanisms of lithospheric mantle and crustal extension. Further, Levell et al. (2010) compiled a wide variety of passive margin petroleum plays that are presently being pursued by the industry which include major and minor deltaic depocenters, slope by-pass systems, carbonate platforms and their margins, deepwater fold belts, syn-rift and pre-rift sections, deepwater far-outboard areas, and frontier Arctic margins.

1.4. THE INDIAN CONTINENTAL MARGIN

The Peninsular India has nearly 7500-km-long coastline and approximately 2.0 million sq km of Exclusive Economic Zone (EEZ) that includes both Lakshadweep and Andaman group of islands. The coastal areas are characterized by vast sedimentary cover with thick pile of sediments both onland and offshore. The continental margin extends on either side of the Indian Peninsular shield and is referred to as the Eastern Continental Margin of India (ECMI) toward the Bay of Bengal and the Western Continental Margin of India (WCMI) toward the Arabian Sea. Further, the region of Andaman–Nicobar islands is located along an active plate boundary where the Indian plate is subducting below the southeast Asian plate. It is the only example of Pacific type of continental margin in India, and is another promising sedimentary basin for hydrocarbon exploration. Despite the vast extent of the sedimentary cover and very promising deepwater area, India still remains as one of the least explored regions of the world.

1.4.1 Tectonic Framework

The ECMI has a strike length of about 2000 km and evolved during the breakup of India from East Antarctica in early Cretaceous (Powell et al., 1988). The conjugate nature of the East Antarctica with ECMI is supported by magnetic lineations and structural lineaments traced into both continents (Johnson et al., 1976; Lawver et al., 1991). The ECMI along the coast is characterized by the presence of major onshore–offshore sedimentary basins, such as the Cauvery–Palar, Krishna–Godavari, Mahanadi, and Bengal basins (Sastri et al., 1973). On the other hand, the WCMI has evolved through rifting and breakup of India–Madagascar during mid-Cretaceous and India–Seychelles during late Cretaceous (Storey et al., 1995). The second breakup event is also associated with large magmatic episode and gave rise to the Deccan Continental Flood Basalt province as a result of emplacement by the Reunion mantle plume. The WCMI comprises five sedimentary basins, such as the Kutch, Saurashtra, Mumbai, Konkan, and Kerala basins.

1.4.2 Sedimentary Basins of India and Their Classification

As plate tectonic setup provides varying habitats of petroleum, in any exploration program, it is very important to understand the tectonic setting of the individual basins under which they have developed. The classification of Indian sedimentary basins in terms of their tectonic regime became known after the publication of the Tectonic Map of India by Eremenko and Negi (1968). Subsequently, some preliminary attempts were made to classify the basins, according to their time of formation (Sastri et al., 1974) or based on their tectonic setting (Raju, 1979). A comprehensive classification of Indian sedimentary basins was published for the first time by Biswas et al. (1993) using the Dickinson (1976) scheme of basin classification. According to this scheme, four factors such as the geometry and size, nature of stratigraphic fill, structural style, and thermal history are important facets of basin evolution and their interrelationship under various tectonic settings provides valuable insights in the categorization of basins. Based on the above scheme with due modifications to Indian subcontinent, Biswas et al. (1993) classified 38 Indian sedimentary basins and have grouped them under three main settings with further subclassification: (i) intracratonic, (ii) rifted, and (iii) orogenic settings (Fig. 1.8A). Out of these basins, in almost 26 selected basins, exploration activities for hydrocarbon were initiated in mid-1950.

Presently, these basins were categorized by the Directorate General of Hydrocarbons (DGH) based on their status and success ratio of three decades of exploration and have been grouped under four major categories (source: DGH) (Fig. 1.8B):

Category I: Proven commercial productivity (Cambay, Mumbai Offshore, Assam Shelf, KG, Cauvery, Tripura–Mizoram, and Rajasthan basins)
Category II: Identified prospectivity (Kutch, Andaman–Nicobar, and Mahanadi basins)
Category III: Prospective basins (Himalayan foreland, Ganga, Vindhyan, Konkan–Kerala–Lakshadweep, and Bengal basins)
Category IV: Potentially prospective (Karewa, Spiti Zanskar, Satpura, South Rewa, Damodar, Narmada, Deccan Syneclise, Cuddappah, Bhima-Kalagdi, Pranhita-Godavari, Bastar, and Chhattisgarh basins).

Further, the deepwater area beyond 400 m water depth up to the EEZ forms a separate category and can be considered highly prospective. From the Indian

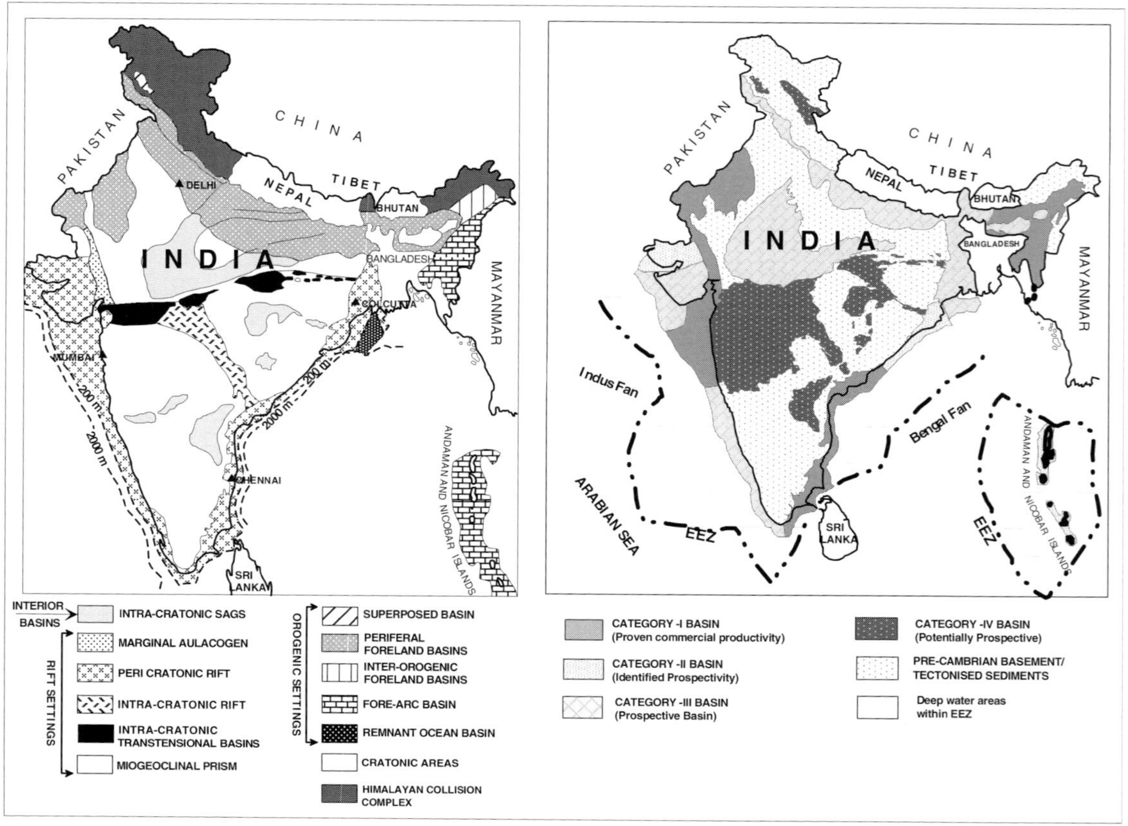

FIGURE 1.8 Maps showing the classification of Indian sedimentary basins (A) based on the tectonic setting (left panel) (redrawn from Biswas et al., 1993) and (B) based on the potential for occurrence of hydrocarbons (right panel) *(Source: DGH, http://www.dghindia.org/Images/Sedimentary_Map.jpg).*

context, the Tertiary sequences account for the major portion of prognosticated hydrocarbon resources; however, considering the success in several other deepwater basins of the world, the hydrocarbon resource potential of the country is expected to increase substantially as the exploration for hydrocarbons is aggressively taken up in the frontier areas, particularly deepwater, and pre-Tertiary sequences like Mesozoics. Based on the ongoing studies and analysis, the potential hydrocarbon resources in various offshore and onland basins of India (source: DGH) are shown in Fig. 1.9.

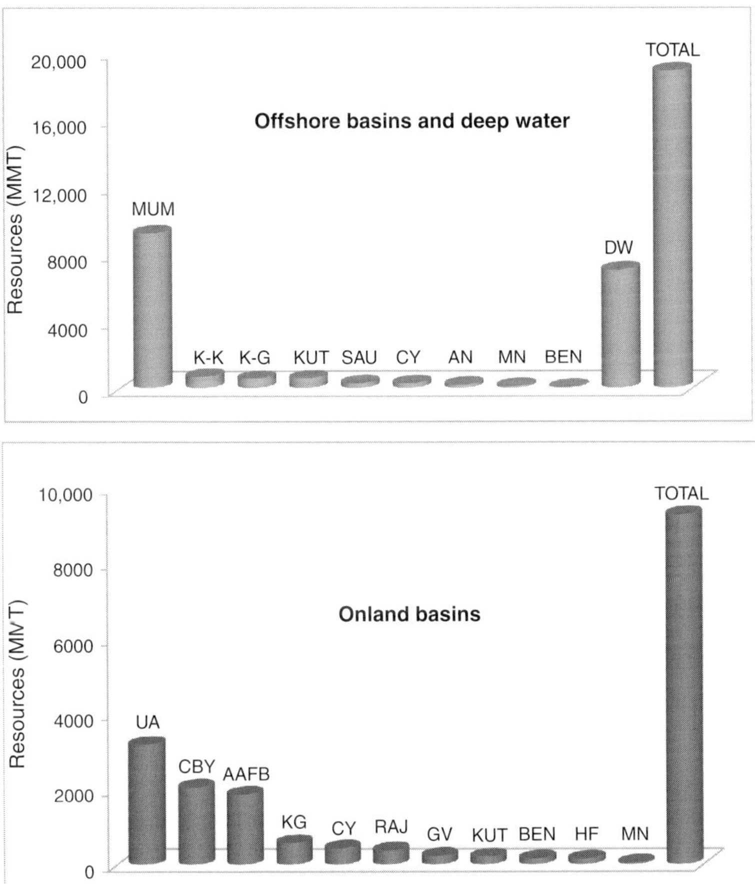

FIGURE 1.9 Offshore and onshore basins—total hydrocarbon resources *(data source from DGH, 2010). MUM, Mumbai; KK, Kerala–Konkan; KG, Krishna–Godavari; KUT, Kutch; SAU, Saurashtra; CY, Cauvery; AN, Andaman; MN, Mahanadi; BEN, Bengal; DW, deepwater; UA, upper Assam; CBY, Cambay; AAFB, Assam Arakan fold belt; RAJ, Rajasthan; GV, Ganga valley; HF, Himalayan foreland.*

1.5. GROWING ENERGY REQUIREMENT

Due to continuous rise of population and the income levels, the energy needs of the country are increasing at a much faster rate. It is known that the nation is heavily dependent on the fossil fuel reserves; therefore, we need to ensure that these resources are exploited to their maximum in terms of recovery in a sustainable way. Figure 1.10A demonstrates the demand distribution of various energy sources that we expect to see in the coming two decades. If focus is only on the hydrocarbon resources, the demand for these resources will be almost doubled. The gradual increase between the production and supply–demand gap of oil in the country is reflected in the U.S. Energy Information Administration (Fig. 1.10B) survey. To obtain a clear picture on various initiatives taken

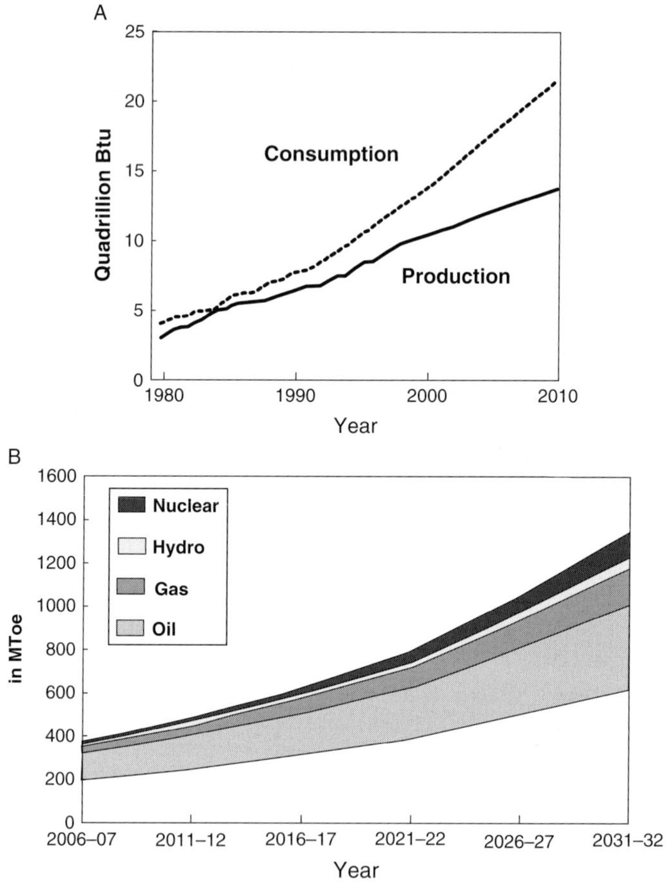

FIGURE 1.10 (A) India's oil supply and demand gap (source: International Energy Agency—IEA, www.iea.org and www.eia.gov). (B) Projected demand growth curve of fossil fuel and nuclear energy for India up to 2030 *(source: Planning Commission Report, August 2006; Reference US Energy Information Administration Survey, Washington, DC at www.eia.gov/forecasts).*

by the Indian Government and the petroleum industry from time to time, in order to address the above demand versus supply gap, a brief overview of the petroleum exploration scenario in the country from its early times until now is given below taking excerpts from ONGC (2006) and the DGH:

1.6. PETROLEUM EXPLORATION SCENARIO IN INDIA

1.6.1 Historical Development

Though petroleum exploration was initiated in India as early as 1867 and subsequently led to the discovery of oil in Digboi, Assam in the year 1889, the exploration activity in the country remained insignificant during the preindependence period till 1947. However, during the postindependence era, the Government of India realized the importance of oil and natural gas resources and gave impetus to the exploration activity by setting up the Oil and Natural Gas Commission (ONGC) in the year 1956. As a result, a number of onland oil and gas-bearing structures were discovered by ONGC in Gujarat and Assam in 1960s and a major offshore oil discovery in Mumbai High in 1974. Subsequently, due to several new discoveries during 1970s and mid-1980s in both onland and shallow water areas such as the Bassein, Tapti, KG–Cauvery basins, Cachar (Assam), Nagaland, and Tripura led India to achieve the self-sufficiency level of 70% in petroleum products. However, due to gradual decline in the oil production by the end of 1980s, the Government of India recognized the need to expand the domestic exploration and production activity which subsequently gave rise to the initiation of New Exploration Licensing Policy (NELP) in 1997.

1.6.2 Developments After the NELP Initiative

Fueled by this paradigm shift in the Indian oil and gas exploration regime, the exploration activity took a big leap forward because of the competitive environment and participation of several major private oil companies. The impact of NELP bidding rounds since its initiation in 1997 is quite significant and is evident from the comparison of explored versus nonexplored areas from 1995–1996 to 2009–2010 status (DGH, 2011) in terms of total percentage area that has undergone exploration in recent times (Fig. 1.11). At the same time, this tremendous growth in blockwise exploration activity had a negative impact with regard to the integration of regional basinwide data sets.

1.6.2.1 Increased Deepwater Exploration Activity

Obviously, a number of new challenges came in picture and gave rise to new frontier areas during various NELP bidding rounds, such as the deepwater exploration as well as the sub-basalt exploration. All these challenges forced the oil and gas industry to look forward for solutions that could address the challenges associated with new areas. As can be seen from Fig. 1.12, the

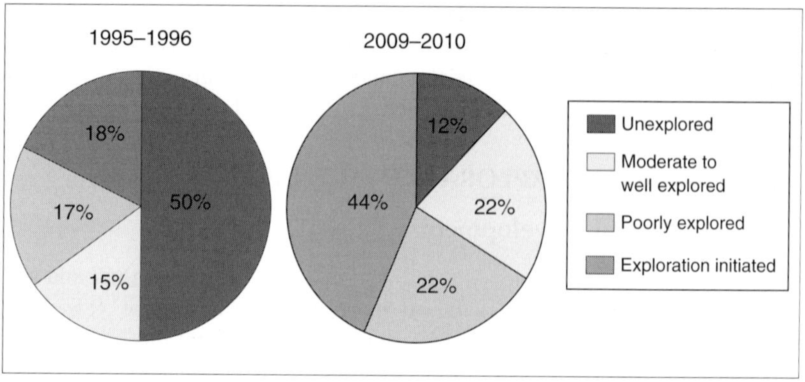

FIGURE 1.11 Exploration status of acreages for hydrocarbons in India *(source: DGH, 2011).*

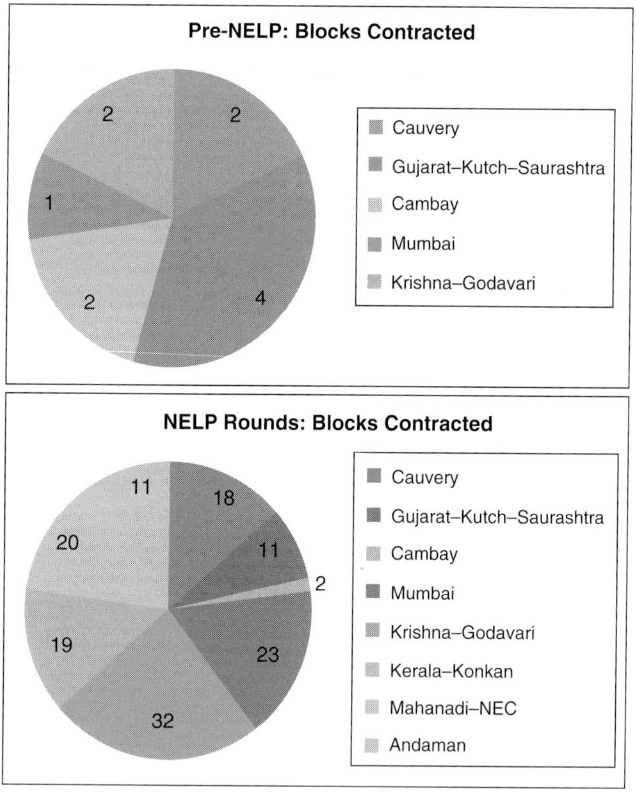

FIGURE 1.12 Pre-NELP and the NELP I-VIII awards scenario *(data source: DGH, 2011). (For color version of this figure, the reader is referred to the online version of this chapter.)*

deepwater exploration activity had significantly increased after the introduction of NELP rounds. While during pre-NELP era, only 11 shallow water offshore blocks were contracted, during the NELP rounds, a total of 136 offshore blocks that include 80 deepwater were contracted (DGH, 2011). A majority of these are in the KG, Mahanadi–NEC, and frontier areas such as Kerala–Konkan and the Andaman basins (Fig. 1.8B).

1.7. CHALLENGES AND TECHNOLOGICAL ADVANCES

Historically, hydrocarbon exploration activity in India for a long time mainly comprised of regional gravity and magnetic data integrated with the analysis of close-spaced 2D seismic reflection and refraction data. Though modern seismic data acquisition was initiated by ONGC in the pre-NELP era, during the post-NELP rounds, tremendous progress has taken place with the consistent utilization of high-resolution 3D seismic data which made a major impact on the exploration. The deepwater exploration activity certainly brought out many new challenges to the petroleum industry in India. Further, successful development of the deepwater blocks clearly required the application of advanced exploration technology, and the industry quickly made use of the expertise which was already available with several international oil companies.

In any deepwater exploration, application of various advanced seismic imaging techniques is very critical for reducing the geological risk to tolerable levels. Advanced seismic data acquisition techniques such as the single-sensor recording, steerable streamer capabilities, larger streamer configurations, and geo-streamer acquisition are found to be extremely useful to understand the reservoir characteristics. For a better prediction of the presence of deepwater reservoirs, techniques like the direct hydrocarbon indicators (DHIs), amplitude-versus-offset (AVO), pre-stack time migration (PSTM), and pre-stack depth migration (PSDM) are useful.

The role of modern technology is an integral part of any exploration and production (E&P) organization and covers almost all areas of its operation. It is very well established that the application of PSDM became very important in the Indian offshore, and such superior technologies will have major impact on exploration and development of offshore projects in India in the near future. On the other hand, developments in the basin modeling as well as interpretation techniques have resulted in the generation of multiple and robust geological models for better understanding of the subsurface basin evolution through geological time.

We present here a summary of various key challenges that need solutions as well as application of new technologies in the major offshore basins of India. Along the western offshore, in the Kutch and Saurashtra basins, the carbonate buildup and pinch-outs certainly need to be properly mapped, whereas in the Mumbai offshore basin, the fractured carbonates in tilted fault blocks in

Mumbai High region need to be addressed through application of advanced techniques. Further, the shallow water multiples and high absorption make the seismic imaging difficult and pose a major geophysical challenge in this basin. The Konkan–Kerala offshore is a major frontier area of exploration that requires definite improvement in resolution due to potential depth imaging issues as the basin is inferred to be a major sub-basalt province and contains pre-trap Mesozoic sediments below the widespread volcanic layer above which the Tertiary sediments are overlying.

In the eastern offshore, the Cauvery basin is credited with minor oil discoveries both offshore and onshore; however, the area closer to Konkan–Kerala is affected by sub-basalt with complex geologic structure and poses a major imaging issue. The Krishna–Godavari (KG) offshore is a major gas-producing basin in India, and thin sand beds in this basin require the application of precise seismic resolution techniques. In fact, high-resolution seismic data in the deepwater basins significantly helped in resolving thin beds for improving the facies evaluation of amalgamated channel reservoirs and of distributaries in fans. The Mahanadi–NEC basin is another important basin having gas discoveries. The complex geologic structure, issues related to velocity modeling, and the delineation of deepwater channels in this basin pose a major geophysical challenge. In the case of Andaman offshore which is lying along the trench-arc system, the fore-arc as well as deepwater back-arc basin areas still needs to be understood in greater detail in terms of the basin architecture and total petroleum system.

1.8. SCOPE OF THE PRESENT BOOK

A detailed evaluation of petroleum systems developed under the divergent margin setting would require knowledge of regional geological history and various factors that contributed to the basin formation in both space and time. The main objective of this book is to provide a concise review of the status of hydrocarbon exploration and its potential of occurrence along the continental margins of India. For a general understanding, a brief overview on the tectonic framework of passive continental margins and mechanisms of their formation is discussed in Chapter 1. A general description of different categories of Indian sedimentary basins as well as the scenario of petroleum exploration in India is also provided in this chapter. The knowledge on rifting and breakup history of the eastern Gondwanaland is critical to decipher the geological history and development of Indian continental margins as it is very well known that India had played an important role in the breakup of Gondwana and subsequent evolution of the Indian Ocean. Some of these aspects have been explained in Chapter 2, in which a detailed synthesis of various constraints available for the reconstruction of eastern Gondwanaland continents, breakup history, plume-related volcanism, and the evolution of the Indian Ocean are discussed. In Chapter 3, we presented a detailed compilation of various geological and geophysical investigations with a view to understand the tectonic framework and crustal structure along the east and west coasts of India as well as the adjoining offshore areas. The collision history of the Indian landmass with Asia during the northward movement of India has important implications in terms of the uplift of Himalayas as well as development of large submarine fan systems on either side of the Peninsular India. The morphology, internal architecture, and sedimentation history of the Indus and the Bengal fans are presented in detail in Chapter 4. The rift–drift processes, sedimentation, and various other geological events have led to the development of major sedimentary basins (both onshore and offshore) along the eastern and western continental margins of India. The tectono-stratigraphic development and burial history of these basins are important for understanding the generation, migration, and the entrapment of oil and natural gas. A detailed description of the geology, depositional history and the development petroleum system in various basins located along the eastern and western margin of India are provided in Chapters 5 and 6, respectively. In the offshore exploration program, particularly in the deepwater areas, the E&P industry is heavily dependent on the application of modern state-of-art methodologies of imaging the seabed as well as the subsurface for reducing the geological risk and for accurate prediction of reservoir characteristics. In Chapter 7, we discussed some of the major issues related to imaging and logging with particular reference to the analysis of thin beds, estimation of pore pressure prediction, and various shallow geohazards, and wherever possible, we presented few examples and case studies from the Indian offshore.

Chapter 2

Reconstruction of Eastern Gondwanaland Breakup and Evolution of the Passive Continental Margin of India

2.1. INTRODUCTION

The global tectonic model for the development of the Indian Ocean has relevance to the geological history and development of Indian continental margins as the Indian plate has played an important role during the breakup and dispersal of Gondwanaland. Oceanic magnetic anomaly identifications, fracture zone, and transform fault trends revealed by bathymetry maps, and the paleomagnetic data of continental areas provide very useful information required to reconstruct the continental fragments of the Gondwanaland (Fig. 2.1) so that the early breakup history and subsequent development of its constituent margins could be deciphered (Reeves and de Wit, 2000). Some of the earliest publications (e.g., Smith and Hallam, 1970) describing the breakup of Gondwanaland and northward drifting of India are loosely reconstructed due to lack of adequate magnetic data (Reeves et al., 2002). During the 1970s, acquisition of new shipborne bathymetry and magnetic data in the Indian Ocean provided better understanding of the evolution of the Indian Ocean since the Late Cretaceous (McKenzie and Scalter, 1971; Sclater and Fisher, 1974) and to some extent during the Early Cretaceous times (Besse and Courtillot, 1988; Norton and Sclater, 1979; Powell et al., 1988). Subsequently, developments in the satellite altimetry based ocean floor mapping provided clarity to the topography of world oceans (Sandwell and Smith, 1997) which revealed the trends of transform faults and fracture zones over large distances, much better than the earlier bathymetric surveys, allowing scientists to reconstruct the drift paths of the continents in greater detail (Royer and Sandwell, 1989; Royer et al., 1989). Reeves and de Wit (2000) pointed out that such highly resolved transform and fracture zone data offer a more robust system of constraints than that available from magnetic anomaly data alone and help in bridging the gap in the magnetic record caused by the Cretaceous Magnetic Quiet Period (~118–84 Ma). While knowledge of

Developments in Petroleum Science, Vol. 59. http://dx.doi.org/10.1016/B978-0-444-53604-4.00002-8

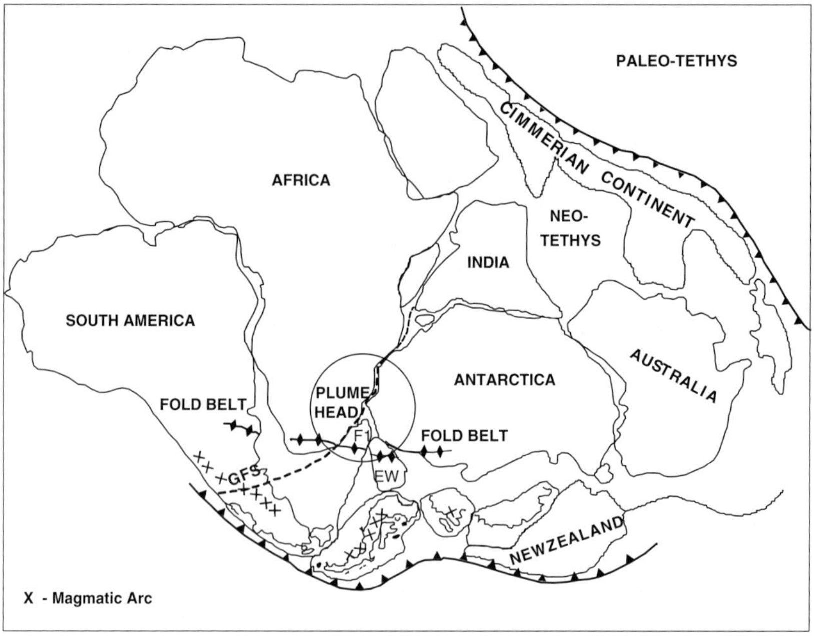

FIGURE 2.1 Pre-breakup Early Mesozoic Gondwanaland reconstruction depicting the subduction on either side of the continent and the Gondwana fold belt *(Reproduced from Storey et al., 1992)*. Location of the plume head (circle) is from White and McKenzie (1989). GFS, Gastre Fault system; FI, Falkland islands; EWM, Ellsworth Whitmore Mountains.

the reconstruction of the Indian Ocean until the Cretaceous had tremendously enhanced in recent times, the role of India in the Gondwanaland dispersal remained very uncertain in the Mesozoic global plate circuit due to complex breakup history and lack of paleomagnetic as well as marine geophysical data in certain critical areas (Coffin et al., 1992; Powell et al., 1988; Gaina et al., 2007).

According to Reeves et al. (2002), the Gondwanaland breakup was initiated in the mid-Jurassic (about 167 Ma) when the East Gondwanaland comprising the Antarctica, Madagascar, India, and Australia began to separate from Africa, and during the Early Cretaceous around 140 Ma, South America slowly started drifting westward from Africa and opened the South Atlantic Ocean. Further, the Eastern Gondwanaland also became fragmented about 130 Ma when India started moving southward less rapidly than the combined Antarctica-Australia. Thus, the formation of the South Atlantic and Indian Oceans commenced with the breakup of Gondwanaland in the Mesozoic and as the subduction activity in the Indian Ocean is minimal and limited to the eastern boundary along the Sunda trench, most of the Indian Ocean evolution was preserved (Norton and Sclater, 1979) and hence could be deciphered from the mapped fracture zones and magnetic anomaly identifications.

The complex evolution of India, that formed an integral part of Gondwanaland, into a freestanding continent, has led to various sections of Indian passive margins evolve into different ways and at different times (Reeves, 2003). In this chapter, we present a comprehensive evaluation of paleogeographic reconstruction of the Eastern Gondwanaland continents and in the process retrieve valuable insights with regard to the evolution of Indian continental margins, the adjoining oceanic areas as well as their conjugate counterparts. As almost all structural and tectonic features of the Indian Ocean were formed during the northward movement of India as well as due to the complex breakup processes between various Eastern Gondwana continents, it is necessary to understand the present day plate setting and morphology of the Indian Ocean.

2.2. PRESENT DAY PLATE GEOMETRY AND MORPHOLOGY OF THE INDIAN OCEAN

The topography of the Indian Ocean is dominated by three major active mid-ocean ridge systems (Fig. 2.2) which mark the present boundaries of the India, Arabia, Africa, Australia, and the Antarctica plates (Wiens et al., 1985). These three ridge segments meet at the Rodriguez triple junction at 25°S, 70°E in the Central Indian Ocean. Previous geophysical investigations along these ridges show marked variation in the spreading rates as well as ridge-transform geometry. The half-spreading rates obtained from magnetic anomaly identifications indicate 12–13 mm/yr for Carlsberg Ridge (McKenzie and Sclater, 1971), 16–25 mm/yr for the Central Indian Ridge (Fisher et al., 1971; DeMets et al., 1990), around 30 mm/yr for the Southeast Indian Ridge (Sclater et al., 1981), and 8 mm/yr for the Southwest Indian Ridge (Sclater et al., 1981). Further west in the Gulf of Aden, the Sheba and the Aden ridges have formed due to westward progression of the oceanic rift toward the Afar region (Laughton et al., 1970; Manighetti et al., 1997), cutting through several Jurassic and/or Cretaceous basins (Bosence, 1997). The estimated half-spreading rates for this ridge system indicate 8–11 mm/yr for the Aden Ridge (Drake and Girdler, 1964) from west to east and 10 mm/yr for the Sheba Ridge (Laughton et al., 1970; LeRoy et al., 2004).

The plate motion models for the Indian Ocean provided valuable constraints on the geometry and kinematics of the present-day plate boundaries in the Indian Ocean (Fig. 2.2). Wiens et al. (1985) stated that the observed plate motion along the Owen fracture zone is negligible and hence it may not define the plate boundary. However, later plate motion models (DeMets et al., 1994; Gordon and DeMets, 1989; Royer and Chang, 1991) suggest that despite low slip rate along this fracture zone, separate India and Arabia plates are consistent with the observed plate motion data. The location of the Arabia–India–Africa (Somalia) triple junction was variously interpreted. While some investigators located it at the intersection of Sheba Ridge, Owen

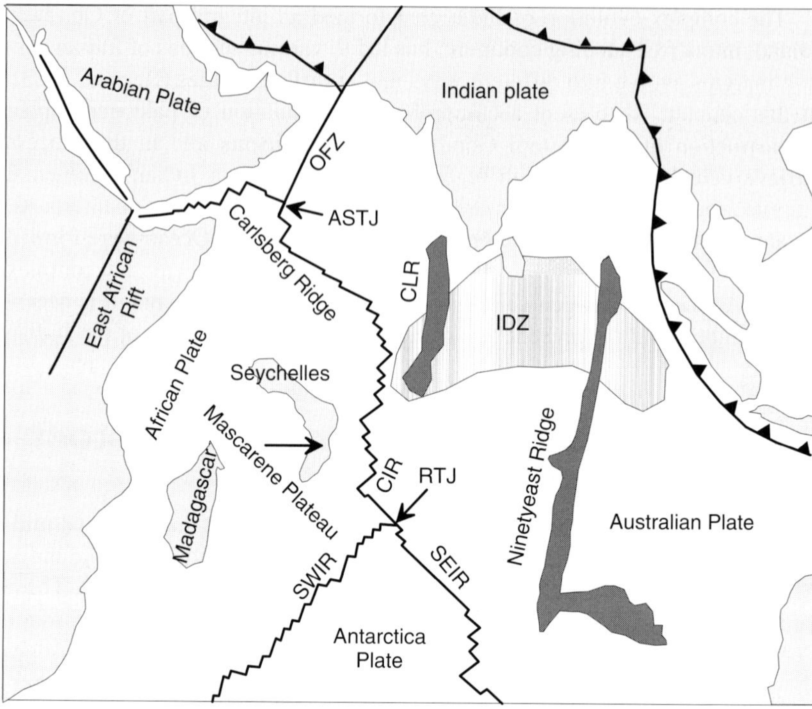

FIGURE 2.2 Present-day five plate geometry of the Indian Ocean showing the mid-Indian ocean ridge system and the Owen fracture zone *(Redrawn from Wiens et al., 1985)*. The extent of the diffuse plate boundary separating the Indian and Australian plates is marked in gray. IDZ, Intense Deformation Zone; ASTJ, Arabian Sea Triple Junction; RTJ, Rodriguez Triple Junction.

transform fault, and the Owen fracture zone (Gordon and DeMets, 1989; DeMets et al., 1990), Fournier et al. (2001) have relocated the triple junction at the bend of the Sheba Ridge near 56°E, with the Owen fracture zone being connected to the Sheba Ridge by an ultra-slow divergent boundary.

Wiens et al. (1985) proposed a diffuse equatorial plate boundary zone between India and Australia in the central Indian Ocean which extends eastward from the Central Indian Ridge, through the Chagos Bank, to the Ninetyeast Ridge. Gordon et al. (1990) proposed that this zone intersects the Central Indian Ridge near the equator and located the Africa–Australia–India triple junction between 6 and 3°S. However, DeMets et al. (1994) located the triple junction between 9 and 6°S along the Central Indian Ridge considering the boundary between India and Australia as discrete. The above observations confirm that the present-day geometry of the Indian Ocean is defined as a mosaic of India, Arabia, Africa, Australia, and Antarctica plates with much improved estimates of relative motion between these plates coming from continuously updated plate motion models from NUVEL 1-A (DeMets et al., 1990) to MORVEL (DeMets et al., 2010). These plate motion models describe well the geologically averaged present-day plate kinematics and form a very important data set for the plate reconstructions.

2.2.1 Aseismic Ridges, Plateaus, and Fracture Zones

Apart from the active mid-oceanic spreading ridge system that dominates the entire Indian Ocean morphology, the floor of the Indian Ocean (Fig. 2.3) comprises of numerous basins separated by aseismic ridges or swell like features.

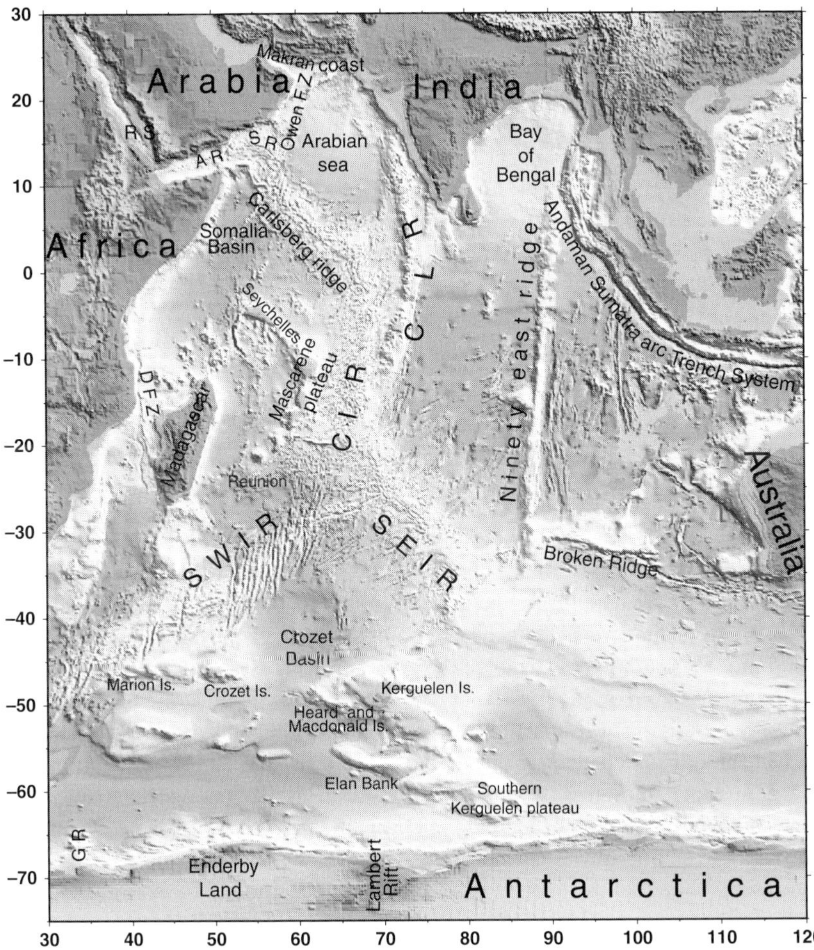

FIGURE 2.3 Shaded relief map of the topography in the Indian Ocean and the adjoining regions. Various morphological and tectonic features in the Indian Ocean such as the mid-oceanic ridges, Aseismic ridges, plateaus, and fracture zones can be noticed. SWIR, Southwest Indian ridge; SEIR, Southeast Indian ridge; CIR, Central Indian ridge; CLR, Chagos-Laccadive Ridge; DFZ, Davie fracture zone; GR, Gunnerus Ridge; RS, Red Sea; AR, Aden Ridge; SR, Sheba Ridge.

An excellent review on the morphology and genesis of these features is provided in Bhattacharya and Chaubey (2003) and Ramana et al. (2003). The ridges and plateaus of intermediate depth appear on either side of the spreading ridges. The size, number, distribution as well as their nature and origin are very important to understand their role in the continuing development of the Indian Ocean (Royer et al., 1989). Heezen and Tharp (1966) pointed out that some of these aseismic ridges/plateaus could be continental fragments. The most prominent topographic features are the Chagos-Laccadive and the Ninetyeast ridges which show considerable elongation in the N–S direction. Based on the presence of fracture zones, McKenzie and Scalter (1971) proposed northward movement of India along these two ridges. While many topographic rises are wholly submarine, few rise above the sea level and appear as islands. Numerous seamounts and Guyots having considerable relief above the ocean floor are also present. Apart from Madagascar Island which is one of the constituents of Eastern Gondwanaland fragmentation, the other minor continental fragments (slivers) in the Indian Ocean include the Seychelles islands, Laxmi ridge, the Chagos Bank, and the Elan Bank of Kerguelen plateau (Ben-Avraham and Bunce, 1977; Ingle et al., 2002; Naini and Talwani, 1982). The submarine Mozambique ridge and Agulhas Plateau consist partly of continental crust (Ben-Avraham et al., 1995). However, few other subaerial islands such as the Laccadives and Andaman-Nicobar islands are also debated as continental in nature (Acharyya, 2007; Chaubey et al., 2002a).

2.3. EVOLUTION OF THE INDIAN OCEAN: GEOLOGICAL AND GEOPHYSICAL CONSTRAINTS

As stated earlier, the high-resolution satellite altimetry based maps of ocean floor topography (Sandwell and Smith, 1997) provide the well-preserved record of the seafloor created during the separation and dispersal of continents. Such a dataset can be integrated with other geological and geophysical constraints to build models that attempts to solve the geometrical reassembly problem for better reconstruction of the Eastern Gondwanaland (e.g., Reeves et al., 2002). However, building up of a tighter reconstruction becomes very difficult due to lack of full understanding on the nature of crust in certain critical areas of the continental margins as well as below large aseismic ridges/plateaus, and uncertainties in magnetic anomaly identifications. The large-scale magmatic activity and plume tracks/volcanic traces at the margins as well as in the oceanic areas obliterate the signatures of crust that had formed by normal seafloor spreading processes. Further, lack of clear understanding on past major plate reorganizations, ridge jumps, and fossil spreading segments complicates the reconstruction of Eastern Gondwanaland. In the following sections, we present a brief account of various geological or geophysical constraints obtained through deep sea drilling efforts in the Indian Ocean, key morphological/tectonic features, Mesozoic magnetic anomaly identifications, and the role of plumes and associated volcanism, for better understanding of the early breakup history of Eastern Gondwanaland as well as the evolution of the Indian Ocean.

2.3.1 Synthesis of Results from DSDP and ODP Investigations in the Indian Ocean

The information regarding continental crustal fragments, extent of volcanism, and the nature of crust/crustal transition form important constraints and are essential for better reconstruction of the Eastern Gondwanaland as well as the early breakup history of Gondwana continents. The scientific drilling campaigns of Deep Sea Drilling Project (DSDP) and Ocean Drilling Program (ODP) have been one of the most successful International scientific efforts aimed at understanding the geological record/history of the ocean basins by drilling through sediments as well as the basement rocks in certain critical or ambiguous areas as decided by the world geoscientific community from time to time (Duncan et al., 1992). As a part of this global scientific venture, deep drilling was carried out at several selected sites in the Indian Ocean (Fig. 2.4). Many salient results of the drilled sediments and rock samples in the Indian Ocean have been published in various reports and later synthesized in several dedicated special issues (Duncan et al., 1992; von der Borch, 1978). The results have provided the most valuable ground truth data for validating the inferences of geophysical investigations. These sites have been carefully selected after thorough scientific deliberations with an objective of answering some of the puzzling questions for understanding the evolution and most of the sites are located over some of the major basins, aseismic ridges, plateaus, islands, and fracture zones in the Indian Ocean region (Duncan et al., 1992). We present here the salient results of the above investigations summarizing information about some of the drill sites where inferences about the basement as well as deeper sediment layers (Table 2.1) could be obtained.

The results of the DSDP site 248 (Schlich, 1982) suggest that the oldest basement rocks in the Mozambique basin are of Mesozoic age. The DSDP site 223 in the easternmost Owen basin (Whitmarsh et al., 1974) revealed basal igneous units of trachy basalt and breccia overlain by late Paleocene tuffaceous claystone. In the ODP initial reports, Mountain et al. (1989) suggested that the Owen Basin had formed during the separation of East and West Gondwana along with the Mozambique, north and west Somali Basins. The ODP Legs 122 and 123 (von Rad et al., 1992) drilled a complete transect of the old and sediment starved passive margin of the Exmouth Plateau and the Argo abyssal plain and provided information with regard to early Mesozoic rifting and breakup history of the northwest Australian margin of Eastern Gondwanaland facing southern Tethys Sea. Perhaps, one of the outstanding results of ODP drilling (site 1137, Leg 183) activity that changed our earlier perceptions on the Eastern Gondwanaland fragmentation with regard to India–Antarctica breakup was the unequivocal evidence of continental crust

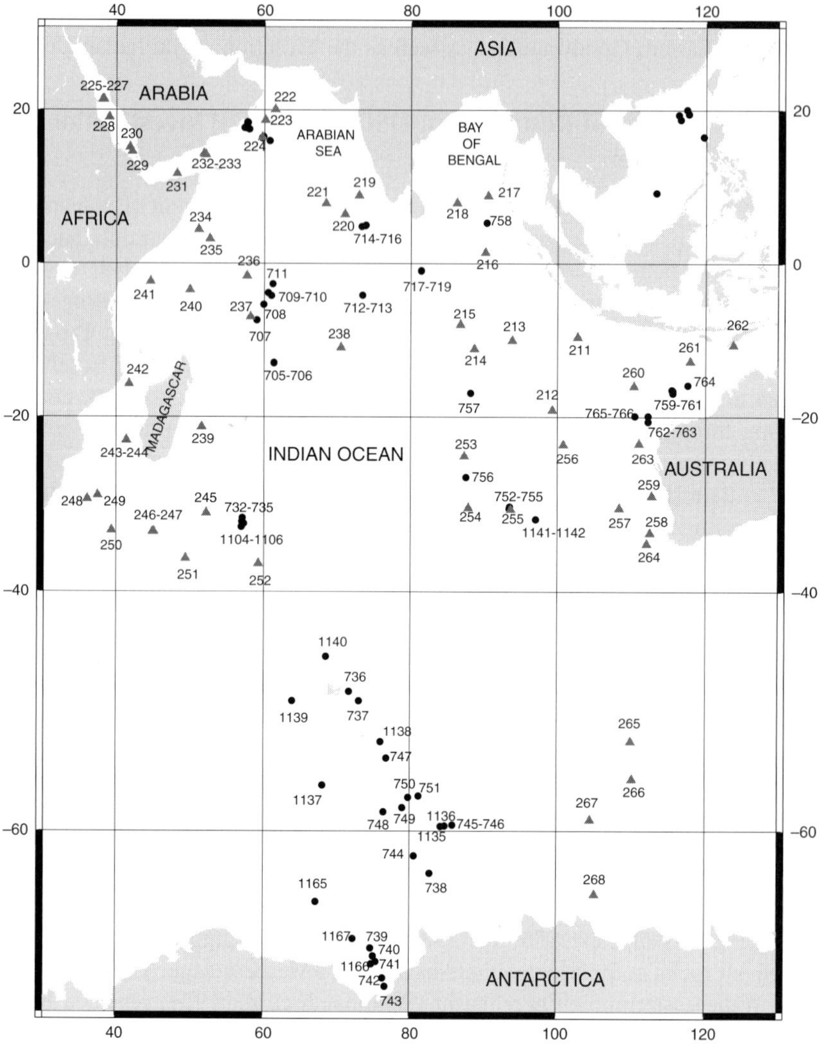

FIGURE 2.4 Location of the DSDP (triangle) and ODP (circle) drilled sites in the Indian Ocean. *(Source: http://www-odp.tamu.edu/sitemap/sitemap.html).*

below the Elan Bank of Kerguelen Plateau with affinity to Eastern Ghat rocks of the East Coast of India (Nicolaysen et al., 2001; Weis et al., 2001) and this aspect will be discussed in subsequent sections.

2.3.1.1 *Plume Tracks and Volcanism*

In the Indian Ocean, volcanism has occurred in a large variety of tectonic settings such as the continental, intraplate, and mid-oceanic ridges and gave rise

TABLE 2.1 Geological Information (Age and Nature of Deeper Layer) Derived from the Scientific Drilling Campaigns (DSDP and ODP Sites) in the Indian Ocean (Compiled from Various Publications of the DSDP and ODP Drilling Results, Duncan et al., 1992; Bhattacharya and Chaubey, 2003)

Location	DSDP/ ODP Site	DSDP/ ODP Leg	Position	Basement Reached (age of basement)	Nature of basement	References
Between Laccadive (Lakshadweep) Plateau and Maldive Ridge	DSDP (219)	23	9°01.75′N 72°52.67′E	No	–	Whitmarsh (1974)
Western flank of the Laccadive Plateau	DSDP (220)	23	6°30.97′N 70°59.02′E	Yes (undated)	Basalt	Whitmarsh et al. (1974)
Arabian Basin	DSDP (221)	23	20°5.49′N 61°30.56′E	Yes (undated)	Basalt	Whitmarsh et al. (1974)
Western edge of the Owen Ridge	DSDP (223)	23	18°44.98′N 60°07.78′E	Yes (undated)	Trachy Basalt	Whitmarsh et al. (1974)
Owen Fracture Zone	DSDP (224)	23	16°32.51′N 59°42.10′E		Lamprophyre	Whitmarsh et al. (1974)
Eastern flank of the Chain Ridge	DSDP (235)	24	3°14.06′N 52°41.64′E	Yes (Cretaceous?)	Basalt (diabase)	Holmes and Watkins (1992)
Southern Mascarene Basin	DSDP (239)	25	21°17.67′S 51°40.73′E	Yes (>71 Ma)	Basalt	Simpson et al. (1974), Holmes and Watkins (1992)
Western Somali Basin	DSDP (240)	25	3°29.28′S 50°3.42′E		Basalt	Simpson et al. (1974), Holmes and Watkins (1992)

Continued

TABLE 2.1 Geological Information (Age and Nature of Deeper Layer) Derived from the Scientific Drilling Campaigns (DSDP and ODP Sites) in the Indian Ocean (Compiled from Various Publications of the DSDP and ODP Drilling Results, Duncan et al., 1992; Bhattacharya and Chaubey, 2003)—Cont'd

Location	DSDP/ODP Site	DSDP/ODP Leg	Position	Basement Reached (age of basement)	Nature of basement	References
Mozambique Basin	DSDP (248)	25	29°31.78′S 37°28.48′E	Yes (72.3±7 Ma)	Basalt	Simpson et al. (1974), Holmes and Watkins (1992)
Mozambique Ridge	DSDP (249)	25	29°56.99′S 36°4.62′E		Basalt	Simpson et al. (1974), Holmes and Watkins (1992)
Mascarene Plateau	ODP (706)	115	13°06.85′S 61°22.26′E	Yes (36 Ma)	Vesicular basalt	Backman and Duncan et al. (1988)
Saddle between Seychelles and Saya de Malha	ODP (707)	115	7°32.72′S 59°01.01′E	Yes (62 Ma)	Massive basalt	Backman and Duncan et al. (1988)
Northern margin of the Chagos Bank	ODP (713)	115	4°11.58′S 73°23.65′E	Yes (48 Ma)	Basalt	Backman and Duncan et al. (1988)
Eastern margin of the Maldive Ridge	ODP (715)	115	5°04.89′N 73°49.88′E	Yes (55–60 Ma)	Olivine basalts	Backman and Duncan et al. (1988)
Kerguelen Plateau	ODP (738)	119	62°42.54′S 82°47.25′E	Yes ? (>91 Ma)	–	Backman and Duncan et al. (1988)
	ODP (747)	120	54°48.68′S 76°47.64′E	Yes (101–110 Ma)	Basalt	Schlich et al. (1989)

	ODP (749)	120	58°26.45′S 78°58.89′E	Yes (110 Ma)	Basalt	Schlich et al. (1989)
	ODP (750)	120	57°35.54′S 81°14.42′E	Yes (101 Ma)	Basalt	Schlich et al. (1989)
Naturaliste Plateau	DSDP (264)	28	34°58.13′S 112°02.68′E	? (>91 Ma)	Volcaniclastic rocks	Hayes et al. (1975)
Ninetyeast Ridge	DSDP (214)	22	11°20.21′S 88°43.08′E	Yes (59 Ma)	Vesicular basalt	Von der Borch, Sclatter et al. (1974)
	DSDP (215)	22	8°07.30′S 86°47.50′E	Yes	Basalt	Von der Borch et al. (1974a,b)
	DSDP (216)	22	1°27.73′N 90°12.48′E	Yes (81 Ma)	Basalt	Von der Borch et al. (1974a,b)
	DSDP (253)	26	24°52.65′S 87°21.97′E	Yes ? (>44 Ma)	–	Davies et al. (1974)
	DSDP (254)	26	30°58.15′S 87°53.72′E	Yes (38 Ma)	Olivine basalt	Davies et al. (1974)
	ODP (756)	121	27°21.330′S, 87°35.805′E	Yes (43 Ma)	Basalt flows	Peirce et al. (1989)
	ODP (757)	121	17°01.458′S 88°10.899′E	Yes (58 Ma)	Porphyritic basalt	Weissel et al. (1991)
	ODP (758)	121	5°23.049′N 90°21.673′E	Yes (82 Ma)	Slightly Plagioclase phyric to aphyric basalt	Weissel et al. (1991)

to emplacement of several long linear aseismic ridges or major volcanic islands (Fig. 2.5), and some of these were believed to have been originated due to the mantle plume activity. One of the most important objectives of the ODP drilling initiative was to investigate the life cycle of plume related volcanism by drilling along well-preserved plume tracks (Fig. 2.6) such as the Kerguelen and Reunion plumes (Duncan and Storey, 1992) in the Indian Ocean. The Chagos-Maldives-Laccadive ridge and the Ninetyeast ridges are the prominent hotspot tracks and the volcanism associated with these plume tracks such as the Deccan Traps, Rajmahal Traps, Kerguelen, and the Mascarene Plateaus would provide valuable data having both spatial and temporal resolution for investigating the compositional evolution of mantle sources (Weis et al., 1992). During the five legs of the ODP drilling (Legs 115, 119, 120, 121,123), nearly eight wells were drilled and along with two

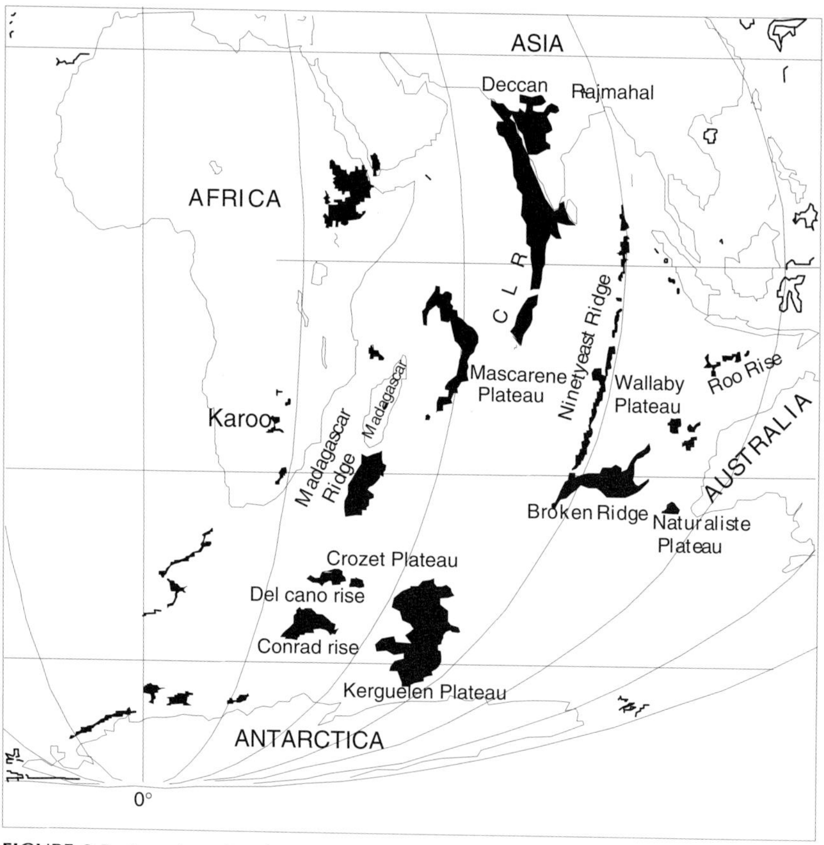

FIGURE 2.5 Location of major and known volcanic provinces of the Indian Ocean and the adjoining continents considered from the global distribution of large igneous provinces compiled by Coffin and Eldholm (1994) (redrawn from Coffin and Eldholm, 1994).

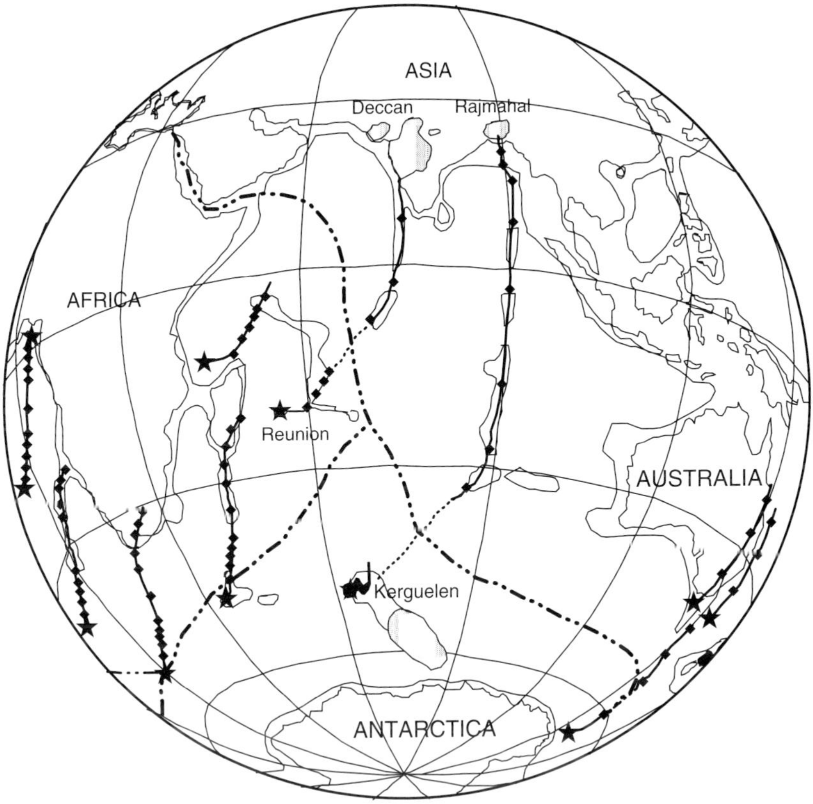

FIGURE 2.6 Modeled hotspot tracks in the Indian and Atlantic Oceans *(Reproduced from Duncan and Storey, 1992)*. It is assumed that hotspots are stationary. Initiation of Kerguelen and Reunion hotspots is characterized by Deccan and Rajmahal volcanism on the Indian subcontinent. Thick dots along the track indicates an age increment of 10 Ma for the volcanic rocks.

industry wells, these sampled the Reunion Plume track (Fig. 2.6), on the other hand, eight drilled wells along with the dredged samples at four DSDP sites sampled the Kerguelen Plume track (Duncan et al., 1992). All these sampled sites combined with the subaerial samples from the Deccan and Rajmahal traps, Kerguelen archipelago, etc., proved the volcanic nature of the basement as well as the age progression of the volcanic ridges since 65 Ma for the Reunion and 117 Ma for the Kerguelen plume (Duncan and Storey, 1992).

However, despite the concerted efforts made during the DSDP and the ODP investigations, there are still many features in the Indian Ocean whose genesis could not be clearly established. For example, two drilled holes, site 237 (Fisher et al., 1974) on the Seychelles-Saya de Malha saddle and site 246 (Simpson et al., 1974) over the Madagascar ridge did not reach the basement, and hence their origin and nature of crust underlying them could not be resolved.

2.3.2 Key Elements During the Separation of East and West Gondwanaland

The Gondwana began to breakup and disperse in Mesozoic times with rifting and the associated volcanism few tens of millions of years preceding the breakup (Royer and Coffin, 1992). The first separation of Gondwanaland into the eastern and western Gondwanaland took place sometime in the Late Jurassic (Reeves et al., 2002). Due to paucity of geophysical data and poorly defined continent-ocean boundaries, the overall plate configuration at the time of breakup is uncertain. We noticed four key elements that have relevance to the early breakup disposition of East and West Gondwana continents for tighter reconstruction, and understanding these elements has implications with regard to the evolution of southwest margins of India and Sri Lanka.

2.3.2.1 Davie Fracture Zone

Marine geophysical data and the DSDP drilling results off the eastern African margin in the Mozambique channel (Heirtzler and Burroughs, 1971; Scrutton, 1978; Simpson et al., 1974) revealed the presence of a steep fault-controlled structural feature that extends from 5°S in offshore Kenya to 26°S in southwest of Madagascar (Fig. 2.7). A topographic high, called the Davie Ridge, coincides with the above structural feature in its central part; hence, it was named as the Davie fracture zone (Scrutton, 1978). The fracture zone is about 30 km wide having steeper side toward the west and can be very well seen on the bathymetry and free air gravity map of the region. According to Bassias (1992), the fracture zone occurs over crystalline continental basement covered by deformed flysch sequence, alkaline lavas, tuffs, and breccias and by a thin succession of carbonate oozes ranging in age from Cretaceous to Pleistocene. In the plate tectonic reconstructions, this feature is considered as a major transform fault which guided the southward drift of Madagascar–India–Antarctica during Late Jurassic to Early Cretaceous splitting of East Gondwana and formation of West Somali basin (Droz and Mougenot, 1987). Though, there is a general agreement on such a movement, considerable ambiguity still exists with regard to original position of the Madagascar and timing of the movement (Embleton and McElhinney, 1975; Scrutton, 1978; Scrutton et al., 1981; Tarling and Kent, 1976).

2.3.2.2 Mozambique Ridge

The Mozambique Ridge is about 2000 km long and 300 km wide topographic feature extending from 35°S to 25°S along the southeast margin of Africa (Fig. 2.7). The ridge has strikingly different flanks, characterized by a gentle western flank and a steep eastern flank having a 3000 m deep scarp descending into the Mozambique Basin (Maia et al., 1990). Results from DSDP Site 249 located near the ridge crest (Simpson et al., 1974) recovered basalts;

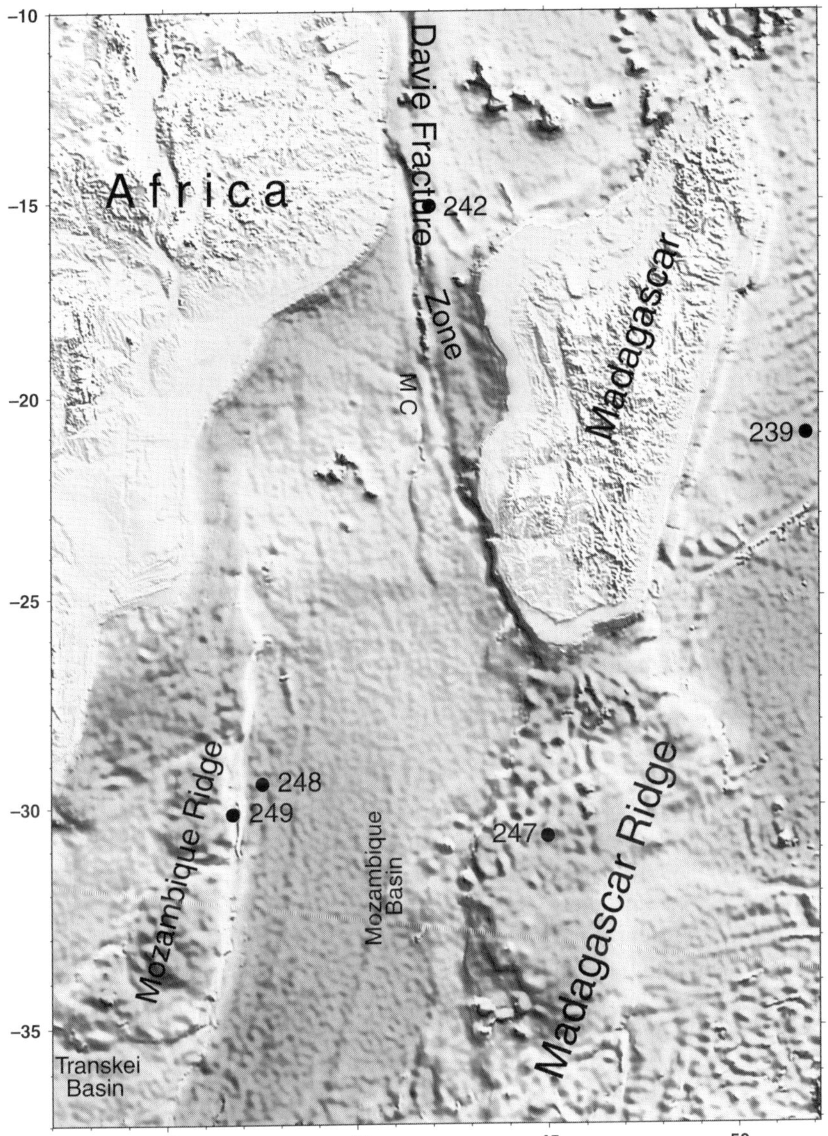

FIGURE 2.7 Disposition of Davie Fracture zone, Madagascar ridge, and Mozambique ridge off the coasts of Africa and Madagascar as seen from the topographic relief map. Location of DSDP sampling sites is also shown.

however, isotopic data on it are equivocal as to whether the ridge can be regarded as continental or oceanic (Erlank and Reid, 1974). Geophysical studies indicate that the ridge is underlain by 22 km thick crust and probably formed between M10 and M2 times by an anomalous activity of a spreading axis linking the northern Mozambique and Transkei basin accretion centers (Maia et al., 1990; Schlich, 1982).

2.3.2.3 *Madagascar Ridge*

The Madagascar Ridge located south of Madagascar is a N–S elongated feature between 26°S and 36°S (Fig. 2.7) and separates the Mozambique and Madagascar basins in the western Indian Ocean. While, the crust in east, in the Madgascar basin is of 53–80 Ma age (Schlich, 1975), the basalts recovered from DSDP site 248 in the west in the Mozambique basin are of 72 Ma age (McKee, 1974). Further west, the identified magnetic anomalies show anomaly 34 toward southwest in the Mozambique basin (Bergh and Norton, 1976) and M2 to M16 anomalies toward north in the Mozambique channel (Segoufin, 1978). However, nature of crust characterizing the ridge is not clearly distinguished and is very important due to the position of Madagascar in the Gondwana reconstruction. Geophysical investigations carried out along the ridge (Goslin et al., 1981; Schlich, 1982) indicate that the ridge could be made up of two domains, having an anomalous crust, neither purely continental nor oceanic, in the northern part and oceanic crust in the southern part. Seismic refraction data combined with gravity modeling over the ridge indicated thicker than normal oceanic crust formed by hot spot related volcanism during Cretaceous (Sinha et al., 1981).

2.3.2.4 *Gunnerus Ridge*

The Gunnerus Ridge located between 32°E and 35°E in the Indian sector of the Antarctic margin is a prominent structural feature (Fig. 2.8) covering an area of 25,000 km^2 and separates the deep sea plains of the Riser-Larsen Sea to the west and the Cosmonaut Sea to the east (Gersonde et al., 1998). Northern extension of the Gunnerus ridge is the Kainan Maru seamount with a steep E–W trending valley separating the two features. The multichannel seismic surveys conducted in the region revealed the presence thin sediments over the ridge and indicates that both Gunnerus ridge as well as the Kainan Maru seamount are of continental nature flanked by non-volcanic passive margins (Roeser et al., 1996; Saki et al., 1987). Based on detailed compilation of results obtained from geophysical data, Gohl (2008) indicated the ridge to be an extended continental crust.

2.3.3 Mesozoic Magnetic Anomalies in the Indian Ocean

It is now well established that the south Atlantic and the Indian Oceans commenced with the breakup of Gondwanaland in the Mesozoic (Norton and Sclater, 1979). Identification of the earliest seafloor spreading magnetic anomalies and the fracture zones along the constituent Gondwanaland

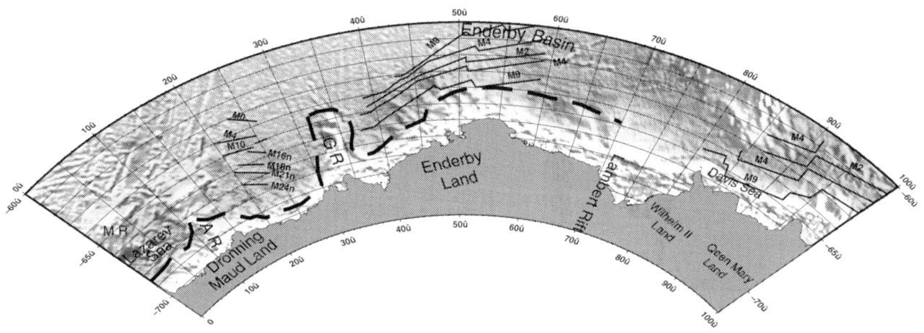

FIGURE 2.8 Bathymetry relief map of the Antarctic margin between Dronning Maud Land and the Queen Mary Land. The mapped Mesozoic magnetic anomaly identifications along the margin are shown along with the Ocean–Continent Boundary. Source of identifications is area west of Gunnerus ridge (after Nogi et al., 2004), in the Enderby Basin (after Gaina et al., 2007), off Queen Mary Land (after Gaina et al., 2007). GR, Gunnerus Ridge. Dashed line shows the location of Ocean–Continent Boundary, west of GR after Gohl (2008) and east of GR after Gaina et al. (2007).

continental margins provides valuable constraints with regard to understanding the early breakup history of the continents. Simpson et al. (1979) identified Mesozoic magnetic anomalies M2 through M22 in the northern Mozambique Basin (Fig. 2.9A) and suggested Early Jurassic or older separation of Africa and Antarctica, with Mozambique Basin and the Mesozoic seafloor off Droning-Maud land of Antarctica (Bergh, 1977) forming the conjugate segments. Using more recent dataset, Jokat et al. (2003) further refined these anomaly identifications and suggested that the first oceanic crust between the Mozambique Basin and Riser-Larsen Sea formed at around M24 (155 Ma). Identified magnetic anomalies indicate the presence of Mesozoic oceanic crust along the northwest margin of Australia (Fig. 2.9B) showing M26–M16 anomalies in the Argo abyssal plain (Fullerton et al., 1989; Larson, 1975) and M10 anomaly sequence in the Gascoyne abyssal plain and Cuvier abyssal plain regions (Fullerton et al., 1989). Along the southwest margin of Australia, the Mesozoic anomaly sequences showing M11–M0 anomalies in the Perth abyssal plain and M4–M0 anomalies in the Naturalist Plateau region were mapped by Markl (1974). The timing of breakup and early spreading history between the East Coast of India and the Enderby Basin of Antarctica is not clearly understood due to lack of clarity in the identification of Mesozoic anomalies. This has resulted in two equivocal models, one model suggests the presence of Mesozoic anomalies M11–M0 (Fig. 2.9C) in the Bay of Bengal as well as southeast of Sri Lanka (Desa et al., 2006; Ramana et al., 1994) with similar anomalies and spreading rates at the conjugate Enderby Basin (Ramana et al., 2001), the other model suggests younger to M0 (120 Ma) crust below Bay of Bengal (Banerjee et al., 1995; Gopala Rao et al., 1997). In addition to this controversy, the discrepancies in identification

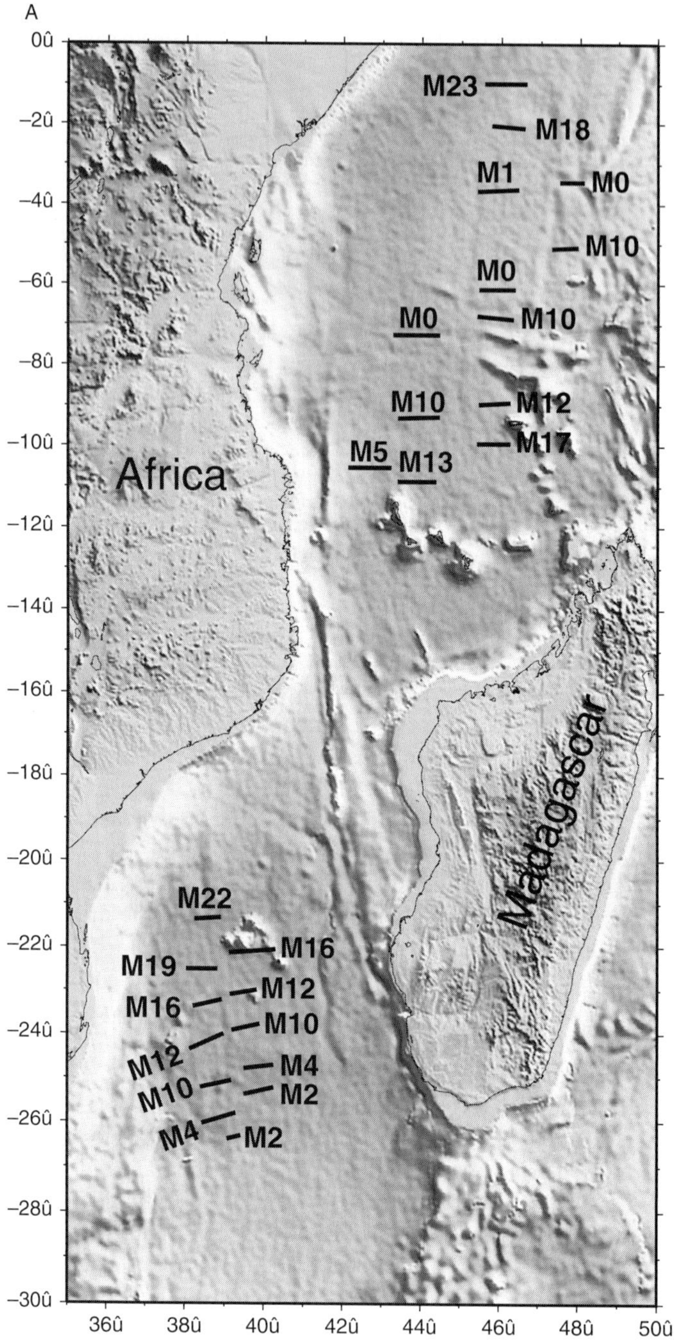

FIGURE 2.9—Cont'd

of Mesozoic sequences also exist along the Enderby Basin. Nogi et al. (1996) identified Mesozoic sequence from M9 with a general NE–SW trend in the central Enderby Basin. Gaina et al. (2003, 2007) identified M9o–M2o

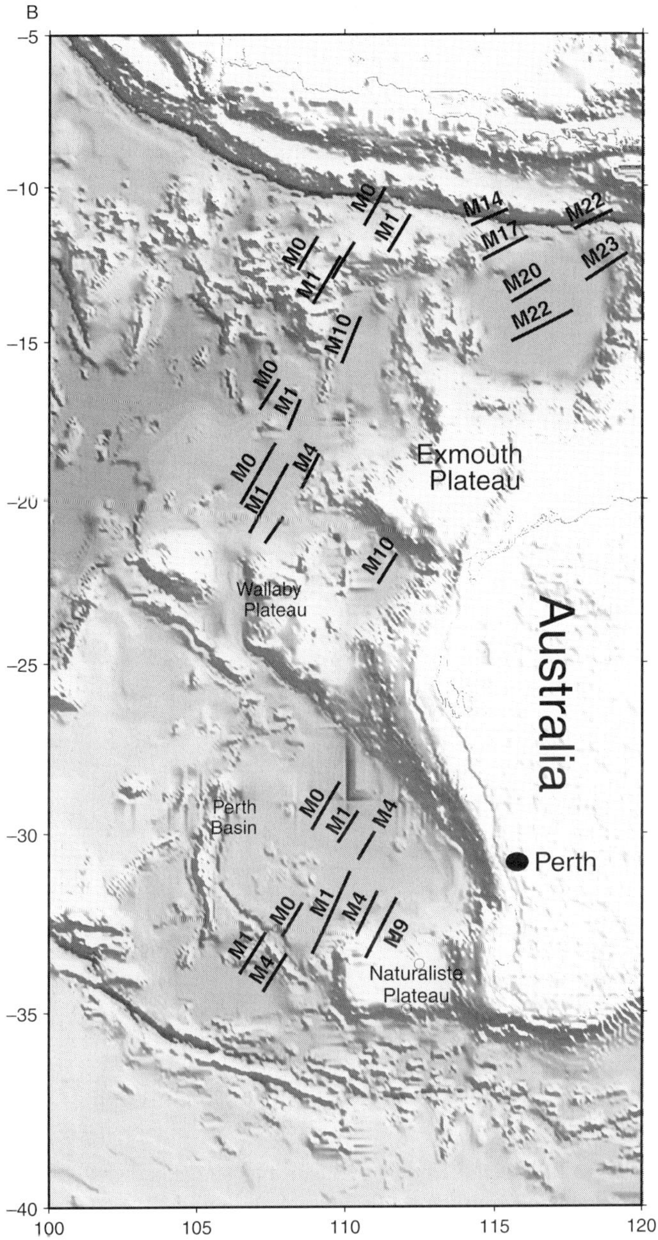

FIGURE 2.9—Cont'd

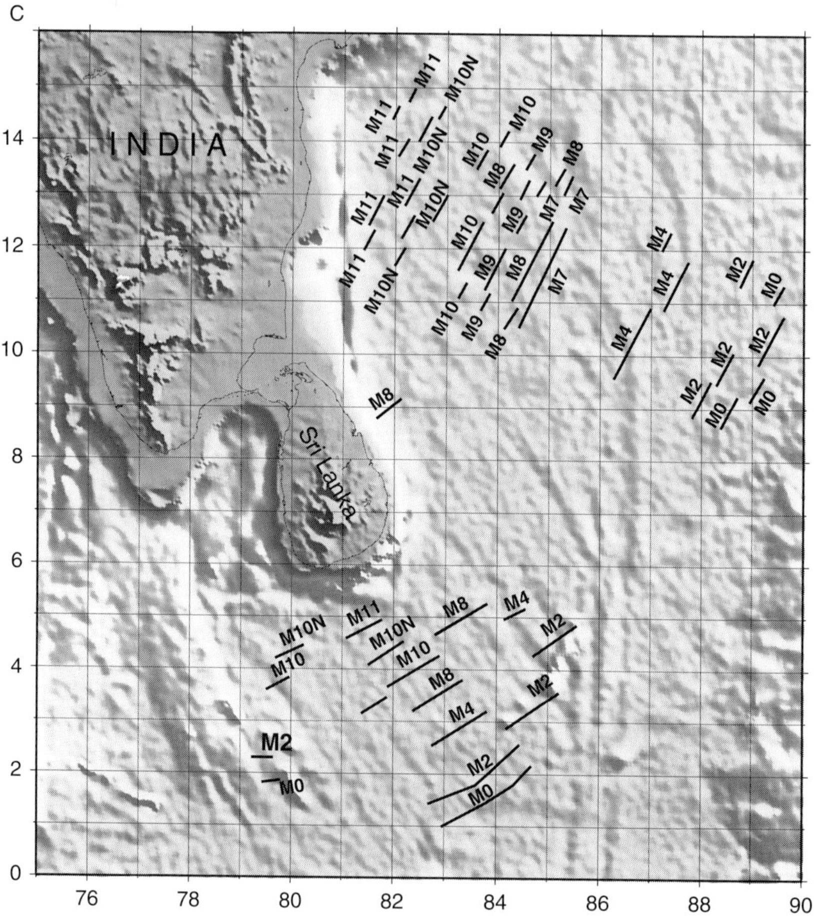

FIGURE 2.9 Mesozoic magnetic anomaly identifications along (A) off the coast of Africa in the Mozambique Basin (after Simpson et al., 1979; Norton and Sclater, 1979) and in the Somali Basin (after Segoufin and Patriot (1980); (B) off the west coast of Australia, north of Exmouth Plateau (after Fullerton et al., 1989), west of Exmouth Plateau (after Johnson et al., 1980), and off Perth ((Markl, 1974, 1978); and (C) off the southeast coast of India (after Ramana et al., 1994) and Sri Lanka (Desa et al., 2006).

anomalies (Fig. 2.8) in the Enderby Basin and suggested that the spreading was contemporaneous with the Perth abyssal plain along the southwest margin of Australia. Further, they identified a fossil spreading ridge between Elan Bank and the Enderby Basin due to a ridge jump around M2 which triggered the separation of Elan Bank from the East Coast of India. This observation had a considerable implication with reference to the breakup history of the eastern continental margin of India.

2.4. TECTONIC RECONSTRUCTION AND DEVELOPMENT OF INDIAN CONTINENTAL MARGINS

The continental breakup processes between India, East Antarctica, Madagascar, and the Seychelles have left their imprints on the newly evolved continental margins of India as well as on the adjoining deep ocean floor. According to Courtillot et al. (1999), the continental breakup in Gondwanaland is believed to have resulted from the interaction of series of hotspots or mantle plumes such as the Kerguelen, Marion, and the Reunion plumes in different times (Fig. 2.10). The eastern margin of India had been affected by the Kerguelen plume during the breakup as evidenced by the widespread volcanism such as (e.g., Rajmahal traps) at ~118 Ma (Coffin, 2002). The interaction of combined India–Madagascar–Seychelles block with the Marion mantle plume resulted in the separation of Madagascar from Seychelles–India block around 88 Ma (Storey et al., 1995). Similarly, the separation of Seychelles and India is contemporaneous with the widespread magmatic event of the Reunion mantle plume around 65 Ma and led to the formation of Deccan Continental Flood Basalt Province (Courtillot et al., 1986; Subba Rao and Sukheswala, 1981). Though such causal link between the plume interaction and continental breakup events needs to be further substantiated, as India moved further north, the influence of the Reunion Plume caused the emplacement of the Chagos-Laccadive Ridge and reorganization of nearby spreading centers in the oceanic areas.

The Precambrian shield of India comprises of the Archean cratons, Proterozoic mobile belts and shear zones, and continental rifts such as the Narmada-Son, Godavari, and the Mahanadi rifts are (Fig. 2.11) found to be closely linked to similar structural features in the constituent continents, the Antarctica and Madagascar of the Eastern Gondwana (Yoshida et al., 1999). Presentation of a complete and unified reconstruction of the constituent Eastern Gondwanaland continents and the evolution of the Indian Ocean is extremely complicated due to several inconsistencies observed by many workers. Although reconstruction of the Indian Ocean till the Late Cretaceous could be confidently made, due to lack of critical information particularly in the time window of Mesozoic, further understanding becomes difficult. As such an exercise is out of scope for this book, we intend to provide here all those aspects of reconstruction that have relevance for understanding the breakup history and development of continental margins of India. It is now very well established that both eastern and western continental margins of India have distinctly different tectonic and breakup histories as well as their subsequent development (Subrahmanyam and Chand, 2006). In order to simplify the reconstruction as well as to resolve the events, we present here the breakup events of the Eastern Gondwanaland in three different time windows: (i) the earliest breakup between Africa and the combined India–Madagascar during 250–140 Ma which is believed to have some association with certain parts of the present-day Western Continental Margin of India (WCMI),

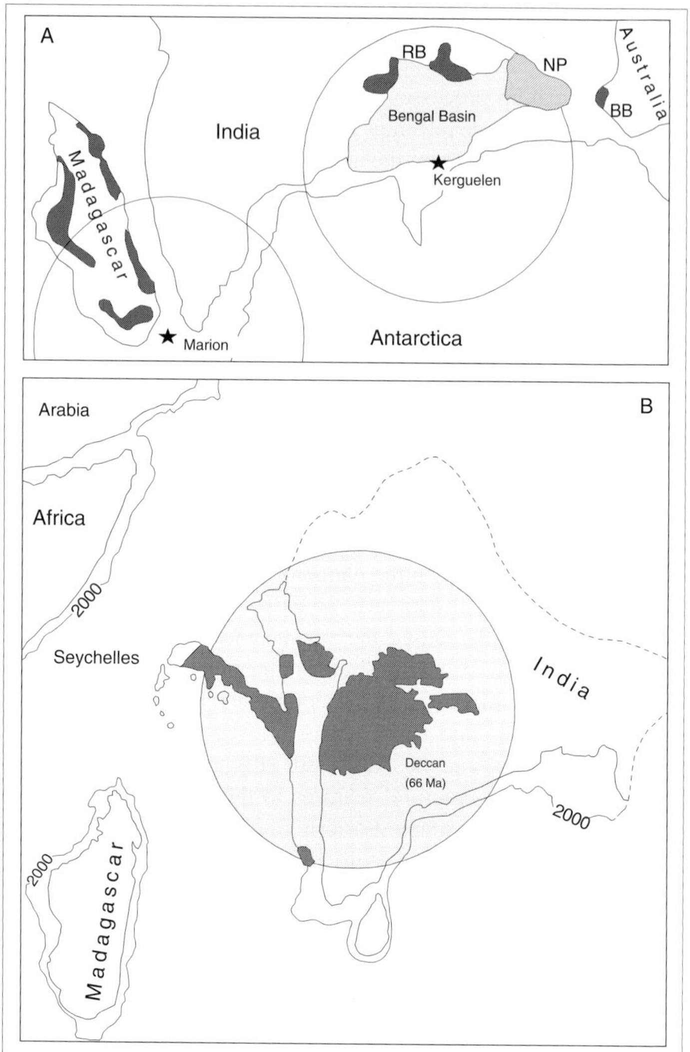

FIGURE 2.10 (A) Location of the Kerguelen plume around 118 Ma and the proximal volcanic regions. RBJ, Rajmahal Basalts; BB, Bunbary basalts; NP, Naturaliste Plateau and the position of Marion Plume around 88 Ma *(Reproduced with permission from Storey, 1995 © Nature Publishing Group)*. Original reconstruction is from Kent (1991). (B) Extent of Deccan volcanism (circle) before the separation of Seychelles from India (around 66 Ma) (Reproduced from White and McKenzie, 1989).

(ii) the India–Antarctica breakup history during 140–100 Ma and the development of passive margin along the East Coast of India, (iii) the India–Madagascar–Seychelles breakup history during 100–50 Ma and the development of the WCMI. While the Indian passive margins were taking shape during the above three stages, the Indian plate along the eastern boundary is transforming into

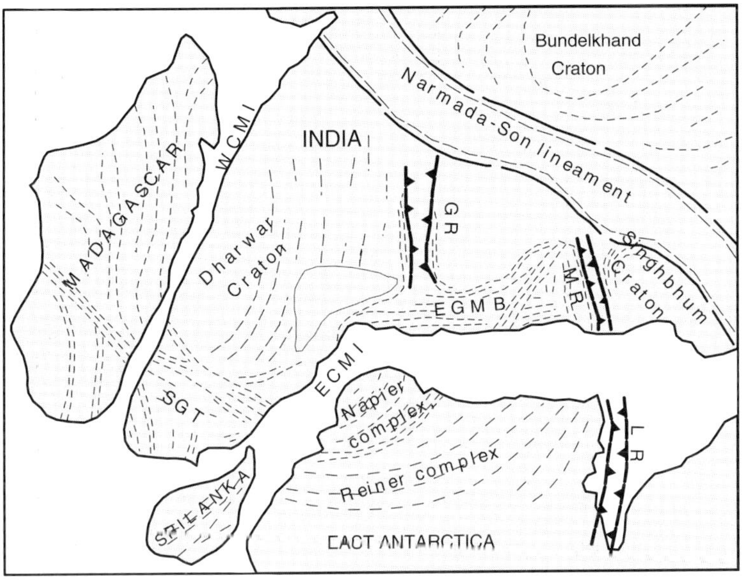

FIGURE 2.11 Reconstruction of India and the constituent eastern Gondwanaland continents (Antarctica, Madagascar, and Sri Lanka) showing the structural trends (Reproduced from Subrahmanyam and Chand, 2006). EGMB, Eastern Ghats Mobile Belt; GR, Godavari rift; LR, Lambert rift; MR, Mahanadi rift; SGT, Southern Granulite Terrain. The Western (WCMI) and the Eastern (ECMI) Continental Margins are also marked.

an active plate boundary at the present-day Sunda and Andaman-Burmese arc that led to the subduction of the Indian plate below the Eurasian plate. Apart from describing various stages of reconstruction, we also provide here a brief synthesis of tectonic history of this margin with an emphasis on the development of Andaman trench and back arc system in the eastern Indian Ocean as it is the only segment of Indian continental margin that is of Pacific type.

2.4.1 Karoo and Gondwana Rift System: Initiation of the Indian Ocean (250–140 Ma)

Gondwanaland reconstruction depicting major geological events during the period (250–140 Ma) have been presented by Reeves and his co-workers in several publications (Bastia et al., 2010e; Reeves, 2003, 2009; Reeves and de Wit, 2000; Reeves et al., 2002, 2004, and the references therein). According to these workers, before 200 Ma, while Helmand departed into the Tethys Ocean from between India and Arabia, marine passive margin conditions were developing along the northwest India, southeast Arabia, and the eastern most Somalia, and the continued rifting through the Karoo and Gondwana rift system bisected the Gondwanaland. At about 167 Ma, the Somali, Weddell,

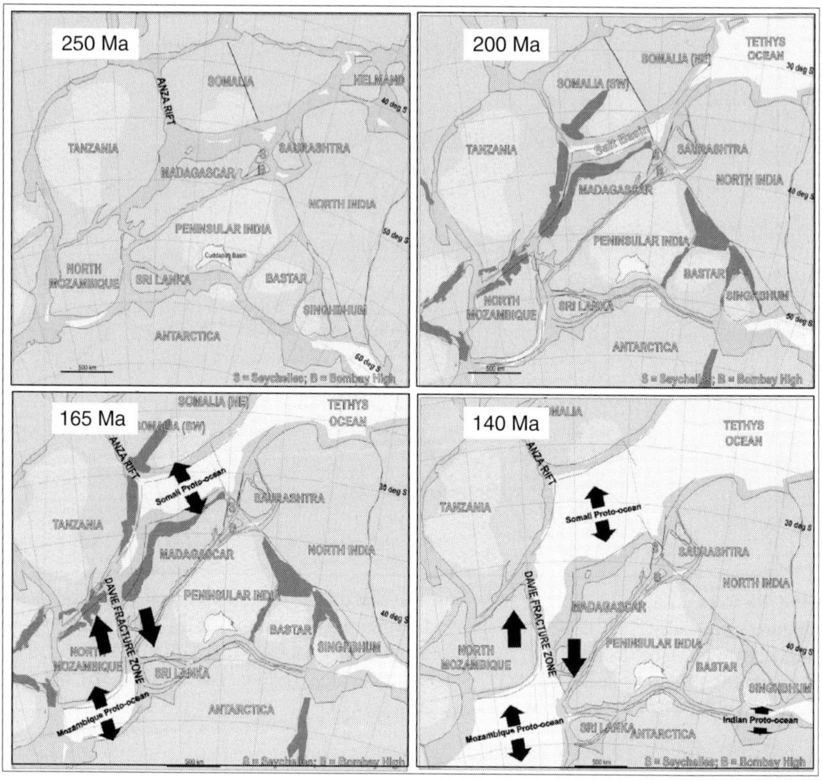

FIGURE 2.12 Reconstruction of the eastern Gondwanaland during 250–140 Ma (after Bastia et al., 2010e) depicting the formation of proto-oceanic crust and the activity of Davie fracture zone. (For color version of this figure, the reader is referred to the online version of this chapter.)

and Mozambique proto-oceans were created with the formation of a salt basin due to intermittent marine access to the Somali proto-ocean (Fig. 2.12). König (2005) infers that the earliest episode that might have initiated the passive evolution of the West Coast may be when Madagascar first moved south from Africa at about 167 Ma, and this would mean that the panel of ocean between Madagascar and Somalia grew at a slightly slower pace than that between India and Somalia. These two panels of ocean crust might be separated by a transform that would have grown between the Narmada-Son lineament in India and the prominent fault that crosses Somalia (Bastia et al., 2010e; Reeves, 2003).

2.4.2 India-Antarctica Breakup History and Development of the East Coast Passive Margin of India

Though the evolution of India's East Coast from the incipient rift zones to passive margin started later, it evolved more quickly than the corresponding events along the West Coast (Reeves, 2003). The early separation between the Australia and Greater India was well documented by the identification of Mesozoic magnetic anomalies in the northwest margin basins of Australia (Larson, 1977; Markl, 1974; Robb et al., 2005). In most of the Eastern Gondwana reconstructions (e.g., Powell et al., 1988; Smith and Hallam, 1970), the East Coast of India is placed against the East Antarctica. In this scenario, the Enderby Land, east of Gunnerus ridge to Lambert Rift (Fig. 2.8) sector of the eastern Antarctica (Gohl, 2008) becomes conjugate segment to the East Coast of India. The crustal and sedimentary structures of this segment of the east Antarctic margin was studied by detailed seismic investigations (Solli et al., 2007; Stagg et al., 2004, 2005). In the Indian sector of the East Antarctic margin, the Enderby Basin can be divided into two zones of distinct character (Stagg et al., 2004), the western Enderby Basin (west of 58°W) having characteristics of mixed-transform setting with fracture zones terminating in an oblique sense at the margin, while the eastern Enderby margin zone has more of a normal rifted margin setting. The Mac Robertson Coastal Anomaly (MCA) which is the most prominent magnetic anomaly in the eastern segment correlates with the northern limit of the continent ocean transition in this zone (Gaina et al., 2007). The seismic investigations (Stagg et al., 2004, 2005), in general, revealed absence of major SDR sequences in the Enderby Basin margin, suggesting that during the breakup at about 130 Ma (Gaina et al., 2007), no major mantle plume interference exists at the margin. However, according to Gaina et al. (2007), the margin off central and eastern Prydz Bay is affected by both the inherited structure of the Paleozoic-Mesozoic Lambert Rift system and the magmatic events of the Kerguelen Plateau nearly 10–15 Ma after the initial breakup between India and Antarctica.

Most of the earlier plate reconstructions (Powell et al., 1988; Royer and Sandwell, 1989) attempted for India-Antarctica segment of the Indian ocean have remained conjectural for a long time due to lack of identified magnetic lineations of Mesozoic age both in the Bay of Bengal and its conjugate Enderby Basin. A first attempt of identifying Mesozoic magnetic lineations in the Bay of Bengal was made by Ramana et al. (1994) on the basis of several NW–SE-trending ship track data. These authors inferred anomalies ranging from M11 to M0 with half spreading rates of 3.6–3.4 cm/yr during anomalies M11–M9 and M4–M0 and a faster spreading rate of 10.0 cm/yr during anomalies M9–M4. Subsequently, Ramana et al. (2001) reported Mesozoic magnetic lineations M11–M0 in the Enderby Basin similar to those observed in the Bay of Bengal, suggesting conjugate nature of these two segments. They further interpreted NNE–SSW trending fracture zones from satellite gravity anomalies between the Gunnerus ridge and the Conrad Rise and assumed that the fracture zones continue in the seafloor close to Antarctica. Based on vector magnetic anomalies in the region east of Gunnerus ridge, ENE–WSW and E–W Mesozoic magnetic anomaly lineations accompanied by NW–SE and NNW–SSE trending fracture zones were identified by Nogi et al. (1996). Further,

Nogi et al. (2004) interpreted the initiation of spreading in the western Enderby Basin before 127 Ma and introduced the extinct spreading centre between the Gunnerus ridge and Conrad Rise. These interpretations have been considerably modified after the observation of Elan Bank microcontinent during the Leg 183 of the ODP drilling activity in the Indian Ocean.

2.4.2.1 Elan Bank: Its Place in Breakup History

The Kerguelen Plateau is a structurally complex region and forms a large igneous province (LIP) representing voluminous Cretaceous volcanism associated with arrival of the Kerguelen plume head (Duncan and Storey, 1992; Storey, 1995). The Elan Bank encompassing nearly 140,000 km^2 of seafloor west of the central Kerguelen Plateau became geologically significant after the recovery of continental crustal rocks (Borissova et al., 2003). The basement underlying the Elan Bank was first sampled during ODP Leg 183 of Site 1137 (Coffin et al., 2000). The rocks recovered were clasts of garnet-biotite gneiss in a fluvial conglomerate intercalated with basaltic flows (Nicolaysen et al., 2001). The gneiss clasts shows an affinity to crustal rocks from India, particularly those of the Eastern Ghats Belt (Ingle et al., 2002), and represents another example of a continental fragment in the Indian Ocean. The clasts of continental crustal material shed new light on the origin of geochemical heterogeneities in basalts of the Kerguelen Plateau (Weis et al., 2001). Gaina et al. (2003) suggested that the history of early seafloor spreading is complicated by ridge—hotspot interaction during the growth of the Kerguelen Plume and the Elan Bank microcontinent was most probably isolated from the Indian continent by one or several ridge jumps associated with the Kerguelen Plume. This is in tune with the observation of identified M-sequence anomalies by Gaina et al. (2003), where M9o–M2o anomalies with a fossil spreading ridge in the central Enderby Basin have been mapped. Based on the requirement of accommodating the Elan Bank microcontinent, these authors proposed a two-stage rifting history for the ECMI, first, the India–Antarctica separation during Early Cretaceous, second, the separation of Elan Bank from the present day ECMI at M2. The observation further corroborates the inference of previous workers with regard to the segmentation of ECMI into northern rifted margin and southern sheared margin (Chand et al., 2001; Krishna et al., 2009; Subrahmanyam et al., 1999). Taking this scenario further, Gaina et al. (2007) proposed a reconstruction of the breakup events at the margin (Fig. 2.13).

From the above new information and arguments, two most important observations have significantly changed the earlier perceptions on the plate reconstruction between India and Antarctica during the early Cretaceous. They are (i) the continental nature of Elan Bank with an affinity to rocks belonging to Eastern Ghat Mobile Belt (EGMB) of the East Coast of India (Ingle et al., 2002) and (ii) observation of M9o–M2o anomalies with a fossil spreading ridge in the Enderby basin of Antarctica (Gaina et al., 2003, 2007).

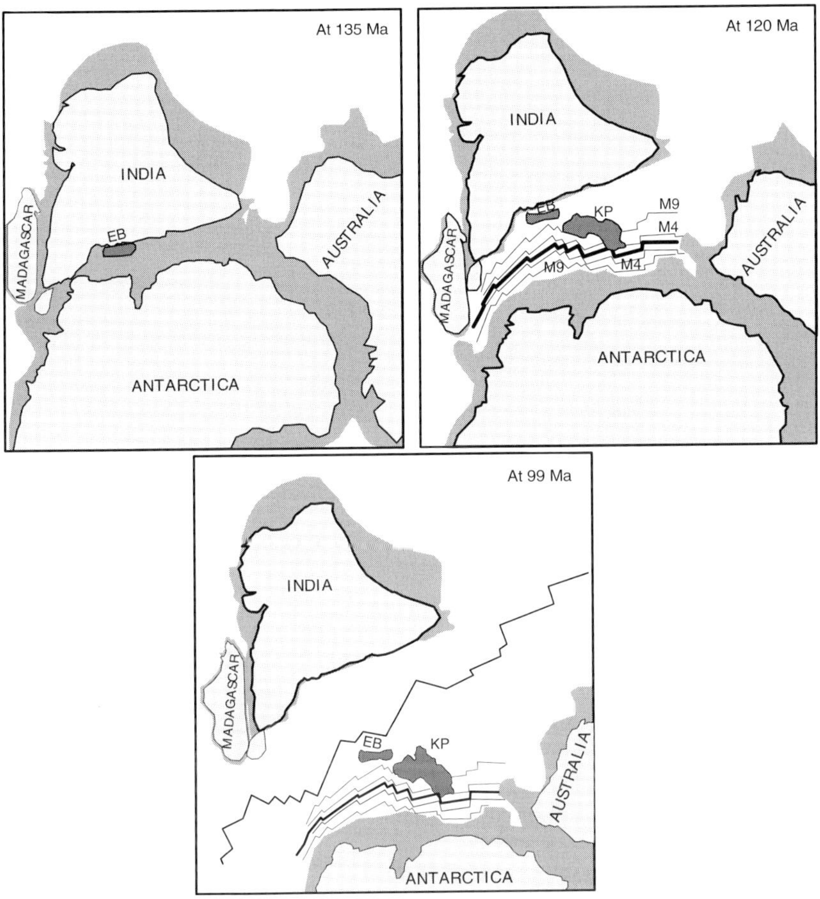

FIGURE 2.13 Early breakup plate reconstruction models for the Eastern Continental Margin of India (ECMI) showing the double breakup events due to the ridge jump at M2 *(Redrawn from Gaina et al., 2007)*. EB, Elan Bank; KP, Kerguelen Plateau.

These observations indicate a double breakup sequence for the evolution of the East Coast of India: the first event between southern part of East Coast with the conjugate western Enderby Basin during which the seafloor spreading (during M9o–M2o) was taking place between Elan Bank and the eastern Enderby Basin and the second breakup event occurred between northern part of East Coast of India and the Elan Bank microcontinent by a northward ridge jump at M2. However, timing of breakup between India and the conjugate counterparts of the eastern Gondwanaland and subsequent evolution of the lithosphere in the Bay of Bengal and the Enderby Basin still remain an unresolved issue because magnetic and gravity data of both the regions do not provide unambiguous interpretation.

2.4.3 India-Madagascar-Seychelles Breakup: Development of the WCMI

It is now very well established that the juxtaposition of the West Coast of India and the East Coast of Madagascar (Fig. 2.14) is consistent with Precambrian trends, lithologies, and age provinces (Katz and Premoli, 1979; Raval

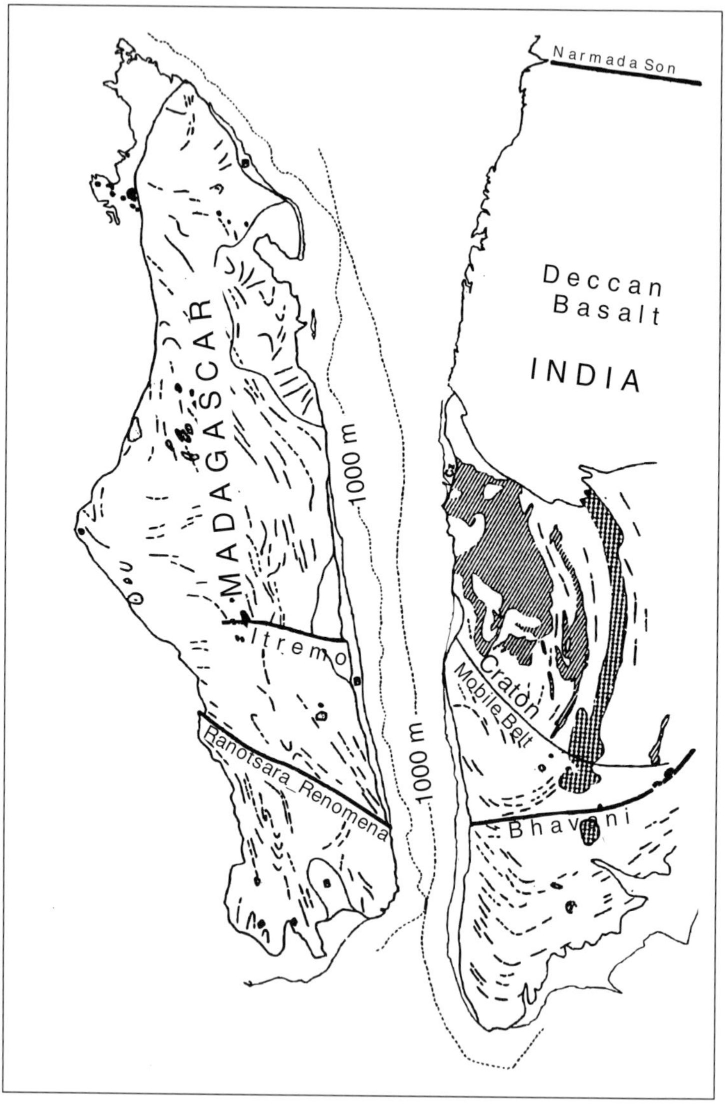

FIGURE 2.14 Fit of India and Madagascar at 1000 m isobath and matching of the Precambrian Structural Trends *(Reproduced with permission from Katz and Premoli, 1979 © Nature Publishing Group).*

and Veeraswamy, 2003a; Yoshida et al., 1999). Though most of the plate reconstructions consider the Madagascar between the West Coast of India and the Africa, its position during the Gondwana times is not very clear (Embleton and McElhinney, 1975; Green, 1972). The Precambrian geology as well as geophysical characteristics between India and Madagascar compare very well (Agarwal et al., 1992). Some of the earlier reconstructions place Madagascar (i) juxtaposed to the West Coast of India (Katz and Premoli, 1979; Yoshida et al., 1999), (ii) close to the Somalia, Kenya, and Tanzania (Coffin and Rabinowitz, 1987; McElhinney et al., 1976), and (iii) adjacent to the Mozambique margin (Green, 1972). It is relevant to note that geologically, both India and Madagascar comprise extensive alkaline mafic intrusions and structural continuation (Yoshida et al., 1999). While the Axial Shear Zone of Madagascar is compared with the dextral Moyar–Bhavani shear zone of southern part of India by Windley et al. (1994) and Yoshida et al. (1999), the E–W trending dextral Palghat–Cauvery Shear Zone is correlated with either the western edge of Axial Shear Zone or the Ranotsara Shear zone by Katz and Premoli (1979). The India and Madagascar fit was further supported by the presence of Upper Cretaceous volcanic rocks, mafic dykes, and the geophysical signatures of conjugate nature (e.g., Chand and Subrahmanyam, 2003; Kumar et al., 2001). The steep and straight edge of the East Coast of Madagascar is generally considered as an evidence of transform motion between India and Madagascar (Barron, 1987; Lawver et al., 1999). Yatheesh et al. (2006) presented a detailed synthesis of all available reconstructions between India and Madagascar attempted by various workers (e.g., Besse and Courtillot, 1988; Morgan, 1981; Müller et al., 1993; Norton and Sclater, 1979). Further, they identified two bathymetric features, one, the Terrace off Trivandrum at the Indian margin and, the other, a bathymetric notch at the southern margin of Madagascar and proposed that these two form the conjugate segments at the time of breakup. The Paleogeographic reconstruction (Fig. 2.15) attempted by these authors provides a close fit between India and Madagascar with gaps and spaces accommodated by the microcontinents or continental slivers that existed during the pre-drift scenario.

The separation of Madagascar and Seychelles from India and subsequent seafloor spreading resulted in the formation of Arabian Sea (Chaubey et al., 1993; McKenzie and Scalter, 1971; Naini and Talwani, 1982; Norton and Sclater, 1979; Whitmarsh, et al. 1974); however, the actual timing of opening of the Arabian Sea is debated as considerable ambiguity in the identifiable magnetic anomalies along the western margin still exists. The knowledge on the location of Ocean Continent Boundary (OCB) and the earliest seafloor spreading anomalies along this margin are critical for constraining the tectonic evolution of the Arabian Sea, particularly the initial opening history of the Arabian Sea and eastern Somali basin as well as the separation of Seychelles from India. Recent investigations in the Arabian Sea have mainly focused on some of these aspects (Bhattacharya et al., 1994; Chaubey et al.,

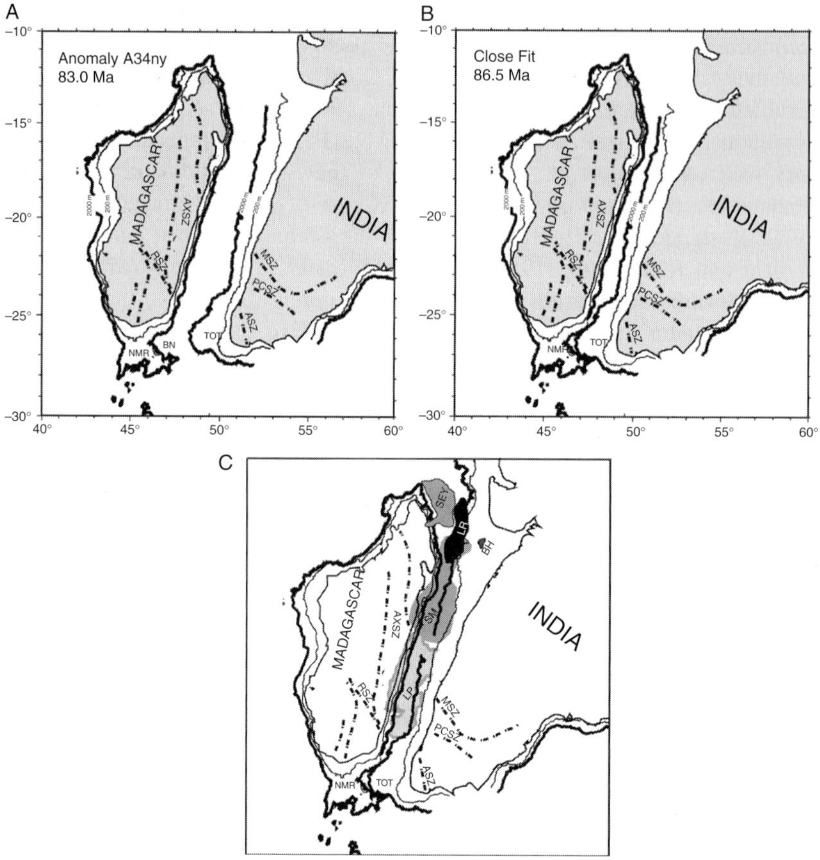

FIGURE 2.15 Paleogeographic reconstruction models of India and Madagascar showing the possibility of accommodating various continental slivers mapped in the western Indian Ocean *(Reproduced from Yatheesh et al., 2006)*. (A) Reconstruction at 83 Ma using the rotation parameters of Norton and Sclater (1979), (B) close fit reconstruction using the rotation parameters of Yatheesh et al. (2006), and (C) close fit model (schematic) accommodating various continental fragments in the gap between 2000 m isobath. TOT, Terrace off Trivandrum; BN, Bathymetric notch; NMR, Northern Madagascar Ridge; ASZ, Achankovil Shear Zone; PCSZ, Palghat-Cauvery Shear Zone; MSZ, Moyar Shear Zone; RSZ, Ranotsara Shear Zone; AXSZ, Axial Shear Zone; SEY, Seychelles Bank; LP, Laccadive Plateau; LR, Laxmi Ridge; SM, Saya de Malha Bank.

2002b; Krishna et al., 2006; Malod et al., 1997; Miles et al., 1998; Radhakrishna et al., 2002; Talwani and Reif, 1998; Todal and Eldholm, 1998). Based on the identification of sea floor spreading magnetic anomalies in the Laxmi basin by Bhattacharya et al. (1994), Talwani and Reif (1998) proposed a reconstruction which moves Seychelles original location closer to India and a new rotation pole between anomalies 28 and 34. While Talwani and Reif (1998) inferred the crust below the Laxmi basin as representative of an initial oceanic crust, several other investigators support continental type of crust below this basin (e.g., Krishna et al., 2006; Miles et al., 1998; Todal and Eldholm, 1998). On the basis of magnetic anomaly identifications, Todal and Eldholm (1998) proposed a reconstruction (Fig. 2.16) of the Seychelles microplate with reference to India which suggests that the continental extension is followed by fan-shaped spreading between India and the Seychelles during A29-27 time. Subsequently, the spreading between India and Seychelles took place and the passage of Reunion plume might have affected the southern part of western margin.

2.4.4 Geological History of Sunda–Andaman–Burmese Subduction Plate Boundary and the Development of Andaman arc and the Back-arc Andaman Basin

The Andaman–Nicobar Islands region which is the northward continuation of the Sumatra–Java arc system has been evolving through mainly subduction-related processes which responded kinematically to the plate reorganizations and other tectonic adjustments during the evolution of the Indian Ocean. The Andaman arc along with the Burmese arc serves as a transitional tectonic link between Himalayan Collision zone to the north and Sunda arc–trench system in the south (Hamilton, 1979). The Andaman–Nicobar Islands are the subaerial expressions of the outer arc ridge and separates the Andaman Sea from the Bay of Bengal (Curray et al., 1982). Rodolfo (1969) inferred that the islands have formed during Oligocene–Miocene times due to E–W compression of sediments derived from the Malayan shelf. Lay et al. (2005) observed about 1.4 cm/yr of convergence along the Andaman arc and suggested that the age and thickness of the subducted oceanic crust as well as the rate of convergence increase from the Andaman toward the Java along the arc. The Andaman back-arc spreading ridge-transform system which joins with the Sumatran fault in the south accommodates some of the plate motion. Curray et al. (1982) inferred that the formation of the Andaman Sea has a definite genetic linkage with the evolution of the Bay of Bengal. During this evolution, modifications to the subduction zone geometry had produced considerable mass anomalies at depth.

The complex evolution of the subduction zone in the eastern Indian Ocean took place since the Cretaceous during the northward flight of India. Curray (2005) synthesized all available geological and geophysical data and proposed different stages of opening of the Andaman Sea region. Some salient aspects of his reconstruction (Fig. 2.17) are presented below and more details have been outlined in Curray (2005). According to his reconstruction, subduction was initiated along the eastern margin of Asia at least during Cretaceous when the separation of India from Australia and Antarctica was taking place. Some of the older sediments found in Myanmar and the Andaman–Nicobar Islands indicate that prior to the initiation of subduction, the region could be a passive

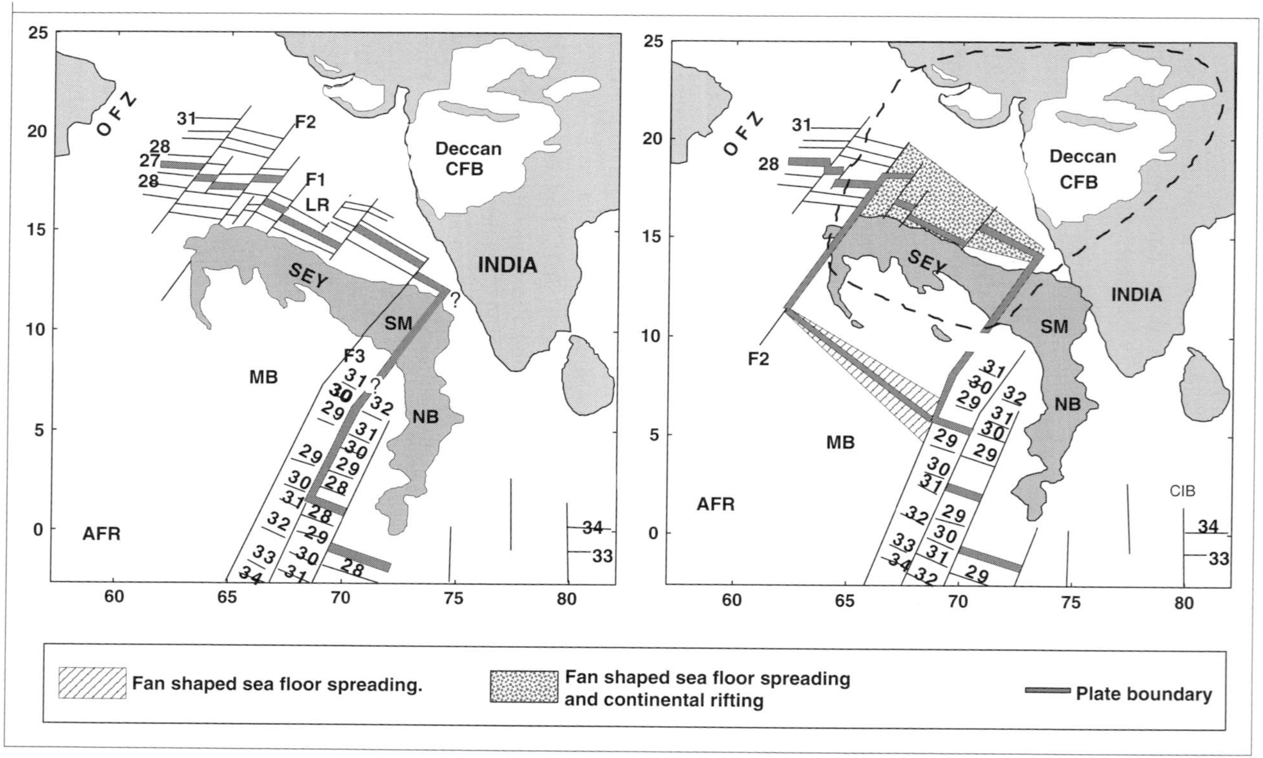

FIGURE 2.16 Plate reconstruction of Seychelles (SEY) and India (IND) depicting the fan-shaped spreading and breakup of Seychelles from India *(Reproduced from Todal and Eldholm, 1998)*. AFR, African plate; OFZ, Owen fracture zone; MB, Mascarene basin; CIB, Central Indian basin; NB, Nazareth bank; SM, Saya de Malha bank; LR, Laxmi ridge. Thick dashed line in the right diagram encircles the Deccan Large Igneous province.

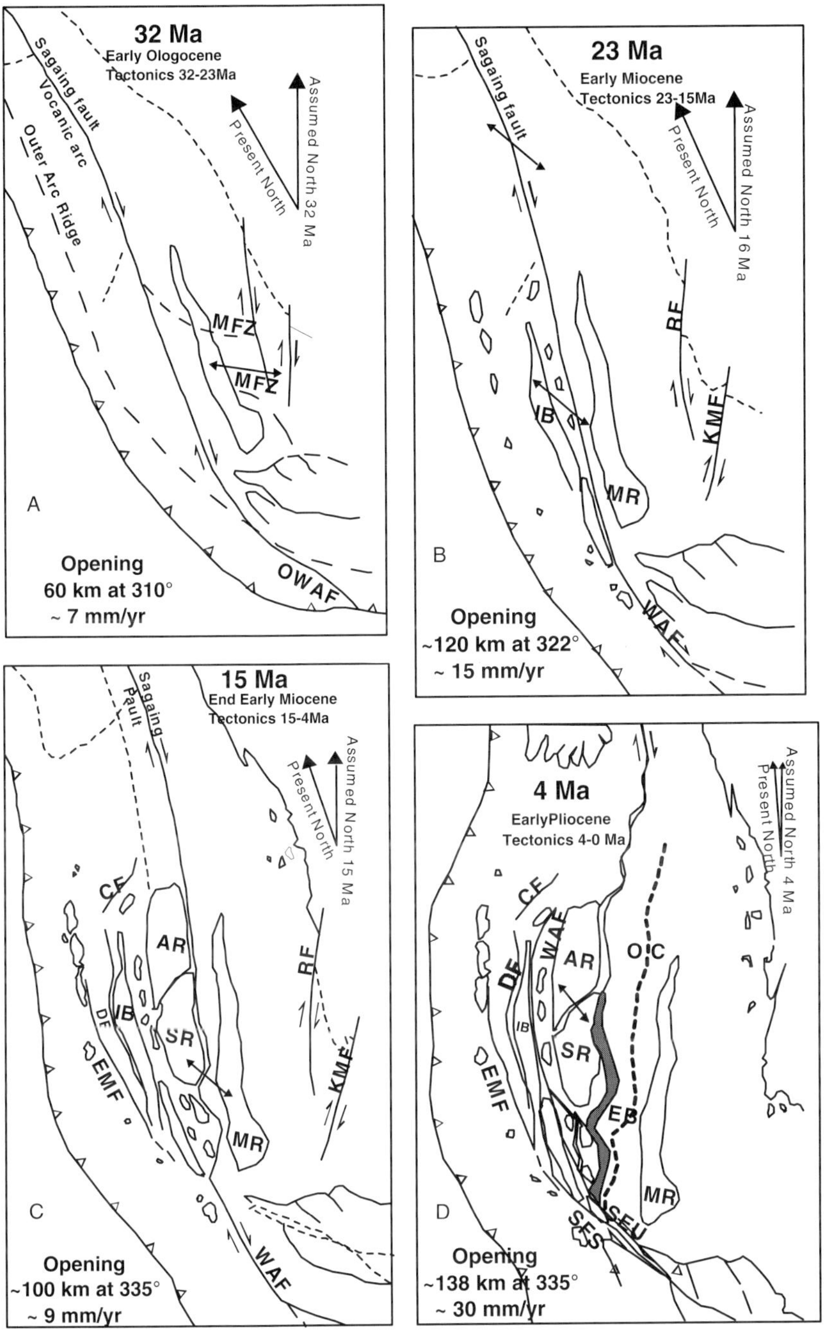

FIGURE 2.17 Reconstruction of the Andaman arc region and the evolution of the back-arc Andaman Sea basin since early Oligocene *(Reproduced from Curray, 2005)*. Details are discussed in the text. OWAF, Old West Andaman Fault; AR, Alcock Rise; SR, Sewell Rise; EB, East Basin; MR, Mergui Ridge; IB, Invisible Bank; CF, Cocos Fault; EMF, Eastern Margin Fault; DF, Diligent Fault; BY, Bago Yoma.

continental margin. Around 59 Ma, the northeast corner of greater India hit this subduction zone (called as soft collision) which was followed by counter clockwise rotation of India until 55 Ma. From this time up to 44 Ma, India was indenting the Asian margin causing the subduction zone to rotate in clockwise direction and led to oblique convergence. During middle to late Eocene, about 44 Ma, a sliver fault had formed which can be inferred to be the forerunner of the old West Andaman and Sagaing Fault systems. The northern strand of the Mergui fault may have crossed the Mergui Ridge as a splay of the Sagaing Fault (Fig. 2.17A), and the Mergui Ridge could have been part of the original volcanic arc. In early Miocene at about 23 Ma (Fig. 2.17B), due to oblique plate convergence, the extension and back-arc sea floor spreading moved westward to the sliver fault, and the seafloor spreading process had formed the rock masses comprising Alcock and Sewell Rises. At the end of early Miocene, about 15–16 Ma (Fig. 2.17C), the Alcock and Sewell Rises started rifting away from the edge of continental crust thus forming the East Basin. About 4 Ma (Fig. 2.17D), the plate edge again migrated and separated the Alcock and Sewell apart and subsequently formed the present plate edge between the Southeast Asian and the Burma Sliver Blocks.

Chapter 3

Regional Tectonic Framework and Crustal Structure

3.1. INTRODUCTION

The passive continental margins of India have evolved as a result of rifting and breakup of India from East Antarctica, Madagascar, and the Seychelles during different geological times. On the basis of plate reconstructions, seafloor spreading magnetic lineations at the conjugate margins, and timings of magmatic episodes, the approximate timing on the respective breakup events could be inferred, such as India–East Antarctica (~130 Ma), India–Elan Bank (~124 Ma), India–Madagascar (~88 Ma), and finally India–Seychelles (~65 Ma). The breakup processes and subsequent geological events have left their imprints on the newly formed Indian margins as well as the surrounding oceanic areas. The seafloor created by spreading processes from the early Cretaceous period to the present, and the aseismic ridges that are believed to be plume traces, document the tectonic history of the region. Several other factors, such as major plate reorganizations, directions and rates of plate motions, interactions of hot spots with spreading centers, intraplate deformation, and collision of India with Eurasia and the Himalayan orogeny, have influenced the evolved crust in different ways. Therefore, an understanding of the evolution of Indian continental margin would require critical assessment of the crustal structure and tectonics of the entire region surrounding them. The passive margin of India seems to have evolved through a variety of tectonic processes ranging from passive to active rifting, shearing, plume activity, and so on with wide variations along different segments (Subrahmanyam and Chand, 2006). In this chapter, salient results compiled from previous investigations along the Indian margins as well as the adjoining oceanic areas are presented particularly with reference to seismic imaging, magnetic anomaly identifications, and modeling from potential field data, which helped in deciphering the age, nature, and structure of the crust in the region for better understanding of the present-day continental margin configuration.

Developments in Petroleum Science, Vol. 59. http://dx.doi.org/10.1016/B978-0-444-53604-4.00003-X

3.2. REGIONAL GEOLOGICAL SETTING

The Precambrian shield of India comprising Archean cratons and Proterozoic mobile belts is surrounded by ocean basins of the Bay of Bengal (BOB) and the Arabian Sea. Some of the major continental rifts, such as the Godavari, Mahanadi, and the Narmada-Son, can be prominently noticed on the regional tectonic map of India, and these are found to be closely linked to similar rifts in their Gondwana counterparts (Mishra et al., 1999; Subrahmanyam et al., 1999). Similarly, the Neoproterozoic shear zones and granulite terrains of south India are also important as they are found to be continuous with similar zones in Madagascar (Janardhan, 1999; Raval and Veeraswamy, 2003a; Yoshida et al., 1999) and East Antarctica (Acharyya, 2000; Yoshida et al., 1992). The structural setup and basin configuration at the margins on either side are partially controlled by some of these onshore trends that continue offshore. On both sides of the Indian shield, the continental margins prograde into deep-sea fans, the Bengal Fan in the east and the Indus Fan in the west, both of which are characterized by thick pile of sediments eroded and transported from the Himalayas. The topographic shaded relief map of the region (Fig. 3.1) reveals that the Indus Fan in the Arabian Sea is bounded by the Carlsberg ridge in the south, and the Owen fracture zone and Murray ridges in the west, while the Bengal Fan located in the BOB is terminated eastward by the Ninetyeast ridge and the Andaman trench–arc system. These bathymetric features have limited the lateral growth of the fan systems. While the Owen fracture zone extending roughly from offshore Karachi to the Carlsberg ridge forms the boundary between the Indian and Arabian plates in the west, in the eastern part, the plate boundary is characterized by the convergent margin setting along the Andaman arc where eastward subduction of the Indian plate takes place below the Eurasian plate (Fig. 3.1). The region lying between the two extreme plate boundaries contains several structural features/aseismic ridges, apart from numerous horst–graben structures at the margin; the most prominent among them are the Chagos-Laccadive, Laxmi, 85° E, Ninetyeast, and the Comorin ridges. These are believed to have formed either during or subsequent to the development of eastern and western continental margins. The nature of crust, the mode of emplacement of these ridges, and the structure of the intervening basins have been a subject matter of debate for several decades.

3.3. BATHYMETRY AND GRAVITY ANOMALIES IN THE INDIAN OFFSHORE AND CONTIGUOUS AREAS

The free-air gravity-shaded relief map of the Arabian Sea and BOB region is presented in Fig. 3.2. This map is prepared on the basis of the high-resolution $1' \times 1'$ grid of satellite-derived gravity anomalies from the DNSC08 Global bathymetry and gravity database (Andersen and Knudsen, 2008) in the oceanic areas. The topography relief map presented in Fig. 3.1 shows that in the western

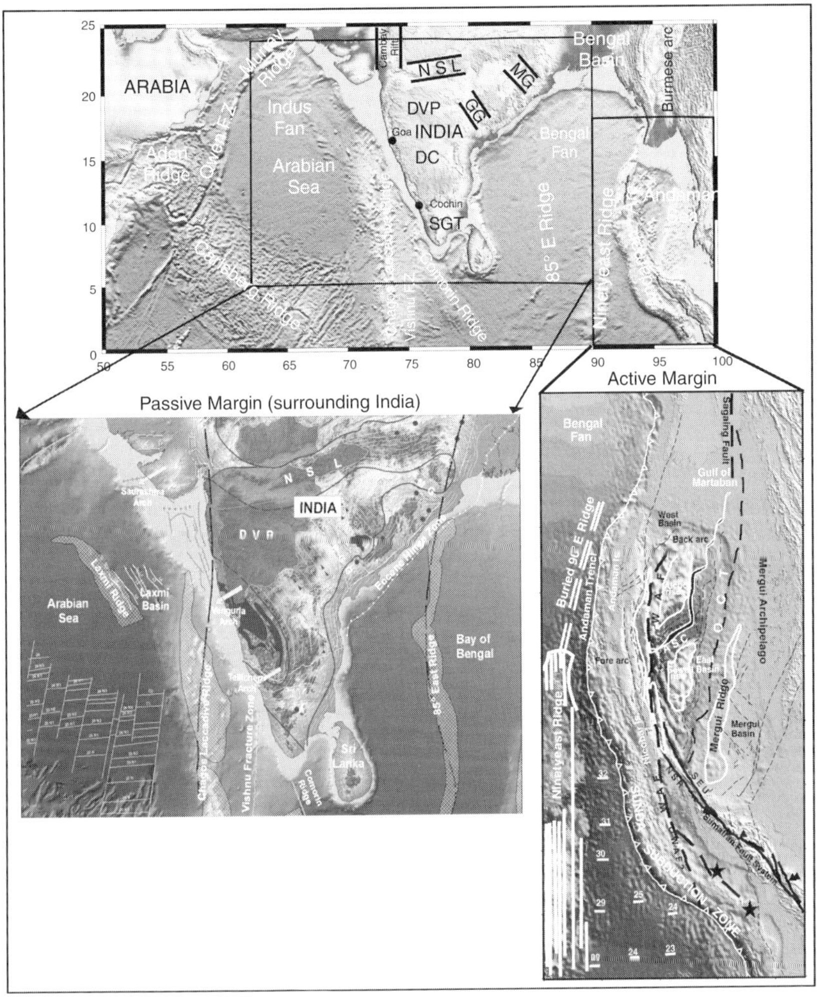

FIGURE 3.1 Topographic relief map of the Peninsular India and the adjoining oceanic areas showing various important structural and tectonic features. Simplified tectonic framework of the passive continental margin surrounding Indian landmass (bottom left panel) and the tectonic map of the active continental margin along the eastern margin of the Bay of Bengal (bottom right panel) are also shown (compiled from several previous works). Details are described in the text. (For color version of this figure, the reader is referred to the online version of this chapter.)

offshore, the width of the shelf varies from nearly 150 km in the north off the Gulf of Kutch, is 300 km in the Bombay offshore region, and gradually narrows down to about 50–60 km in the region off Cochin toward south. A normal continental slope-rise can be recognized north of Goa; however, further north, the morphology of the margin has been modified by the presence of the Indus Fan (Naini and Kolla, 1982). Most of the deep offshore area is characterized by smooth topography of the Arabian basin due to Indus Fan sedimentation.

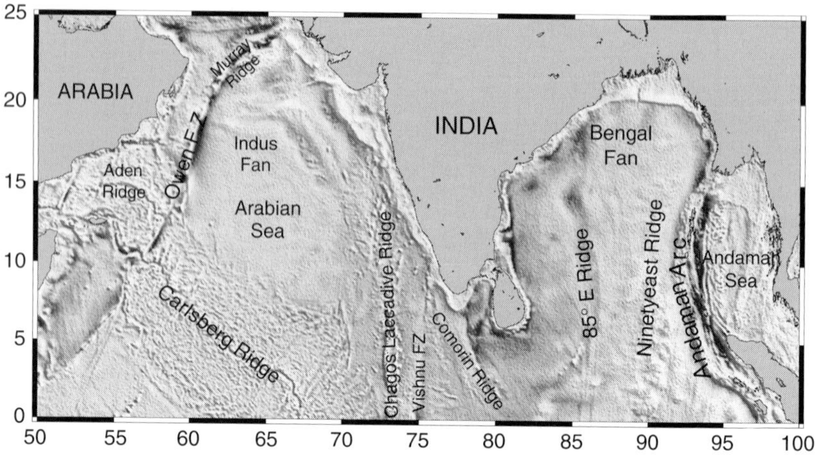

FIGURE 3.2 Shaded relief map of satellite-derived free-air gravity anomalies based on the DNSC08 database (Andersen and Knudsen, 2008) for the Arabian Sea and the Bay of Bengal region. (For color version of this figure, the reader is referred to the online version of this chapter.)

The western boundary of the Indus Fan is characterized by the Murray and Owen ridges along the Owen fracture zone. The Indus River is the most dominant sediment supplier to the Indus Fan; however, in the eastern Arabian Sea, the Narmada and the Tapti rivers have also significantly contributed sediments at the margin (Kolla et al., 1981). The most prominent topographic feature along the Western Continental Margin of India (WCMI) is the Laccadive Ridge (LR), which runs in N–S direction. The region between the LR and the margin south of about 16° N consists of a complex bathymetry with numerous topographic highs and lows (Fig. 3.1). East of the LR, two prominent topographic expressions with limited strike length are observed. These are the Vishnu fracture zone (Norton and Sclater, 1979) along the eastern flank of LR and the Comorin ridge west of Sri Lanka (Kahle et al., 1981). The fracture zone system of the slow-spreading Carlsberg ridge characterized by short-wavelength topography is also seen on the map.

Prominent morphological expressions in the northeast Indian Ocean include the shelf–slope part of the Eastern Continental Margin of India (ECMI), the smooth Bengal Fan floor, the Ninetyeast ridge (NER), Andaman arc–trench, and the Andaman Sea. While the width of the shelf along the eastern margin of India gradually increases from south to north, the base of the slope deepens from north to south (Mukhopadhyay and Krishna, 1991; Rao and Rao, 1985). The depth contours of the eastern continental margin nearly follow N–S trend in the southern part, and further north, they take a NE–SW trend up to Mahanadi coast in the north, thus running subparallel to the coastline (Murthy et al., 1993). The Bengal Fan underlying the northeast

Indian Ocean has a smooth southward slope, where water depth varies from 200 m in the northern part of the Fan to more than 3500 m near the Ten Degree Channel at about 10° N latitude (Curray and Moore, 1971). The BOB region west of the NER shows relatively smoother topography except few isolated topographic features related to seamount volcanism in the distal Bengal Fan areas. The most outstanding morphological feature in the eastern Indian Ocean is the NER; the ridge forms the eastern limit of the Bengal Fan against the Andaman arc, where it separates the Bengal and Nicobar Fans. To the north, the ridge is exposed to about 10° N; beyond that, the ridge plunges beneath the Bengal Fan sediments. The morphological features of the Andaman arc-sea region (after Rodolfo, 1969 and Curray, 2005) can be broadly defined as (i) the outer arc sedimentary complex comprising of Andaman–Nicobar islands between the Burmese coast in north and the Sumatra in south over a length of 1100 km, (ii) the Nicobar Deep toward east of the islands with a prominent bathymetric depression having 2 km of relief, and (iii) the Andaman volcanic arc consisting of volcanoes and seamounts. The Andaman Sea, which can be considered as a back-arc basin, lies east of the Andaman arc and extends up to the Mergui terrace at the Malayan margin. The basin is mostly demarcated by the N–S trending 2-km isobath, while a narrow bathymetric depression in the middle of the Andaman Sea characterizes the Andaman back-arc spreading center.

The gravity anomaly map (Fig. 3.2) shows that except in the inner shelf region, free-air anomalies in general are negative in most part of the eastern Arabian Sea. The bipolar edge effect anomalies, that is, a gravity high in the inner shelf region and a low in the slope, can be seen all along the western margin, except between 12° N and 16° N, where the gravity field seems to be disturbed perhaps due to the presence of several basement ridges and isolated bathymetric features (Radhakrishna et al., 2002). The gravity trends of fracture zones along the Carlsberg ridge can be seen in the southwest corner of the map and a similar grain is noticed on the topography map indicating a close match between the bathymetry and free-air gravity anomalies. A NW–SE trending gravity low of ~−40 mGal correlates well with that of the Laxmi ridge. The Laxmi basin as a whole is characterized by a broad regional gravity high with a prominent NNW gravity low at the center associated with the crest of the Panikkar ridge except in areas where the ridge is elevated and manifested as isolated seamounts (Krishna et al., 2006). North of Laxmi ridge, the gravity high has been inferred by Malod et al. (1997) as due to the presence of a basement high. Between LR and Sri Lanka, prominent N–S trending gravity lineaments along 75° E join the west coast at 10° N and are probably the trace of the Vishnu fracture zone (Norton and Sclater, 1979). Several lineaments having NW–SE orientations can be noticed on the gravity map in the oceans surrounding India and Sri Lanka and reflect the spreading episode between India and East Antarctica (Desa et al., 2006; Krishna et al., 2009; Subrahmanyam and Chand, 2006).

In the BOB, the north-striking gravity anomalies are predominantly observed along the NER, Andaman arc, and the ECMI. It is interesting to note that despite the huge sediment thickness, the Bengal Fan shows a depressed / subdued gravity field where the free-air anomalies range from −20 to −30 mGal. The most prominent gravity signature associated with the ECMI is the bipolar gravity anomaly: a gravity high over the inner shelf that changes to a gravity low at the shelf edge and slope region, and this low is traceable all along the coast off Sri Lanka in south up to about 20° N in the Mahanadi offshore over a strike length of 1700 km (Mukhopadhyay and Krishna, 1991). In the Mahanadi offshore, the shelf-edge gravity anomaly coincides with the gravity anomaly low of the 85° E ridge. Further north, the gravity high at the shelf break is followed by a gently decreasing gravity anomaly toward the deep sea. On the basis of the gravity and bathymetry signatures along the ECMI, Chand et al. (2001) divided the ECMI into two segments, the northern rifted segment and the southern sheared segment, suggesting the nature of breakup at the margin. Faint traces of NW–SE fracture zones have been inferred between the 85° E ridge and the ECMI, revealing early seafloor spreading history between India and East Antarctica (Bastia et al., 2010a; Krishna et al., 2009) with their counterparts in the offshore regions of East Antarctica (Ramana et al., 2001; Rotstein et al., 2001). Detailed mapping of basement features from multichannel seismic (MCS) data in the BOB by Radhakrishna et al. (2012) confirmed the presence of many NW–SE trending fracture zones. It is noteworthy that Sandwell (1992) also identified several NW trending linear features in the gravity images of the offshore region of East Antarctica, which were inferred as representative of the spreading episodes between India and East Antarctica. However, as already discussed in Chapter 2, the breakup and subsequent spreading history between India and Antarctica are not properly resolved because of uncertainties in the magnetic anomaly identifications as well as ambiguity in the position of Elan Bank during the breakup.

The gravity anomaly map (Fig. 3.2) brings out a N–S trending negative anomaly associated with the 85° E ridge, which is quite prominent between 5° N and 15° N; the peak amplitude of 80 mGal occurs at about 14° N. This negative anomaly trend seems to merge with the shelf-edge anomaly at the ECMI, though with less certainty; the present gravity map indicates that the 85° E ridge is abutting the east coast of India at about 18–19° N and further north the extension of the ridge is not visible from the gravity map. Further east, the en-echelon block structure of the NER is very well reflected on the gravity anomaly map. One can notice from the map that positive gravity anomaly of the NER extends further north up to 16° N and possibly beyond (Maurin and Rangin, 2009; Mukhopadhyay and Krishna, 1995; Subrahmanyam et al., 2008a). Apart from the linear N–S positive trend of gravity anomalies, the individual blocks of the NER display NE–SW oriented trends in both bathymetry and gravity (Petroy and Wiens, 1989; Subrahmanyam et al., 2008a). The eastern boundary of the NER is characterized by steep gravity gradient against the Andaman trench/arc zone. However, a distinct offset in the NER gravity trend can be noticed at about 8° N, where it blends with the trench gravity low (Subrahmanyam et al., 2008a). While, a strong gravity low coincides with the Andaman trench, a comparable gravity high outlines the volcanic arc in the overriding Burma plate (Mukhopadhyay, 1988). The gravity anomaly high seen all along the Andaman fore arc also shows a prominent break in its continuity at about 7–8° N (Radhakrishna et al., 2008; Subrahmanyam et al., 2008a). The most prominent gravity anomaly in the entire region is the gravity low related to the fore arc basin, where the anomaly drops below −200 mGal level. This gravity low is reflected as a curvilinear belt of negative gravity contours running close to and just behind the fore arc and seems to fork out with one arm of the fork continuing behind the fore arc, while the other veers out and blends with the trench negative gravity anomaly in front of the Sumatra arc (Subrahmanyam et al., 2008a).

3.4. STRUCTURE AND TECTONICS OF THE WESTERN OFFSHORE

3.4.1 Carlsberg Ridge

The Carlsberg ridge, located in the northwest Indian Ocean, separates the African and Indian plates. It extends from 2° N to the Owen transform at 10° N, which offsets the Carlsberg and Sheba ridges. It falls under the category of slow-spreading mid-ocean ridge system, which is spreading at a half-spreading rate of 12 mm/year characterized by rugged topography along its flanks and a central median valley (McKenzie and Sclater, 1971). The ridge was probably formed during the late Paleocene (~61–63 My) by a WNW trending spreading center between the Seychelles Bank and the Laxmi ridge (Chaubey et al., 1995; McKenzie and Sclater, 1971; Mercuriev et al., 1996; Norton and Sclater, 1979) and gave rise to the conjugate Arabian and Eastern Somali basins.

3.4.2 Owen Fracture Zone

The northern part of the Carlsberg ridge is discontinuous and characterized by the presence of the major NE–SW trending Owen transform at 58° E. The Owen Transform lies on the India–Somalia plate boundary and offsets the Carlsberg ridge from the Sheba ridge by nearly 300 km characterized by sinistral motion (Gordon and DeMets, 1989). Two seismic profiles across the transform have imaged the uplifted mantle rocks (Bonatti and Hamlyn, 1978). It can be seen that the Owen transform extends on either side as a fracture zone; however, the northern limb of the fracture zone shows considerable seismicity and relative motion, due to which, Gordon and DeMets (1989) considered it as a plate boundary separating the Arabian and Indian plates. Another interesting aspect of this fracture zone is that the triple junction between the Arabia–India–Somalia plates has been identified by Fournier et al. (2001) at the bend of the Sheba ridge near 56° E, with the Owen fracture zone being connected to the Sheba ridge by an ultraslow divergent boundary.

3.4.3 Makran Coast and Gulf of Oman Basin

Along the continental margin off Pakistan and Iran, the offshore extension of Makran accretionary complex is an active plate margin in which the oceanic crust of Gulf of Oman has been subducting under the Asian continent since late Cretaceous (Arthurton et al., 1982). Fruehn et al. (1997) observed extremely low angle of subduction and suggested that subduction continued at least since 45–55 Ma. The Gulf of Oman basin is an oceanic basin covering an area of 300-km wide and 950-km long lying between Oman and southern Pakistan–southern Iran, bordered on the west by Arabia and on the east by the Murray ridge (Fig. 3.3) (Uchupi et al., 2002). Progressive accretion onto the Asian plate resulted in a coast-parallel topography characterized by uplifted basins and ridges along the Makran coast (White, 1982). The topography of the accretionary wedge ranges from more than 3000 m below sea level to

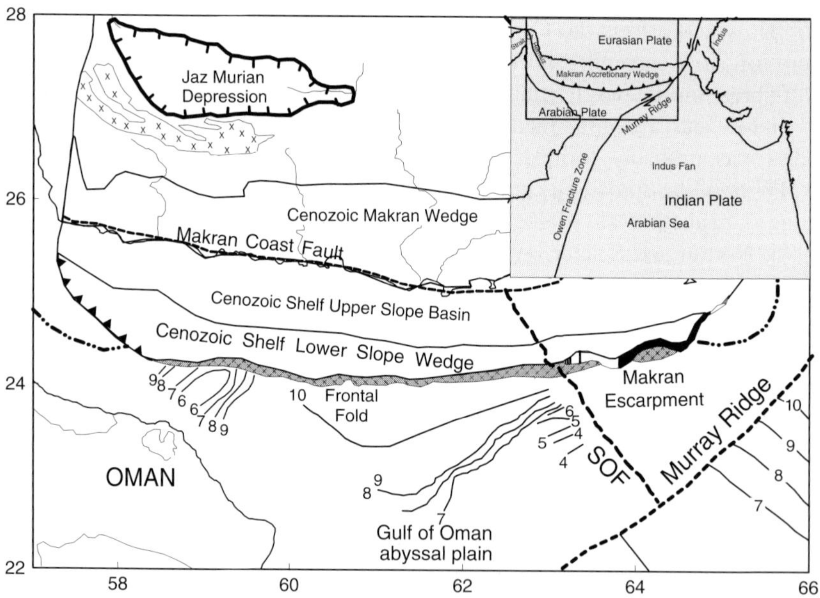

FIGURE 3.3 Tectonic map of the Gulf of Oman basin region (Redrawn from Uchupi et al., 2002). Contour values represent depth to oceanic basement. Inset shows major plate boundaries in the northern Arabian Sea (after Gaedicke et al., 2002a).

heights of 1500 m above sea level (Gaedicke et al., 2002a). In the Oman abyssal plain, around 7 km of thick sediments are present at the toe of the accretionary wedge. According to Uchupi et al. (2002), the accretionary wedge can be morphologically divided into three segments, the northern, central, and the offshore segments; while, the northern wedge developed in Cretaceous–Paleocene time is characterized by microcontinents, the central segment along the Makran coast is made up of Eocene, Oligocene, and middle Miocene distal and proximal turbidites derived from the east, and upper Miocene–Holocene strata derived from the north, and the offshore segment belonging to Eocene–Holocene wedge consists of middle Miocene and older turbidites derived from the east and Makran sands of late Miocene and younger derived from the north. The presence of submarine canyons and the left lateral Sonne fault (Kukowski et al., 2000) in the offshore (Fig. 3.3) further complicates the morphology and tectonic fabric of the offshore accretionary wedge.

3.1.3.1 Murray Ridge

The Murray ridge extending from the continental slope south of Karachi to the northernmost part of the Owen fracture zone is part of the plate boundary separating the Indian and Arabian plates (see inset Fig. 3.3). The ridge is not a single topographic feature and forms part of several subparallel sediment-covered ridges, seamounts, basins, and troughs (Burgath et al., 2002). Seismic reflection profiles and earthquake focal mechanisms show many extensional structures in the Murray ridge and Dalrymple trough region (Edwards et al., 2000; Gordon and DeMets, 1989). The region contains more than 8 km of thick Cenozoic sediments and is characterized by the presence of submarine volcanism near the Murray ridge crest (Gaedicke et al., 2002b). Geophysical studies in the region suggest that the southern Murray ridge and the Dalrymple trough region are underlain by faulted and tilted blocks of thinned continental crust having thick volcanic rock associations (Edwards et al., 2000; Gaedicke et al., 2002b; Malod et al., 1997). The ridge divides the northern part of the Arabian Sea into two, the north–south trending present-day Indus Fan in the Arabian basin and an east–west trending Proto-Indus Fan along the Makran coast. According to Whitmarsh (1979), the Indian–Arabian plate boundary has been in its present position, that is, along the Murray ridge since 90 Ma, whereas Mountain and Prell (1990) proposed that the plate boundary was initially located along the Oman margin in the west and jumped to its present position during Oligocene–early Miocene time, which was further supported by Barton et al. (1990) on the basis of Oman margin being a former transform boundary. The geophysical data interpreted from the Gulf of Oman (Edwards et al., 2000) could not support either of the above propositions. However, detailed seismic investigations in the Murray ridge region (Clift et al., 2001; Gaedicke et al., 2002) suggest that the ridge was brought to the present level by two major uplift events in the geological past. While, the initial uplift of the ridge took place during the Eocene–Oligocene period, the second phase of significant uplift took place along with the Owen ridge in early Miocene time during the regional plate reorganization in the western Indian Ocean.

3.4.4 Arabian Basin

The Arabian basin forms deep oceanic areas of the Arabian Sea and is covered by Indus Fan sediments. The basin is bounded by the Owen fracture zone in the northwest, Carlsberg ridge in the southwest, and the LR and Laxmi ridge along the east and northeast (Fig. 3.1). Magnetic anomaly identifications in the Arabian basin indicated the presence of continuous sequence of seafloor spreading isochrons 28ny to 18ny (Fig. 3.4) between the Owen fracture zone near the vicinity of Laxmi ridge (Chaubey et al., 1998). The oldest oceanic crust mapped 28ny is close to the Laxmi ridge. The oblique offsets in magnetic lineations between anomalies 28ny and 20ny (Chaubey et al., 1998, 2002b; Dyment, 1998; Royer et al., 2002) represent the pseudo-faults associated with paleo-propagating ridges. The thickness of sediments is variable in the basin and ranges from 1.3 to 4.2 km with velocity variations from 1.7 to 3.8 km/s below which typical oceanic crust underlies having average velocities of 5.5 and 6.6 km/s for the oceanic layers 2 and 3, respectively (Naini and Talwani, 1982).

3.4.5 West Coast of India and the Adjoining Continental Margin

The WCMI has evolved in different stages, first, through rifting and subsequent seafloor spreading between India and Madagascar during mid-Cretaceous ($\sim$88 Ma) and later through rifting between India and Seychelles during late Cretaceous (Norton and Sclater, 1979). The margin comprises several surface/subsurface ridges, such as the Chagos-Laccadive ridge (CLR), Prathap ridge, and Laxmi ridge, and a belt of numerous horst–graben structural features (Fig. 3.4), which are believed to have been controlled by the Precambrian structural grain of the Indian shield (Biswas, 1987; Kolla and Coumes, 1990). Many of the structures and horst–graben features have been related to the rifting history between India and Madagascar. Hence, it is necessary to examine the geology and tectonics of the western part of the Indian shield to unravel the basement control on the breakup. Further, the northern part of the margin encountered a major flood basalt episode and another breakup event between India and Seychelles. In order to understand the evolutionary history of the WCMI, it is important to study these two breakup events properly. In the following sections, we present the morphotectonics and geology of the western Indian shield region and the geophysical characteristics of major tectonic features of the WCMI so as to discuss the structure and nature of crust and development of the margin.

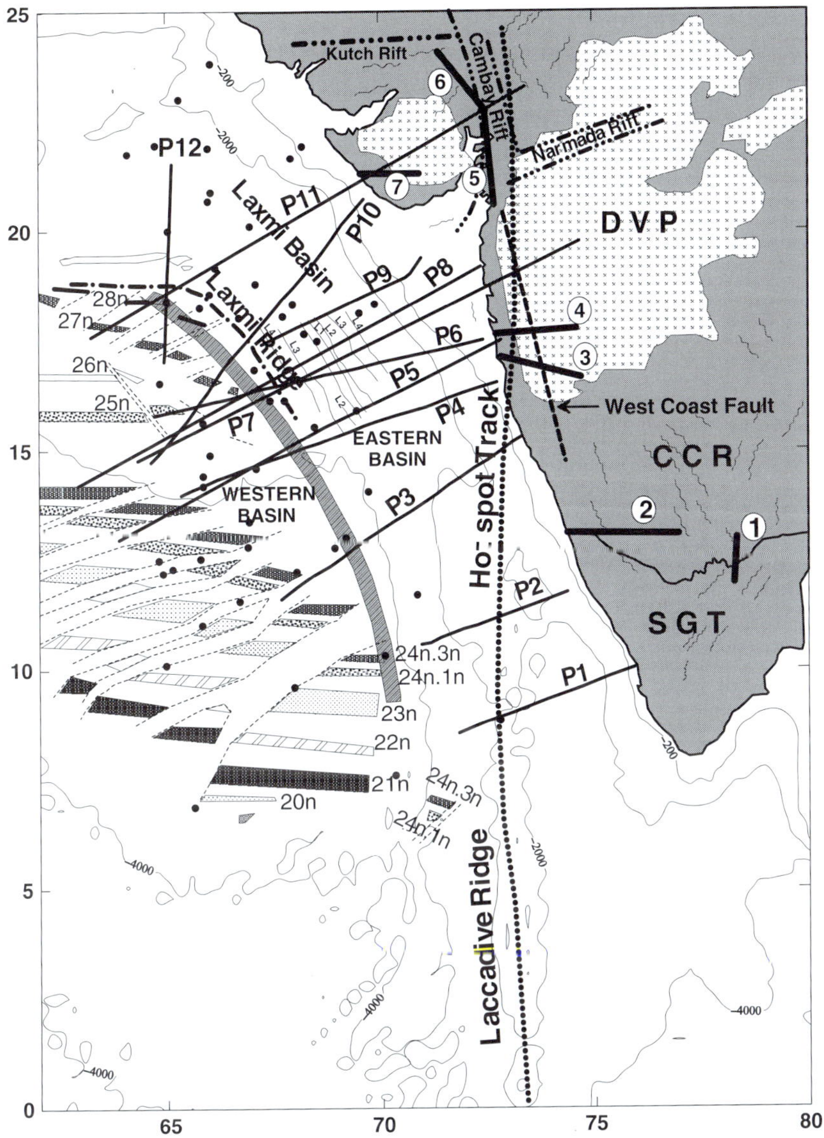

FIGURE 3.4 Regional tectonic framework of the western Indian shield and the contiguous eastern Arabian Sea showing magnetic anomaly identifications in the western basin (Chaubey et al., 1998, 2002b) and in the Laxmi basin L1–L4 (Bhattacharya et al., 1994). Dots are seismic refraction stations (Naini and Talwani, 1982). Thick dotted line indicates the track of the reunion plume. Thick lines with numbers encircled along the West Coast show the location of Deep Seismic Sounding (DSS) profiles. Lines P1 through P12 show the location of the MCS lines, crustal transects, or gravity models discussed in the text. DVP, Deccan Volcanic Province; CCR, Central Cratonic Region; SGT, Southern Granulite Terrain.

3.4.5.1 Regional Structural Trends

The western Indian shield region consists of Archean gneisses, schists, charnockites, and metamorphosed sedimentary rocks with most of the northern part, north of Goa, covered by the Deccan traps (Das and Ray, 1976). Qureshy (1981) identified a number of linear features mostly having N–S and NW–SE trends on the basis of landsat imagery data over the Indian Subcontinent. Krishnan (1953) described persistent regional trends such as regional foliation strikes, fold axes of ancient mobile belts, and strike faults noticeable in the Precambrian rocks of India which manifest as major tectonic lineaments. The dominant Precambrian structural and tectonic trends observed along the West Coast (Fig. 3.5) are the NW–SE to NNW–SSE Dharwar trend; NE–SW Aravalli trend; and ENE–WSW to E–W Satpura trends (Biswas, 1982). These trends are seen as metamorphic belts representing three major orogenic cycles; of these, the Aravalli and Dharwar represent the oldest cycles followed by the Satpura (Biswas, 1982). Along the coast, the Dharwar trend dominates up to Bombay and further north, in the Gulf of Cambay region, the Satpura trend dominates the structural style, whereas in the Kutch–Saurashtra region, the Aravalli trend is predominant. The northern component of Aravalli, the trend of the Delhi fold belt, swings E–W and continues into the Kutch region across the Cambay graben. According to Biswas (1982), this Precambrian structural grain has a bearing on the rifting style from north to south during the breakup of eastern Gondwanaland.

3.4.5.2 Deccan Volcanic Province

The Deccan Volcanic Province (DVP) is one of the world's largest Continental Flood Basalt (CFB) provinces similar to those of Columbia River, Parana, Karoo, and Etendeka and is known to be associated with the rifting and continental breakup of India and the Seychelles during Paleocene (Chandrasekharam, 1985; Courtillot et al., 1986; Norton and Sclater, 1979; Todal and Eldholm, 1998; Vandamme et al., 1991). The CFBs are believed to be originated from the mantle plumes, though dissenting views exist (e.g., Sheth, 2005). It is well known that the DVP has resulted from the Reunion mantle plume activity; however, its association with the breakup event between India and Seychelles is not clear (Fig. 3.6). The Deccan volcanic rocks have erupted over the Precambrian crust of Dharwar, Aravalli, Bastar, and Bundelkhand cratons, and the Satpura mobile belt, and are at places underlain by Cretaceous sediments that are described as infra-trappeans. The volcanic rocks of the DVP, popularly known as Deccan traps, presently cover an area of 5×10^5 km^2 in the western and central India and constitute the best Large Igneous Provinces (LIP) of the world (Mahoney and Coffin, 1997). Considering the area prior to erosion and the concealed portion in the offshore, the original extent of the traps could be 1.5×10^6 km^2 (Deshmukh, 1982). The intrusive phases of the traps extend into many neighboring areas

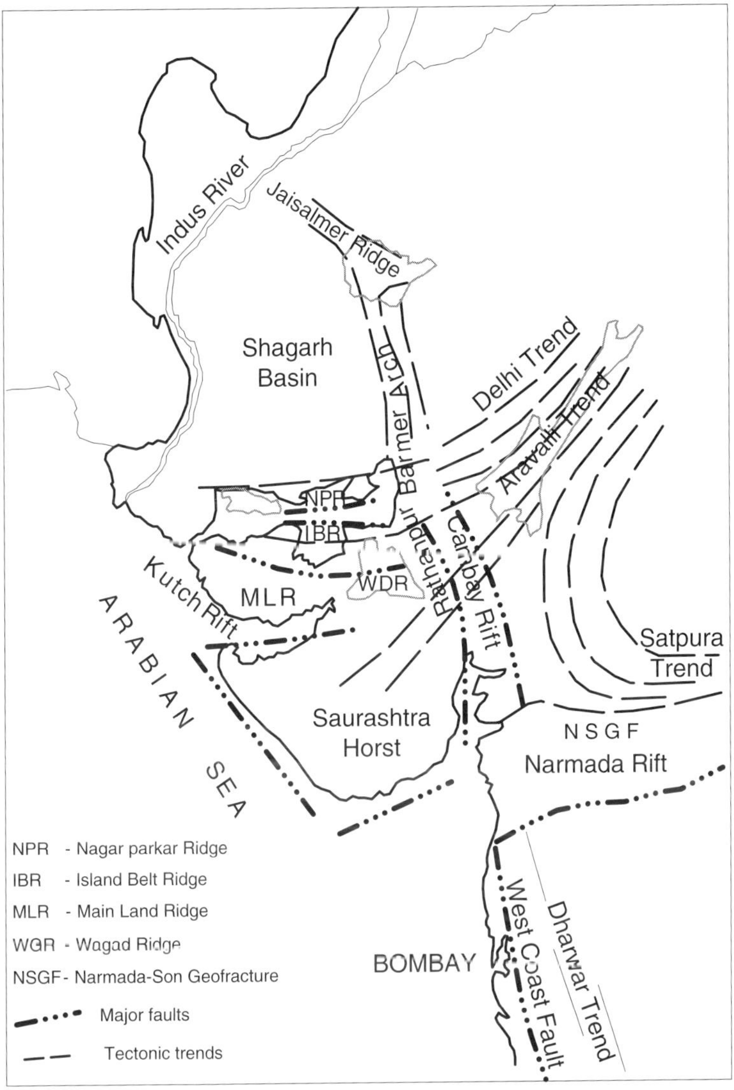

FIGURE 3.5 Tectonic map of the northwest Indian shield showing the pericratonic rift basins and major structural lineaments *(adapted from Biswas, 1987)*.

as far as Karnataka and Kerala dykes in the south (Kumar et al., 2001; Radhakrishna et al., 1994) and Rajahmundry in Krishna–Godavari basin (Baksi, 2001) in the east coast (Fig. 3.6). The volcanic rocks of the province are dominantly subaerial flows of tholeitic basalts with subordinate picrites, picritic basalts, and alkaline basalts. The trap thickness estimated from Deep Seismic Sounding (DSS) profiles (Kaila, 1988) in the DVP indicates that they

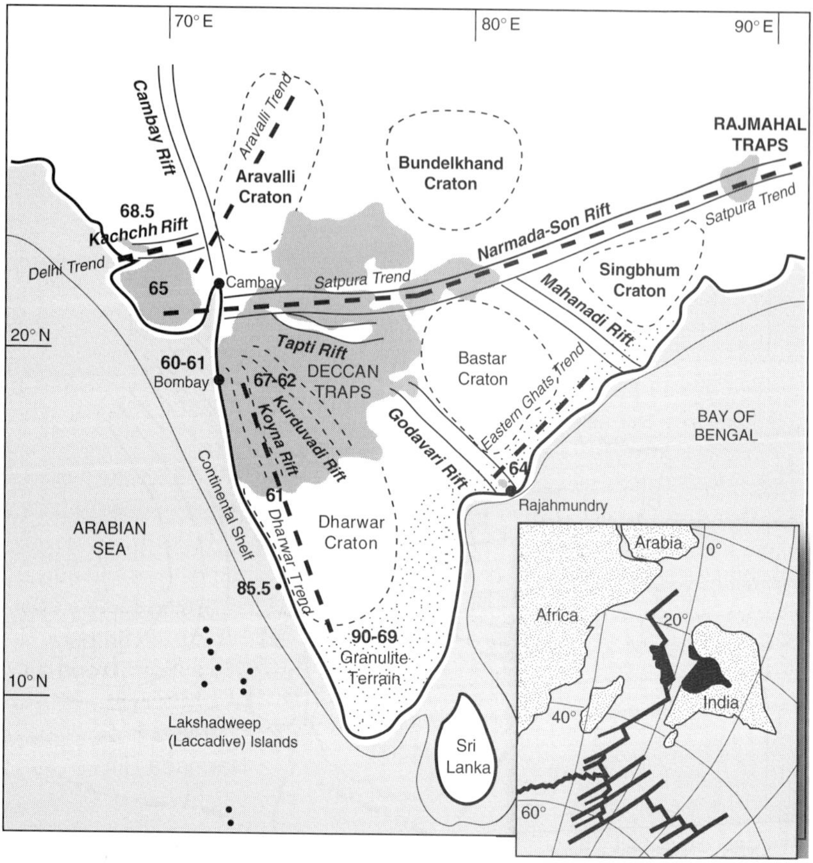

FIGURE 3.6 Map showing the Deccan Volcanic Province, Precambrian cratons, and structural trends in the Indian shield region. Dots with numbers are age of volcanics. Breakup of Seychelles from India is shown in the inset *(Reproduced with permission from Sheth, 2005 © Geological Society of America).*

are the thickest over the Western Ghats, especially along the Kalsubai–Mahabaleshwar axis and is estimated to be of the order of 1.5 km, reaching a maximum of 2 km locally. However, the trap cover thins down eastward, and within some 500–600 km, the Precambrian basement is exposed. In the low Konkan plains, west of the erosional scarps of the Western Ghats, the trap thickness is <1 km, the upper part of the trap having been removed by erosion (Widdowson and Cox, 1996). Owing to the westerly dip (Sukheswala and Poldervart, 1958), the thickness of traps increases to 1.2 km along the coastal tract and because of erosion, the flood basalts are best exposed over the Western Ghats region. The exposed Deccan Basalts at many places contain mantle nodules and emplacement of volcano–plutonic complexes (Chandrasekharam, 2003). There is considerable ambiguity regarding the age of volcanics and time span of emplacement of flood basalts. The emplacement time window for the traps ranges between very rapid, catastrophic events having 1.0–0.5 My span (Allegre et al., 1999; Courtillot et al., 1988; Duncan and Pyle, 1988; Hofmann et al., 2000) to wider time span of 4–5 My (Pande, 2002; Venkatesan et al., 1993).

3.4.5.2.1 Panvel Flexure

Several prominent CFB provinces of the world such as the Karoo in South Africa, Parana in South America, the East Greenland, and the Deccan province in India exhibit coastal monoclinal flexures (Cox, 1988; Nielsen and Brooks, 1981). The monoclinal flexure in the Deccan CFB province along the west coast of India (Fig. 3.7) was noticed long ago by Blanford (1867). Auden (1949) described the structure in detail and called it the "Panvel flexure." He ascribed the flexure to simple monoclinal bending of the lava pile. Later, several other workers proposed the flexuring due to warping of crust in response to rifting (Devey and Lightfoot, 1986), uplift of the Western Ghats, and consequent subsidence of the western margin (Watts and Cox, 1989). More recent studies of this region, on the other hand, doubt this feature as a monoclinal flexure and suggest it as an extensional fault structure comprising en-echelon east-dipping faults and westerly tilted fault blocks (Dessai and Bertrand, 1995), a reverse drag structure on an east-dipping master fault that lies offshore Bombay having numerous subsidiary antithetic and sympathetic faults (Sheth, 1998), represent the eastern periphery of large circular structure (Mishra, 2001).

3.4.5.3 Western Ghats Escarpment and Development

The West Coast of India has an important geomorphological feature characterized by an elevated inland plateau, a huge erosionally controlled escarpment, coast parallel monoclinal flexure, and the low-lying coastal plain (Gunnell and Radhakrishna, 2001), and this feature runs parallel to the coast all along its length (Fig. 3.7). The elevated portions are flat topped and covered by laterites. This precipitous terrain is known as the Western Ghats (Gunnell and Radhakrishna, 2001). Significant variation in the lithology and age of rocks forming the escarpment is observed which consists of peninsular gneisses, granulites, and the Deccan basalts (Radhakrishna and Vasudev, 1977). Gunnell and Fleitout (2000) inferred the formation of the escarpment to the post-Cretaceous processes of shoulder uplift and scarp recession. Based on the Apatite Fission Track Analysis, Gallagher et al. (1998) suggested that the escarpments had formed due to uplift during the rifting, followed by a lateral scarp retreat. Differential denudation and flexural isostasy is suggested by Gilchrist and Summerfield (1991) as a mechanism for the formation of Western Ghats. Such geomorphological features have also been observed along other passive margins and at the rifted CFB provinces, such as Parana basin

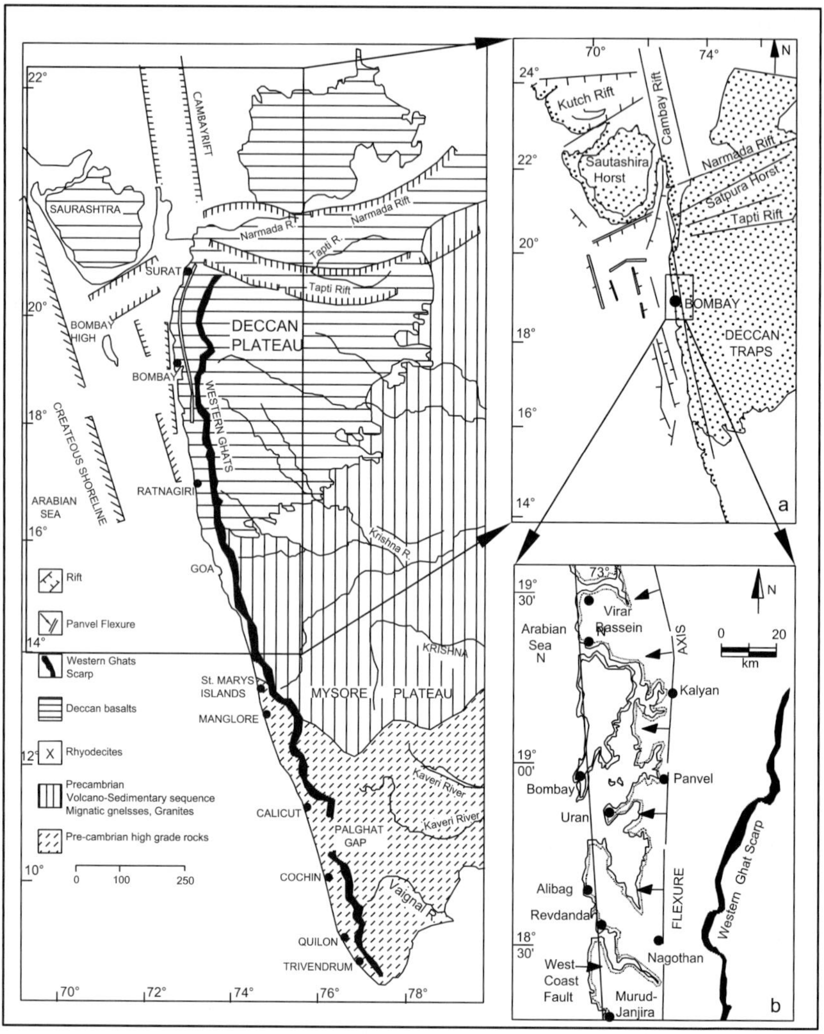

FIGURE 3.7 Generalized geological and tectonic map (left panel) of the West Coast of India *(adapted from Subrahmanya, 1998)*. The Western Ghat scarp and Panvel flexure are indicated. The area indicated in rectangle is further zoomed (right panel) to present first-order tectonic features (A) and the axis of Panvel flexure and the West Coast fault *(adapted from Sheth, 1998)*.

in Brazil, Karoo of SE Africa, and Etendenka of SW Africa (Gallagher et al., 1998). Detailed studies carried out to understand such remarkably consistent morphological expression brought out two basic models: (i) the downwarp model (King, 1967; Ollier and Pain, 1997) and (ii) the scarp retreat model (Gilchrist and Summerfield, 1990; Gilchrist et al., 1994; Kooi and Beaumont, 1994). For the Western Ghats, the studies helped to classify the western margin as a high elevated rifted margin (Chand and Subrahmanyam, 2003; Gunnell and Fleitout, 1998). The above-mentioned studies also stressed the significance of the offshore sedimentary record to understand the landscape development at the coast, because for generating models on flank-uplift topography, the onshore denudation, sedimentary deposition offshore, and the resulting flexural response form important constraints (Brown et al., 1990; Gilchrist and Summerfield, 1991; Rust and Summerfield, 1990). Further, the phenomena of uplift at the margin have been attributed by many workers to magmatic underplating (McKenzie, 1984; Watts and Cox, 1989), transient thermal effects (Cochran, 1983), the rift-related mechanisms of crustal thinning (Royden and Keen, 1980; White and McKenzie, 1988), secondary convective effects associated with extension (Buck, 1986), and flexural unloading (Weissel and Karner, 1989).

3.4.5.4 Onshore–Offshore Tectonic Linkages at the West Coast

The western offshore areas comprise several surface as well as subsurface ridges and horst–graben complexes evolved during the rifting history of the margin. It is believed that some of these structural trends at the margin were mainly controlled by the Precambrian structural grain of the western Indian shield margin (Biswas, 1987; Kolla and Coumes, 1990; Subrahmanyam et al., 1995). As mentioned in Section 3.4.5.1, three major Precambrian structural trends seem to terminate along the West Coast of India such as the NNW–SSE Dharwar trend, the NE–SW Delhi–Aravalli trend, and the ENE–WSW Satpura trend. Several previous workers (Bhattacharya and Subrahmanyam, 1986; Biswas, 1987; Biswas and Singh, 1988; Kolla and Coumes, 1990; Rao, 1984; Singh and Lal, 1993; Subrahmanyam et al., 1993) presented geophysical evidence on the extension of structural and tectonic trends from the Indian subcontinent into shelf areas of the western Indian margin. Some of these tectonic trends controlled the structural styles of intracratonic and pericratonic sedimentary basins (Cambay, Kutch, and Narmada) of the West Coast (Biswas, 1987), while movements along the ancient structural trends greatly influenced the evolution of basins in the offshore. Kolla and Coumes (1990) noticed that the fault trends, basement highs, and structural patterns in the shelf and slope region offshore western India and Pakistan are mainly parallel to the Precambrian structural grain of the western Indian shield (Fig. 3.8A). Analysis of geophysical data in the western offshore areas revealed extension of Narmada-Son lineament into the Mumbai offshore (Fig. 3.8B) area (Bhattacharya and Subrahmanyam, 1986), a NNW–SSE trending mid-shelf basement high, and the presence of several lineaments that extend from onshore into the offshore areas (Fig. 3.8C) between Goa and Kasargod in the central part of WCMI (Subrahmanyam et al., 1993). Singh and Lal (1993) observed that several structures in the shelf region along the Konkan–Kerala basin have continuity from the mainland (Fig. 3.8D).

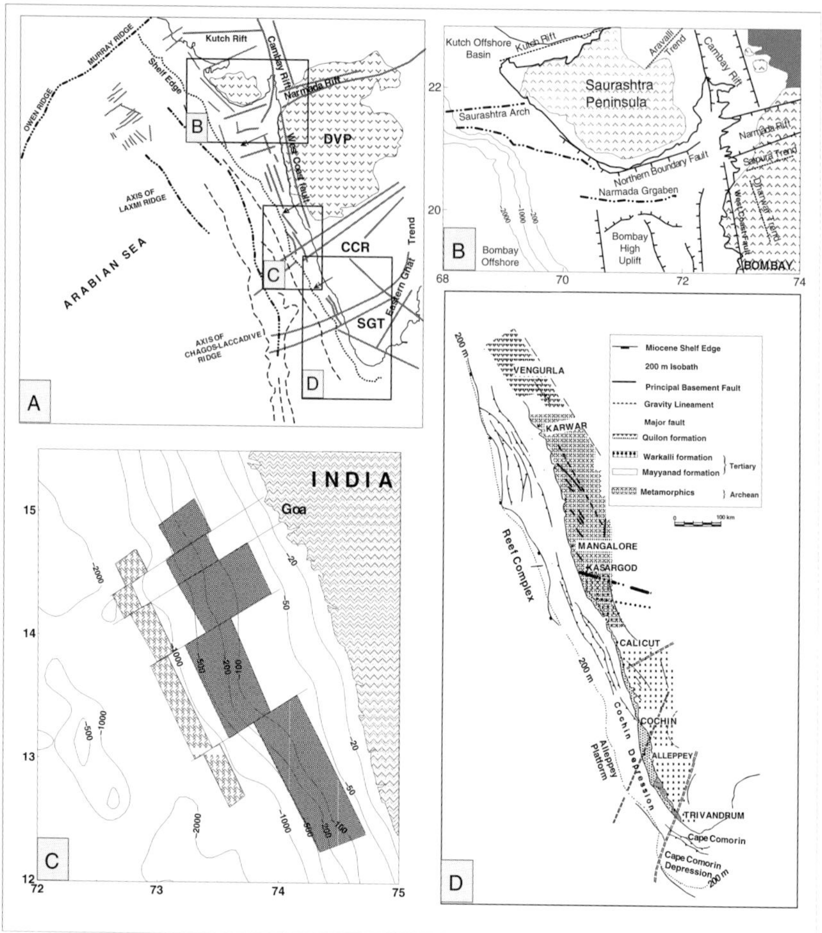

FIGURE 3.8 Map showing the continuation of onshore structural trends of India into the western offshore areas. (A) Structural trends adapted from Kolla and Coumes (1990). (B) Extension of Narmada-Son lineament into the offshore *(adapted with permission from Bhattacharya and Subrahmanyam, 1986 © Springer)*. (C) Structural offsets in the central part of the margin *(adapted from Subrahmanyam et al., 1993)*. (D) Structural continuity into the shelf areas of the Konkan–Kerala basin *(adapted from Singh and Lal, 1993)*.

3.4.5.5 Seismic Refraction Velocities and Crustal Seismic Transects

The WCMI and the adjoining deep oceanic areas of the eastern Arabian Sea consist of numerous structural features namely, the Laccadive ridge, the Laxmi ridge, the Prathap ridge complex, and the Panikkar ridge, which are mostly buried under the sediments (Naini and Talwani, 1982). The Laccadive and Laxmi ridges divide the eastern Arabian Sea into the Western and Eastern basins (Fig. 3.4). While the geophysical studies indicate that the Western basin is underlain by oceanic crust (Chaubey et al., 1995; McKenzie and Scalter, 1971; Naini and Talwani, 1982; Whitmarsh, 1974), the nature of crust below the Eastern basin could not be clearly established.

Seismic data along the WCMI and the deep oceanic parts of the Arabian Sea include wide-angle reflection and seismic refraction surveys carried out by Francis and Shor (1966), Rao (1967), Naini and Talwani (1982), Naini and Kolla (1982), Chaubey et al. (2002a), Krishna et al. (2006), and Minshull et al. (2008). Information on crust/mantle structure along the West coast of India is available from seven refraction DSS profiles collected by National Geophysical Research Institute, Hyderabad, India, along the Kavali–Udipi profile (Kaila et al., 1979; Reddy et al., 2000; Sarkar et al., 2001), Guhagar–Chorochi profile (Koyna-I) and the Kelsi–Loni profile (Koyna-II) in Koyna region (Kaila et al., 1981a,b; Krishna et al., 1991), Mehmadabad-Billimora profile in Cambay basin (Kaila et al., 1981b), Amreli–Navibandar profile in Saurashtra (Kaila et al., 1988), and Kuppam–Palani transect in the Southern Granulite Terrain (SGT) (Reddy et al., 2001; Vijaya Rao et al., 2006), and Dharimana–Gandhinagar profile (Kaila et al., 1990b). The locations of available refraction stations in the eastern Arabian Sea and the DSS profiles in the western Indian shield are shown in Fig. 3.4. The interpreted DSS sections belonging to the west coast are shown in Fig. 3.9.

DSS investigations have clearly shown the unstretched continental crust and its post-rifting features in the western Indian shield. A few inferences from the DSS studies along the west coast (after Radhakrishna et al., 2002) as revealed from Figs. 3.9 and 3.10 are (i) the velocities for Deccan traps range between 4.5 and 5.5 km/s and have maximum thickness of 2.0 km. The traps are underlain by Mesozoic sediments having velocities of 4.0 km/s in Saurashtra; (ii) a two-layered continental crust, an upper crust of 10–20 km characterized by velocities of 5.8–6.5 km/s and a lower crust having velocities of 6.6–6.9 km/s. The Moho is observed at a depth of 38–42 km in the shield regions; (iii) Moho in general shallowing towards the coast; (iv) high lower crustal velocities ($>$7.0 km/s) at 23–25 km depth and a shallow Moho at a depth of 31–33 km below Cambay rift basin; (v) transitional Moho and low-velocity layers in the Koyna region; and (vi) continental type of crust below Saurashtra.

The sonobuoy refraction data shown as velocity–depth profiles in Fig. 3.11 compiled from Naini and Talwani (1982) show distinct variations in crustal structure below the western and the eastern basins in the Arabian Sea. A summary of this data indicates that the sedimentary layers are characterized by velocities varying from 1.7 km/s for top unlithified sediments to 4.9 km/s for the deeper layers above the acoustic basement, the basal sedimentary layers in the Eastern basin and the Laxmi ridge region show velocities in the range of 4.0–4.9 km/s, and the acoustic basement having velocity of 5.0 km/s. For the crustal layers, velocities between 5.2 and 7.3 km/s (with predominantly 6.2–6.4 km/s values) have been observed. A few refraction stations in the Laxmi basin and the Laxmi ridge show velocities $>$7.0 km/s in the lower crust,

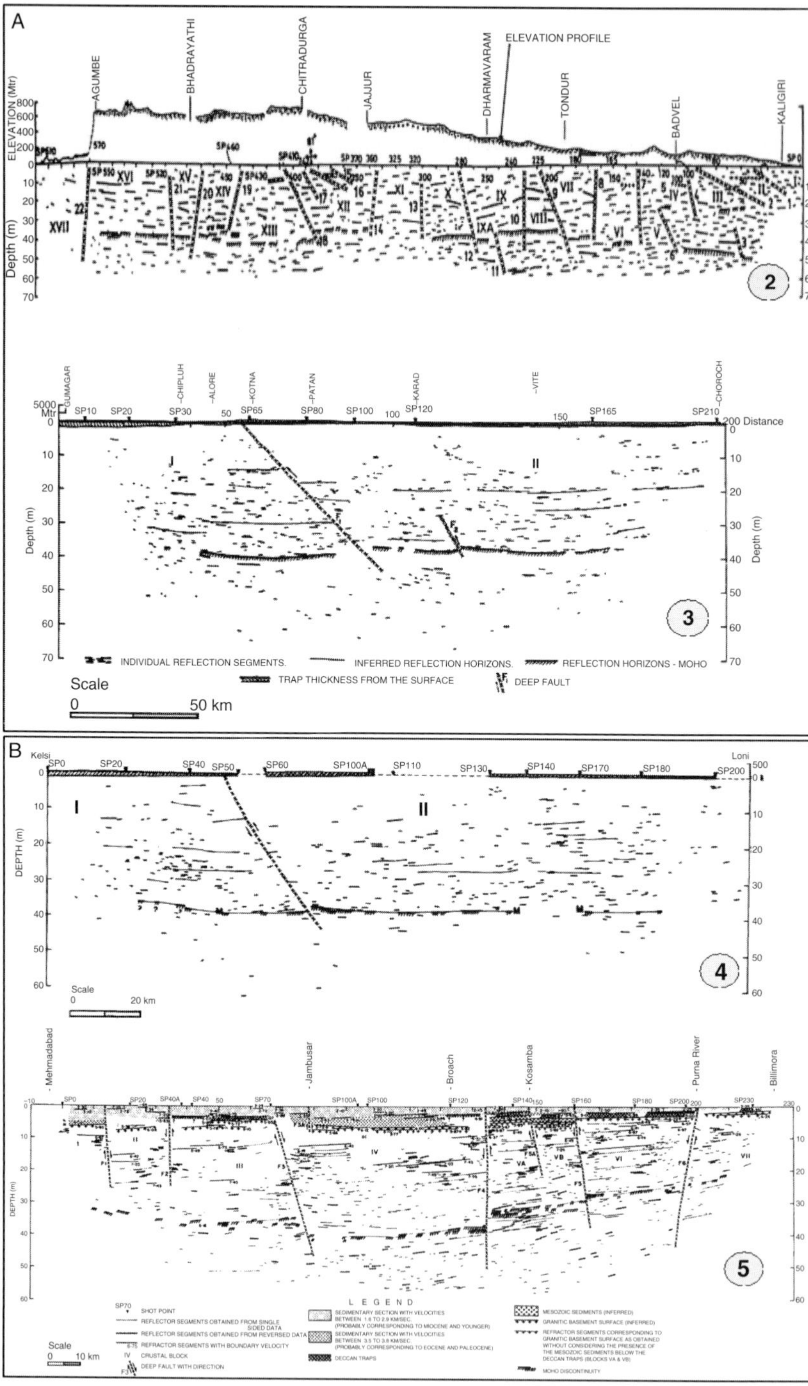

FIGURE 3.9—Cont'd

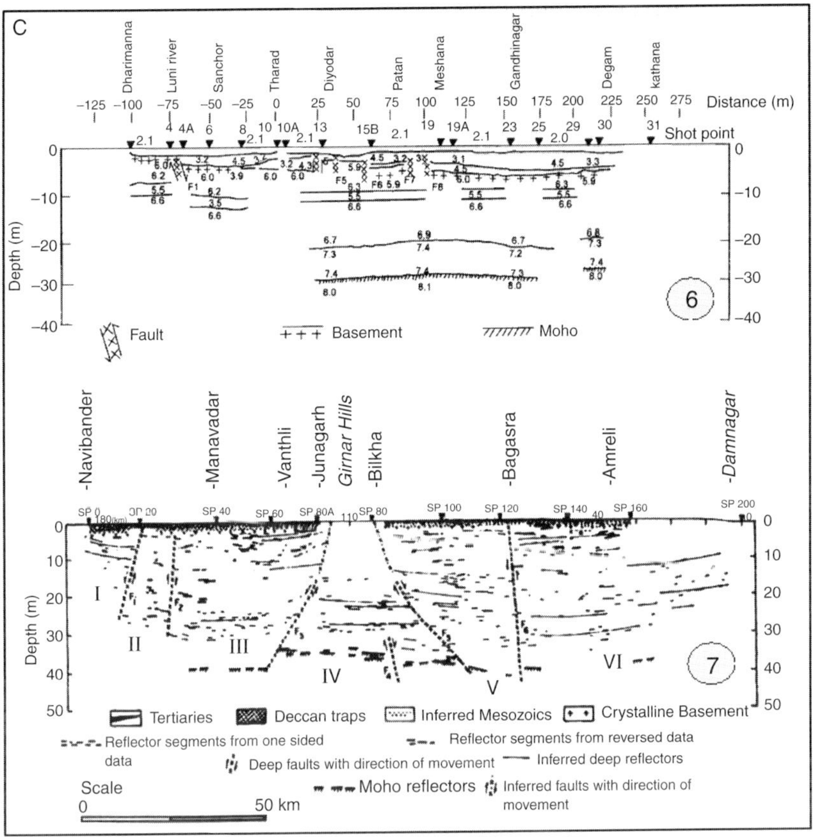

FIGURE 3.9 Various DSS sections (location shown in Fig. 3.4) close to the West Coast of India compiled from different sources. (A) Crustal section (2) along Kavali–Udipi profile *(adapted from Kaila et al., 1979)* and crustal section (3) along Koyna I (Guhagar–Chorochi) profile *(adapted from Kaila et al., 1981a)*. (B) crustal section (4) along Koyna II (Kelsi–Loni) profile *(adapted from Kaila et al., 1981b)* and crustal section (5) along the Mehmadabad to Billimora profile in the Cambay basin *(adapted from Kaila et al., 1981c)*. (C) Crustal section (6) along north Cambay and Sanchor basins from Mehmadabad to Dharimanna *(adapted from Kaila et al., 1990b)* and crustal section (7) along Navibandar-Amreli profile in the Saurashtra peninsula *(adapted from Kaila et al., 1988)*.

although the Moho could not be mapped. However, the Moho having velocities of 7.9–8.3 km/s could be mapped in the western basin at several locations at an average Moho depth of 11.5 km. The refraction data unequivocally indicate that the western basin is underlain by oceanic crust with average velocities of 5.5 and 6.7 km/s for oceanic layers 2 and 3, respectively. While the seismic velocities over the Laxmi ridge indicate continental affinity of the ridge, unfortunately, in the Laxmi basin region, the velocity data did not provide any clues regarding the nature of crust. Miles et al. (1998) interpreted the seismic velocities

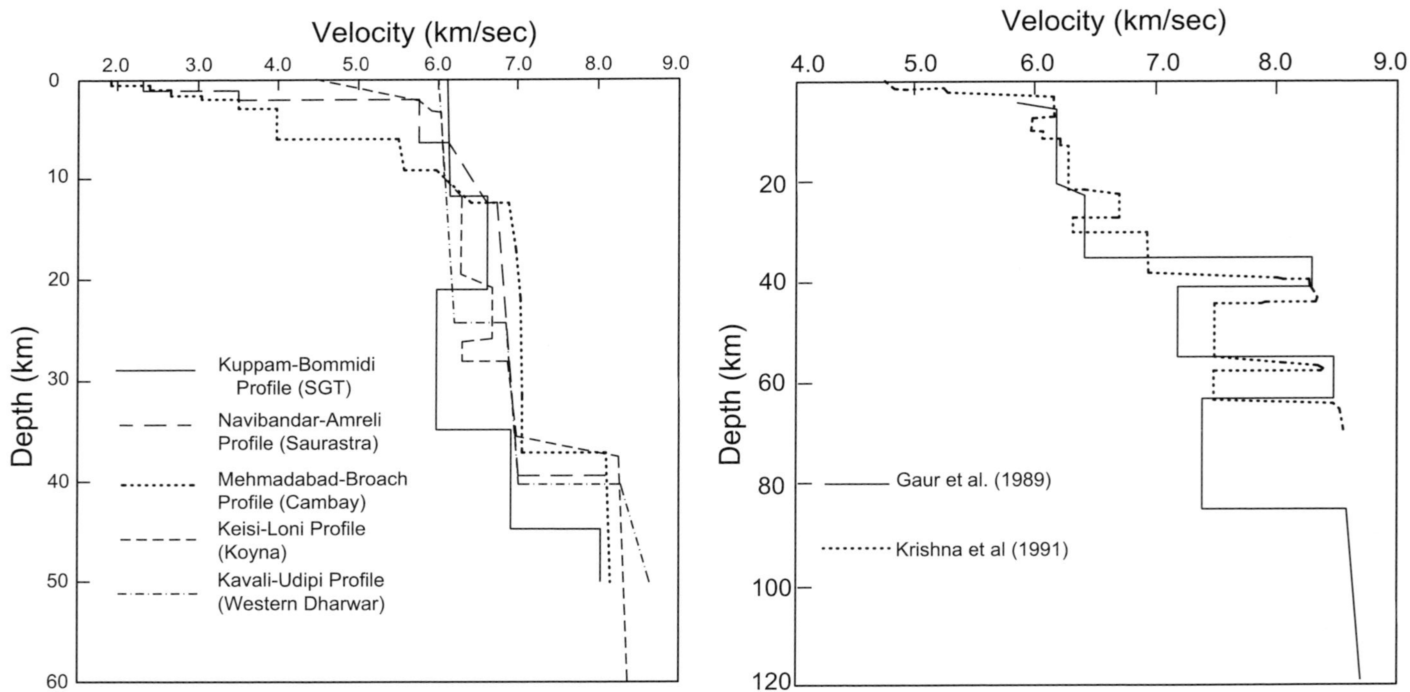

FIGURE 3.10 The compiled velocity models for different DSS profiles along the western Indian shield region and the velocity models for lithosphere below the West Coast in the DVP *(adapted from Radhakrishna et al., 2002)*.

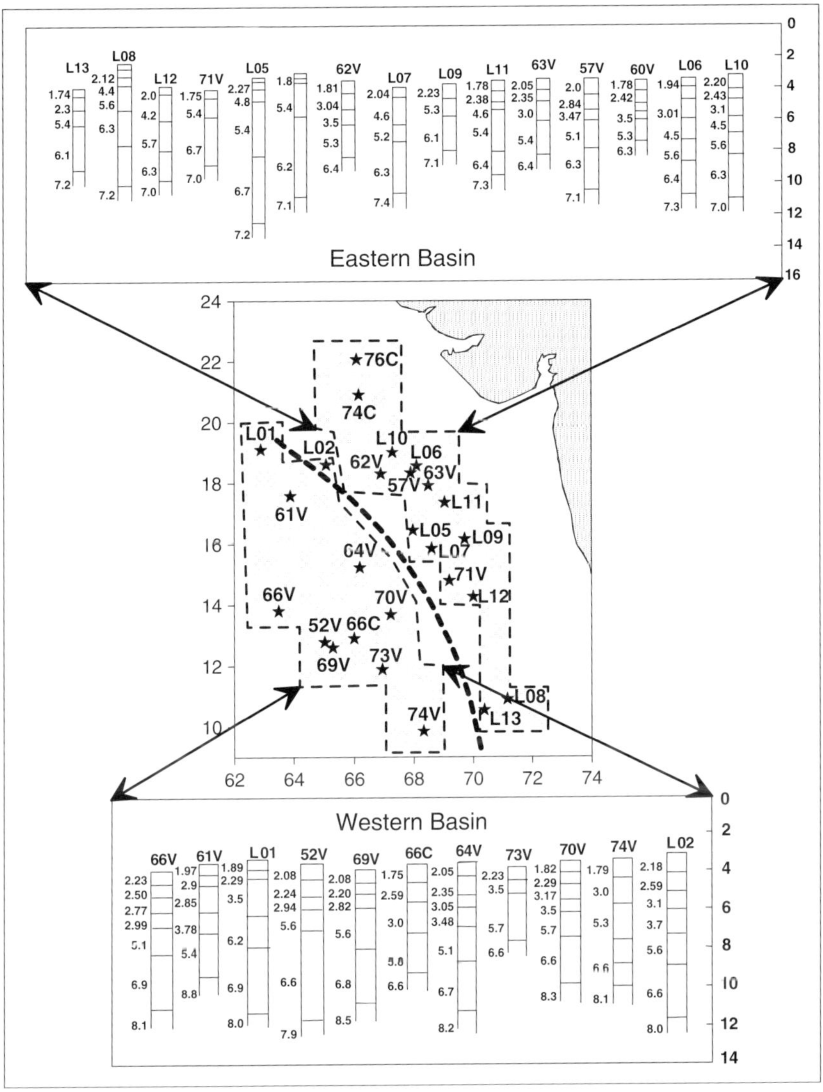

FIGURE 3.11 The velocity–depth profiles obtained from the sonobuoy refraction data in the eastern Arabian Sea (redrawn with permission from Naini and Talwani, 1982 © AAPG Publication).

(as low as 6.3 km/s) in the Laxmi basin as representing the continental crust, whereas Talwani and Reif (1998) attributed the velocities greater than 7.0 km/s to the presence of initial oceanic crust. Krishna et al. (2006) infer the basin to be underlain by rifted continental crust.

Further, for better understanding of crustal geometry across the western margin, crustal transects have been earlier constructed by few workers (Radhakrishna et al., 2002; Todal and Eldholm, 1998) by projecting all available seismic information in the nearby areas onto the transect lines (Fig. 3.12). While Todal and Eldholm (1998) constructed the sections in the offshore (P5 and P8 in Fig. 3.4), Radhakrishna et al. (2002) extended them into the coastal areas of western shield (P7 and P11 in Fig. 3.4) and projected the upper and lower crustal boundary as well as the Moho variations known from the DSS sections in the West Coast which provided information on the stretched as well as unstretched continental crust at the margin.

3.4.5.6 MCS Reflection Profiles

Most of the earlier inferences on structural features in the shelf–slope region along the WCMI are from the analysis of single-channel seismic reflection profiles combined with potential field data. However, the MCS reflection imaging of the margin covering the shelf to deep water areas provided significant new information with regard to the nature of the crust, volcanism, sedimentation, and rift architecture at the margin. Location of few MCS profiles considered here is shown in Fig. 3.4. Owing to the ambiguity in the nature of crust below Laxmi basin and the widespread impact of Deccan volcanism in the offshore, the northern part of WCMI became quite complex. Proper geological understanding of this region is very important due to its implications on paleogeographic reconstructions of the western Indian Ocean (Krishna et al., 2006). Few MCS profiles (P4, P6, P9, and P10) in the northern part of WCMI have thrown some light on the above aspects; however, the interpretations derived from these profiles are equivocal and the main ambiguity could not be resolved (Fig. 3.13). While Krishna et al. (2006) suggested the presence of stretched continental crust below Laxmi basin on the basis of shallowness of basement (in comparison to the western basin) with intrusive features observed in MCS data, Corfield et al. (2010) suggested normal oceanic crust below Laxmi basin between the continental Laxmi ridge and the west coast. However, both these investigators identified several features in their seismic data that are related to the Deccan magmatic event along this part of margin. In the central part of WCMI, the MCS profile (P3) by Chaubey et al. (2002a) across the shelf and the deeper oceanic areas covering the LR and the Arabian Sea (Fig. 3.14) revealed six major sedimentary sequences (H1–H6) ranging in age from Paleocene to Holocene. On the basis of the tilted fault blocks, volcanic intrusives, and other geophysical characteristics, these authors inferred the presence of thinned/stretched continental crust extending up to the LR, and the Continent–Ocean Boundary (COB) west of LR. The southern part of WCMI is not very well understood because of lack of seismic information for a long time. However, Ajay et al. (2010) carried out detailed analysis of the MCS profiles in the Konkan–Kerala offshore, and we present here two of their profiles (P1 and P2) in which the Seaward Dipping Reflector (SDR) sequences along the western flank of the LR have been identified (Fig. 3.14). On the basis of SDRs, they proposed that the

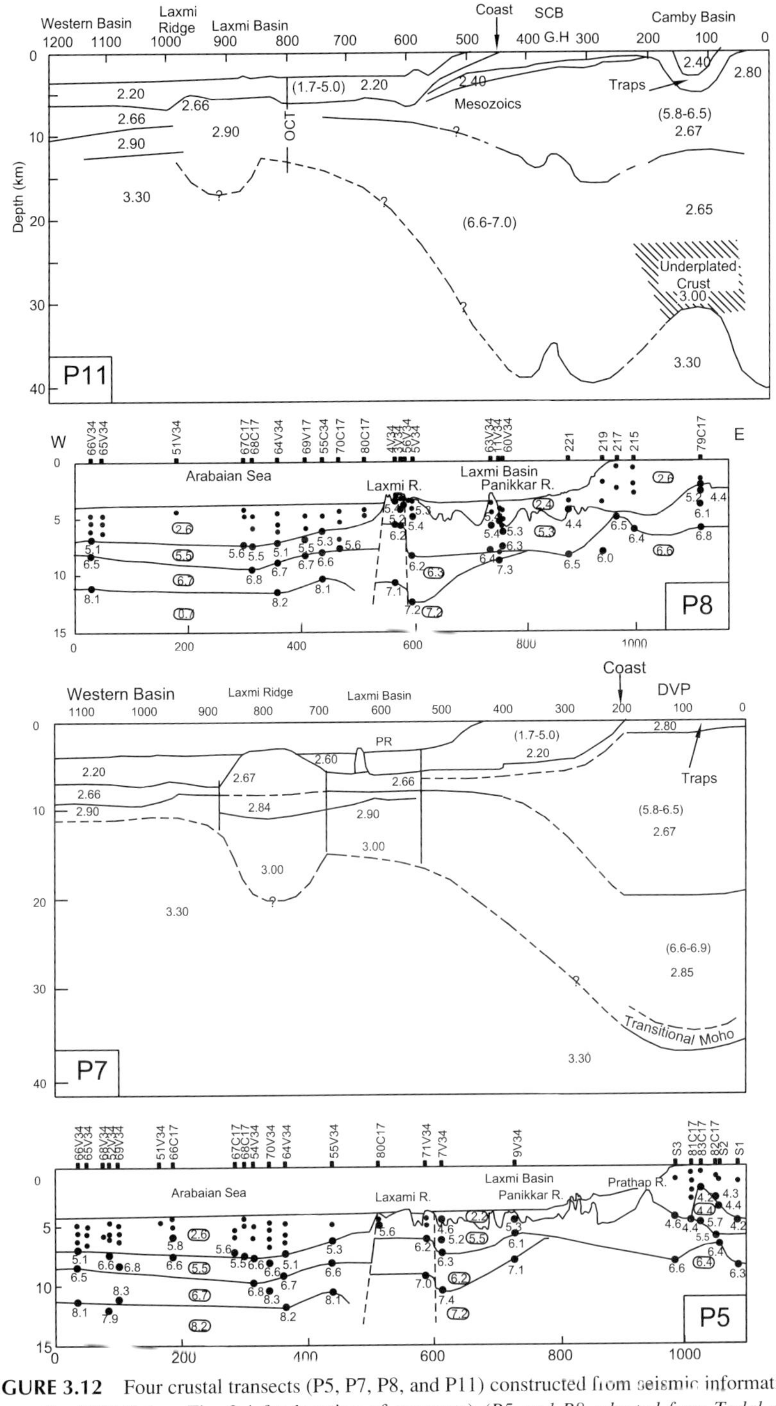

FIGURE 3.12 Four crustal transects (P5, P7, P8, and P11) constructed from seismic information across the WCMI (see Fig. 3.4 for location of transects) *(P5 and P8 adapted from Todal and Eldholm, 1998; P7 and P11 adapted from Radhakrishna et al., 2002).*

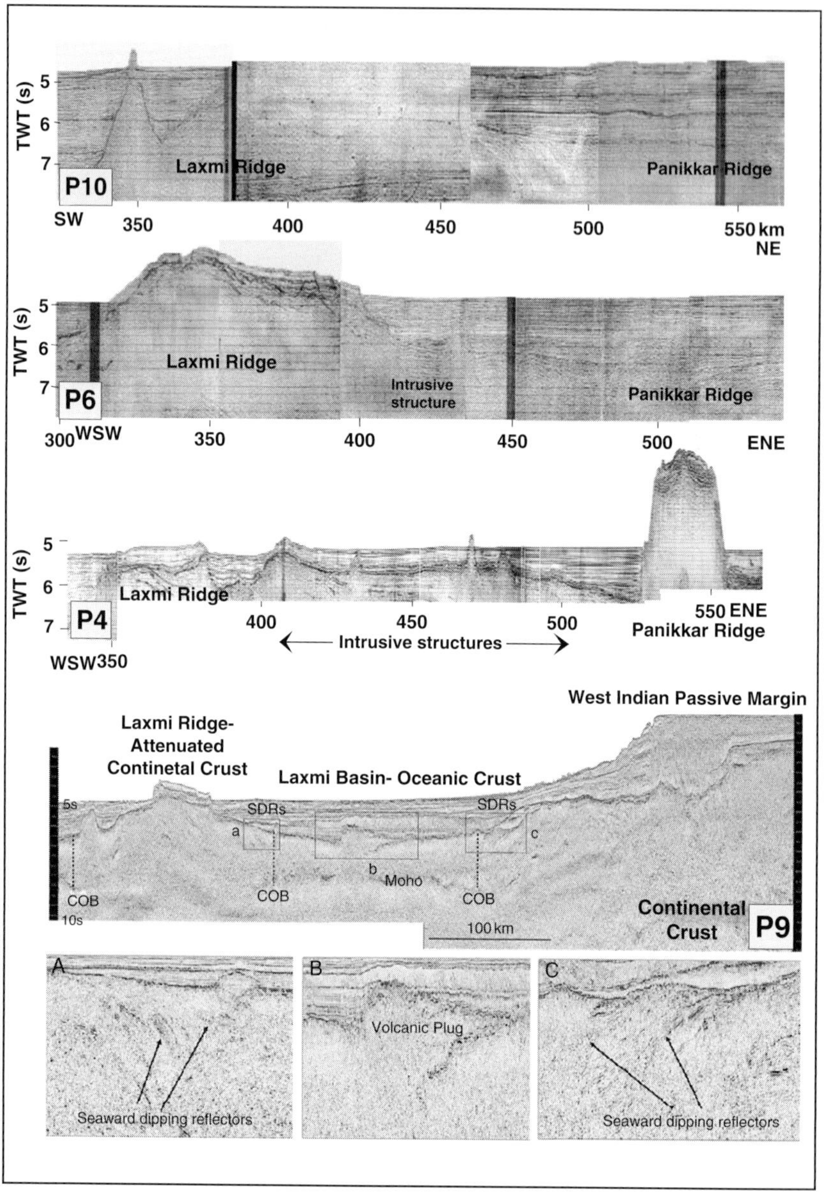

FIGURE 3.13 Regional seismic sections (P4, P6, P9, and P10) across the Laxmi basin and Laxmi ridge area in the northern part of the WCMI (see Fig. 3.4 for location of transects). P4, P6 and P10 (Reproduced from Krishna et al., 2006) and P9 (Reproduced with permission from Corfield et al., 2010 © Geological Society of London). (For color version of this figure, the reader is referred to the online version of this chapter.)

southern part of WCMI could be the volcanic passive margin. Gopala Rao et al. (2010) identified several basement features in the shelf–slope area in their reflection data and discussed the rift characteristics of this part of the margin.

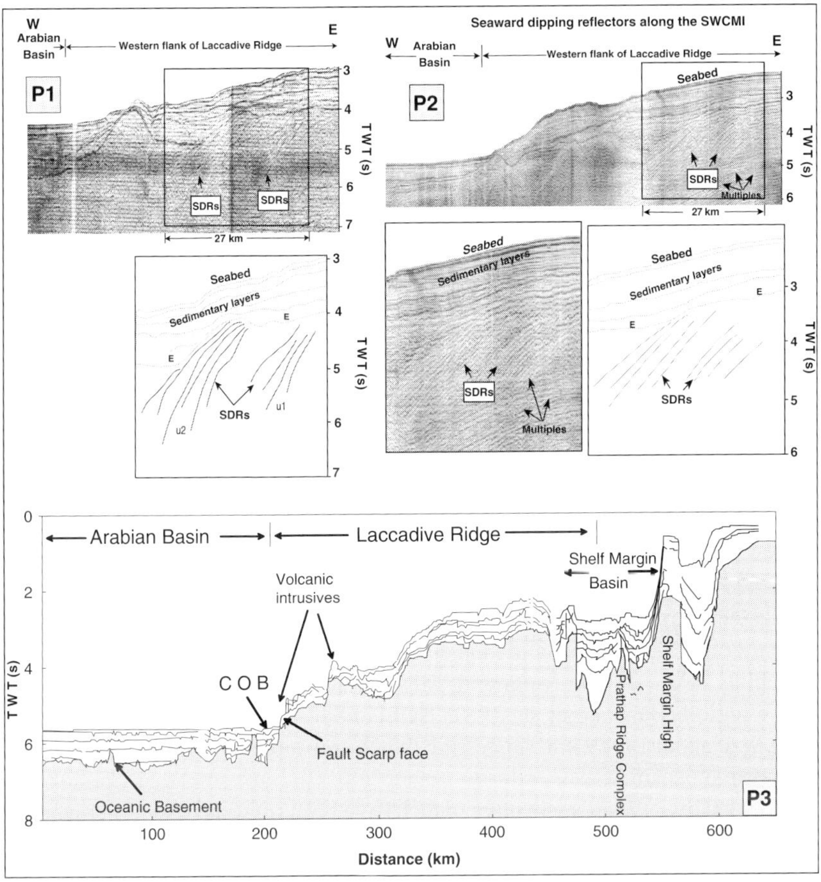

FIGURE 3.14 MCS sections (P1 and P2) and the respective interpreted line diagrams in the Konkan–Kerala basin *(adapted from Ajay et al., 2010)*. Interpreted line drawing of MCS line (P3) across the central part of WCMI *(adapted from Chaubey et al., 2002a)*. See Fig. 3.4 for the location of sections.

3.4.5.7 Laccadive Ridge

The LR is the northern segment of the N–S trending CLR complex (Ben-Avraham and Bunce, 1977) and lies almost parallel to the west coast of India between 8° N and 16° N (Fig. 3.1). The ridge is characterized by nearly 20 islands and banks, together known as Laccadive Islands, oriented along the N–S trend of the CLR (Bhattacharya and Chaubey, 2003). The ridge is distinctly noticed in the seismic profiles as a basement raise feature bounded on either side by fault scarps, and separates the shelf margin basin (Laccadive basin) in the east and Arabian basin in the west (Ben-Avraham and Bunce, 1977; Chaubey et al., 2002a). The free-air gravity anomalies replicate the morphological pattern of the ridge and show relatively positive gravity anomalies with respect to the surrounding regions (Naini and Talwani, 1982). The low and subdued gravity field of the Arabian Sea sharply rises toward east over the CLR and further eastward along its boundary, and the prominent gravity low correlates with the Vishnu fracture zone (Fig. 3.2). Magnetic anomalies over the ridge indicate that the magnetic field is subdued in the eastern part, whereas along its western part, prominent magnetic anomalies related to igneous intrusives and flows are observed (Chaubey et al., 2002a; Gopala Rao et al., 1987; Naini and Talwani, 1982). The origin of the ridge is not clearly established because of the ambiguity in the nature of crust underlying the ridge. Naini and Talwani (1982) observed that the LR and Laxmi ridge divide the eastern Arabian Sea into two provinces: the Western and Eastern basins. It is widely agreed that the western basin is underlain by oceanic crust (McKenzie and Sclater, 1971; Naini and Talwani, 1982) and formed by propagating spreading ridge segments during the Paleogene (Chaubey et al., 1998, 2002b; Dyment, 1998; Miles et al., 1998; Royer et al., 2002). However, east of LR, the region encompassing the Laccadive basin is characterized by the presence of either transitional type of crust formed during the initial phase of rifting (Naini and Talwani, 1982) or a failed rift of stretched continental crust associated with volcanic intrusives (Chaubey et al., 2002a). The crust underlying the LR is variedly interpreted as a transitional crust (Narain et al., 1968; Radhakrishna et al., 2002) and a thinned continental crust (Chaubey et al., 2002a; Naini and Talwani, 1982). Seismic refraction studies in the region (Babenko et al., 1981; Naini and Talwani, 1982) image Moho at a depth of about 18–19 km, suggesting that the crust below the LR is thicker than the normal oceanic crust.

Several hypotheses are in place, which have been suggested by earlier workers with regard to the mode of emplacement or formation of the CLR. Some of the earliest workers suggested that the ridge had formed along a transform fault during the northward movement of India (McKenzie and Sclater, 1971; Sclater et al., 1971). Later, Ben-Avraham and Bunce (1977) suggested that along the CLR, the LR and Chagos Bank are composed of volcanic features associated with leaky transform faults, whereas the Maldive ridge has characteristics of a microcontinent that had separated away from India. Further, the LR is also interpreted as a continental fragment (Naini and Talwani, 1982), as a continental fragment intermixed with volcanics (Chaubey et al., 2002a), or as a transitional crust affected by the passage of Reunion plume (Radhakrishna et al., 2002). Several studies also suggest its origin as due to the trace of the Reunion hot spot formed during the northward motion of the Indian plate (Duncan, 1990; Richards et al., 1989), which was later supported by the DSDP and ODP results in the region (Whitmarsh, 1974).

3.4.5.8 Laccadive Basin

The narrow basin lying between LR and the continental slope of southern India is referred to here as the Laccadive basin (Whitmarsh, 1974), though various nomenclatures exist in the literature (Bhattacharya and Chaubey, 2003). The basin widens from 16° N at the WCMI toward south and opens up into the central Indian Ocean (Fig. 3.1). The seismic reflection studies in the basin reveal that it is characterized by the presence of sediment accumulation ranging in TWT around 3.5 s in the north to 2.5 s in the south (Chaubey et al., 2002a; Naini and Talwani, 1982; Singh and Lal, 1993). Some earlier workers reported the presence of a NNW–SSE trending Prathap ridge in the basin (Chaubey et al., 2002a; Naini and Talwani, 1982; Subrahmanyam et al., 1995), whose origin is not yet established. It could be a feature manifested by rift-related volcanism (Subrahmanyam et al., 1995), uncompensated basement high over a thinned continental crust (Radhakrishna et al., 2002), or be related to failed rift and volcanism of the stretched continental crust (Chaubey et al., 2002a).

3.4.5.9 Laxmi Ridge

The Laxmi ridge is a NW–SE trending basement high feature (relief of ~2 km) lying between 16° N and 18° N in the northeastern Arabian Sea (Naini and Talwani, 1982). The ridge is partly buried under a thin sediment cover and separates the Arabian basin in the west and the Laxmi basin in the east. Similar to the 85° E ridge in BOB, this ridge is also an enigmatic structural feature due to the characteristic gravity low (~40–50 mGal) over it unlike other oceanic aseismic ridges, which are generally associated with gravity highs. Sonobuoy refraction measurements in the eastern Arabian Sea (Naini and Talwani, 1982) indicate that the ridge comprises a 2.4-km thick upper crustal layer with a velocity of about 2.3 km/s, 4.5-km thick middle crustal layer with a velocity of about 6.3 km/s, and a lower crust having a velocity of about 7.2 km/s. Unfortunately, Moho could not be mapped in their refraction data; however, other geophysical constraints indicate that it may be at a depth of 21 km, suggestive of continental affinity (Naini and Talwani, 1982), which is in general agreement. Further, the geophysical signatures also establish the presence of crustal underplating below the ridge (Singh, 1999).

3.4.5.10 Laxmi Basin

While the velocity structure clearly indicates the presence of oceanic crust below the western basin, in the eastern basin, several contrasting views exist with regard to the nature of crust in the Laxmi basin: whether it is the oceanic crust (Bhattacharya et al., 1994; Malod et al., 1997), initial thick oceanic crust (Talwani and Reif, 1998), or the thinned extended continental crust (Krishna et al., 2006; Miles et al., 1998), and some of these aspects are also discussed in Section 3.4.5.5. Refraction measurements in this basin indicate the Moho at a depth of about 17 km below sea level, implying that the crust below the Laxmi basin is thicker than normal oceanic crust and nearly half of the normal continental crust, which suggest that it could be a transitional type crust.

However, according to Malod et al. (1997), the crust in the region north of Laxmi ridge is of oceanic which formed before 64 Ma due to spreading around a triple junction connecting the Laxmi basin, Narmada-Son lineament, and the Gop rift. On the other hand, Talwani and Reif (1998) proposed that opening of the Laxmi basin predates the emplacement of Deccan volcanism and suggesed the COB east of the Laxmi basin. In contrast, the thinned continental crust along with an underplated crust below the Laxmi basin as well as Laxmi ridge region was proposed by Miles et al. (1998). They further suggest that the basement features in the Laxmi basin may be due to large-scale intrusions.

3.4.5.11 Gravity-Derived Crustal Models Across the Margin

The seismically constrained gravity modeling has been carried out along different profiles (along P1, P3, P4, P11, and P12) covering the WCMI by various workers (Ajay et al., 2010; Chaubey et al., 2002a; Krishna et al., 2006; Radhakrishna et al., 2002; Yatheesh et al., 2009), and these models are shown in Fig. 3.15. It must be remembered that variations exist in terms of basic assumptions as well as the density layering pattern while carrying out the modeling. However, the models generated by these workers provide valuable information on variation in the crustal configuration along as well as across the WCMI.

3.4.5.12 Comorin Ridge

While compiling the echosounding data acquired during the International Indian Ocean Expedition, Heezen and Tharp (1964) reported for the first time a NNW–SSE trending ridge in the region west of Sri Lanka and south of India between latitudes 1.5° N and 6.5° N with an average width of 150 km, and this feature is named as the Comorin ridge. The ridge is associated with low-amplitude positive gravity anomalies. Kahle et al. (1981) carried out a detailed analysis of the marine geophysical data acquired over this ridge and suggested that it had formed on an oceanic crust with COB located along its eastern boundary. Desa et al. (2006) interpreted this feature as a transform ridge formed during the early phase of opening of the Indian Ocean. From the existing knowledge, it is not clear whether the ridge had formed during India–Madagascar breakup or during an earlier breakup of Gondwanaland at the southwest margin of Sri Lanka. Geomagnetic induction models reveal the presence of an anomalous conductive structure below the Comorin ridge (Agarwal and Weaver, 1989; Mareschal et al., 1987; Thakur et al., 1986). Sheena et al. (2007) carried out coherence analysis of gravity and bathymetry over the ridge and inferred a Te value of 10 km. Sreejith et al. (2008) made admittance analysis of this ridge for comparing Te value along the ridge between southern and northern part. The gravity modeling carried out by them revealed that the southern part of the ridge was emplaced on the relatively

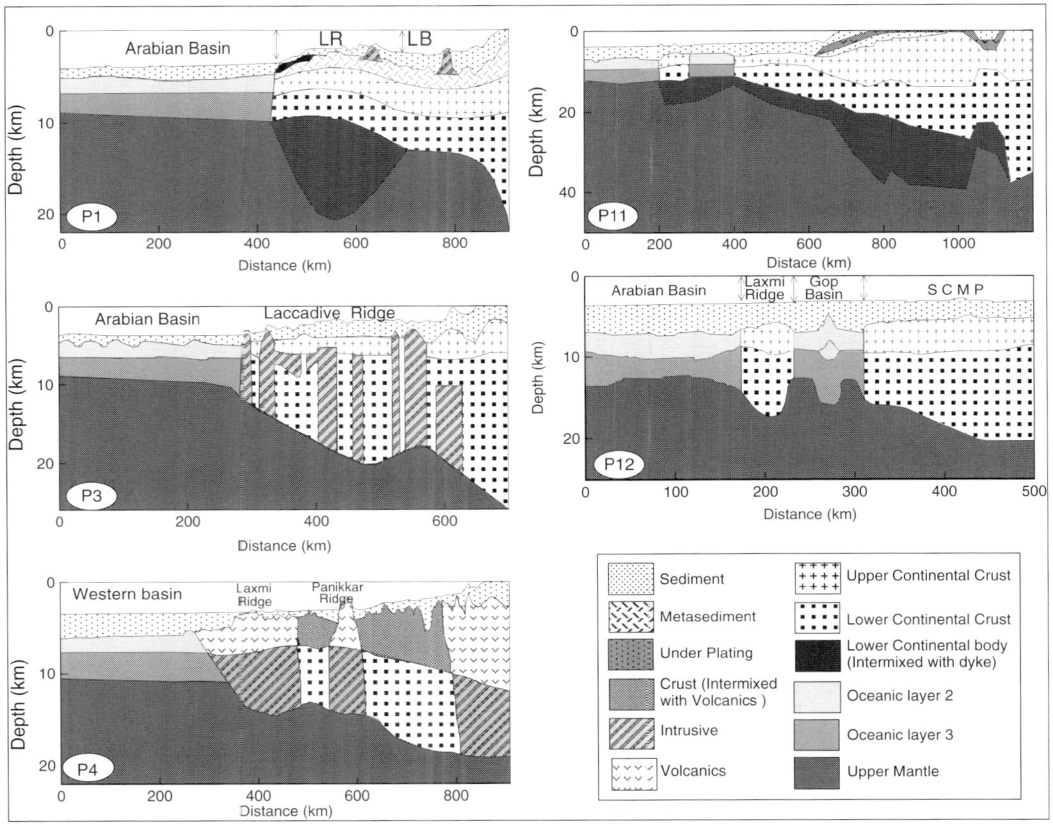

FIGURE 3.15 Seismically constrained gravity derived crustal models across the Western Continental margin of India compiled and redrawn from earlier studies (see Fig. 3.4 for location) (P1, Ajay et al., 2010; P3, Chaubey et al., 2002a; P4, Krishna et al., 2006; P11, Radhakrishna et al., 2002; P12, Yatheesh et al., 2009).

weak oceanic crust characterized by both surface and subsurface loading, while the northern part was emplaced on the continental crust.

3.5. STRUCTURE AND TECTONICS OF THE EASTERN OFFSHORE

The breakup of eastern India from East Antarctica seems to have been accommodated along the Proterozoic Eastern Ghat Mobile Belt (EGMB), which extends for over 1000 km along the east coast. The ECMI is a divergent type of margin that evolved as a consequence of breakup of eastern Gondwanaland between India and east Antarctica during early Cretaceous (Powell et al., 1988). The major geological features underlying the ECMI and the BOB are inherited from the breakup of eastern Gondwanaland and subsequent spreading of the Indian Ocean lithosphere. The Kerguelen plume activity beneath east Gondwanaland might have given rise to rifting and associated magmatism at least in the northern parts of the ECMI (Kumar et al., 2003; Kent, 1991; Kent et al., 1997). The ECMI has a strike length of about 2000 km extending from the Sri Lanka shelf in the far south to the Bengal basin in the north and consists of several major river-based sedimentary basins: the Bengal, Mahanadi, Krishna–Godavari, Palar, and Cauvery basins (Fig. 3.16), which are believed to have originated during the rifting and breakup (Talukdar, 1982). In the following sections, we present the morphotectonics and geology of the east coast and the geophysical characteristics of ECMI and discuss the structure, nature of crust, and the development of the margin.

3.5.1 Structural Trends and the Breakup

The dominant rock types along the east coast of India belong to the EGMB in the northern part and the SGT in the south. It is observed that the Precambrian structural trends, viz., the Godavari and Mahanadi rifts, shear zones, and granulite terrains, at the east coast of India (Fig. 3.16) have close linkages with similar features along the east Antarctica (Fedorov et al., 1982; Yoshida et al., 1992). Reconstruction models considering the prerift and drift extensions (Reeves et al., 2004; Veevers, 2009) provided further constraints on the rifting scenario between India and the Antarctica. The NE–SW trending ridge-depression basement configuration underlying the ECMI basins typically demonstrates that they developed over a highly faulted and rifted continental crust probably during the Cretaceous time. Ship-borne as well as the satellite-derived gravity and magnetic investigations along the eastern Indian offshore and the conjugate Enderby basin, east Antarctica, provided valuable information on the structural fabric as well as the nature of the crust (Chand et al., 2001; Golynsky et al., 2002; Krishna et al., 2009; Ramana et al., 2001; Rotstein et al., 2001). Geophysical investigations along the ECMI

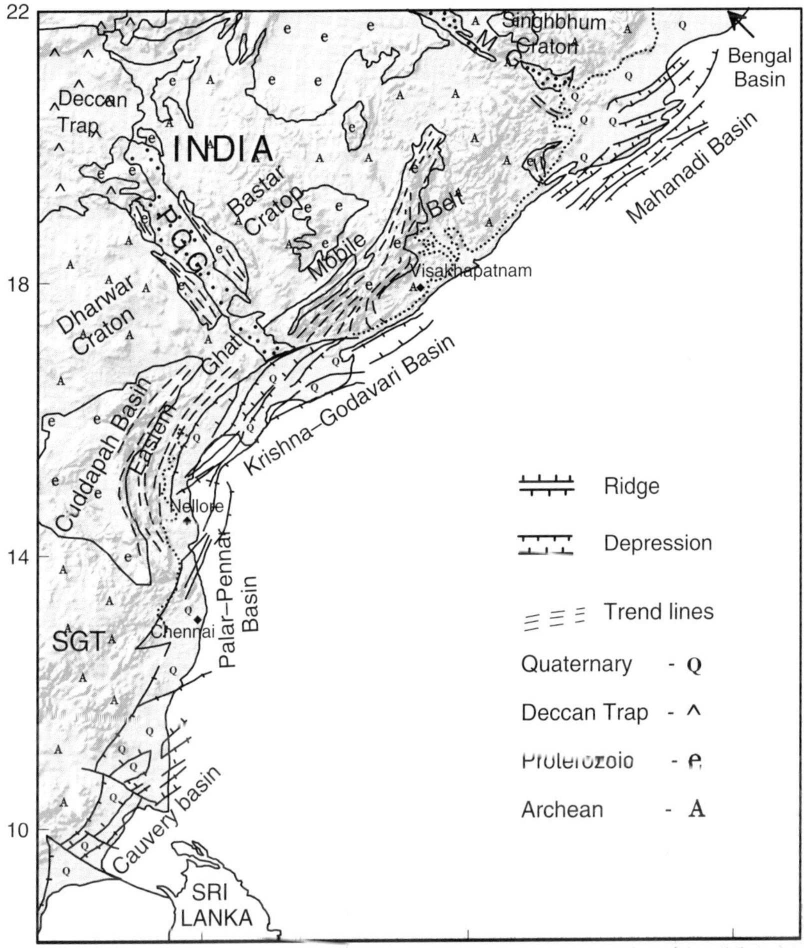

FIGURE 3.16 Shaded relief map of the eastern Indian Shield margin showing tectonic details of the East Coast of India. Geological formations and tectonic details are from Gupta et al. (2001). (For color version of this figure, the reader is referred to the online version of this chapter.)

suggest that the margin can be considered as a composite of two segments (Fig.3.17), the northern part (north of 16° N) having rifted margin character and the southern part developed as a consequence of shearing in the early stages of continental separation (Chand et al., 2001; Subrahmanyam et al., 1999). The initial breakup seems to have occurred at the bight of the present-day Krishna–Godavari (K–G) basin (Ramana et al., 2001), in two major stages along two different segments: the northern (K–G) rifted segment and the southern (Cauvery) sheared or transform segment (Subrahmanyam et al., 1999). On the basis of the requirement of accommodating the Elan Bank

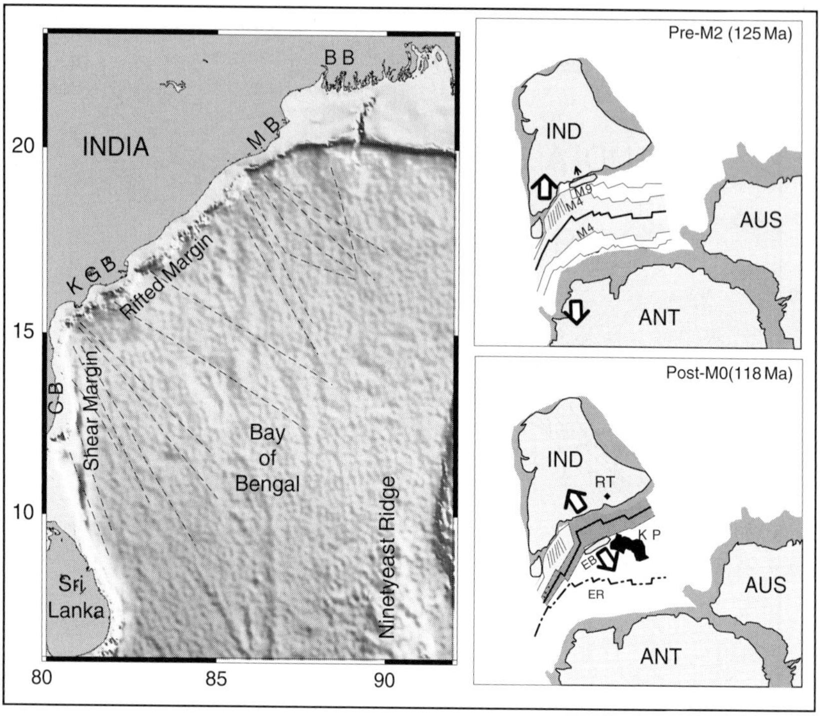

FIGURE 3.17 Map shows the morphology of the western BOB region (left panel). Dashed lines indicate the identified fracture zone trends (after Radhakrishna et al., 2012). Schematic plate tectonic reconstruction models (right panel) showing the double breakup scenario between East coast of India and East Antarctica (adapted from Radhakrishna et al., 2012). (For color version of this figure, the reader is referred to the online version of this chapter.)

microcontinent in the western margin of Kerguelen Plateau (Borissova et al., 2003; Gaina et al., 2003, 2007; Ingle et al., 2002), Krishna et al. (2009) proposed a two-stage rifting history for the ECMI: first, an Australia–Antarctica separation from greater India during early Cretaceous; second, the separation of Elan Bank from the present-day ECMI at about 120 Ma. Analysis of basement structural features from several MCS profiles in the BOB region by Radhakrishna et al. (2012) revealed several NW–SE trending fracture zones (Fig. 3.17) along the ECMI, which show two different sets of fracture zones belonging to these two breakup events. However, they suggested the timing for the breakup events as around M4 for the first event and M2 for the second event, as depicted in Fig. 3.17.

3.5.1.1 Geophysical Data analysis

Different geophysical data sets mainly comprising the ship-borne bathymetry, gravity and magnetic, high-resolution satellite-derived gravity, and single and multichannel seismic reflection data covering the shelf and slope regions of the eastern Indian offshore have been collected and analyzed by many previous workers (Bastia et al., 2010a–c; Curray et al., 1982; Gopala Rao et al., 1997; Krishna et al., 2009; Mukhopadhyay and Krishna, 1991, 1992; Murthy et al., 1993; Radhakrishna et al., 2012; Rao and Rao, 1985; Subrahmanyam et al., 1999, 2006, 2008b). These studies are useful to delineate structural features, morphotectonic elements, and lineaments in the eastern offshore. Bastia et al. (2010a) presented a detailed analysis of the high-resolution satellite-derived gravity data (Andersen and Knudsen, 2008) along with nearly 33,000 line km of ship-borne gravity and 18,000 line km of magnetic data of National Institute of Oceanography (NIO) marine trackline database and the NGDC trackline data along with the regional multichannel seismic data in the ECMI region. Some salient results obtained from this study are given here to facilitate better understanding of the structural fabric and the basement trends along the eastern offshore. The free-air, Bouguer, and the band-pass-filtered (10–200 km) maps of the eastern offshore prepared from the above database are shown in Fig. 3.18. For resolving various features such as the basement highs, faults, and the edges of geological features, the band-pass-filtered Bouguer anomaly map is further reduced by calculating first vertical derivative (FVD), second vertical derivative (SVD), horizontal gradient (HG), and FVD of HG, and the resultant maps are shown in Fig. 3.19. From the ship-borne magnetic data, the total field magnetic intensity (TMI) and the reduction-to-pole (RTP) maps have been prepared by them (Fig. 3.20). The TMI map shows considerable variation in the wavelength and amplitude of anomalies and suggests numerous magnetic sources at different depths. In the inner shelf region, presence of high-frequency anomalous zones of magnetic field parallel to the coast has been observed which are interpreted as intrusive bodies associated with the rift-phase volcanism (Murthy et al., 1993; Ramana et al., 2003). The comparison of TMI and RTP maps shows that the anomalies appear to have migrated in a certain direction. High-low pairs of anomalies tend to exhibit dominant positive anomaly features. Using these enhanced maps along with vast amount of seismic data, Bastia et al. (2010a) identified several lineaments, faults, and gravity trends along the margin which suggests that in general, the ENE–WSW trending faults dominate the structural configuration in the eastern offshore (Fig. 3.21).

3.5.1.1.1 Basement Depth Map and Continent–Ocean Boundary Along the ECMI

Magnetic anomalies in Fig. 3.20 show the linear magnetic trends following the boundary of continental/proto-oceanic to oceanic crust matching with the gravity enhancements, viz., Bouguer gravity band-pass (10–200 km) and its vertical and horizontal derivative maps (Figs. 3.18 and 3.19). Bastia

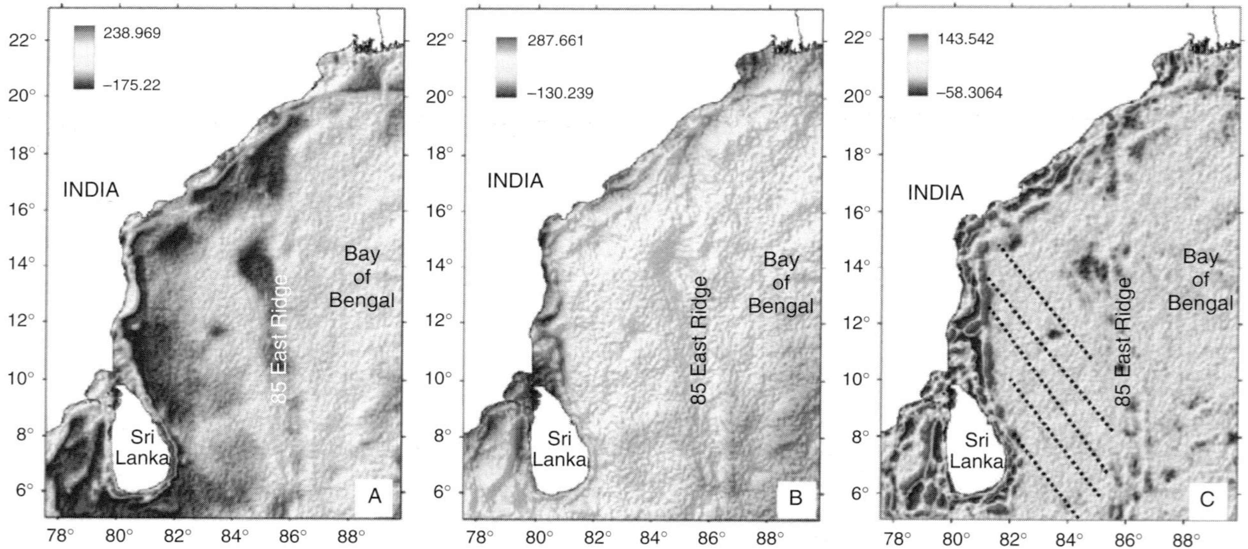

FIGURE 3.18 Shaded relief maps showing (A) free-air gravity anomalies, (B) Bouguer gravity anomalies, and (C) band-pass filtered (10–200 km) Bouguer gravity anomalies along the ECMI and the adjoining Bay of Bengal region. High-resolution satellite-derived gravity data, KMS-02 (Andersen and Knudsen, 2008), were used for this purpose. Inferred fractures zones are from Krishna et al. (2009) *(adapted from Bastia et al., 2010a)*. (For color version of this figure, the reader is referred to the online version of this chapter.)

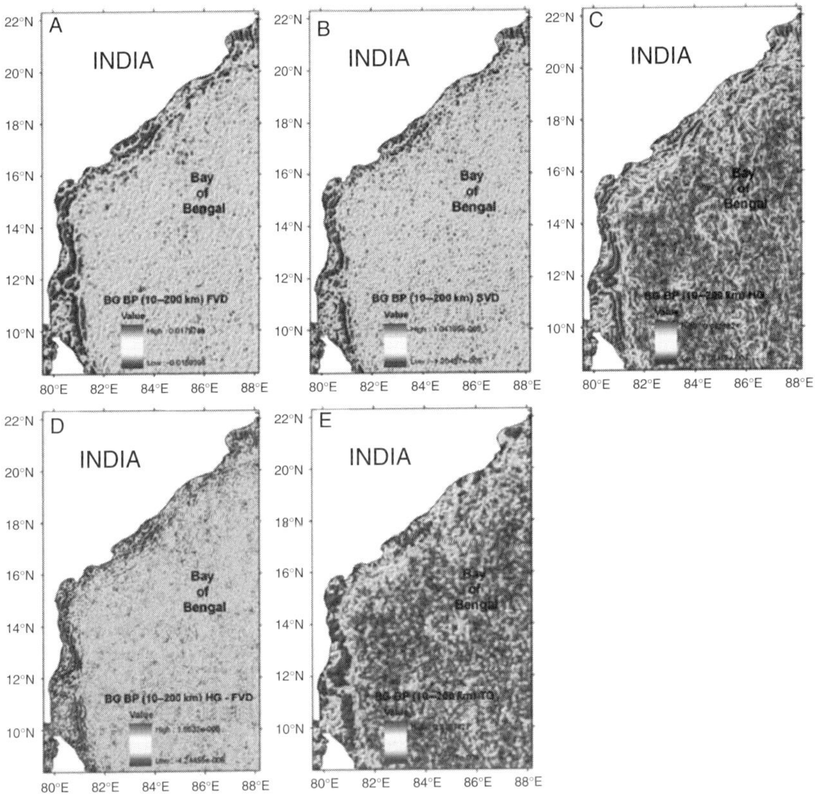

FIGURE 3.19 Image enhanced maps prepared from the band-pass filtered Bouguer anomaly map shown in Fig. 3.18C. (A) First vertical derivative (FVD), (B) second vertical derivative (SVD), (C) horizontal gradient (HG), (D) HG of first vertical derivative, and (E) total gradient (TG) (*adapted from Bastia et al., 2010a*). (For color version of this figure, the reader is referred to the online version of this chapter.)

et al. (2010a) identified the COB from the potential field enhancements integrated with the seismic data along the ECMI (Fig. 3.21). The band-pass-(10–200 km) filtered map (Fig. 3.18) exhibits a linear Bouguer gravity high trend in the offshore which reflects the clear density transition in the region of attenuated continental crust. Both TMI and RTP maps (Fig. 3.20) indicate the intensity of crustal attenuation to be maximum with gradually fading magnetic anomalies in the Mahanadi, Northeast Coast (NEC), and northern part of Krishna–Godavari basins and suggest normal extension or rifting, whereas the southern part of Krishna–Godavari, Palar and Cauvery basins show abruptly ending magnetic anomalies reflecting minimal crustal attenuation. The abrupt transition between continental and oceanic crust characterizing oblique transform movement indicates that the southern part is a sheared margin. The COB along the ECMI has been discussed by many earlier workers (Gopala Rao et al., 1997; Krishna et al., 2009; Mukhopadhyay and Krishna, 1991, 1992;

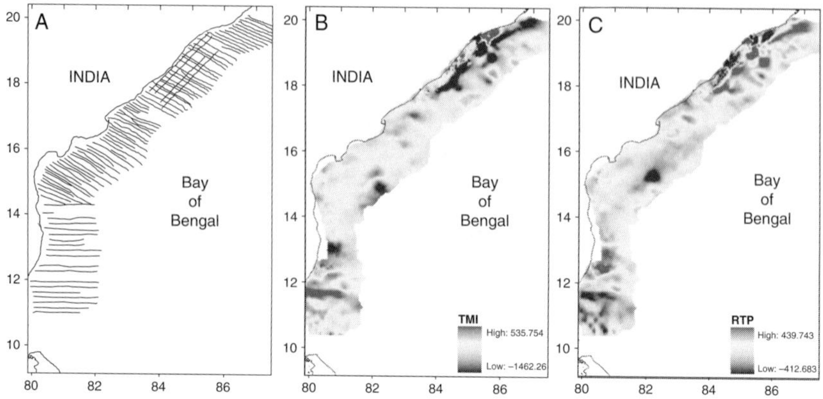

FIGURE 3.20 The ship-borne magnetic profiles (A) used to prepare the color-coded images of total magnetic intensity (TMI) anomalies in (B) and the Reduction-to-Pole (RTP) map in (C) *(after Bastia et al., 2010a)*. (For color version of this figure, the reader is referred to the online version of this chapter.)

Murthy et al., 1993; Subrahmanyam et al., 1999), and, in general, it is agreed that it lies along the foot of the continental slope showing similar characteristics as mentioned above. The magnetic data are further analyzed by Bastia et al. (2010a) to present the basement depth and sediment thickness in order to interpret the structures in specific regions. Figure 3.22 depicts the basement depth estimated from the above analysis and the COB identified along the margin. The depth to the basement ranges from 1 to 12 km below sea level and deepens toward the Bengal Fan in the north. In the Cauvery basin, close to Point Calimere in the south, two E–W trending basement lows truncated at the COB are observed. The basin is characterized by a more N–S depositional axis in the north. The COB is much closer to the onshore area at this location. The NE–SW strike-slip faults are the predominant structures in the basin which continue across the COB. In the northernmost part of the basin, a WNW–ESE trending basement high exists. In the Krishna–Godavari basin, the NE–SW strike-slip faults are predominantly observed which continue into deeper oceanic areas. The COB follows these faults in some areas and is close to the coastline in the Machilipatnam bay area, and the main basement low areas may be lying on the oceanic crust. The abrupt change in the basement trends from N–S in the Cauvery basin to ENE–WSW in the Krishna–Godavari basin may be marking the basin boundary. Further north, the deepest portion of magnetic basement is characterized by ENE–WSW trending basement low (>10 km). Such local depressions at the margin could have formed due to the subsidence caused by sediment loading. In the Mahanadi basin, the basement contours are parallel to the coastline, reflecting a probable early syn-rift geometry of the basin prior to the breakup. Basement highs and lows in the northern part of the basin are oriented in NW–SE. The structures truncate against the COB indicating that these axes have been perhaps controlled by

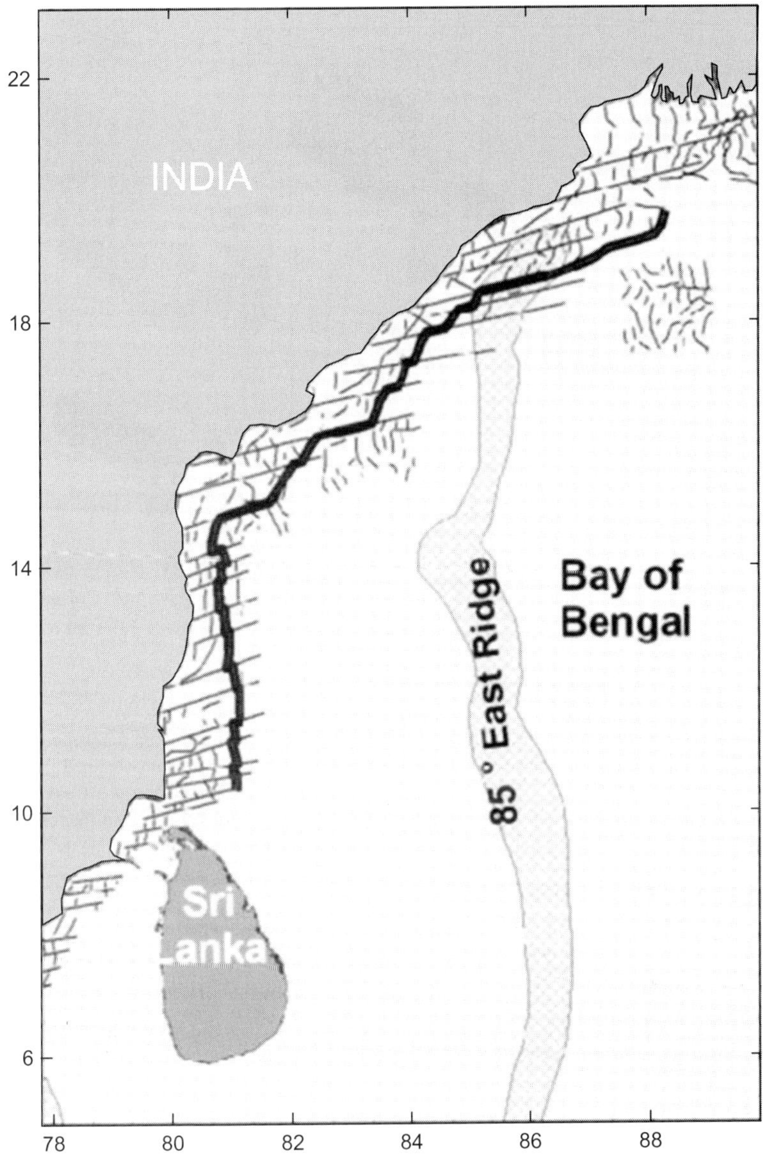

FIGURE 3.21 The inferred offshore lineaments/faults from the interpretation of potential field data along the ECMI (after Bastia et al., 2010a). Thick line indicates the location of inferred COB.

the pre-existing rift geometry. In the NEC basin, the N–S trending features dominantly observed probably are related to the Bengal Fan. The depth to basement increases toward the Bengal Fan throughout the basin. For

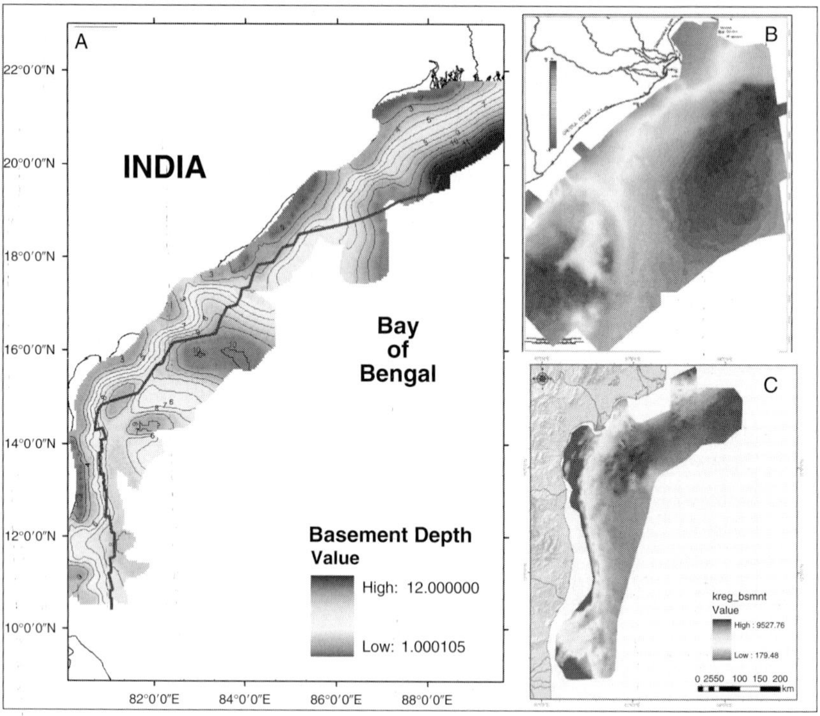

FIGURE 3.22 Map showing (A) depth to magnetic basement obtained from magnetic data for the ECMI. Inferred COB is shown in red line on the map. Depth to basement obtained from offshore seismic data is also presented for (B) Mahanadi and NEC basins and (C) Cauvery and Krishna–Godavari basins *(adapted from Bastia et al., 2010a)*. (For interpretation of the references to color in this figure legend, the reader is referred to the online version of this chapter.)

comparison, the basement depth maps prepared for the northern part (Mahanadi and NEC basins) and southern part (Cauvery and Krishna–Godavari basins) of the ECMI using large volume of seismic data are presented in Fig. 3.22. Both these maps show the presence of horst–graben features related to rifting. A comparison of basement depth estimates derived from the magnetic and the seismic data indicates that basement trends and depths in general are comparable in Cauvery and Krishna–Godavari offshore. However, in the Mahanadi offshore, the basement structure over the 85° E ridge has been better resolved in the seismic data.

3.5.1.2 Onshore–Offshore Tectonic Linkages at the East Coast

The Precambrian structural grain of the Indian shield to a larger extent controlled the continental fragmentation and the subsequent tectonic reactivations that led to the present-day basement trends as well as margin segmentation. The ECMI is characterized by the presence of five major sedimentary basins in the onshore areas, and detailed geophysical studies over these basins reveal that the basins extend offshore and merge with the deep-sea Bengal Fan (Curray et al., 1982). The underlying basement in the basins is typically dominated by alternating set of NE–SW trending ridges and depressions (Bastia, 2006; Fuloria, 1993; Prabhakar and Zutshi, 1993; Sastri et al., 1981). Detailed studies on the land–ocean tectonics along the eastern offshore of India are made through systematic acquisition of marine magnetic, bathymetry, and single-channel seismic data in the nearshore as well as shelf–slope areas of the margin by many previous workers (Murthy et al., 1993, 1995, 2010; Subrahmanyam et al., 1995, 2006, 2010). Marine magnetic studies in the nearshore regions of the margin revealed offshore extension of structural trends and dyke intrusions related to rift-phase volcanism (Murthy et al., 1993). Subrahmanyam et al. (2006) reported offshore extension of two major Precambrian mega-lineaments: the Palghat–Cauvery Lineament and Moyar–Bhavani–Attur Shear zone following 11° N and 12° N, respectively (Fig. 3.23). Differential tectonic movements that resulted in subsidence of

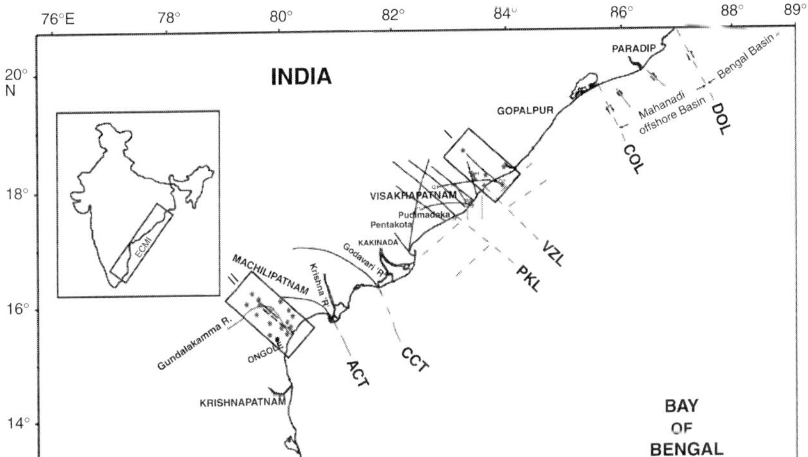

FIGURE 3.23 Map showing the onshore–offshore structural trends across the East Coast of India *(Reproduced from Murthy et al., 2010)*. Blocks (I, II, III) show the areas of seismic activity associated with the Land Ocean Tectonics (LOTs) at the coast. CCT, Chintalapudi Cross Trend; ACT, Avanigadda Cross Trend; PKL, Pudimadaka Lineament; VZL, Vizianagaram Lineament; COL, Chilka Offshore Lineament; DOL, Dhamra Offshore Lineament; MBA, Moyar–Bhavani–Attur Lineament; PCL, Palghat–Cauvery Lineament. (For color version of this figure, the reader is referred to the online version of this chapter.)

the Pondicherry coast along two NE–SW trending subparallel faults are reported by Ramasamy et al. (2001). Extension of four major NW–SE trending cross trends (Fig. 3.23) perpendicular to the NE–SW oriented horst–graben basinal structures from the coast toward the offshore areas is observed along the K–G basin. These are the Ongole, Chintalapudi, Avanigadda, and Pithapuram cross trends (Murthy et al., 1995; Venkatarengan and Ray, 1993). Subrahmanyam et al. (2008b) reported extension of structural features and lineaments of the land into the Mahanadi shelf that represent the rift-related tectonics of the Gondwana graben. Some of these lineaments show signatures of reactivation in terms of occurrence of minor tremors at the coast, and at least three such areas (blocks) along the east coast (indicated in Fig. 3.23) where frequent tremors are occurring have been reported (Subrahmanyam et al., 1995, 2006; Murty et al., 2002; Sarma et al., 2009). A detailed analysis of the land–ocean tectonics of the east coast with special emphasis on the seismogenesis of these three active tectonic blocks is presented in Murthy et al. (2010).

3.5.2 Crustal Structure Below the East Coast and the Adjoining BOB

The Bay of Bengal (BOB) comprising of the deep-sea Bengal fan is an interesting area to undertake studies related to early crustal evolution and the sedimentary basin research. The BOB primarily receives sediments from the north; however, the east-flowing river systems of Peninsular India also contribute significant amount of sediments from the west into it. This led to the formation of major offshore sedimentary basins along the ECMI. As the crust in the BOB region had evolved because of rift–drift processes of continental separation between India, East Antarctica, and the Elan Bank, a synthesis of crustal structure as obtained from the seismic reflection and refraction data as well as from gravity inversion would be very useful to understand the tectonic evolution of the region.

3.5.2.1 Crustal Configuration as Evidenced by the DSS Data

Post India–Antarctica rifting, on the basis of intrinsic differences between various sub-crustal blocks (Lal et al., 2009; Radhakrishna, 1989), the East Coast of India is tectonically and geomorphologically segmented into several pericratonic rift basins such as the Cauvery, Palar, Krishna–Godavari, Mahanadi, and Bengal basins. Geophysical studies over the East Coast basins (Fuloria, 1993; Prabhakar and Zutshi, 1993; Rangaraju et al., 1993; Rao, 2001; Sastri et al., 1973, 1981) revealed the presence of several NE–SW trending horst–graben structural features which were formed during rifting. The DSS data provide information on the velocity distribution, geometry about the basement configuration, and also the crustal section down to the Moho. As prominent velocity discontinuities are generally reflected in the DSS data, their velocity distribution and structural details help to understand the basic nature of the crust.

We present here eight DSS profiles (Fig. 3.24) that include both refraction and wide-angle reflection data available for the eastern Indian shield margin (Fig. 3.25) in the K–G and the adjoining Cuddapah basin (profiles a–d), in the Mahanadi basin (profiles e-g), and in the Bengal basin (profile h). The compiled profiles are (a) Kavali–Parnapalle profile (Kaila and Tewari, 1985; Kaila et al., 1979), (b) Alampur–Koniki–Ganapeswaram profile (Kaila and Tewari, 1985), (c) Kallur–Polavaram profile (Kaila et al., 1990a), (d) Paloncha–Narsapur profile (Kaila et al., 1990a), (e) Konark–Mukundpur profile (Behera et al., 2004), (f) Baliamba–Jagannathpur profile (Behera et al., 2004), (g) Paradip–Kabatabandha profile (Behera et al., 2004), and (h) Beliator–Bangoan profile (Kaila et al., 1992; Mall et al., 1999). The refraction velocities obtained along these profiles falling in the region of the east coast gave rise to the layered velocity structure for the eastern Indian shield crust with Moho mapped at a depth of 35–39 km in the south below the Cuddapah basin, 41–43 km in the K–G basin, 33–37 km in the Mahanadi basin, and 26–36 km in the Bengal basin regions. In the K–G and Cuddapah basin region, the upper crustal layer with seismic velocities between 5.5 and 6.2 km/s represents the crystalline basement. In the K–G basin, the crystalline basement is underlain by two crustal layers with velocities in the range of 6.5–6.6 and 6.7–6.8 km/s having the Moho boundary marked by a velocity jump from 6.9 to 8.1 km/s (Kaila et al., 1990a). The crust beneath the Cuddapah basin also has similar velocity structure with individual layers showing minor variations; however, this region is characterized by the presence of a high velocity layer (>7.2 km/s) in the deeper part of the crust above the Moho. In the Mahanadi basin, Behera et al. (2004) have observed a five-layered crustal structure with velocities ranging between 6.0 and 7.5 km/s. Most importantly, they mapped a mid-crustal low-velocity zone (6.0 km/s) and a 10-km thick high-velocity (7.5 km/s) layer at the crustal base, both suggestive of typical rift-related evolution of the basin and plume-related magmatic underplating. A similar high-velocity (7.5 km/s) layer is also mapped in the lower crust for the profile across the Bengal basin by Mall et al. (1999). The presence of underplating material at the crustal base in both Mahanadi and the Bengal basins suggests widespread influence of plume-related magmatism at the northeast coast of India.

3.5.2.2 *Seismic Refraction Velocities in the Bengal Fan*

The seismic P-wave velocity structure underlying the northern part of the Bengal Fan was first given by Curray and Moore (1971). The velocity structure beneath the Bengal Fan was also presented using the thickness and depth relations obtained from the wide-angle reflection and spot refraction data by Naini and Leyden (1973). Based on the velocity–depth structure beneath the Bengal Fan, five lithologic sections were prepared by them for delineating

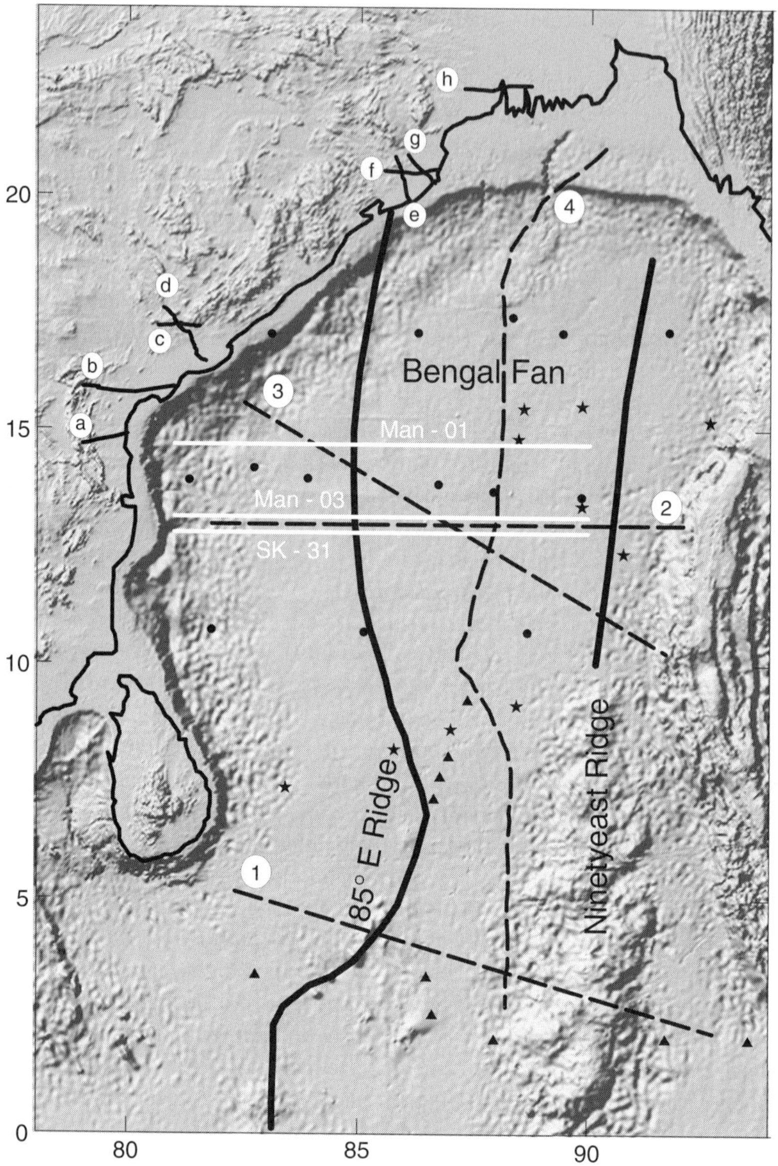

FIGURE 3.24 Shaded relief map of the Bay of Bengal showing the location of DSS profiles along the East Coast of India and regional MCS lines and selected crustal sections from 3D gravity interpretation in the offshore. Filled circles and stars, seismic refraction stations; triangles, OBS measurements *(adapted from Radhakrishna et al., 2010)*.

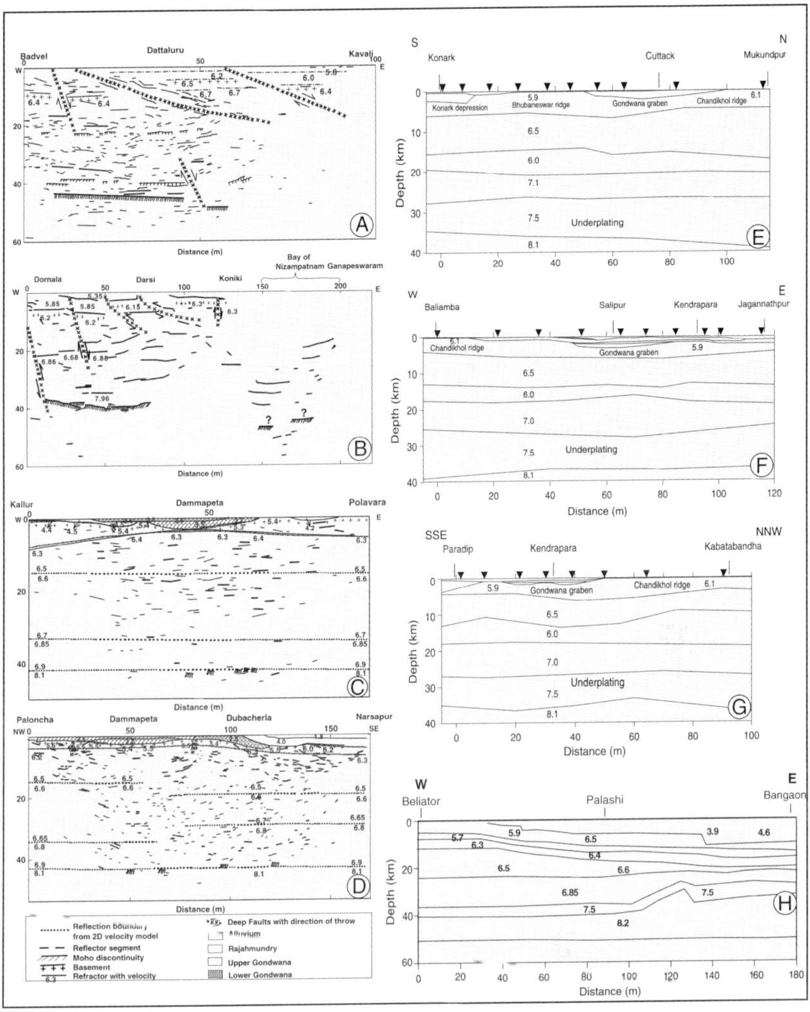

FIGURE 3.25 Interpreted crustal velocity–depth layering along different DSS profiles (A–H) for the East Coast of India compiled from different sources. Profile locations are shown in Fig. 3.24. Profiles A–D are from Cudappah and K–G basin region, (A) Kavali–Badvel section (Kaila et al., 1979), (B) Koniki–Ganapeswaram section (Kaila and Tewari, 1985), (C) Kallur–Polavaram section (Kaila et al., 1990a), (D) Paloncha–Narsapur section (Kaila et al., 1990a), (E) Konark–Mukundpur section (Behera et al., 2004), (F) Baliamba–Jagannathapur section (Behera et al., 2004), (G) Paradip–Kabatbandha section (Behera et al., 2004), and (H) Baliator–Bangoan section (Mall et al., 1999).

the oceanic basement and the overlying sediments. From these sections, a total of nine reflecting layers with velocities ranging between 2.0 and 5.78 km/s and the oceanic basement characterized by a velocity of 6.2 km/s have been identified. Later, Curray et al. (1982) provided detailed velocity structure of the deeper sedimentary layers beneath the BOB in the north based on seismic refraction studies and several regional multichannel seismic reflection sections across the BOB. They considered a P-wave velocity of 5.7 and 6.0–7.5 km/s to represent and distinguish the oceanic basement from the oceanic igneous crust, respectively. Overlying the oceanic basement, a layer of Cretaceous–Paleocene age with velocities of 4.0–4.5 km/s was mapped throughout the Bengal Fan. The observed crustal velocities beneath the Ninetyeast ridge where a thin cover of sediment exists range from 6.2 to 7.0 km/s. The Ninetyeast ridge is progressively buried northward under thick sediments. In the southern part of BOB (south of 10° N), Subrahmanyam and Singh (1992) presented the refraction data collected during Russian cruises from ocean bottom seismograph stations, and the velocity measurements from this data suggest velocities of 2.0–5.7 km/s for the sediments, 6.1–7.7 km/s for the oceanic crust, and 8.3–8.5 km/s for the oceanic upper mantle. The velocities of 7.3 km/s represent the oceanic layer 3 and are overlain by the layer with velocities of 5.7–6.8 km/s. The above-mentioned velocity information obtained from each refraction station in the BOB region (see Fig. 3.24 for location) is compiled and presented by Radhakrishna et al. (2010) in map view (Fig. 3.26). From this data, the sediment and crustal seismic velocities can be very well distinguished at many stations. However, there is a large uncertainty on the representative seismic velocity which ranges between 5.4 and 6.2 km/s that demarcates the sediment–basement interface or in other words the top of the oceanic volcanic layer. It may be noted that while the crust beneath southern BOB can be classified as normal oceanic crust based on the velocity structure, toward north, the basement is overlain by progressively thicker sediments and this fact must be taken into account. In view of the above, the seismic velocities in BOB were reinterpreted by Curray (1991), and the sedimentary structure has been redefined to include an additional 6-km-thick, high-velocity (6.0–6.4 km/s) metasedimentary layer at a depth greater than 16 km. The inclusion of this additional sedimentary layer in the deepest part gave rise to a maximum thickness of 22 km of sediments in offshore Bangladesh (Fig. 3.27) in northern BOB. Thus, the estimated sediment thickness from seismic data varies significantly in the BOB region from almost 2–4 km near the equator to nearly 22 km in northern most part of BOB in the shelf region. In the central part of BOB, the sediment thickness is about 8 km except in the linear regions encompassing the 85° E and Ninetyeast ridges. While the 85° E ridge is seen to be buried beneath 2–4-km-thick sediments throughout, a thin sedimentary cover exists over the Ninetyeast ridge up to 10° N. However, further north, the ridge is seen to be progressively buried under the thick BOB sediments. The 85° E and the Ninetyeast ridges divide the BOB into the sub-basins, and the deeper part of these sub-basins is characterized by thick pre-Bengal Fan sediments overlying the basement (Curray et al., 1982). Further, the crustal average velocity–depth profile along a longitudinal section (Fig. 3.27) from the equator to the Bengal basin (south

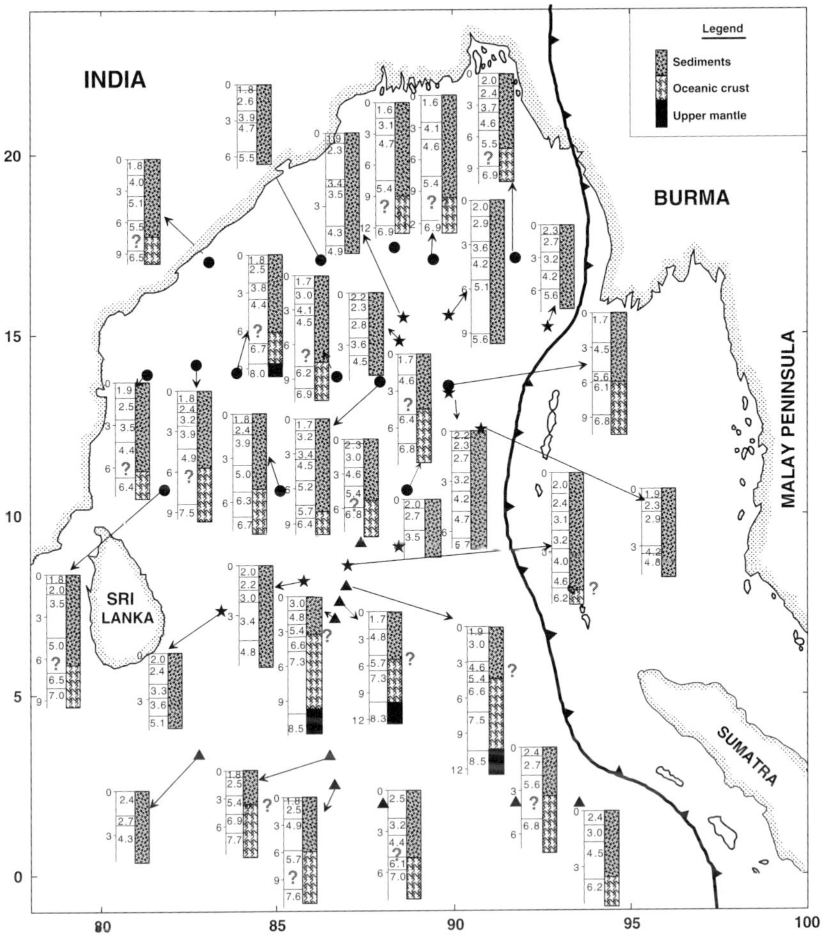

FIGURE 3.26 Seismic refraction station plots showing velocity interfaces with depth in the BOB in plan view *(adapted from Radhakrishna et al., 2010)*. Sea floor is considered as the reference surface. Filled circles are from Curray et al. (1982), stars from Naini and Leyden (1973), and triangles are Russian OBS measurements from Subrahmanyam and Singh (1992). Question marks near refraction velocities indicate uncertainty in identified sediment–basement interface. (For color version of this figure, the reader is referred to the online version of this chapter.)

to north) exhibits the deeper burial effect of sediments as well as the crust toward north (Curray, 1991). The section indicates that the thickness of the metasedimentary layer is variable and decreases toward south along with a decrease in the thickness of the overlying sediments. A detailed analysis of sediment velocity–depth relation in BOB (Radhakrishna et al., 2010) reveals that the southern part of BOB has a different burial relationship when compared to the northern part where the sediments are much thicker. This difference was ascribed to the progressive decrease in the sedimentation rate and

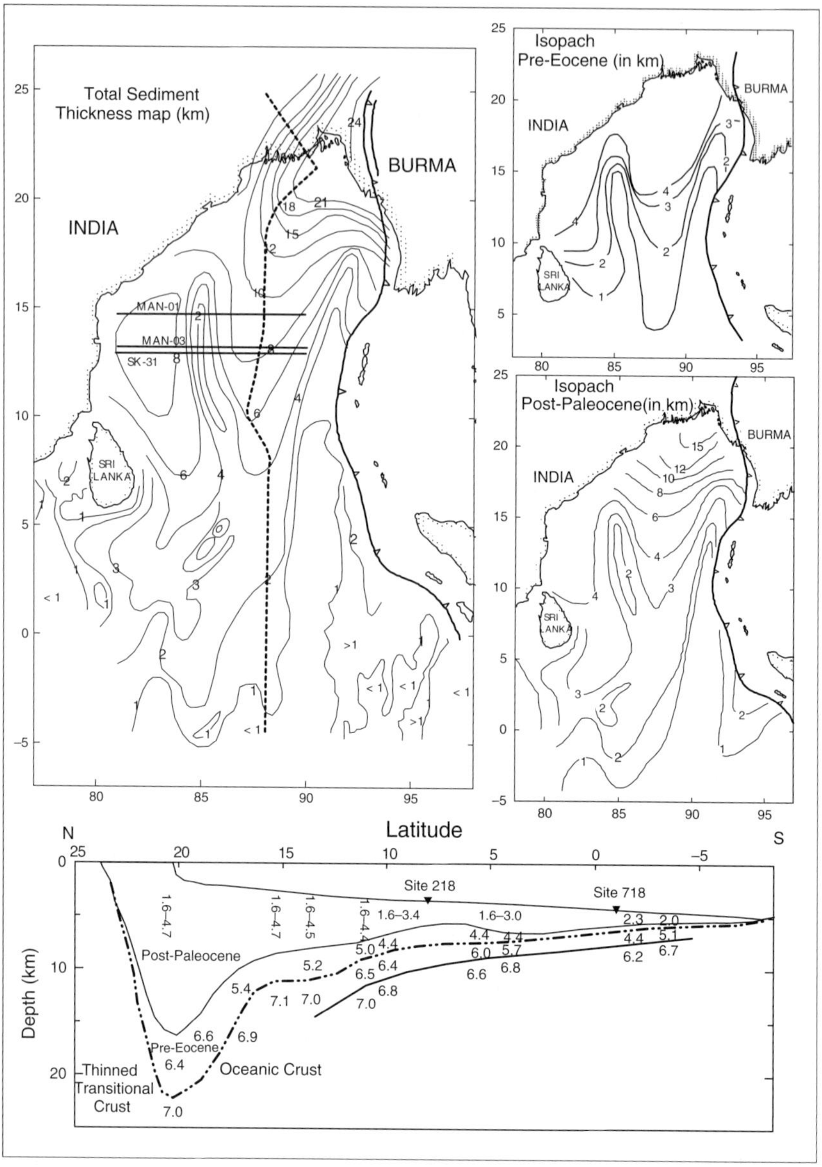

FIGURE 3.27 Isopach map of total sediment thickness in the BOB and Bengal basin (left panel), and the isopach maps of pre-Eocene and post-Paleocene sediments (right panel) *(Reproduced from Curray, 1991)*. The N–S dashed line along the BOB (left panel) is the crustal section (shown below) constructed by Curray (1991).

the sediment character in the Bengal Fan from north to south. Curray (1991) suggested that entire sedimentary section of the BOB region can be divided into two parts (Fig. 3.27) separated by widespread early Eocene unconformity which is related to the timing of the India–Asia collision. They are (i) precollision section representing the Cretaceous and Paleocene continental rise sediments up to 6 km off the Indian margin and (ii) post-collision Bengal Fan section having a maximum sediment thickness of 16 km derived mainly from the denudation of the uplifted Himalayas after collision.

3.5.2.3 *MCS Reflection Profiles Across the BOB*

The internal sedimentary architecture and the basement configuration across the BOB along several E–W oriented lines starting from the east coast to the Andaman trench region are available from the previous investigations (Curray et al., 1982; Gopala Rao et al., 1997; Pateria et al., 1992). We present here the detailed sedimentary strata identified along three such E–W seismic sections across the BOB covering the central and northern parts (Gopala Rao et al., 1997). These sections have clear information from the seafloor to the basement and reveal eight seismic layers H1–H8 (Fig. 3.28) with maximum total thickness of around 5.0 s TWT. The observed depositional sequence boundaries

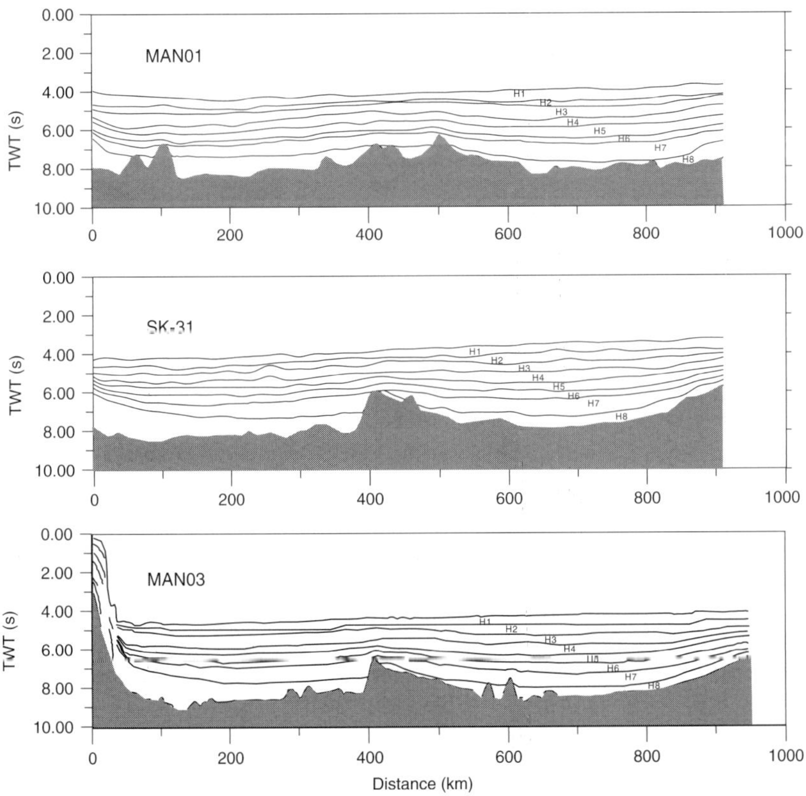

FIGURE 3.28 Interpreted line drawings showing the acoustic basement and the overlying sedimentary layers for three regional MCS profiles across the BOB. See Fig. 3.27 for their location *(Redrawn from Gopala Rao et al., 1997)*.

indicate that the sedimentation in the BOB can be categorized into two distinct phases (Curray et al., 1982; Gopala Rao et al., 1997): an older phase prior to collision along the Himalayan front, and a younger phase marking the post-collision depositional activity related to the erosion of Himalaya and deposition of the Bengal fan sediments.

3.5.2.4 Crustal Thickness Map of BOB

The crustal thickness data for the BOB region are available along the MCS reflection sections through gravity modeling (Gopala Rao et al., 1997; Mukhopadhyay and Krishna, 1991; Subrahmanyam et al., 1999). Utilizing all available seismic information from refraction station velocities and seismic reflection data and the velocity–depth relations, Radhakrishna et al. (2010) carried out 3D gravity analysis in the BOB region by considering the depth-dependent variation of density for the sedimentary layers. After subtracting the gravity effect of BOB sediments, the resultant crustal Bouguer anomalies were inverted to obtain depth to the Moho, from which they estimated thickness of igneous crust in the region (Fig. 3.29). Four crustal transects (see Fig. 3.24 for location) along and across the BOB prepared from the 3D gravity interpretation (Radhakrishna et al., 2010) reveal the presence of a thin crust below the Bengal Fan and thicker than normal oceanic crust below the 85° E and NERs (Fig. 3.30).

The maps presented in Fig. 3.29 bring out new information on regional variations in the thickness of the igneous crust and the Moho depth computed from north to south beneath the BOB. The depth to the Moho varies between 11 and 16 km up to 17° N, and further north, the Moho depth increases abruptly to 32 km near the Bangladesh shelf. Three distinctly different crustal domains were noticed: (i) Moho depth of 11–12 km in the southern BOB, (ii) Moho depth ranging from 14 to 18 km in the central part, and (iii) much deeper Moho depth varying between 18 and 32 km in the north in the offshore Bangladesh region (Fig. 3.29). However, the igneous crustal thickness map, obtained after removing the water layer and sediments, does not reflect the above discussed trends implying that these domains in BOB are due to subsidence caused by large-scale sedimentation in the northern part of BOB. The Moho depths obtained in this study are strikingly similar to the earlier estimates determined through surface wave dispersion measurements, Sn attenuation data, integrated with the reinterpreted seismic data in BOB by Brune et al. (1992). The above crustal thickness estimates, therefore, rule out the presence of anomalously thick crustal layer beneath the BOB inferred previously by Brune and Singh (1986), Brune and Priestley (1989), and Ghosal (2008). The inference of an anomalously thick crust beneath northern BOB by Brune and Singh (1986) may be due to the fact that they neglected the deeper high velocity metasediments, which, however, were later incorporated while inverting the seismological data by Brune et al. (1992). It is seen that moderately thick igneous oceanic

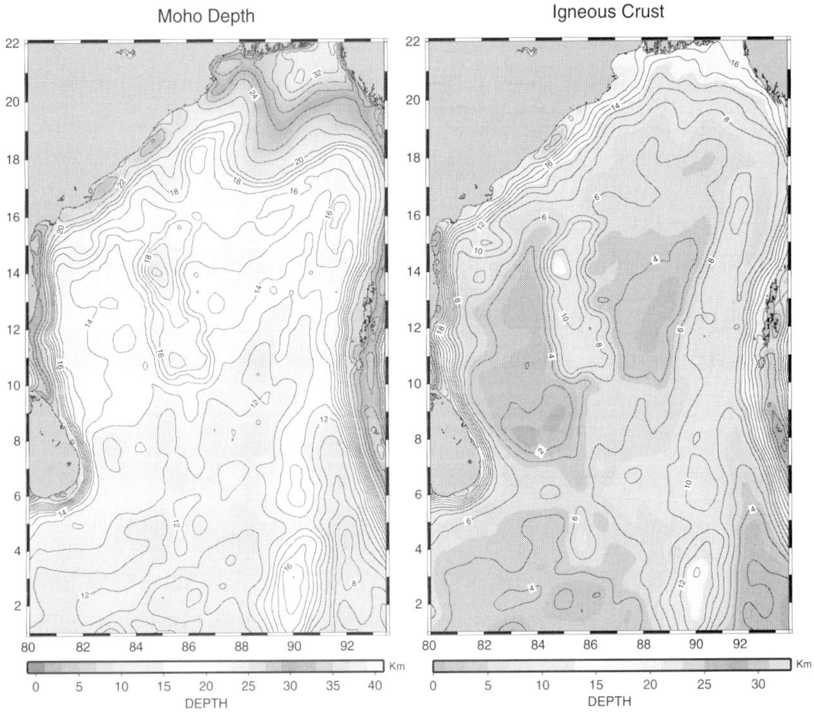

FIGURE 3.29 (A) The Moho depth map and (B) total igneous crust map of the BOB region obtained from 3D inversion of the residual crustal Bouguer anomalies *(adapted from Radhakrishna et al., 2010)*. (For color version of this figure, the reader is referred to the online version of this chapter.)

crust of 10–16 km is present beneath the Bangladesh shelf even after considering an isostatically balanced oceanic crust of 3.0 g/cm^3 beneath 20–22-km-thick sediments suggested by Curray (1994) based on isostatic calculations (see Section 4; Fig. 3.30). The BOB region is underlain by an oceanic crust formed during 120–80 Ma by seafloor spreading. The nature of crust below the Bengal basin is not clearly known, though presence of a denser and thicker oceanic crust was earlier proposed by Verma and Mukhopadhyay (1977). However, Mall et al. (1999) reported a high velocity (7.5 km/s) lower crustal layer which was inferred to be due to underplating. The above observations point toward a thicker and anomalous crust beneath the Bengal basin. Alternatively, the thicker crust in the Bangladesh shelf and the adjoining Bengal basin could possibly represent an initial thick oceanic crust (Holbrook et al., 1994; Roots et al., 1979), or anomalous lower crust/underplating (Trehu et al., 1989) formed during early margin evolution. The igneous crustal thickness map shows 2–6 km of oceanic crust south of 19° N excluding the regions of 85° E and the Ninetyeast ridges. In particular, it is interesting to notice that the deeper basinal parts of the Bengal

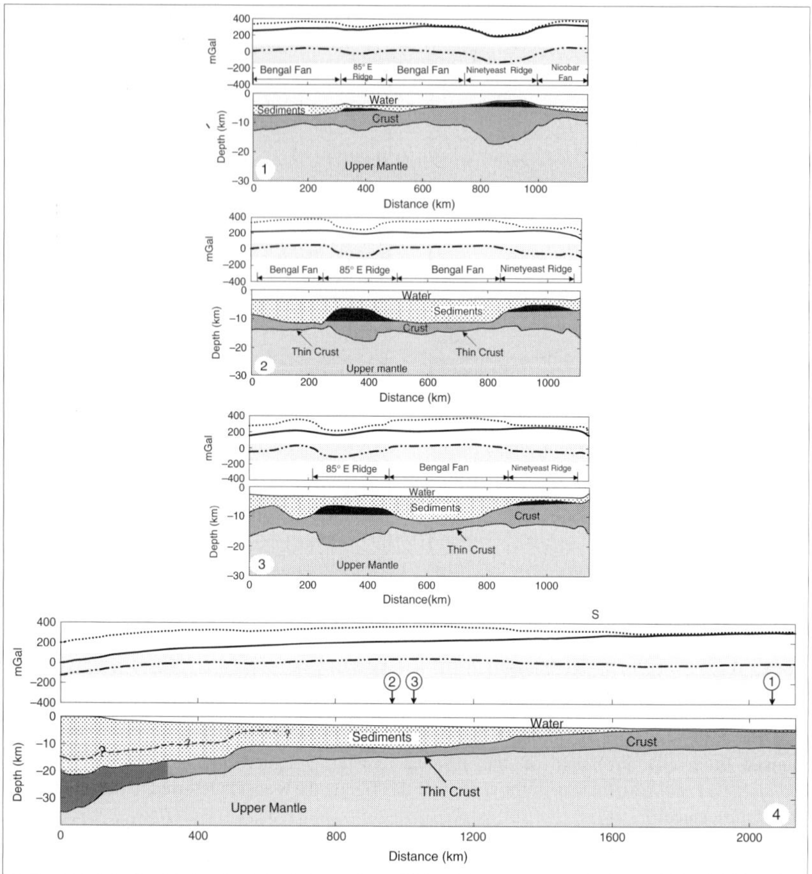

FIGURE 3.30 2D crustal sections along and across the BOB (location shown in Fig. 3.24) constructed from the 3D crustal structure maps (in Fig. 3.29). The upper panel for each section represents gravity anomalies after removing the effect of water layer (continuous line), removing the effect of water and sediment layers (dashed line) and residual anomalies (dashed and dotted lines) *(adapted from Radhakrishna et al., 2010)*. (For color version of this figure, the reader is referred to the online version of this chapter.)

Fan and the Wharton basin, east of the Ninetyeast ridge, are characterized by an anomalously thin oceanic crust of 2–4 km. Such anomalously thin oceanic crust was previously reported along major fracture zones (Detrick and Purdy, 1980; Prince and Forsyth, 1988) and at certain nonvolcanic margins (Hopper et al., 2006). Thin oceanic crust is generally formed only at ultraslow spreading rates below 20 mm a^{-1} through normal seafloor spreading processes (Dick et al., 2003). However, based on seismic reflection and refraction data, Singh et al. (2008) reported the formation of an anomalously thin crust (3.5–4.5 km) at the fast spreading Wharton spreading center. They attributed the formation of such an abnormally thin crust in a fast spreading environment to the interaction between the Kerguelen plume and the Wharton spreading center in the region of depleted or slightly colder than normal mantle. Further, Rodger et al. (2006) also observed such unusually thin crust (<4 km) beneath the Amazon Fan from seismic data and attributed its formation to the ultraslow seafloor spreading process. The deposition of huge sediments of the Amazon Fan over the unusually thin oceanic crust implies a strong upper mantle. Implications of the above analysis are discussed in detail by Radhakrishna et al. (2010).

3.5.2.5 Structure and Emplacement of 85° E and the NERs Below BOB

The sediments of the BOB are underlain by the early Cretaceous oceanic basement, and it is observed that the sediment thickness decreases progressively toward south starting from the Bangladesh shelf toward the distal end of the Bengal Fan (Curray et al., 1982). However, considerable variation in sediment thickness is observed between the ECMI and the Andaman arc region across the BOB due to the presence of N–S trending 85° E and the NERs. The region is divided into major sub-basins by these two ridges which are believed to have been emplaced subsequent to the formation of BOB crust (Curray et al., 1982). Though the exact nature of their emplacement is debated, most of the investigators believe that their origin is the plume-related volcanism. A detailed geophysical data analysis in terms of subsurface geometry and structure is presented here, particularly along their northernmost buried part, so as to obtain a deeper insight into their origin.

3.5.2.5.1 85° E Ridge

The 85° E ridge in the BOB is the most enigmatic aseismic ridge due to the uncertainty about the nature of crustal rocks that form the ridge and the origin of the low gravity anomaly over the ridge. The prominent negative free-air gravity anomaly over the ridge is quite uncharacteristic for an aseismic oceanic ridge having a positive relief. The 85° E ridge extends from the Afanasy Nikitin Seamount (ANS) in the equatorial central Indian Ocean to the Mahanadi offshore in the north. The ridge was initially identified as a basement high on a few seismic profiles (Curray and Moore, 1971), but subsequent seismic studies have mapped the ridge continuously (Fig. 3.31), starting from the ANS up to the East Coast of India. The ridge is buried below thick BOB sediments north of 5° N, whereas toward south, it occasionally rises above the seafloor and finally joins the ANS at 5° S. The ridge is also reported to be associated with complex magnetic signature all along its length (Gopala Rao et al., 1997; Michael and Krishna, 2011; Ramana et al., 1997). The northward continuation of the 85° E ridge beyond 17° N in the BOB has been speculative for a long time. Based on the interpretation of the gravity and magnetic anomalies, several earlier researchers (Nayak and Rao, 2002;

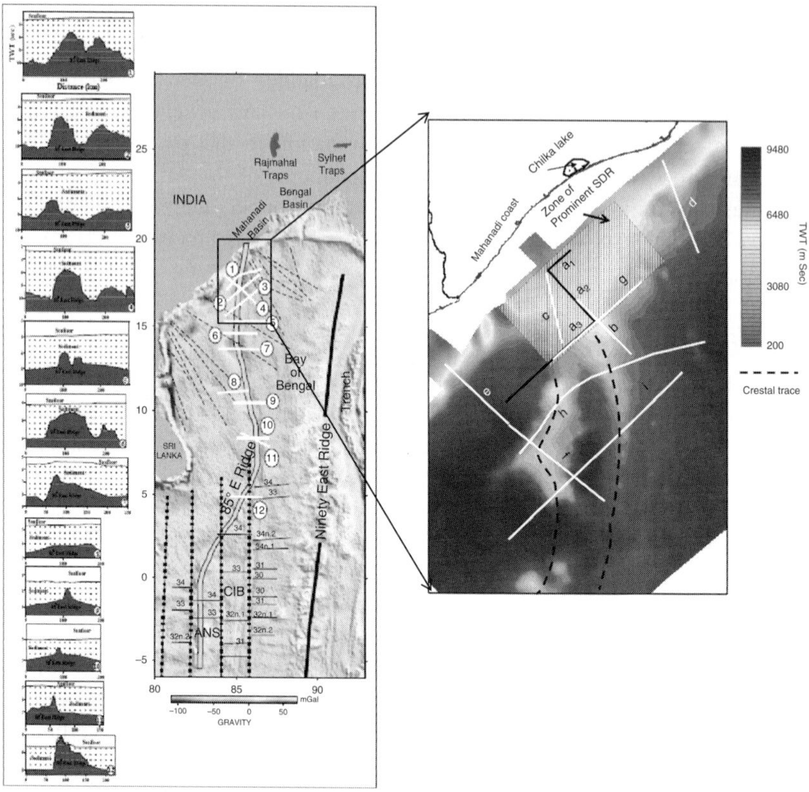

FIGURE 3.31 Shaded relief map of gravity anomalies in the BOB region (centre panel). Several E–W seismic sections show the trace of 85° E ridge in BOB region (left panel). Dashed lines on the gravity map indicate the fracture zones identified by Radhakrishna et al. (2012). The zoomed area in the north (right panel) shows the time–structure map to the top of acoustic basement along with number of MCS profiles pertaining to the northern part of 85° E ridge in the Mahanadi offshore. Shaded area in this map marks the region of prominent SDRs (modified after Bastia et al., 2010b). (For color version of this figure, the reader is referred to the online version of this chapter.)

Ramana et al., 1997; Subrahmanyam et al., 2008b) inferred that the ridge continues up to the Mahanadi coast. Since the characteristic gravity low attributed to the ridge is obscured by the shelf-edge anomaly of the ECMI, it restricts a definitive interpretation of the ridge in the Mahanadi shelf region. Anand et al. (2009) inferred that the ridge terminates at 15° N and ascribed the structure seen along 17° N profile of Curray et al. (1982) to the geomorphic expression of the basement. However, owing to limitations in the gravity and magnetic interpretation at the margin in the absence of further seismic constraints, most of the earlier propositions remained inconclusive. Utilizing vast amount of MCS data, Bastia et al. (2010b) delineated the subsurface geometry of the ridge in the Mahanadi offshore region (Fig. 3.32), which unequivocally demonstrated the northward continuation of the ridge up to the Mahanadi coast. The first significant geophysical study to model the ridge

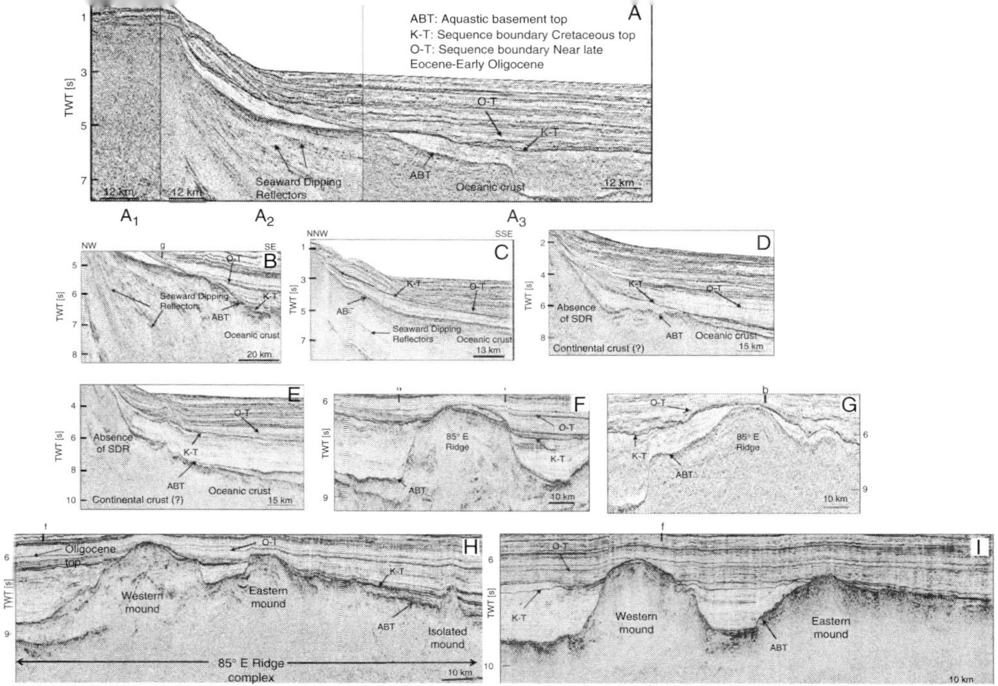

FIGURE 3.32 (A) Zigzag seismic profile showing the morphology and internal seismic signature of the ridge below shelf (a1), slope area (a2), and abyssal plain (a3). (B) Seismic profile showing the presence of prominent SDRs in the slope region. (C) Seismic section across the slope showing the presence of a submerged ridge below the acoustic basement top (ABT) as evidenced by the presence of prominent SDRs. (D) Seismic profile across the shelf slope boundary in the northern part of study area. (E) Seismic profile across shelf slope in southern part of study area. SDRs are absent in profiles D and E, indicating they are away from the trace of the ridge complex. (F) Seismic line across the western volcanic mound rising above the abyssal plain, showing a complex internal geometry. (G) Strike section across the ridge, close to the toe of the slope. Multiple volcanic mounds are present with the overall doming structure. (H) Strike section along the proximal part of the abyssal plain showing the entire ridge complex that includes the western, eastern, and other isolated mounds. (I) Strike section, along the distal part of the abyssal plain, showing a more symmetric nature of the western mound with steeper flanks as compared to the eastern mound which is wider and continuous but with a more complex internal architecture *(Reproduced from Bastia et al., 2010b)*. (For color version of this figure, the reader is referred to the online version of this chapter.)

was by Liu et al. (1982), who explained that the negative gravity field observed over the 85° E ridge was due to the emplacement of the ridge over a weak lithosphere which subsequently attained strength during the sediment loading. Other important hypotheses that were put forward based on the interpretation of geophysical data over the ridge are (i) due to shearing or sagging of crust by horizontal stretching/compressional forces (Anand et al., 2009; Ramana et al., 1997), (ii) emplacement over a weaker lithosphere characterized by underplating (Subrahmanyam et al., 1999), and (iii) emplacement over an older lithosphere with the negative gravity anomaly attributed to downflexing of the oceanic crust as well as the presence of metasediments (Krishna, 2003; Sreejith et al., 2011). Curray and Munasinghe (1991) postulated that the Rajmahal traps, 85° E ridge, and the Afanasy Nikitin Seamounts possibly represent the trace of the Crozet hotspot.

A detailed investigation of the 85° E ridge across the COB in the Mahanadi offshore was carried out by Bastia et al. (2010b) on the basis of several closely spaced high-resolution 2D seismic profiles (Fig. 3.32) as well as the isochronopach maps prepared for the prominent sequence boundaries (Fig. 3.33). This study had provided some valuable insights with regard to the nature of the ridge in the region. Some salient results from this analysis are presented here.

The seismic profiles A_2, B, and C (Fig. 3.32A–C) traversing across the shelf reveal the presence of a prominent seismic event at around 1.5 s (TWT) depth recognized to represent the Acoustic Basement Top (ABT), below which chaotic seismic facies without any discernible seismic reflections are observed. Several wells drilled in the Mahanadi shelf have penetrated this event and correspond to the top of a suite of basic volcanic rocks. The overlying paleontologically dated sediments in correlation with the Rajmahal traps exposed further north (Fig. 3.31) in the Indian continental shield (Fuloria, 1993) indicate that these rocks were of Early Cretaceous age. While typical seismic signatures of rifted passive margins were recorded and drilled wells have penetrated into nonvolcanic crystalline continental crust (Bastia, 2006) in the neighboring Bengal and Krishna–Godavari (K–G) basins, in the Mahanadi shelf region, massive outpouring of lavas inhibits recognition of such typical rift-related features. The presence of a thick pile of volcanic rocks could be ascribed to the passage of the Mahanadi shelf over the foci of volcanic eruption.

In the slope region, thick wedges bounded by SDRs were observed beneath the correlated top of the volcanics (Profiles A_2, B, and C). Toward the basinal side, these wedges are successively younger and increase in thickness significantly. The combined thickness of these wedges along the slope is interpreted to vary sharply based on the observed TWT increase from 2.5 to 4.0 s. The thickness as well as the dips of the wedges is anomalous when compared to the SDRs associated with a normal oceanic crust. Further, it is interesting to notice that away from the main ridge axis, toward north–east as well as south–west, the SDRs are absent in the slope region. However, features showing typical linkage of normal continental curst to oceanic crust through

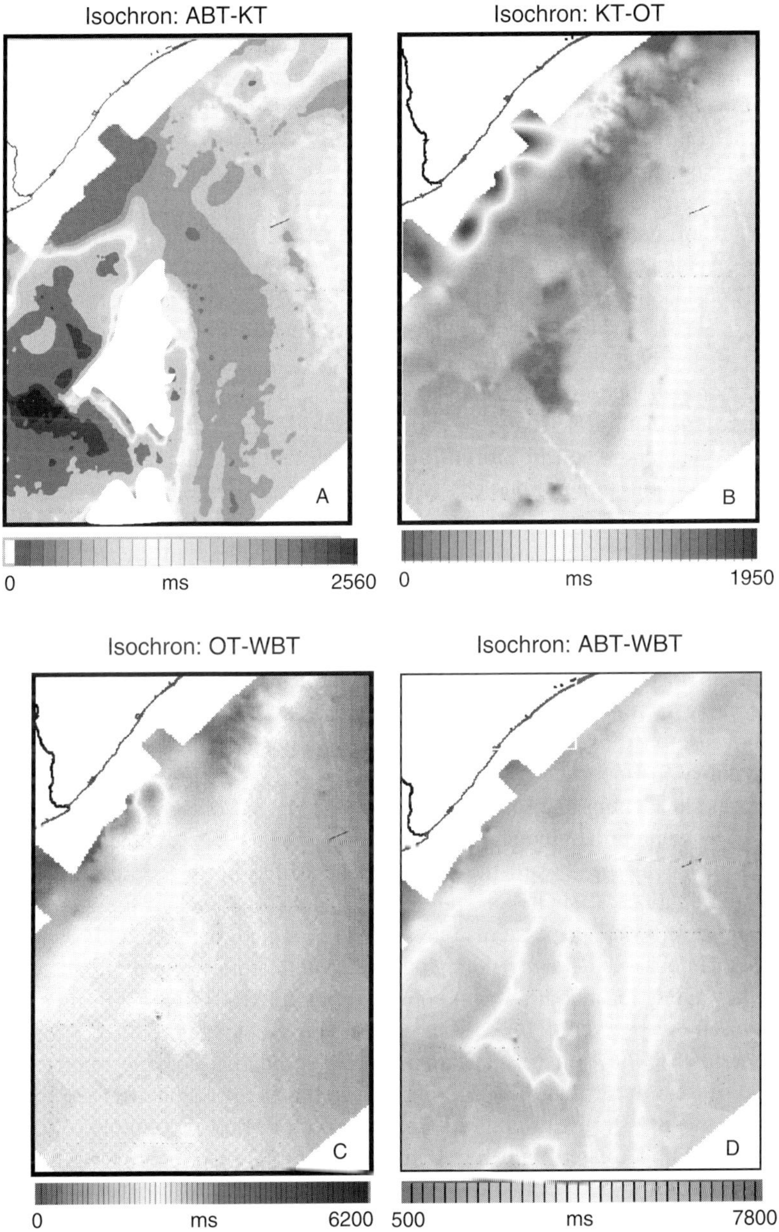

FIGURE 3.33 The isochronopach maps for different stratigraphic sequences over the 85° E ridge. (A) Between the Acoustic Basement Top (ABT) and the KT sequence boundaries. (B) Between the KT and the OT sequence boundaries. (C) Between the OT and the water bottom (WBT). (D) Between ABT–WBT sequence boundaries indicating total sediment thickness in the Mahanadi offshore region *(adapted from Bastia et al., 2010b)*. (For color version of this figure, the reader is referred to the online version of this chapter.)

rifted margin and proto-oceanic crust were observed (Fig. 3.32D and E). Similar type of wedges bounded by SDRs is observed in the Walvis ridge in SE Atlantic Ocean by Elliott et al. (2009). Near the base of the slope, several prominent upward doming events were observed below which the SDRs were absent (Fig. 3.32A_3 and B). This prominent SDR zone corresponds to the broad basin-ward plunging nose seen on the ABT map (Fig. 3.31). It is unclear whether this region in the slope represents the northern termination of the morphological ridge. However, drilled well and seismic information shows evidences of volcanism further northward in the shelf area.

The seismic sections of two dip (Fig. 3.32B and F) and three strike (Fig. 3.32G–I) profiles illustrate the morphology and internal structure of the ridge in the oceanic crust. In the proximal part of the abyssal plain, internal reflection patterns of the ridge suggest the presence of several upward doming bodies, terminating the SDRs on their flanks which are interpreted to be major eruptive centers. The dip profiles across such features show the presence of prominent highs with morphology typical of volcanic seamounts (Fig. 3.32G–I). Seismic images (Fig. 3.32H and I) reveal the presence of two prominent mounds which crop up parallel to each other. Morphology of the ridge reveals two structural highs. While the western high is more prominent and relatively has greater relief, the eastern high is wider and more continuous throughout its trace (Fig. 3.31). The western mound is symmetrical in shape with the steep-dipping western flank. The internal structure of the ridge appears to be complex with upward doming events (Fig. 3.32H) which indicate episodic eruptive activity over a period, as well as build up of prominent volcanoes. The downward concave bright reflector at $\sim$9 s TWT (Fig. 3.32F) indicates lithospheric sagging due to increased load of erupted volcanics. The basement structure map further indicates that the ridge occurs as two linear structural highs which merge toward the basin margin in the north before progressing into the continental shelf. Several subtle offshoots parallel to the main trend are observed on the eastern flank of the ridge. A maximum width of 130 km is observed for the ridge complex in this area, while the highest part of the ridge is located at a depth corresponding to 5100 ms TWT. An isolated sub-basin is observed between the western and eastern highs of the ridge.

From the direction of mapped magnetic anomalies and the trend of the ridge, we can infer that the ridge might have been emplaced after the formation of the BOB crust through different modes of volcanic activity, one oceanic spreading related and the other plume emplaced. Isochronopach maps for the region in and around the ridge are prepared for different stratigraphic intervals (Fig. 3.33) having respective sequence boundaries associated with significant tectonic events such as the K–T boundary and late Eocene–early Oligocene collision event. These maps would be useful to understand the relation between sedimentation and disposition in the vicinity of the ridge in deepwater Mahanadi offshore region and will be discussed in greater detail in Chapters 4 and 5.

Considering its long trace up to the ANS, many complexities are yet to be unraveled with regard to the time and mode of emplacement of the ridge during the evolution of northern Indian Ocean. As mentioned earlier, several theories were suggested with regard to mode of emplacement of the ridge and these are trace of the Crozet hotspot connecting the Rajmahal traps in north and the ANS in south (Curray and Munasinghe, 1991); volcanism through a weak zone emplaced in a short span of time (Chaubey et al., 1991); northward continuation of 86° E fracture zone (Kent et al., 1992). From different geophysical interpretations, subsequent investigators agreed with the hotspot theory for the emplacement of the ridge (Bastia et al., 2010b; Gopala Rao et al., 1997; Krishna, 2003; Krishna et al., 2009; Michael and Krishna, 2011; Radhakrishna et al., 2012; Sreejith et al., 2011; Subrahmanyam et al., 1999). However, the plume model of emplacement must take into account the timing and source for the Rajmahal volcanics, the volcanism at ANS, and the regional plate motion during the emplacement. Based on geochemical studies and the paleo-position of the Indian Plate (Baksi et al., 1987; Frey et al., 1996; Kent et al., 2002), the connection between the Rajmahal traps and the Kerguelen Plume has been established. The DSDP (Legs 22–28) and ODP (Legs 120, 121, 183) drilling data indicate the eruption activity of the Kerguelen Plume in the nascent Indian Ocean after the breakup of Eastern Gondwanaland (India–Antarctica) in the Early Cretaceous. It is relevant to note that the present-day Kerguelen Large Igneous Magmatic Province (KLIMP) spreads over 22° longitude and 15° latitude, of which the older South Kerguelen Plateau encompasses nearly 12° longitude and 7° latitude (Bénard et al., 2010). It is known from the tectonic reconstructions (Royer and Coffin, 1992; Royer et al., 1991) that the Ninetyeast ridge, the Broken ridge, and the KLIMP have originated from a single hotspot in the Indian Ocean. Royer et al (1991) further suggested that the volcanic spur to the west of the Kerguelen Island and its relation to the McDonald Island Volcano in terms of its age and duration of volcanic activity need to be examined. Their reconstruction of the hotspot migration with respect to the Atlantic and Western Indian hotspots, on the plot of Fleitout et al (1989) model, requires presence of a mirror image of the Kerguelen Island/volcanic spur at a position west of the Ninetyeast ridge, which corresponds to the trace of the 85° E ridge. If both Kerguelen Island and the associated volcanic spur—the McDonald Island Volcano—are considered to be representing two heads of the Kerguelen Plume, it alone would support the source for volcanic hotspot-related emplacement for both Ninetyeast and the 85° E ridges (Bastia et al., 2010b).

3.5.2.5.2 Ninetyeast Ridge

The Ninetyeast Ridge (NER) is the longest N–S oriented linear feature in the oceanic areas extending over a length of 5200 km located in the eastern Indian

Ocean. The ridge extends from 30° S, where it intersects the Broken ridge, to almost 18° N in the northern BOB (Fig. 3.34) and can be clearly noticed on the topography map (see Fig. 2.3); however, beyond 10° N, the ridge is buried beneath the Bengal Fan sediments (Curray et al., 1982). The ridge marks the eastern limit of the Bengal fan and separates the Bengal and Nicobar fans

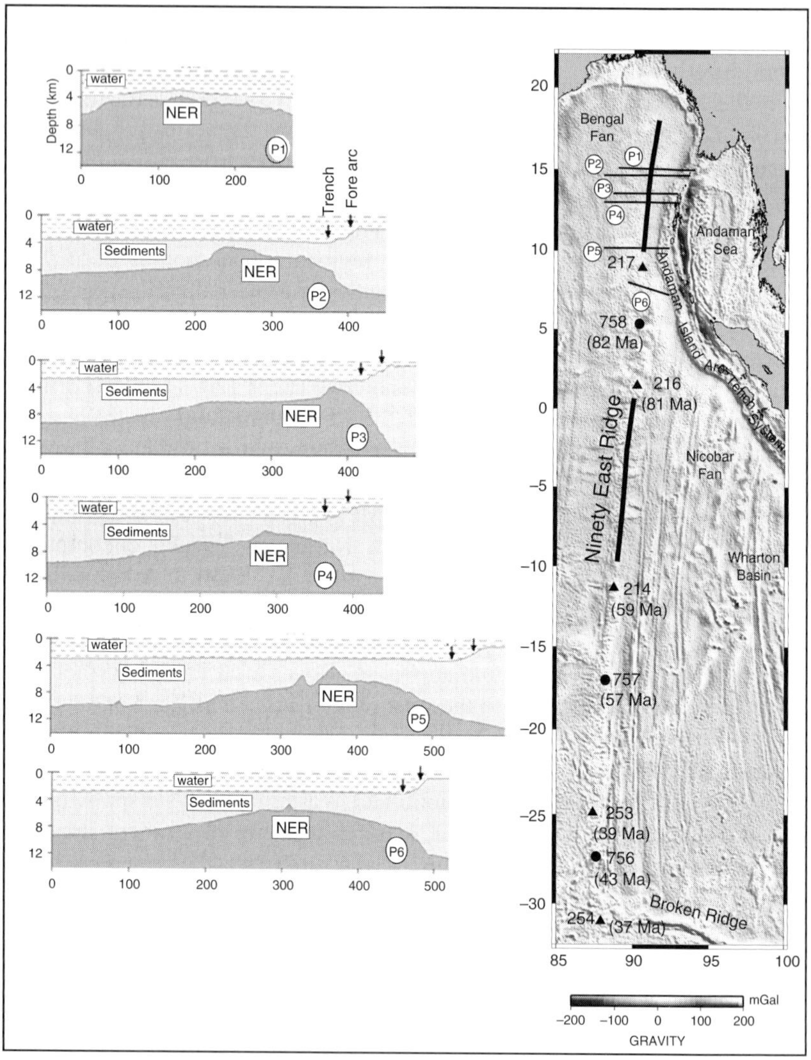

FIGURE 3.34 Map showing the gravity anomaly map of the NER–Andaman convergence zone in the northeast Indian Ocean (right panel). E–W lines (P1–P6) across the northern part of NER (left panel) present structure of the NER (after Subrahmanyam et al., 2008a). Filled circles, ODP; Solid triangle, DSDP sites are indicated over the ridge. (For color version of this figure, the reader is referred to the online version of this chapter.)

(Fig. 3.34). The southern part of the ridge is straight and flat topped and between 5° S and 100° N, it is composed of a series of en-echelon blocks (Curray and Moore, 1974; Sclater and Fisher, 1974). The single and multi-channel seismic data bring out the subsurface picture of the ridge north of 10° N, where it is buried below thick BOB sediments (Curray et al., 1982; Gopala Rao et al., 1997). From 10° N onward, Curray et al. (1982) traced the buried part of the ridge for another 700 km northward to about 17°N below the Quaternary–Eocene sediments of the Bengal Fan; however, recent deep seismic reflection data by Maurin and Rangin (2009) indicate its presence up to 18° N. Many theories have been proposed by earlier workers with regard to the formation of the NER, these are (i) horst-type feature having thickness of the underlying crustal layers similar to that of the ocean basin to the east (Francis and Raitt, 1967), (ii) localized emplacement of gabbro (Bowin, 1973), and (iii) formed by leaky-transform tectonics (Sclater and Fisher, 1974). However, it is generally accepted that the ridge had formed due to the plume-related volcanism near the Wharton spreading ridge. According to the plume model, during the northward drift of India in Late Cretaceous and early Cenozoic period, the hotspot-spreading ridge interaction left a trail of volcanism that formed the ridge (Krishna et al., 2001; Luyendyk and Rennick, 1977; Mukhopadhyay and Krishna, 1995; Royer et al., 1991; Sager et al., 2010). Johnson et al. (1976) attributed the NER volcanism to the combined result of leaky-transform tectonics and the hotspot activity, whereas Luyendyk and Rennick (1977) inferred two hotspots for its formation. The results obtained from several DSDP and ODP drilling sites (Curray et al., 1982; Duncan, 1978; Mahoney et al., 1983; Peirce et al., 1989, Von der Borch et al., 1974a,b; Weis et al., 1992; among many others) located all along the length of the ridge have significantly enhanced our knowledge about it. Some salient results are it has an extrusive volcanic basement, had formed subaerially or in shallow water and subsided with time, moderately evolved tholeitic basalts, and hotspot trace formed in south and transported northward.

Identified magnetic anomalies in the vicinity of the NER (Krishna et al., 1995, 1999; Liu et al., 1983; Peirce, 1978; Royer and Sandwell, 1989; Royer et al., 1991; Sclater and Fisher, 1974) reveal that west of the ridge, the ages decrease from north to south; in contrast, they increase along the eastern flank (Fig. 3.35). Royer and Sandwell (1989) observed that this linear age progression is not in agreement with the age estimates for the oceanic crust under the Wharton basin lying east of the ridge. According to Krishna et al. (1995), the NER is too long to have formed simply by volcanism on the Indian plate because magnetic isochrons of the same age in adjacent basins cover 11° less distance. The identified magnetic anomalies along east of NER in the Wharton basin imply large spreading ridge jumps and may incorporate sections of the ridge formed on the Antarctic plate (Krishna et al., 1999; Pilipenko, 1996; Royer et al., 1991; Sager et al., 2010).

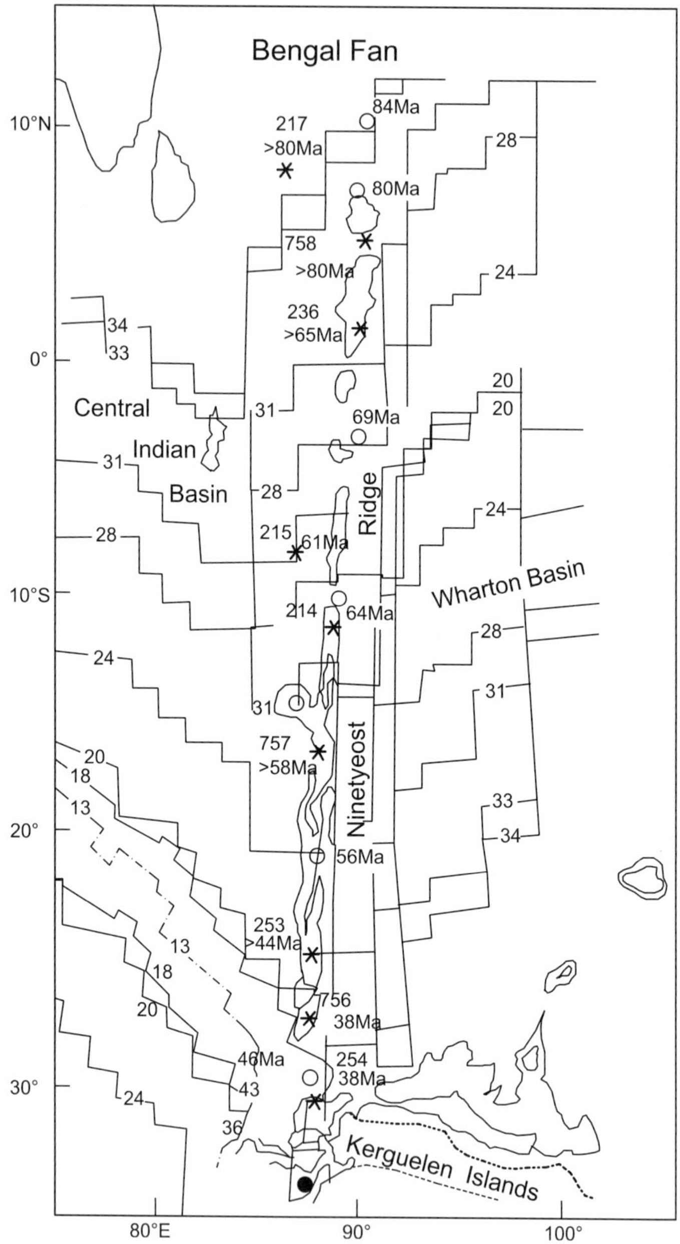

FIGURE 3.35 Plate reconstruction at Chron 13 (36 Ma) for the emplacement of NER in the eastern Indian Ocean (redrawn from Royer and Sandwell, 1989). Interpreted locations of the Kerguelen hotspot (open circles) are shown for different times. DSDP and ODP sites are indicated by stars, with their corresponding ages.

North of the equator, the NER has a known length exceeding 1700 km, including its buried continuation for almost 700 km below the Bengal Fan and gravity interpretation for this part of the ridge is carried out by several workers (Mukhopadhyay and Krishna, 1995; Radhakrishna et al., 2008; Subrahmanyam et al., 2008a). The gravity anomalies over the ridge indicate stronger positive anomalies over the exposed part and subdued gravity field over the buried segment. The gravity models of the ridge from south to north reveal the presence of underplated crust below the ridge and further indicate partial subduction of the ridge below the Andaman trench. The strike-slip and subduction-related tectonic forces gave rise to the en-echelon block structure for the NER in the proximity of the Andaman subduction zone (Radhakrishna and Sanu, 2002; Subrahmanyam et al., 2008a).

3.5.3 Structure and Tectonics of the Andaman Arc–Sea Region

According to Hamilton (1979), the Andaman–Burmese arc system serves as an important transitional tectonic link between the Himalayan Collision zone in the north and Sunda arc–trench system in the south (Fig. 3.1). The Indo-Burman ranges and the Andaman–Nicobar ridges are a northward continuation of the Mentawai ridge of the Sumatran subduction zone. Curray et al. (1982) inferred that the formation of the Andaman Sea has a definite genetic linkage with the evolution of the BOB. They further suggested a relation between thickness of sediments on the subducting plate and height and volume of the outer sedimentary arc and inferred that the Andaman Sea and the central lowlands of Burma are parts of a single structural province. The Andaman Sumatra arc is characterized by oblique plate subduction having predominantly thrust motion in the trench/fore arc region and strike-slip motion in the back-arc region (Dasgupta and Mukhopadhyay, 1993; Radhakrishna and Sanu, 2002; Ravikumar et al., 1996). The strike-slip motion is mainly taken up by the spreading ridge-transform system in the Andaman Sea and the right-lateral Sumatran fault zone in the mainland Sumatra (Curray et al., 1979; Kamesh Raju et al., 2004).

3.5.3.1 Configuration of Andaman Benioff Zone

The geometry of the Andaman Benioff Zone (ABZ) was delineated by several earlier investigators (Dasgupta and Mukhopadhyay, 1993; Dasgupta et al., 2003; Guzman-Speziale and Ni, 1996; Khan and Chakraborty, 2005; Kumar, 1981; Radhakrishna et al., 2008; Ravikumar et al., 1996). Eleven seismologic depth sections (after Radhakrishna et al., 2008) presented in Fig. 3.36 reveal the characteristics of ABZ along the Andaman arc. Their analysis indicates that the dip of the ABZ varies between 38° and 53° along the arc, with values varying between 43° and 53° in north Andaman (sections S1–S3) and between 38° and 50° in south Andaman–north Sumatra (sections: S4–S11).

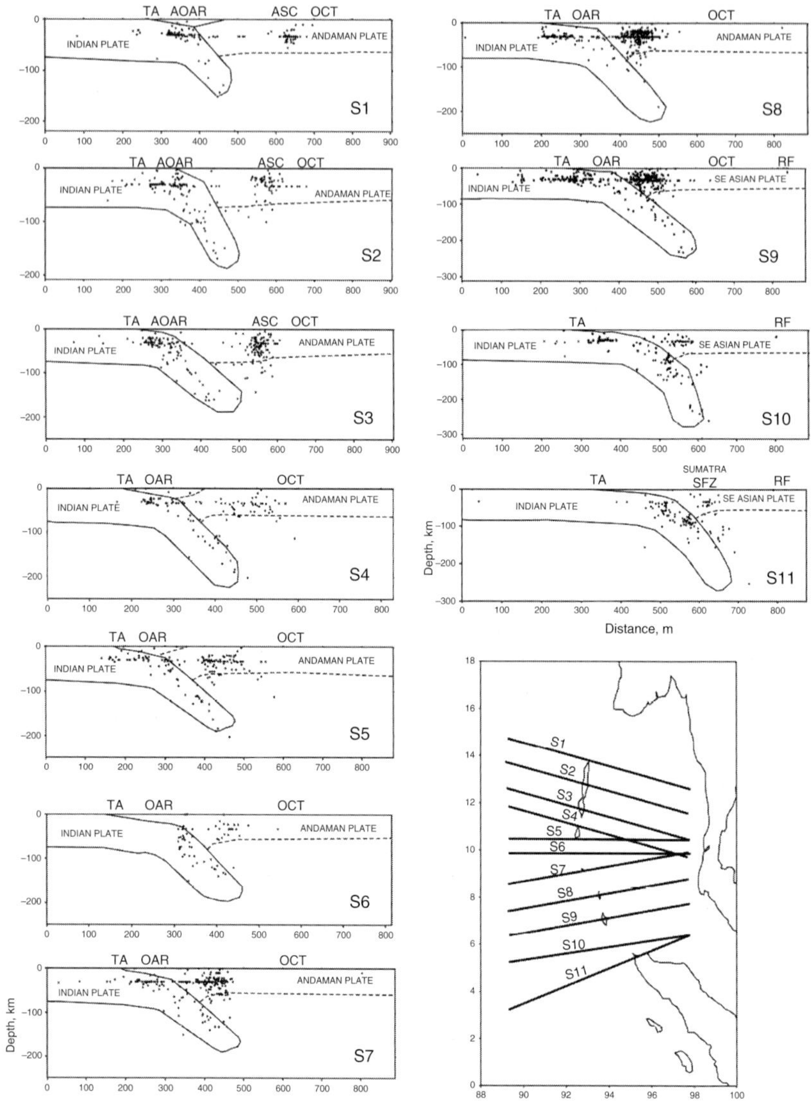

FIGURE 3.36 Seismologic sections showing the Andaman Benioff Zone (ABZ) configuration constructed from the earthquake hypocentral data (period: 1900–2004) in the Andaman arc region. Inset shows location of these sections (S1–S11). TA, Trench axis; OAR, Outer Arc Ridge; ASC, Andaman Spreading Center; OCT, Ocean continent transition; SF, Sumatran fault; RF, Renong Fault *(adapted from Radhakrishna et al., 2008)*.

This observation indicates that the dipping Indian plate does not maintain uniform geometry throughout; rather it is flexed by several transverse faults displaying contortions in the Benioff zone between Andaman and Sumatra (Dasgupta et al., 2003). Previous attempts to map the ABZ have used different procedures, these are Hypocentral Trend Surface (HTS) analysis (Guzman-Speziale and Ni, 1996), closer grid bases of shallow-depth Benioff zone hypocenters (Dasgupta et al., 2003), or smooth curves drawn through hypocenters (Khan and Chakraborty, 2005). The earthquake activity is almost continuous starting from the trench toward east, this is particularly true in the case of south Andaman sections (S5, S7–S11). It can be noticed that the earthquakes are mostly confined to top part of the lithosphere where the down bending of the slab is first initiated. Dasgupta and Mukhopadhyay (1997) observed a seismic gap at a depth of 90–110 km in the central Andaman arc and highlighted its implication on the back-arc volcanism.

3.5.3.2 Gravity Field Along the Arc

The gravity field of the Andaman arc was described by several workers (Mukhopadhyay, 1988; Mukhopadhyay and Krishna, 1991; Peter et al., 1966; Radhakrishna et al., 2008; Subrahmanyam et al., 2008a). We present here the free-air anomaly map prepared for the Andaman arc-sea (Radhakrishna et al., 2008), with superposed tectonic elements (Fig. 3.37), which reveals a broad and distinct gravity low of as much as −200 mGal over the Andaman arc. This low anomaly shifts toward east into the Nicobar Deep at around 8.5° N. Two strong N–S trending highs can be observed over the Invisible Bank and the West Sewell ridge almost close to this low. The Andaman trench is associated with a gravity low having values around −40 to −60 mGal. Between the two gravity lows, a gravity high occurs along the outer arc sediment accretionary prism comprising of Andaman, Nicobar, and Nias Islands. The gravity low observed along the Andaman trench is seen disrupted by the positive gravity field near 8° N and suggests that subduction of the NER partially taking place in this region. As a consequence, the gravity low of Andaman arc is displaced to the east in the Andaman Sea (Fig. 3.37). Subrahmanyam et al. (2008a) interpreted this shift due to the effect of convergence of NER on the Andaman arc. East of the arc, in the Andaman basin, the ASR is characterized by gravity variations ranging between −20 and 20 mGal, and both the Alcock and Sewell rises show gravity highs of around 60 mGal. Further east, the Mergui ridge falls in an area of large and linear gravity high of 50 mGal. Detailed analysis of this region was carried out by Radhakrishna et al. (2008) through the computation of Slab Residual Gravity Anomalies (SRGA) by subtracting the 3D slab gravity anomaly from the free-air anomalies described above (Fig. 3.37). The SRGA map also shows similar anomaly pattern as the free-air anomaly map with an important distinction that anomalies change. We present here two gravity models (AA′ and BB′ in Fig. 3.37) across the Andaman arc region interpreted by Radhakrishna et al. (2008) based on the SRGA map and the calculated mantle Bouguer anomalies (Fig. 3.38). Modeling of SRGA anomalies in the Andaman arc region revealed that below the ANR nearly 40–47-km-thick upper lithospheric

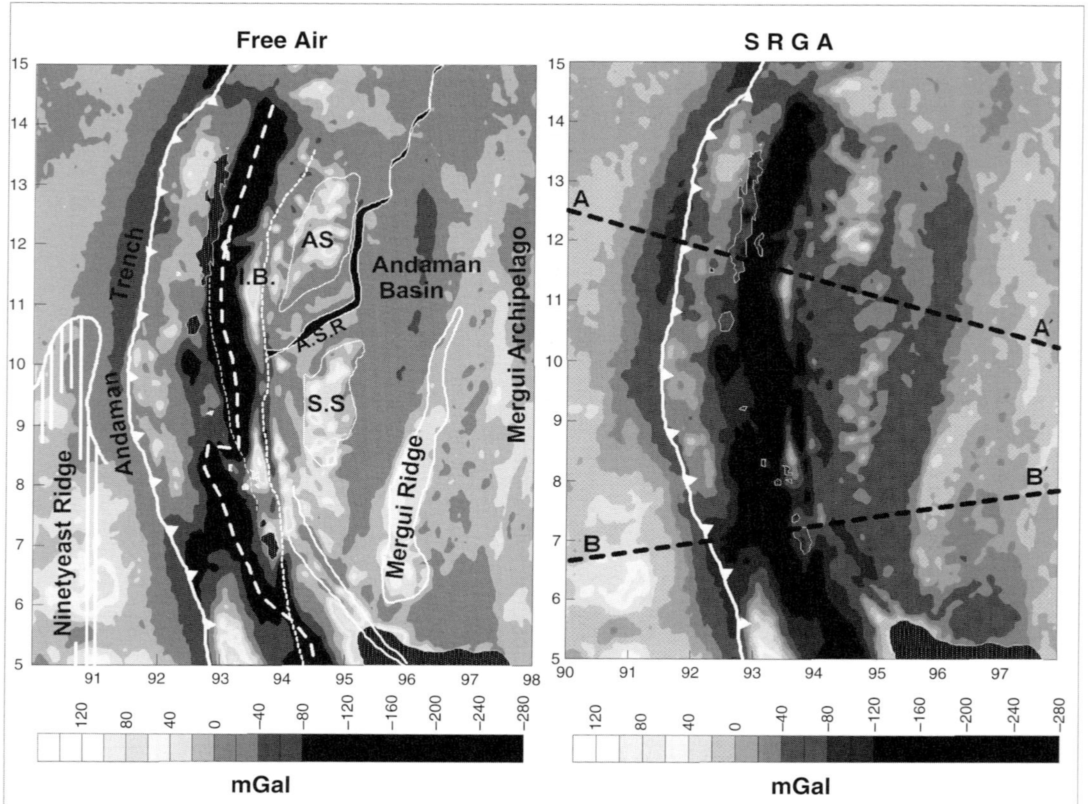

FIGURE 3.37 Satellite-derived free-air gravity anomaly map and the corresponding Slab Residual Gravity Anomaly (SRGA) map of the Andaman arc region (after Radhakrishna et al., 2008). See the correlation of gravity anomalies and tectonic features in the free-air gravity map. Profiles AA´ and BB´ are profiles across the arc used for gravity modeling.

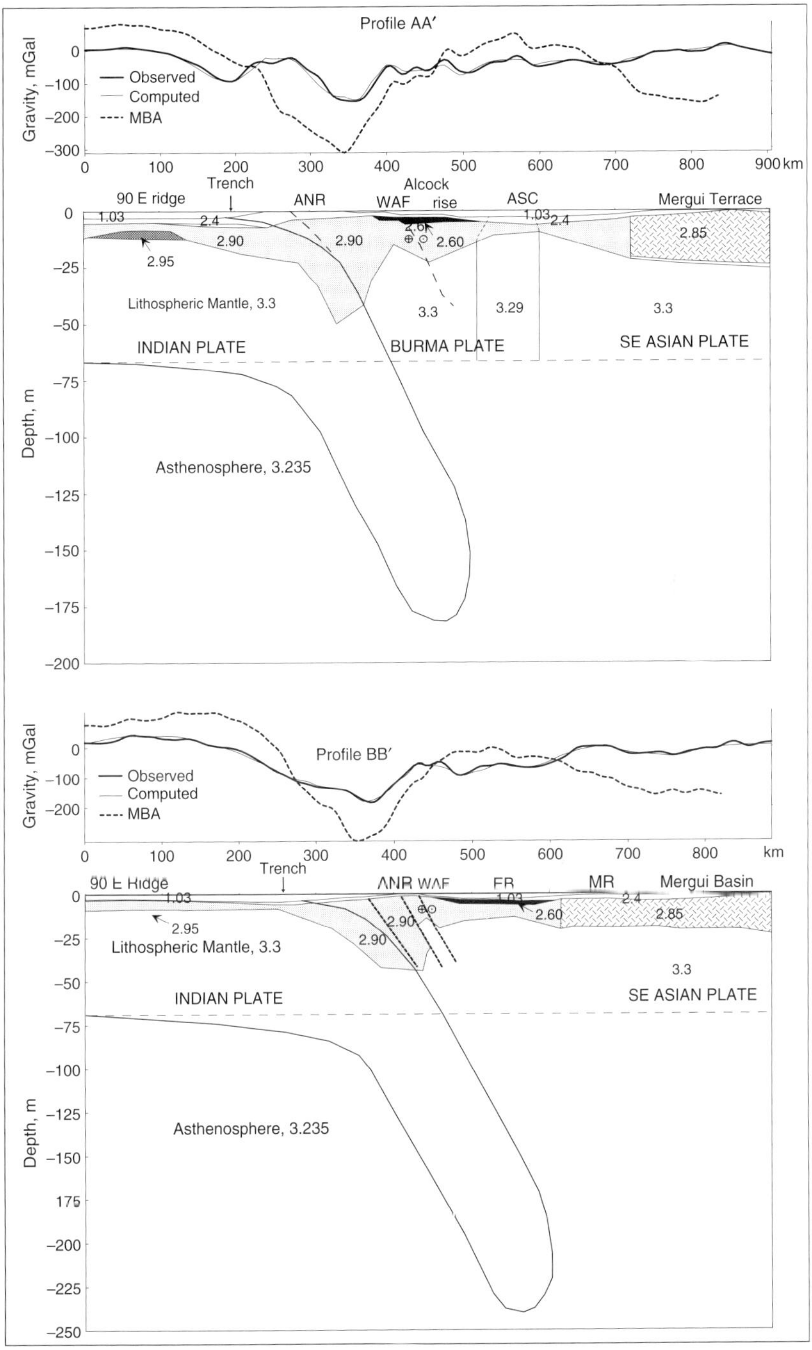

FIGURE 3.38 Seismically constrained gravity-derived lithospheric configuration along profiles AA′ and BB′ in the Andaman arc region (after Radhakrishna et al., 2008). ANR, Andaman Nicobar Ridge; WAF, West Andaman Fault; ASC, Andaman Spreading Center; EB, Eastern basin; MR, Mergui Ridge. Right-lateral shear motion is indicated for WAF.

layer consisting mainly of oceanic crust is present. Such thicker crust along the outer arc may be due to accretion of oceanic crust to the southeast Asian plate during westward shifting of subduction front since the Cretaceous time (Roy, 1992). Further, Roy et al. (2008) suggested that subduction along the arc has probably remained active since the Cretaceous, and repetitive upthrusting of accretionary wedges might have ultimately formed the ANR. Kieckhefer et al. (1980, 1981) suggested that large variations in the thickness of oceanic crust could be observed if imbricate faulting extends downward into the oceanic layer 3; thereby, portions of oceanic crust can actually get incorporated into the accretionary prism.

3.5.3.3 *Andaman–Nicobar Outer Arc Ridge*

The Andaman–Nicobar islands which form part of the fore arc sedimentary complex have evolved during the Oligo-Miocene times (Rodolfo, 1969). The islands act as barrier and separate the Andaman basin from the Bay of Bengal. Mukhopadhyay (1984) observed that the east dipping Benioff zone of the Andaman arc extends downward to about 200 km depth. Geological mapping of the islands reveals that the rock types in the Andaman–Nicobar islands mainly constitute Cretaceous serpentinites, ophiolites with radiolarian cherts, Cretaceous to Eocene cherty pelagic limestone, and a thick section of Eo-Oligocene flysch overlain by Neogene shallow water sediments (Chatterjee, 1967; Eremenko and Sastri, 1977; Roy, 1983). Along the western flank, the ANR is marked by the trench that is filled with the sediments of the Bay of Bengal (Curray et al., 1979). While the structures in the ANR region is dominated by gently folded east dipping nappes, off Sumatra, the folding is tight characterized by intense deformation (Moore and Curray, 1980; Weeks et al., 1967). Eremenko and Sastri (1977) observed that the deformation is more intense in Cretaceous–Oligocene sequences than the younger sequences. Along the ANR and the adjacent eastern offshore, several N–S faults and thrusts have been observed, and Jarwa thrust developed on the main islands (Roy, 1983) and the west Andaman fault (Curray et al., 1979) are the most significant among them. The emplacement mechanism of the ophiolites along Andaman outer arc thrust zones is largely debated as upthrust oceanic crust as a result of subduction since late Mesozoic (Curray, 2005; Curray et al., 1979; Pal et al., 2003), or to the collision of Indo-Burma-Andaman (IBA) microcontinent during late Oligocene and initiation of subduction in late Miocene (Acharyya, 2007; Acharyya et al., 1989). Also, with regard to the nature of crust underlying the Andaman–Nicobar outer arc ridge, two different views (continental vs. oceanic) exist. These are (i) continental crust on the basis of structural and stratigraphic reconstruction of onshore geological data (Acharyya, 2007) and (ii) oceanic crust inferred from a reversed seismic section (showing velocities of 6.3 and 6.9 km/s) in the offshore Indo-Burman Ranges (Curray, 2005). It is relevant to note that Roy (1992) observed the presence of a highly complex basement structure consisting of folded thrust packets successively under thrust from east to west and a tectonic mélange structure in the Cretaceous section embodied in the oceanic basement or ophiolites. This observation is based on well data and multichannel seismic reflection lines in and around the Andaman Islands.

3.5.3.4 *Andaman Basin*

The Andaman basin is a part of back-arc Andaman Sea and lies between Burma in north (Fig. 3.1) and Sumatra in south having an average width of 650 km from the Malay Peninsula in the east to the Andaman–Nicobar Islands in west (Rodolfo, 1969). Several youthful structures dominate the basin which might have originated either due to tensional or from combined tensional and strike-slip movements (Curray, 2005). Rodolfo (1969) presented a detailed description of the Andaman Sea which is marked by the presence of the Nicobar Deep, Barren–Narcondam volcanic islands, Invisible bank, and Alcock and Sewell seamounts. The strike-slip faulting parallel to the trench and the back-arc extension in the Andaman Sea could be the result of oblique subduction of the Indian plate in this region (Curray et al., 1979). According to Eguchi et al. (1979), the collision of NER was initiated in the middle or late Miocene, while this ridge–trench collision had transmitted compressional stresses in the back-arc area, the collision of India with Eurasia exerted a drag in the back-arc region which together caused the opening of the Andaman Sea. The magnetic anomaly identifications by Curray et al. (1979) in the central Andaman Sea indicate a spreading rate of 3.7 cm/year and age of opening of about 13 My or middle Miocene, which then may give rise to total opening of 460 km (Curray et al., 1982). However, Kamesh Raju et al. (2004) inferred that the present full rate of spreading in the Central Andaman basin is about 38 mm/year and that it has opened 118 mm in about 4 Ma, and this observation agrees well with the reconstruction model proposed by Curray (2005). Detailed seismic and compiled drilled well information suggests that the Andaman basin is underlain by oceanic crust with a thick pile of volcanic layer on top below Alcock and Sewell rises (Curray, 2005). Further eastward along the Malayan shelf, the COB occurs at the Malayan shelf. The Mergui north Sumatra basin consists of a series of horsts and grabens, and the basin is underlain by thin rifted continental crust which probably extends further west up to North Sumatra ridge (Curray, 2005).

3.5.3.5 *Back-Arc Volcanism*

The subduction zones are associated with volcanism which mostly occur along a line parallel to the trench and lie above the descending plate (Fowler, 1990). This chain of volcanoes generally results in island arc.

The volcanic arc of the Andaman Sea which has evolved as a result of eastward subduction of the Indian Ocean lithosphere lies between the active volcanic arc of Java and Sumatra, and the extinct or dormant arc in Myanmar (Curray, 2005). Two volcanic islands, the Barren Island and the Narcondam Island, lie in the northern Andaman Sea (Bhattacharya et al., 1993). Barren Island with an area of around 10 km^2 represents the only active volcanic system in the Indian subcontinent, the last event occurring a few months after the great earthquake of December 26, 2004 off Sumatra (Chandrasekharam et al., 2009; Franke et al., 2008; Sheth et al., 2010). Narcondam volcano has no record of activity in the recent times but it appears to have erupted in the Holocene (Simkin and Siebert, 1994). The Alcock seamount about 60 km east of Barren Island and the Sewell seamount about 200 km east of Car Nicobar Island are other imprints of volcanic activity in the Andaman Sea (Curray, 2005). Mitchell (1985) reported the occurrence of arc-related volcanic activity in the Mesozoic and Cenozoic along Burma and Sumatra arc. Banerjee and Shaw (2001) suggest that the volcanic arc structurally continues between Myanmar and Sumatra through the Barren and Narcondam islands and interprets it to be a Cretaceous fold belt of andesite effusive volcanic rocks.

Chapter 4

Sedimentation History and Development of Fan System Along the Continental Margins of India

4.1. INTRODUCTION

Submarine fan systems are known to be potential areas of hydrocarbon exploration (Lopez, 2001) as they form the largest deep-water sediment bodies near continental margins. The depositional history of these sediments also provides useful information with regard to the past land and marine climate, sea-level changes, and local and regional tectonic activity. Submarine channels flanked by levees have been recognized as important components of deep-water systems. During the past three decades, extensive studies were undertaken by researchers throughout the world to understand the depositional processes as well as their hydrocarbon exploration potential along some of the major fans such as the Amazon, Mississippi, Congo, Bengal, and Indus Fans. The architecture of a fan system is mostly dependent on the factors such as the sea-level fluctuations, climate, and nature of sediment input and tectonics (Bouma, 2000; Richards et al., 1998). Shanmugam and Moiola (1988) divided the submarine fans into four types based on their tectonic setting: (i) immature passive-margin fans (North Sea type); (ii) mature passive-margin fans (Atlantic type); (iii) active-margin fans (Pacific type); and (iv) mixed-setting fans (Bay of Bengal). They can also be broadly classified into four types based on the sediment supply and these are mud-rich, for example, Amazon Fan (Flood et al., 1991), sand-rich, for example, Sierra Nevada, California (Busby-Spera, 1985), mud-sand-rich, for example, California deep sea basin Fan (Normark, 1970), and gravel-rich fans, for example, Upper Jurassic system in North Sea (Hurst et al., 2005). In most of the large- and medium-sized submarine fans, the channel–levee systems dominate the upper fan architecture (Bouma et al., 1985; Damuth and Flood, 1985; Emmel and Curray, 1985; Kolla and Coumes, 1987) which act as conduits for clastic sediments to be transported into the deep sea. Therefore, it is very important to

Developments in Petroleum Science, Vol. 59. http://dx.doi.org/10.1016/B978-0-444-53604-4.00004-1

understand the sedimentary architecture as well as depositional and transport processes acting within these turbiditic systems for the purpose of evaluating their potential for hydrocarbon development and trapping.

4.2. HIMALAYAN OROGENY AND SUBMARINE FAN SYSTEM IN THE INDIAN OCEAN

During India's northward drift, the initial soft collision between India and Asia (Fig. 4.1) might have initiated the Himalayan uplift (Curray and Munasinghe, 1989), and subsequently the hard continent–continent collision of the Indian plate with the Asian landmass gave rise to the main Himalayan orogeny which caused significant increase in sediment supply and development of fan systems (Fig. 4.2) along the Indian margins, the Indus Fan in west (Clift et al., 2002) and the Bengal Fan in east (Alam et al., 2003). The timing of initial collision between India and Eurasia is debated (Butler, 1995; Cochran, 1990) and range from Cretaceous–Tertiary boundary (Klootwijk et al., 1992) to different times in the Eocene (Rowley, 1996). An early Eocene collision (referred to as soft collision) at about the time of anomaly 22 (~ 53 Ma) is suggested because at this time, spreading half rates in the Central Indian basin decreased from 8 to 4 cm/yr and the spreading direction changed from nearly N–S to NE–SW (Sclater et al., 1976). A regional unconformity observed throughout the

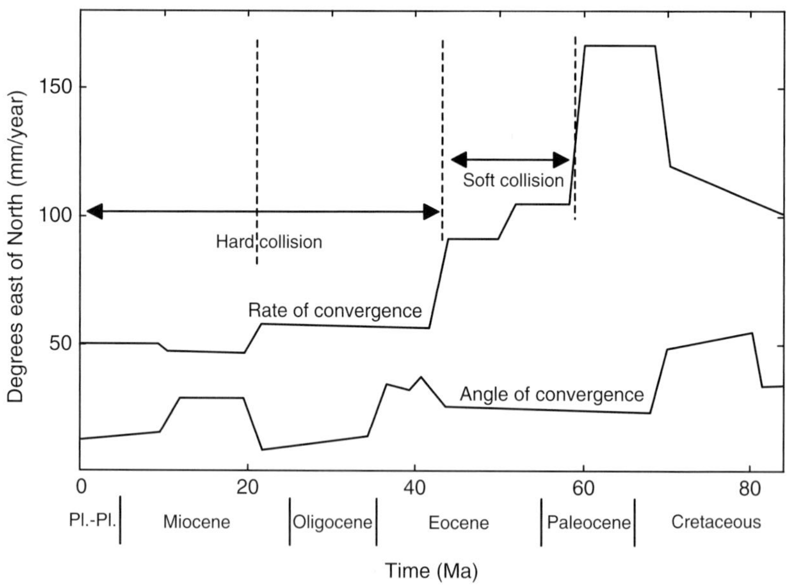

FIGURE 4.1 Rate and angle of convergence between India and Eurasia in different times depicting the associated soft and hard collision *(Reproduced from Alam et al., 2003).*

Bay of Bengal (Curray and Moore, 1971) has also been dated to this time by extrapolation from DSDP Site 217 (Moore et al., 1974). A middle Eocene time is proposed based on similarity of mammalian fossils between India and Mongolia (Sahni and Kumar, 1974) and on paleomagnetic data and plate motions (Dewey et al., 1988). According to Curray and Moore (1971), the formation of the Indus and Bengal fans in the northern Indian Ocean began some time in the Eocene. The hard continent–continent collision occurred in the late Miocene at the time of the initial uplift of the Himalayas (Gansser, 1964; Powell and Conaghan, 1973). Apart from the fan sediments, the land path way that links the Himalaya with these fans, the Indus and Katawaz basins on the west and the Bengal basin in the east contains both pretectonic and syntectonic stratigraphic sequences (Qayyum et al., 1997; Uddin and Lundberg, 1998) and therefore record a significantly older history. Many investigators suggest that the collision might not have taken place simultaneously all along the E–W trending Himalayan belt which is 2500 km long and 300 km wide (Burchfiel, 1993; Dewey et al., 1989; Rowley, 1996; Uddin and Lundberg, 1998) based on stratigraphic evidences from the Himalayan region. According to Uddin and Lundberg (1998), the Himalayan belt is characterized by four lithotectonic units (Fig. 4.2) juxtaposed along north dipping thrust faults and from south to north, these are (i) the sub-Himalaya, consisting of Miocene to Pleistocene molasses deposits of Siwaliks; (ii) the lower or lesser Himalaya, composed of Precambrian and Paleozoic sedimentary rocks, crystalline rocks, and granites; (iii) the higher Himalaya, composed of schists, gneisses, and granites; and (iv) the Tethys Himalaya and Trans Himalaya, representing fossiliferous Cambrian to Eocene sedimentary rocks, batholiths, and volcanic rocks. The detrital record of the Himalayan collision event preserved in both the Bengal and the Indus basins is distinctly different from each other and suggests that continental collision could be diachronous (Rowley, 1996; Uddin and Lundberg, 1998). While Eocene to Oligocene pre- to syncollisional sediments in the Indus basin includes detritus from the developing Himalayas, the correlative deposits in the Bengal basin indicates source from the Indian craton. Based on the above observations, Uddin and Lundberg (1998) proposed that the orogenic uplift was initiated earlier in the western Himalaya and propagated eastward over several tens of millions of years (Fig. 4.3) with a considerable counter clockwise rotation of India during this interval.

The northern Indian Ocean has two world's major submarine fans, the Bengal Fan and the Indus Fan. The Bengal Fan in the Bay of Bengal is the largest deep-sea fan in the world covering an area of $\sim 3.0 \times 10^6$ km^2 and the Indus Fan covering an area of $\sim 1.1 \times 10^6$ km^2 is the second largest (Clift et al., 2002; Curray et al., 2003). The sediments of these fans are largely eroded from the Himalayas and transported by the Indus river system into the Arabian Sea and the Ganga-Brahmaputra river system into the Bay of Bengal. A detailed investigation of sediment depositional history of these two fans would provide valuable information on collision history and different phases of Himalayan

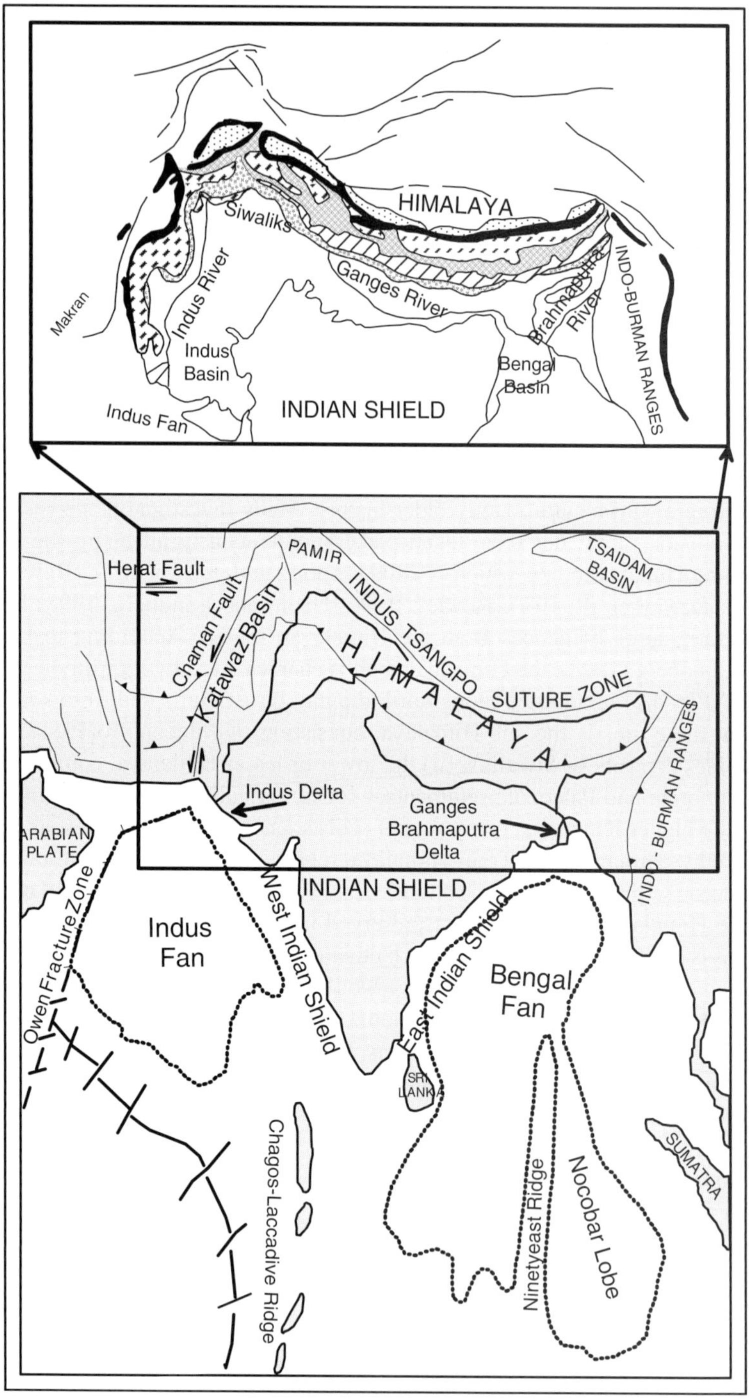

FIGURE 4.2 Tectonic map of south Asia and the surrounding oceanic areas encompassing the Arabian Sea and the Bay of Bengal (Reproduced from Qayyum et al., 2001). The eroded sediments by the Indus river flowing through the western Himalaya and by the Ganges–Brahmaputra river system flowing through the eastern Himalaya reaches the Indus Fan in the west and the Bengal Fan in the east. The upper panel shows the lithotectonic belts of Himalaya (Redrawn from Uddin and Lundberg, 1998).

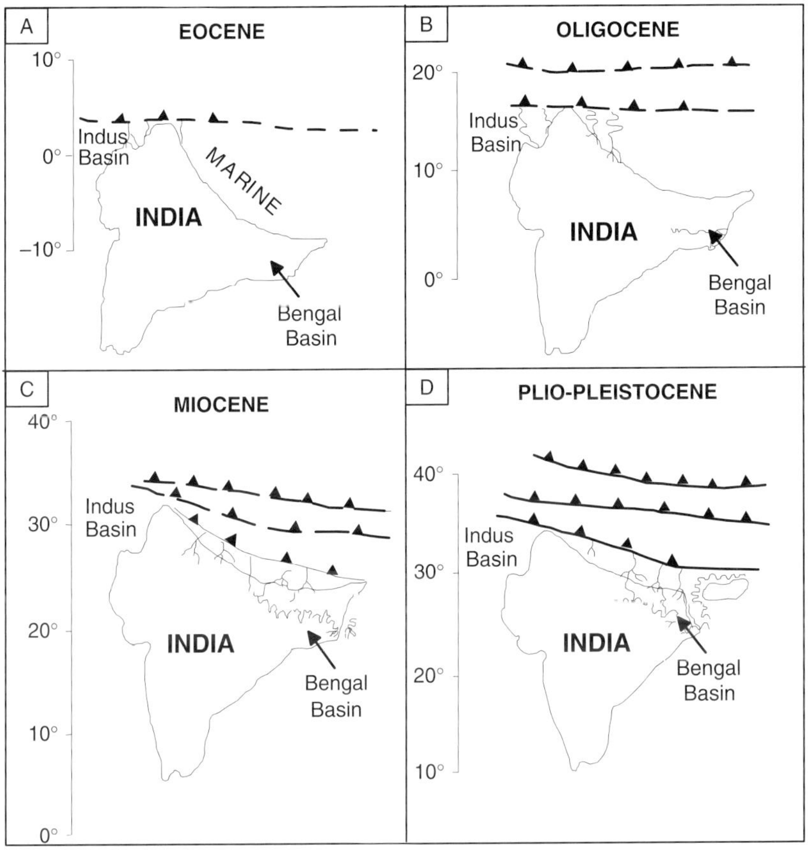

FIGURE 4.3 Eocene to Pleistocene paleogeographic reconstruction (schematic) of the Himalayan foreland and the surrounding areas *(Reproduced from Uddin and Lundberg, 1998)*. Note the eastward propagation of Himalayan collision. Details are discussed in the text.

uplift as well as the architectural development of the submarine fan systems that helps to understand the deep-water hydrocarbon exploration potential along the Indian margins.

4.3. INDUS FAN

4.3.1 Regional Setting

The Indus Fan is the second largest deep-sea fan in the world comprising of material eroded from the western Himalaya, Karakoram, and Hindukush (Clift et al., 2002). The fan developed off the passive continental margin of Pakistan and India covers an area of approximately 1.1–1.25 million km^2 having a length of 1600 km and width of 1000 km (Kolla and Coumes, 1987; McHargue and Webb, 1986). The fan is bounded by the Chagos-Laccadive in the east, Owen fracture zone and Murray ridges in the west, and Carlsberg ridge in the south (Fig. 4.4). These bathymetric features are presently limiting the lateral growth of the fan. The Indus river system is about 2900 km long and flows nearly 1200 km beyond Himalayas in relatively flat plains and has been the dominant sediment source of the fan (Kolla and Coumes, 1987). On its path way, the Indus basin is bounded on the west by Kirthar

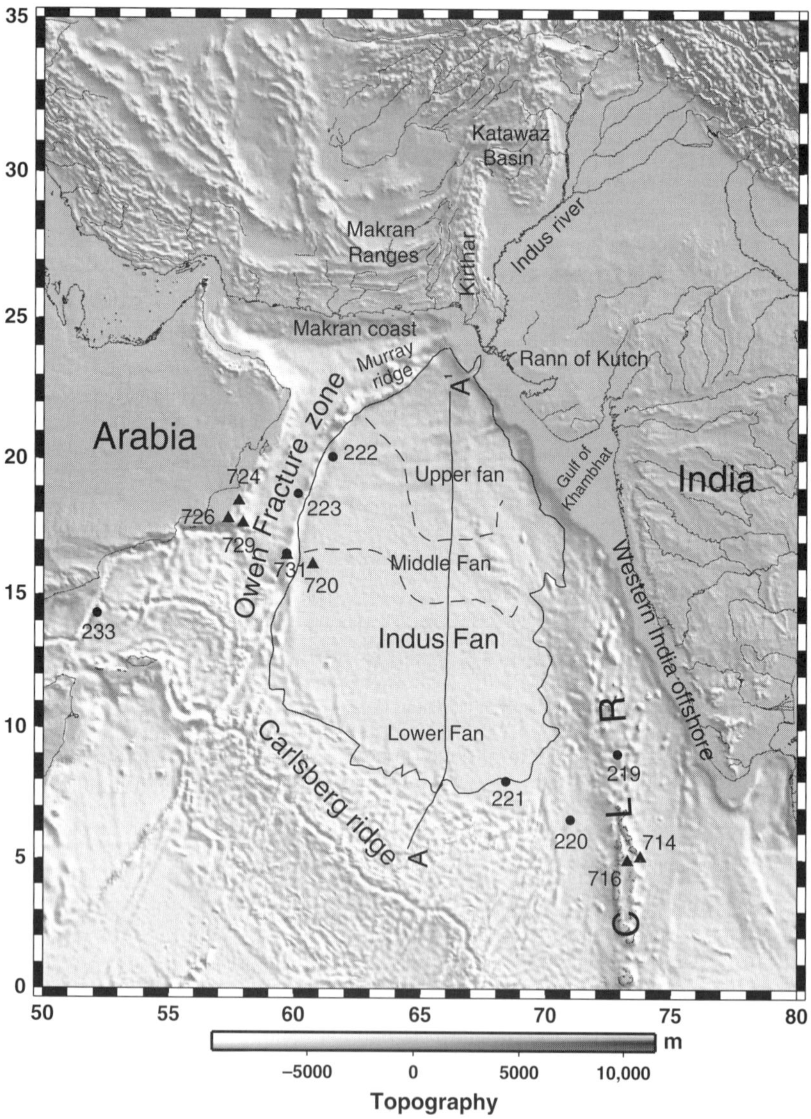

FIGURE 4.4 Topographic relief map depicting morphological setting (after Kolla and Coumes, 1987) of the Indus Fan. The numbered symbols (triangle and circle) show the DSDP and ODP drilling sites. AA′ is a longitudinal section (Fig. 4.7) constructed from the refraction data in the Arabian Sea by Naini and Kolla (1982).

and Sulaiman Mountains (Fig. 4.4) and covers a total drainage area of 966,000 km^2 (Krishnan, 1968).

4.3.2 Morphology and Channel Development

The Indus fan has typical fan morphology in the deep Arabian Sea with water depths range from ~1500 m at the foot of the slope to more than 4500 m at the distal end near the Carlsberg ridge (Fig. 4.4). The width of the shelf varies from 100 to 150 km off Indus Delta to as much as 350 km off Bombay further south (Kolla and Coumes, 1987). Based on the morphological characteristics, Kolla and Coumes (1987) divided the fan into upper, middle, and the lower fan (Fig. 4.4). According to these authors, the upper fan extends from the foot of the continental slope to about 3400 m water depth, while the middle and lower fans extend up to 4000 and 4600 m water depth, respectively. From the upper to lower fan, the seafloor has a relief of several hundred meters with overall gradients greater than 1:500 in the upper fan and less than 1:1000 in the lower fan. The shelf-slope region is characterized by the presence of Indus Canyon having an average width of 8 km and relief of ~800 m, and off Gulf of Kutch, with another lesser magnitude canyon, namely, the Saraswati valley (Fig. 4.5A). Subsequently, high-resolution bathymetric surveys (von Rad and Tahir, 1997)

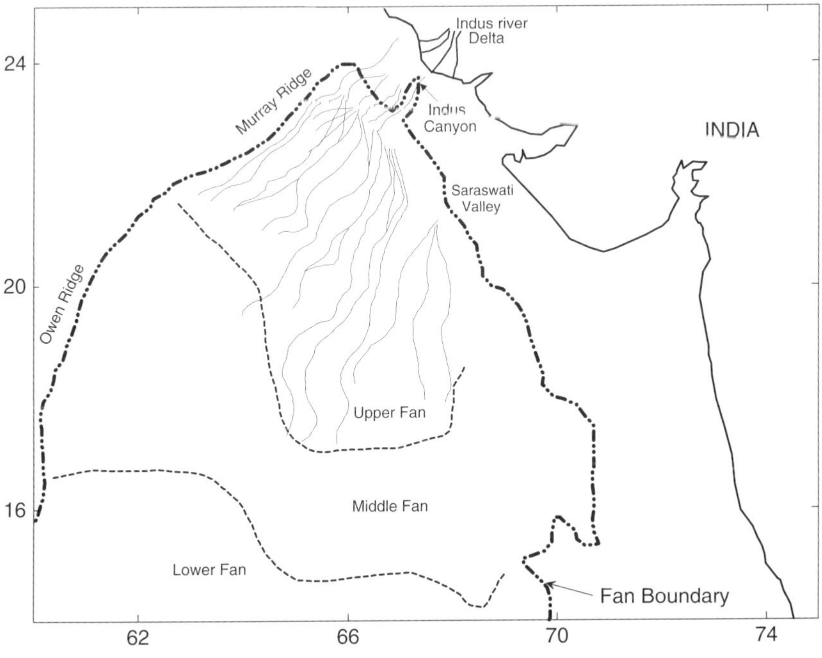

FIGURE 4.5A Channel pattern in the upper Indus Fan region *(Redrawn from Kolla and Coumes, 1987).*

provided detailed morphological characteristics of the Indus canyon. This new data indicate that the canyon is about 170 km long and present from 20 to 30 m water depth in the inner shelf to 1400 m depth at the foot of the continental slope with an eastward bend at about 1000 m water depth (Fig. 4.5B). It shows a gradient of about 1:100 on the shelf and 1:50–1:20 on the slope, while the gradient of continental slope adjacent to canyon is about 1:60–1:30. Kolla and Coumes (1987) proposed a model for origin of Indus Canyon similar to that proposed by Coleman et al. (1983) for Missisippi Canyon. According to them, the Indus Canyon originated mainly by slumping and related mass wasting processes on the continental slope corresponding to huge depocenter near river mouth during Pleistocene low sea level and subsequent retrograde slumping during sea-level rise gave rise to growth of the canyon.

A channel–levee system is defined as a single channel belt bordered by outer levees, whereas, a channel–levee complex is a series of stacked channel–levee systems that are fed by the same canyon (Damuth and Flood, 1985). The Indus Fan is a large river-fed submarine fan (Kenyon et al., 1995) with huge sediment accumulation near the toe-of-slope (Clift et al., 2001). The growth and development of the Indus Fan has been partly controlled by the sea-level changes and the channel–levee systems of the fan were formed during low sea-level stands (Ayub, 1992). The upper Indus Fan consists of some of the largest channel–levee systems which have been studied by several authors using variety of data sets (Deptuck et al., 2003; Droz and Bellaiche, 1991; Kenyon et al., 1995; Kolla and Coumes, 1987; McHargue and Webb, 1986; Naini and Kolla, 1982; von Rad and Tahir, 1997). These studies indicate the presence of three canyons along the Indus Fan margin, one, the modern Indus Canyon and, the other two, ancestral canyons located further west (Fig. 4.6). While canyon marked "1" is the youngest, other two are older (Kolla and Coumes, 1987). Clift et al. (2001) observed that the overlapping and stacked channel–levee system is a characteristic feature of the Indus fan which commenced during the Neogene period. The canyon-channel–levee system has produced a complex seafloor bathymetry in the upper and middle fan areas. The youngest channel–levee complex has been mapped in detail using long range side-scan sonar (GLORIA) data (Kenyon et al., 1995) which brought out 20 channel–levee systems in the Indus fan (Fig. 4.6) of which 6 are large and 14 are small. The data further indicated that the 14 smaller channel–levee systems are distributory complexes of large channel–levee systems A and B. East of the depositional lobe of system A, older small channel–levee systems of system B may have been fed by the same source but through another less prominent canyon, the Saraswati valley (Kenyon et al., 1995; Kolla and Coumes, 1987). Though not completely mapped, out of the main feeder channels emanating from the Canyon 1, channel A is the youngest and channel F is the oldest (Kenyon et al., 1995 ; Prins et al., 2000). The smaller second-order channels A^1–A^7 emanating from the large feeder channel A are the most recent ones as they are covered by only thin veneer of Holocene sediments (Amir et al., 1996; Kenyon et al., 1995; Prins et al., 2000). Kolla (2007) indicated that all the higher-order channels

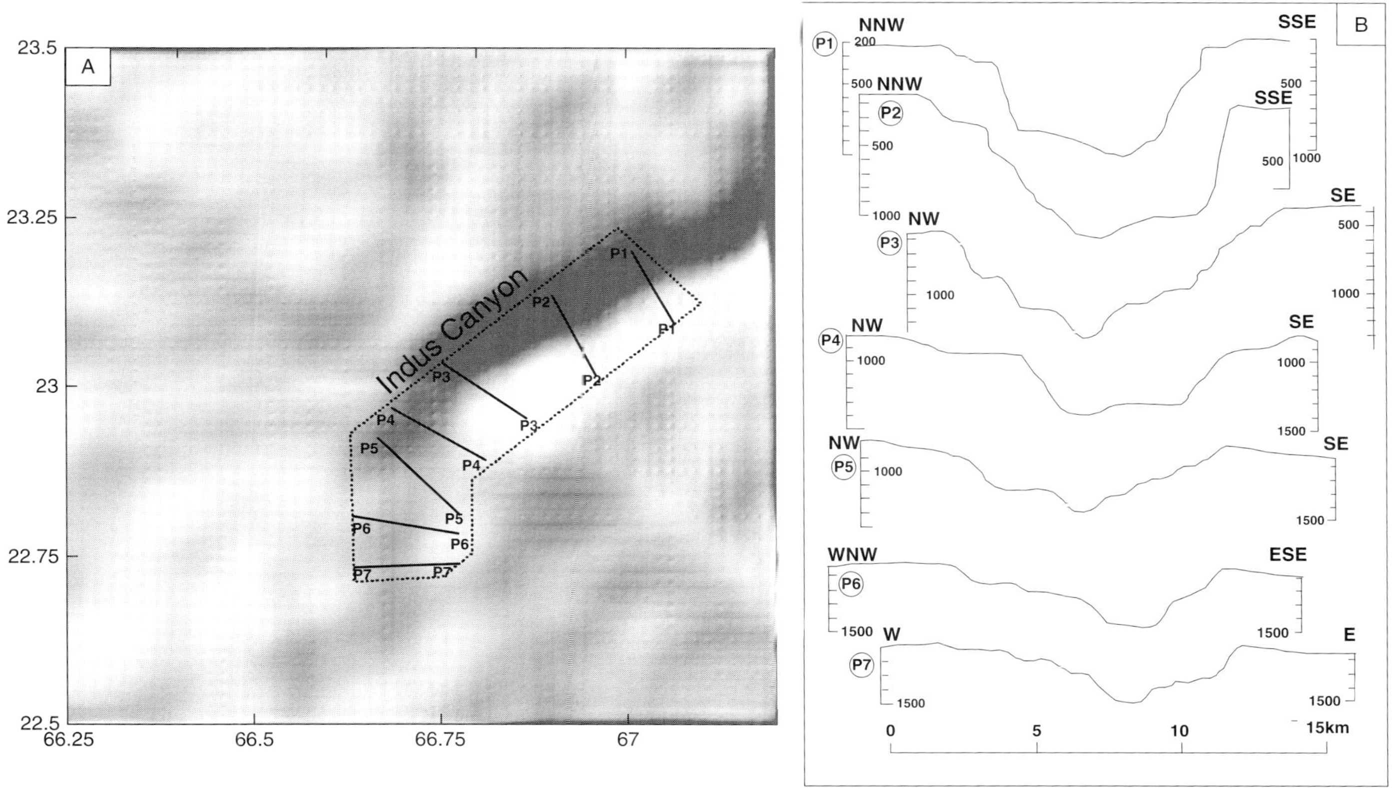

FIGURE 4.5B Representative bathymetry profiles (location in left panel) across the Indus submarine canyon *(adapted from von Rad and Tahir, 1997)*. The transition from the canyon to channel system can be very well seen in the figure.

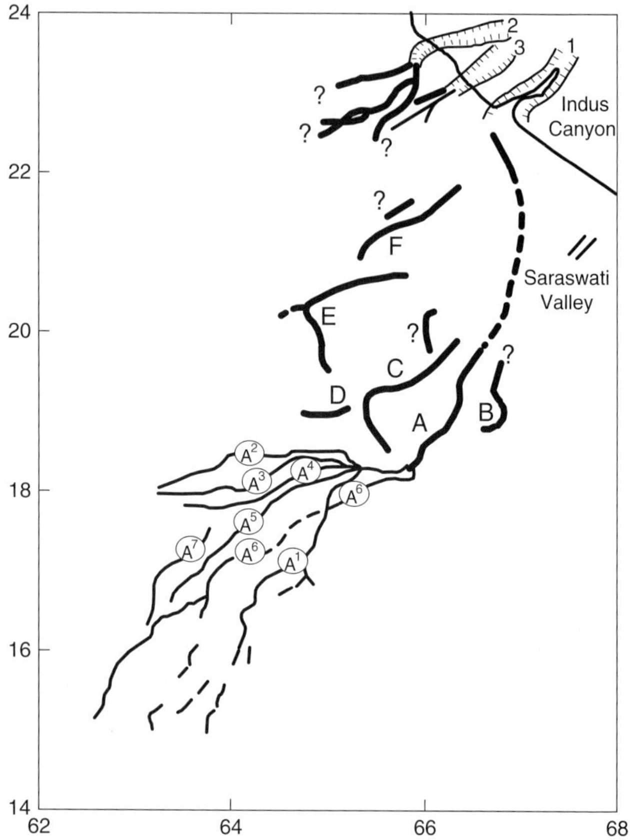

FIGURE 4.6 The most recent channel pattern as well as some older channels in the Indus Fan (after Kenyon et al., 1995). *Modified version adapted from Kolla (2007).*

avulsed in a narrow zone and took radial or divergent courses down-dip and this avulsion zone is the transitional area of the upper to middle Indus Fan where gradients decrease from greater than 2 to 1–2 m/km. Amir (1992) observed that the present thalweg gradients of channel A in the upper fan and higher-order channel A^1 in the middle fan are less than the fan surface gradients. He further noticed that the up-dip to down-dip, cross-sectional area decreases from channel A to A^1 and older channels have been affected by overbank spilling from younger channels and by the hemipelagic sedimentation.

4.3.3 Sediment Structure and Sedimentation History

Though a systematic dataset deciphering the internal structure of the sediment layer overlying the acoustic basement is not available for the entire Indus Fan, the single and multichannel seismic surveys carried out for academic and petroleum exploration purposes provided valuable information on various sedimentary sequences as well as on the sedimentation history of the fan (Clift et al., 2001, 2002; Droz and Bellaiche, 1991; Kolla and Coumes, 1987; McHargue and Webb, 1986; Naini and Kolla, 1982, and many references therein). Out of several DSDP and ODP drilling sites in the Indus Fan region (Fig. 4.4) the Site 221 drilled through the entire fan on the distal southern edge (Whitmarsh, 1974). Seismic refraction station data projected along a longitudinal section passing through the entire fan (Naini and Kolla, 1982) suggests the existence of several sedimentary layers with velocities ranging between 1.7 and 3.8 km/s (Fig. 4.7). The sedimentary layers are also seen gently dipping southward from the proximal to the distal fan. Kolla and Coumes (1987) presented a gross sediment thickness distribution map (Fig. 4.8) that includes both fan and pre-fan sequences for the entire fan which indicates that the sediment thickness (in TWT) ranges from less than 0.5–1.0 s in the distal fan to as much as 6.0 s in the proximal areas of the fan. The map shows three broad basins with thick sediment accumulations separated by ridges or basement highs, the offshore Indus basin having sediment thicknesses exceeding 6 s, the Western basin with thicknesses up to 3 s and the Eastern basin with thicknesses greater than 2 s. The Western and Eastern basins are separated by the Laxmi ridge. The deeper parts of both western and eastern basins forming troughs as well as gaps within the Laxmi ridge served initially as conduits for the turbidity current transport of the Indus River sediments from proximal into the distal areas (Naini and Kolla, 1982). Thickest parts of the Indus Fan lie close to the Pakistan coast off the shelf edge where sediments are thicker than 10 km (Clift et al., 2002). Seismic reflection data clearly depict the existence of older buried channels and channel abandonments similar to that observed for other deep-sea fans (Clift et al., 2002) and indicates that major fan deposits migrated southeastward since the onset of fan sedimentation

4.3.4 Seismic Stratigraphy

Based on a regional single channel seismic section in the mid-fan region, Droz and Bellaiche (1991) observed three layers, an upper Indus Fan mega sequence characterized by channel and levee formation, a ponded parallel-stratified layer, and a basement cover sequence. Multichannel seismic data (Clift et al., 2002; Gaedicke et al., 2002) subsequently confirmed the presence of the above sequences, however, a major unconformity separating the topography-filling series from the basement cover sequence is observed and further indicated that the basement cover sequence to be a pre-breakup sequence. The channel–levee systems observed above the base of middle Miocene (Fig. 4.9) are relatively larger and constructed over a period of 10^5–10^6 years during the late Miocene, whereas, below middle Miocene strata, levees are rare and of low amplitude (Clift et al., 2002). The ODP site 731 on the Owen ridge

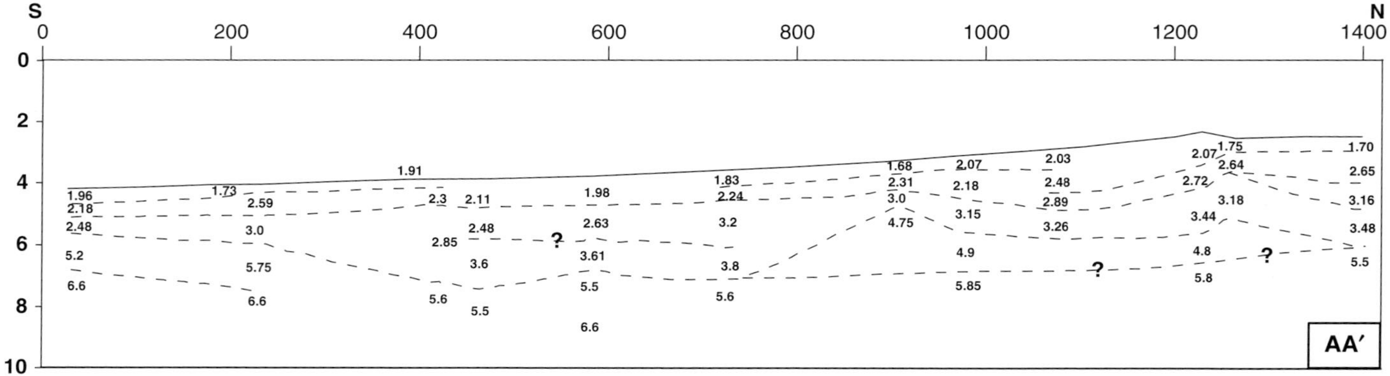

FIGURE 4.7 Longitudinal section showing seismic velocity layering in N–S across the Indus Fan *(adapted from Naini and Kolla, 1982)*.

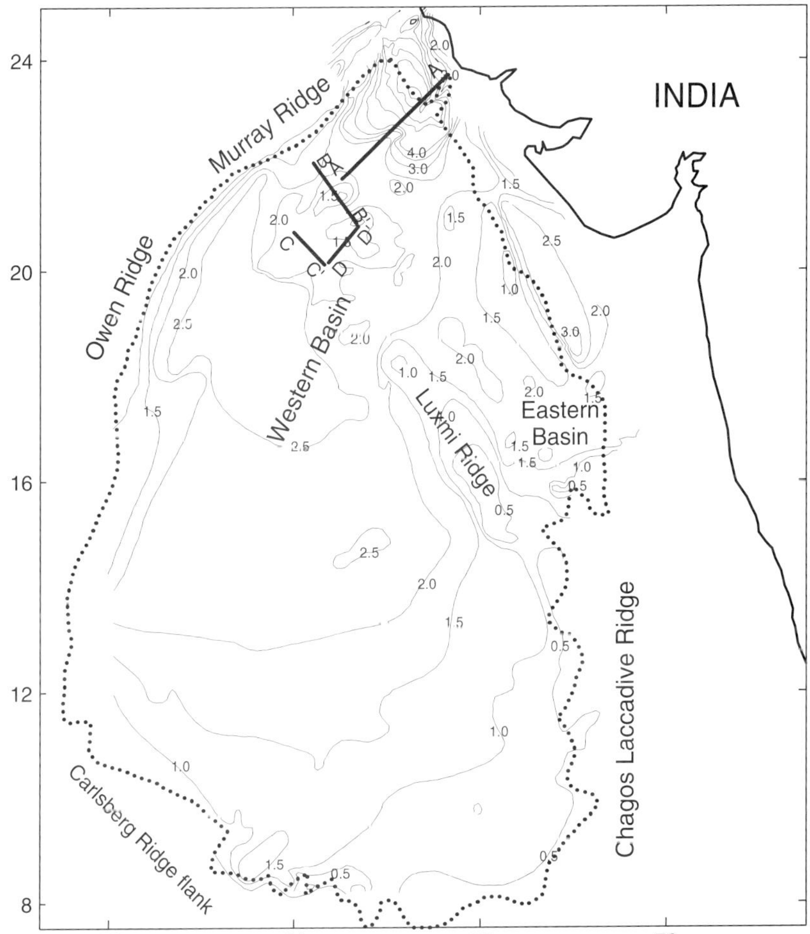

FIGURE 4.8 Time–structure (TWT) map of the sediment thickness in the Indus Fan region *(Reproduced from Kolla and Coumes, 1987)*. Dotted line marks the boundary of the fan. Profiles AA′–DD′ are location of seismic sections shown in Figs. 4.9 and 4.10.

(Prell et al., 1989) revealed fan turbidites, flat bedded distal varieties underlying the Indus Fan megasequence which is of middle Miocene age. According to Weedon and McCave (1991), muddy turbidites of Oligocene and older ages together with Eocene shale overlie the basement along the Owen ridge and these deposits represent the Paleogene Indus Fan in the Arabian Sea. Seismic stratigraphy data (Clift et al., 2001, 2002) reveals that large thickness of Indus Fan sediments were of Paleogene age (Fig. 4.10) and predates the uplift of the Murray ridge. Further, the seismic sections indicate that the Paleogene section thins over the two structural highs of the ridge (Fig. 4.10). It is

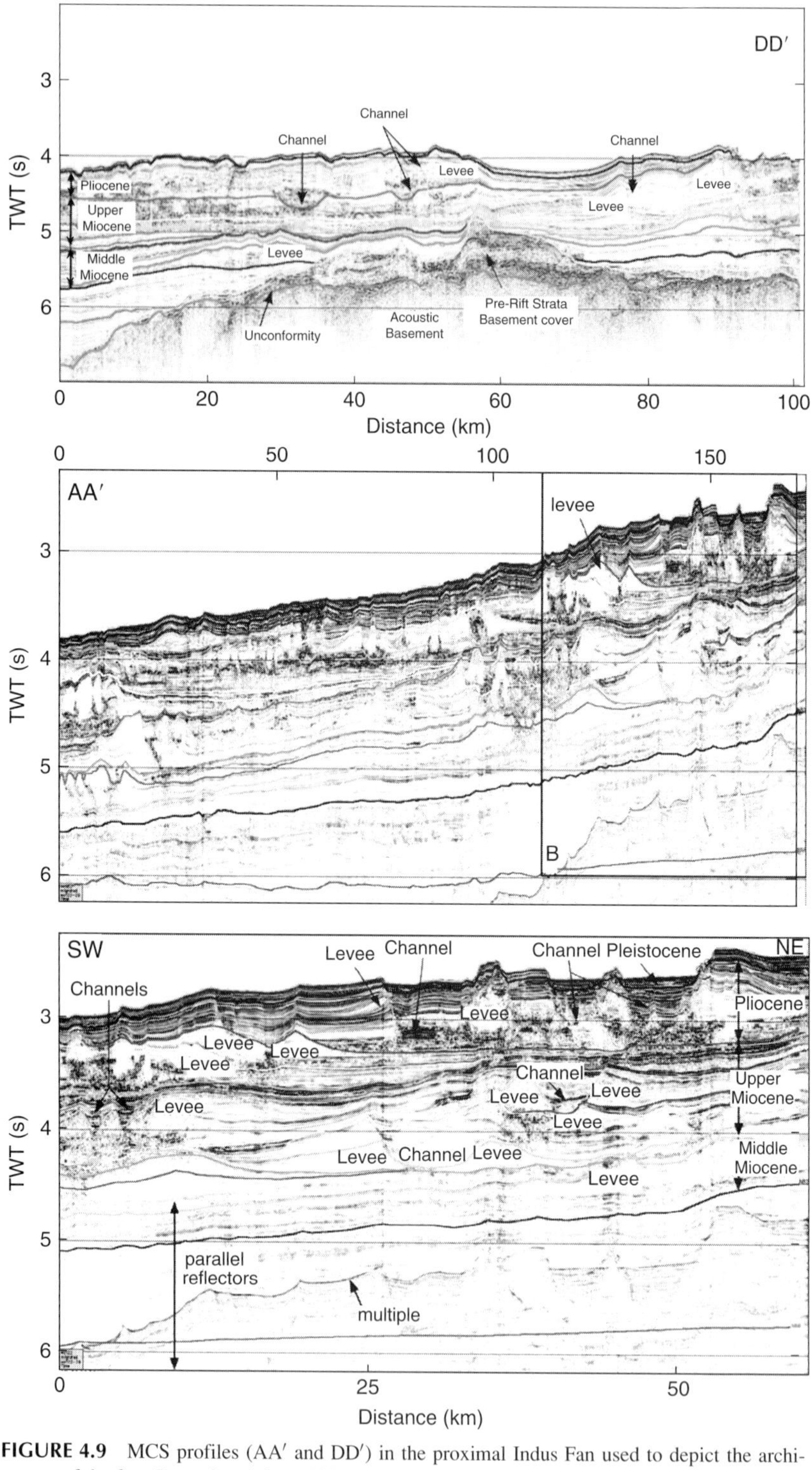

FIGURE 4.9 MCS profiles (AA′ and DD′) in the proximal Indus Fan used to depict the architecture of the fan *(Reproduced from Clift et al., 2002)*. The prerift sediments overlying the acoustic basement and the progressive onlapping of Indus Fan on the structural high can be seen in DD′. The zoomed part (B) in section AA′ shown in the lower most panel depicts well-developed channel–levee systems on the upper part of the Indus Fan.

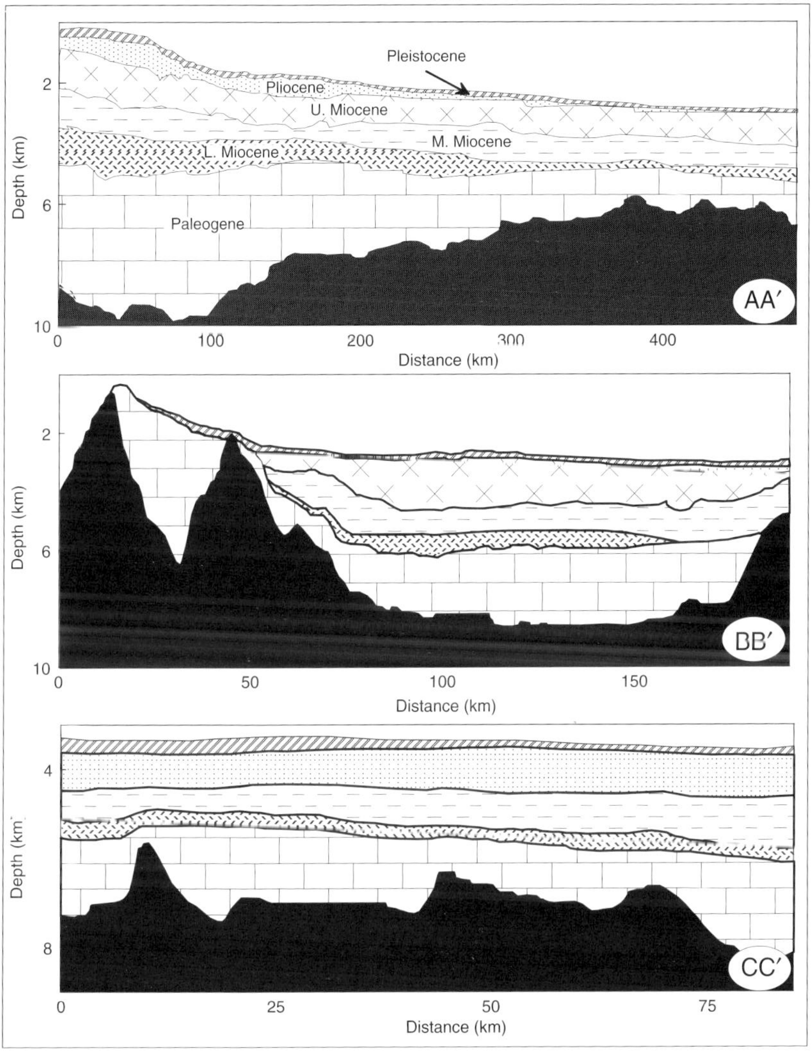

FIGURE 4.10 Depth sections of MCS profiles AA′–CC′ *(Redrawn from Clift et al., 2002).* Details are discussed in the text.

relavant to note here that both Murray and Owen ridges have undergone major uplift during the development of Indus Fan system as a result of stresses induced during plate reorganization (Edwards et al., 2000; Mountain and Prell, 1990). The uplift of the Murray ridge occurred in two stages, the initial uplift, though not well constrained, is inferred to be during Eocene–Oligocene time (Clift et al., 2001) and the second major phase correlates with the early Miocene uplift of the Owen ridge (Mountain and Prell, 1990).

4.3.5 Geologic History

The development of Indus Fan owes its origin to several geological events in the past involving both tectonism and sedimentation in the Arabian Sea as well as in the adjoining land region (Fig. 4.11). It is now well established that the Arabian Sea evolved after the rifting between India and Madagascar during the mid Cretaceous and India-Seychelles in late Cretaceous as well as subsequent seafloor spreading (Norton and Sclater, 1979). The northward drift of India might have led to the establishment of linkages across portions of Tethys Sea by the end of early Eocene (Sahni and Kumar, 1974; Valdiya, 1984) as well as uplift of Himalayas due to collision of the Indian plate with Eurasia in Oligocene. According to Patriat and Achache (1984), around 40 Ma, complete reorganization of the mid-ocean ridge system accommodated the incipient collision with a major change in spreading direction. This was followed by another hard collision during middle to late Miocene which led to major uplift of Himalayas. According to Valdiya (1984), the third major episode of Himalayan uplift took place in late Pliocene to middle Pleistocene. Studies indicate that the Indus River drainage originated by Oligocene time and as the Tethys Sea closed, it built deltas advancing southward along the Indus trough (Kolla and Coumes, 1987). Qayyum et al. (1997) related the thick Eocene–Oligocene turbidites from the onshore Makran accretionary prism to the oceanic fan (paleo Indus deep-sea fan) of fluviatile system known as Katawaz formation and suggested that the modern Indus Fan was not initiated until Early Miocene. Plate reconstructions for the evolution of Arabian Sea since Eocene times by Ellouz-Zimmermann et al. (2007) (Fig. 4.12) depicts the existence of paleo Indus fan before the development of modern Indus fan system. Clift and Gaedicke (2002) attributed much of the middle Miocene influx of sediments to tectonism and rapid exhumation in the Karakoram range which is the dominant source for Indus Fan.

4.4. BENGAL FAN

4.4.1 Regional Setting

The Bengal Fan in the northeast Indian Ocean is bordered by the Indian continental shelf in the west, continental shelf of Bangladesh in the north, and the Andaman–Sumatra trench-arc and the associated accretionary prism in the east (Fig. 4.13). The Bengal Fan along with its onshore counterpart, the Bengal basin, forms a large Meso-Cenozoic basin, called the Bengal geosyncline (Curray and Moore, 1974). Bengal Fan is one of the largest submarine fans in the world covering the entire Bay of Bengal from 20°N in 1400 m water depth to 7°S at 5000 m water depth for almost 3000 km from north to south (Curray and Moore, 1971). The sediments of the fan have been largely contributed by the erosion of the Himalayas and transported by the Ganga–Brahmaputra River system (Curray and Moore, 1971; Curray et al., 2003). However, some of the major rivers from eastern India also feed significant quantum of sediments into the Bay of

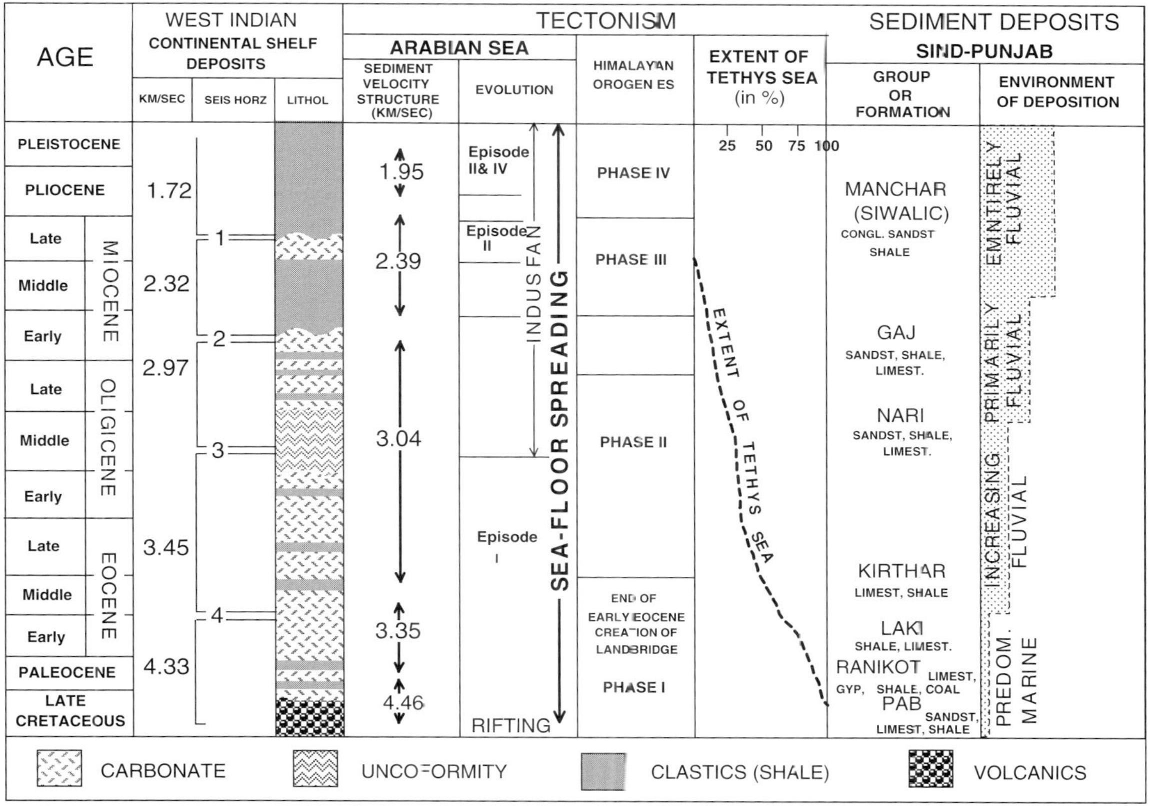

FIGURE 4.11 Geological history of the Arabian Sea, the Incus Fan, and the adjoining land areas compiled from different sources *(Reproduced from Kolla and Coumes, 1987)*.

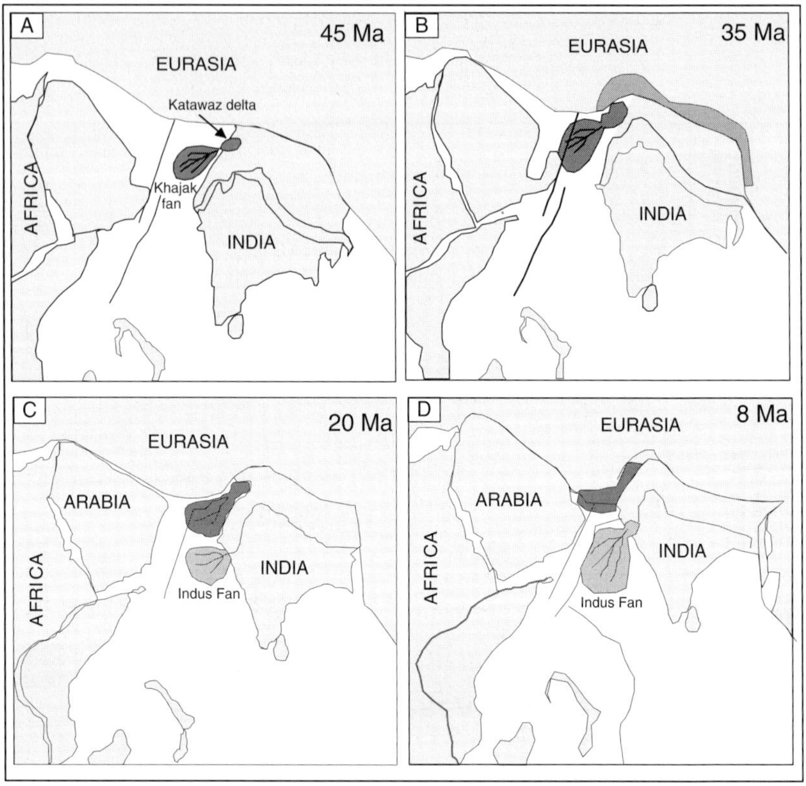

FIGURE 4.12 Plate reconstructions showing evolution of the Arabian Sea since Eocene time *(Redrawn from Ellouz-Zimmermann et al., 2007)*. Note the development of paleo Indus delta (after Qayyum et al., 1997) as well as the modern Indus Fan.

Bengal from the west. The Bay of Bengal is one of the deepest sedimentary provinces of the world. It contains nearly 22-km-thick sediments (includes both pre- and postcollision) having maximum off the Bangladesh shelf with sediment thickness progressively decreasing toward south (Curray, 1994). Considering maximum thickness for postcollision sediments as 16 km at the shelf edge (Curray, 1994; Emmel and Curray, 1985), the total volume of postcollision sediments would be 12.5×10^6 km^3. Two distinct fans have been recognized in the northeast Indian Ocean, separated by the Ninetyeast ridge (Fig. 4.13). The eastern subfan is called the Nicobar Fan while the western lobe is the proper Bengal Fan which encompasses the Bay of Bengal (Curray and Moore, 1971). According to Curray and Moore (1974), the Nicobar Fan has been inactive since mid-Pleistocene as the channel system was cut-off to the fan due to the convergence of Ninetyeast ridge with the Sunda trench. The width of the Bengal Fan varies from 1430 km at 15°N to around 830 km at 6°N and covers an area of $\sim 3 \times 10^6$ km^2 excluding the area of Nicobar Fan (Curray et al., 2003).

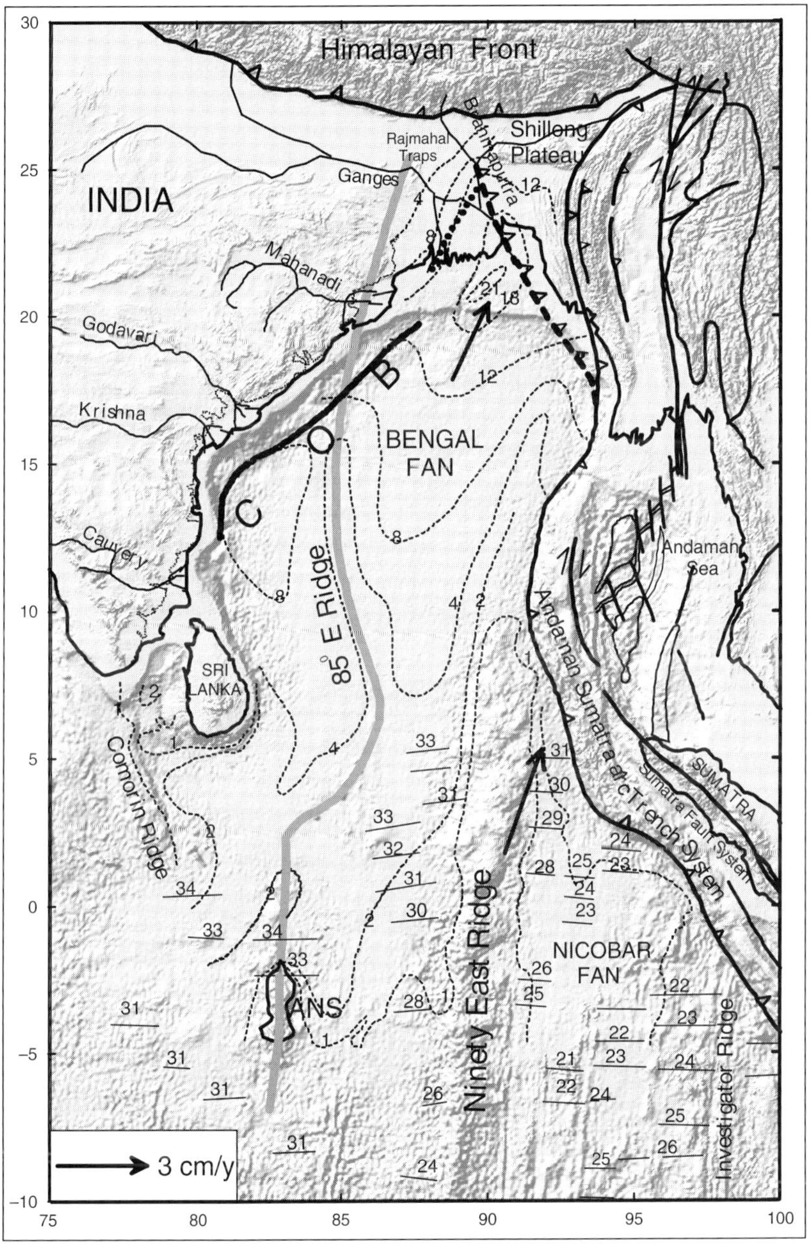

FIGURE 4.13 Shaded relief map of the Bay of Bengal and the adjoining areas. Tectonic details are from Curray (1991). Continent Ocean Boundary (COB) is from Murthy et al. (1993). Sediment thickness in km. ANS—Afanasy Nikitin Seamount. Black filled circles in Bengal basin show position of Eocene shelf edge. Tectonic boundary (marked as thick dashed barbed line) along Burmese arc is from Steckler et al. (2008) *(adapted from Radhakrishna et al., 2010).*

4.4.2 Morphology

The sediments delivered by the Ganga–Brahmaputra river system into the Bengal Fan is more than any other river system on the earth (Milliman and Meade, 1983). From delta to the fan, the sediments are transferred through a canyon (Fig. 4.14) which is referred to as the "Swatch of No Ground" (Emmel and Curray, 1985; Kudrass et al., 1998; Michels et al., 1998). The head of the "Swatch of No Ground" lies in about 38 m water depth and continues for 160 km southward as a long, straight trough up to a water depth of 1400 m, with an average gradient of 8.2 m/km (Curray et al., 2003). The rate of sediment deposition is very high in the subaqueous parts of Bengal Delta, the inner continental shelf (Hübscher et al., 1997; Kuehl et al., 1989; Michels et al., 1998), and also at the inner end of the canyon floor as sediments are trapped and mobilized by storms and tidal currents (Kudrass et al., 1998). However, since the Holocene, supply of sediments to the canyon has been greatly reduced (Curray et al., 2003). The head of the Bengal Fan occurs at the continental slope around 1400 m water depth. The surface of the fan appears smooth with gradients of 5.7 m/km on the upper most fan to less than 1 m/km on the lower reaches (Curray et al., 2003); however, all along its length, the fan is marked by a series of channels that are open for various distances (Fig. 4.15). The channels have been identified as the active channel (AV), channel

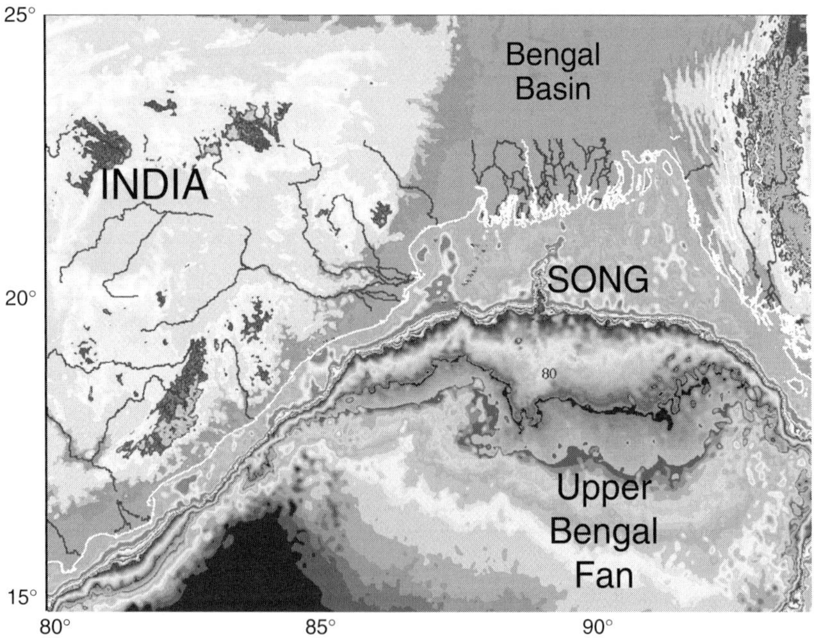

FIGURE 4.14 Bathymetry map of Bengal shelf area showing the submarine canyon Swatch of No Ground (SONG). (For color version of this figure, the reader is referred to the online version of this chapter.)

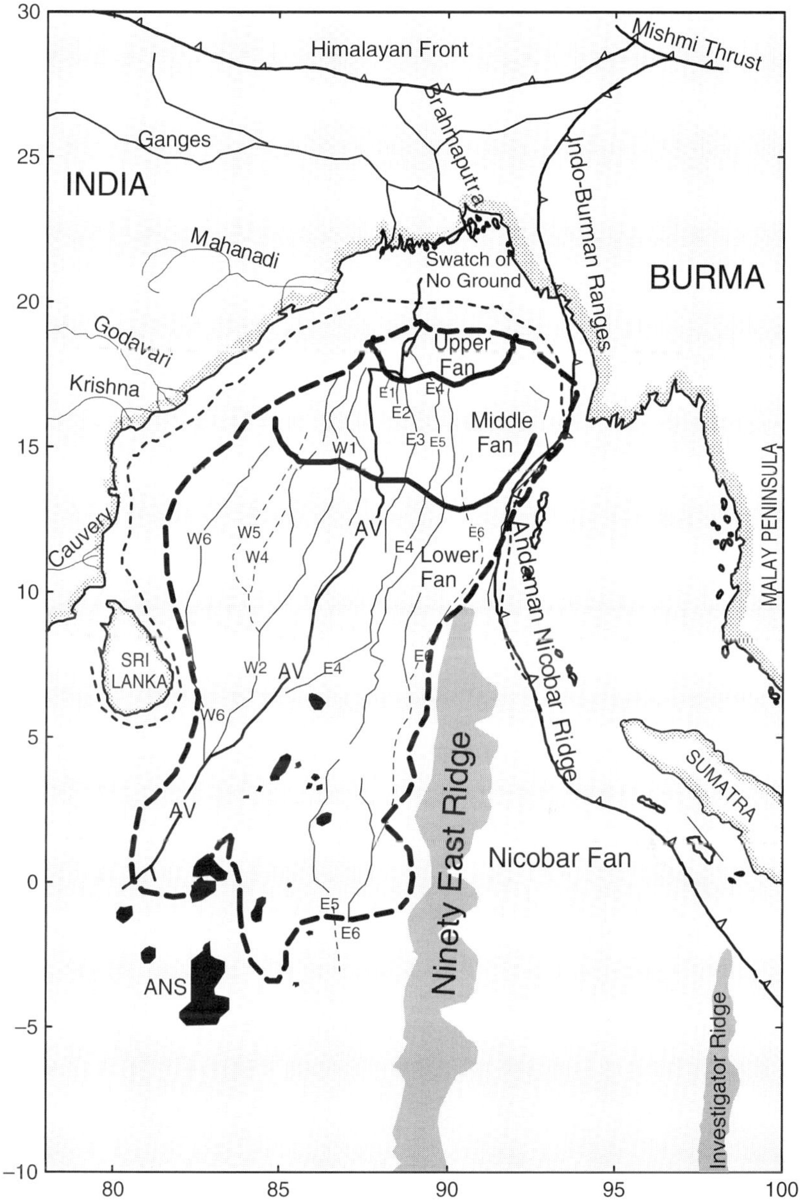

FIGURE 4.15 Morphology of the Bengal Fan showing the distribution of channels E1–E6 and W1–W6 on either side of the active channel AV *(adapted from Curray et al., 2003)*. Thick dashed line indicates the outline of Bengal Fan and thin dashed line is the present-day shelf edge. The subdivision of the Bengal Fan into upper, middle, and lower fans from north to south is indicated with thick east west lines.

systems lying to the west of AV (W1–W6) and toward east (E1–E6) from oldest to youngest (Curray et al., 2003). The number of channels decreases southward due to reoccupation of older channels by the active channel leading to tributary pattern of channels (Curray et al., 2003; Schwenk et al., 2005), in contrast to the distribution pattern observed in other submarine fans (e.g., Amazon Fan). Previous investigations suggest that these channels could have formed in the past 125,000 years (Curray et al., 2003), while channel AV from 12.8 to 6 ka (Hübscher et al., 1997; Weber et al., 1997). The thalweg gradients in the upper fan are about 2.39 m/km and in the middle fan, the gradients are about 1.68 m/km. High-resolution multibeam bathymetry images (Fig. 4.16) along a segment of the active channel system show highly meandering nature of the channel with a sharp bend near 18°N. The sediments supplied through the active fan are deposited in the deeper part in the abyssal plain through the active channels. The Bengal Fan had many canyon sources throughout the Quaternary (Fig. 4.17) and probably during Tertiary (Curray et al., 2003). Avulsions on the upper fan caused the formation of new channel–levee systems and led to the abandonment of the former active channel. However, at any point of time, only a single channel (consequently a single fan)

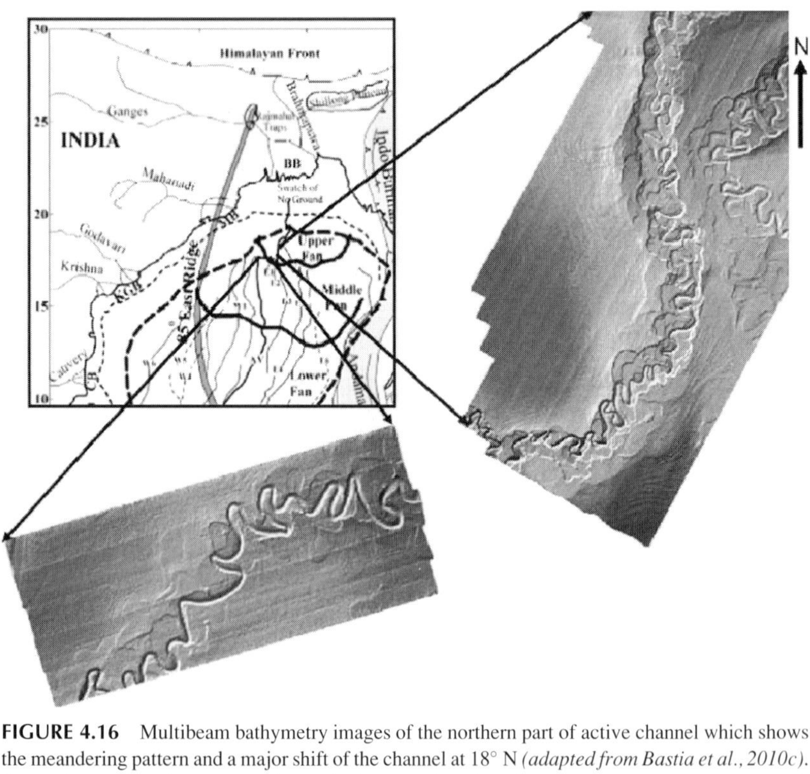

FIGURE 4.16 Multibeam bathymetry images of the northern part of active channel which shows the meandering pattern and a major shift of the channel at 18° N *(adapted from Bastia et al., 2010c)*. (For color version of this figure, the reader is referred to the online version of this chapter.)

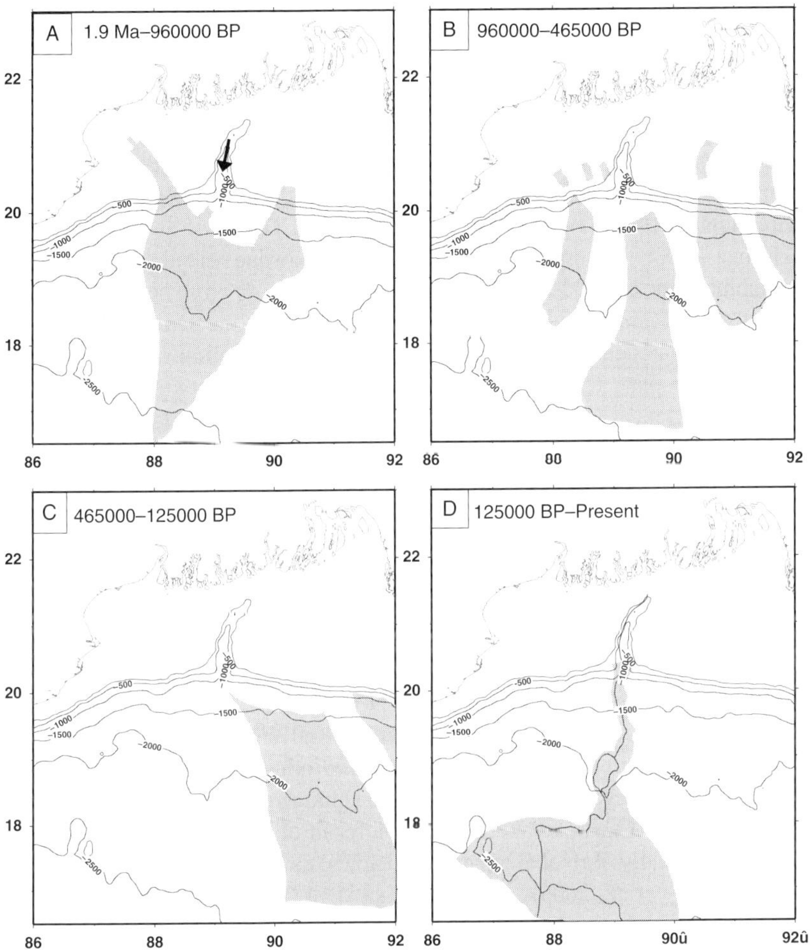

FIGURE 4.17 Quaternary subfans showing shifting of the submarine canyons and fan valley systems *(adapted from Curray et al., 2003)*.

was active; in addition, significant shifting of active canyon was also observed (Curray et al., 2003). The above-mentioned observations suggest that the present-day geometry of the Bengal fan is a collage of several subfans that were active during different geological times. Based on the present-day bathymetric (fan surface) gradient and fan architecture, the Bengal Fan is divided into three subenvironments (Fig. 4.15): the upper, middle, and lower fan (Emmel and Curray, 1985). While the boundary between upper–middle fans occurs at 2250 m water depth, the middle to lower fan boundary occurs at 2900 m. The grain size, structure, and morphology of channel–levee complex in each of these three subdivisions vary considerably (Curray and Moore, 1974).

4.4.3 Tectonic History

The tectonostratigraphic succession as well as detailed marine geophysical investigations in the Bay of Bengal and the adjoining margins gave considerable insight on the early rift–drift history, India–Asia collision scenario, sedimentation, and the growth of the Bengal Fan, apart from understanding the formation of the 85° E and the Ninetyeast ridges (Curray, 1994; Curray and Munasinghe, 1989, 1991; Curray et al., 1982; Gopala Rao et al., 1997; Krishna, 2003; Krishna et al., 2009; Radhakrishna et al., 2010, 2012; Subrahmanyam et al., 1999, 2008a). The formation of crust beneath the Bay of Bengal was initiated during early Cretaceous with the separation of India from the Antarctica in a NW–SE direction. This orientation changed to N–S with a major plate reorganization around 90–80 Ma (Klootwijk et al., 1992). The Bay of Bengal is dissected by the two almost N–S trending ridges namely the 85° E and the Ninetyeast ridges which are believed to have been related to the hotspot emplacement on the Bengal fan crust (Curray and Munasinghe, 1991; Curray et al., 1982; Krishna, 2003; Subrahmanyam et al., 1999). A strong regional unconformity recognized as the Paleocene–Eocene hiatus by Curray et al. (2003) was traced throughout the fan. This is considered as the time of initiation of the Bengal Fan and indicates a period of nondeposition separating the pre-Bengal fan sediments from the Bengal fan sediments. The intraplate deformation of the oceanic lithosphere as evident from the compressional stresses related to the continuing collision of India and Asia has been noticed from the Late Miocene (~8 Ma) records (Cloetingh and Wortel, 1985). The Tertiary deposition in the Bengal Fan was significantly influenced by the early Eocene Himalayan orogeny which had continued through India–Asia collision until the present day (Curray et al., 2003). As a result of this massive Tertiary sedimentary influx, the Bay of Bengal is recognized as the largest fan system in the world (Curray et al., 2003).

4.4.4 Sediment Structure and Stratigraphy

The sediment structure of the BOB is known from the single and multichannel seismic reflection, wide angle reflection, and spot refraction surveys undertaken in the northeast Indian Ocean (Bastia et al., 2010a–c; Curray, 1991; Curray and Moore, 1971; Curray et al., 1982; Gopala Rao et al., 1997; Naini and Leyden, 1973) and some salient results of these studies are discussed in Chapter 3. The seismic reflection data in the BOB shows three different horizons each separated by distinct unconformities (Curray and Moore, 1971). Curray et al. (1982) presented several refraction stations and three east–west multichannel seismic reflection sections across the fan lying between the Indian continental margin and the Andaman arc which provided substantial information on seismic structure of deeper sedimentary layers. They considered 5.7 km/s velocity as representative of oceanic basement with oceanic igneous crust in the velocity range from 6.0 to 7.5 km/s. This gave rise to maximum sediment thickness of

16 km in the northern Bengal Fan. The 85°E and the Ninetyeast ridges divide the BOB into three sub-basins; the basin between eastern margin of India, the depression lying between the two ridges, and the basin lying east of the Ninetyeast ridge and the Andaman arc. Subsequently, Curray (1991) reinterpreted the earlier seismic data and further refined the sedimentary structure that include an additional 6 km thick high velocity, high-density metasedimentary rocks at depth greater than 16 km characterized by 6.0–6.4 km/s velocities which then gave rise to maximum thickness of 22 km of sediments in offshore Bangladesh in northern BOB.

The onset of Bengal Fan can be very well understood by the sedimentary sequences that have been reasonably well mapped from the above-mentioned seismic reflection data as well as the DSDP results. According to Curray (1994), sediments in the BOB principally consist of two units separated by a regional unconformity of early Eocene age (Fig. 3.27). While, the lower pre-Eocene unit represents precollision sediments, the upper unit is interpreted as the post-Paleocene postcollision turbidite section shed by the river system from north. Though drilling through this unconformity was not yet achieved, the nature and areal extent of the unconformity suggest that the hiatus is longer with increasing distance from the apex of the fan (Curray, 1994, Curray et al., 2003) and separates the sediments into the pre-Eocene precollision fan and the post-Paleocene postcollision fan. This is further confirmed by the DSDP Leg 22 in the distal Bengal Fan. Apart from this, the DSDP and ODP investigations in the region (Cochran, 1990; Stow et al., 1990; Von der Borch et al., 1974a) further reveal the upper Miocene and upper Pleistocene unconformities and shallow water carbonates over the Ninetyeast ridge. Haq et al. (1987) reported lowered sea-levels during the upper Oligocene, upper Miocene, and upper Pleistocene. Multichannel seismic reflection data in the central and northern Bay of Bengal (Fig. 3.28) by Gopala Rao et al. (1997) provided some additional constraints on the sedimentary sequences and stratigraphic development of the fan. Utilizing vast amount of high-resolution 2D as well as 3D seismic data obtained for the exploration purpose, Bastia et al. (2010c) presented the isochronopach maps of pre- and postcollision sediments in the upper and middle Bengal Fan and discussed the depositional history. A detailed account of these results is presented below.

4.4.5 Depositional Pattern in the Upper and Middle Bengal Fan

The Isochronopach maps for the total sediment thickness in the region (Fig. 4.18) show development of larger sediment accumulation in the northern as well as in the western part of the Bengal Fan (Bastia et al., 2010c). As multiple systems were active in the region during the development of the basin, we need to properly understand the depositional pattern in different times.

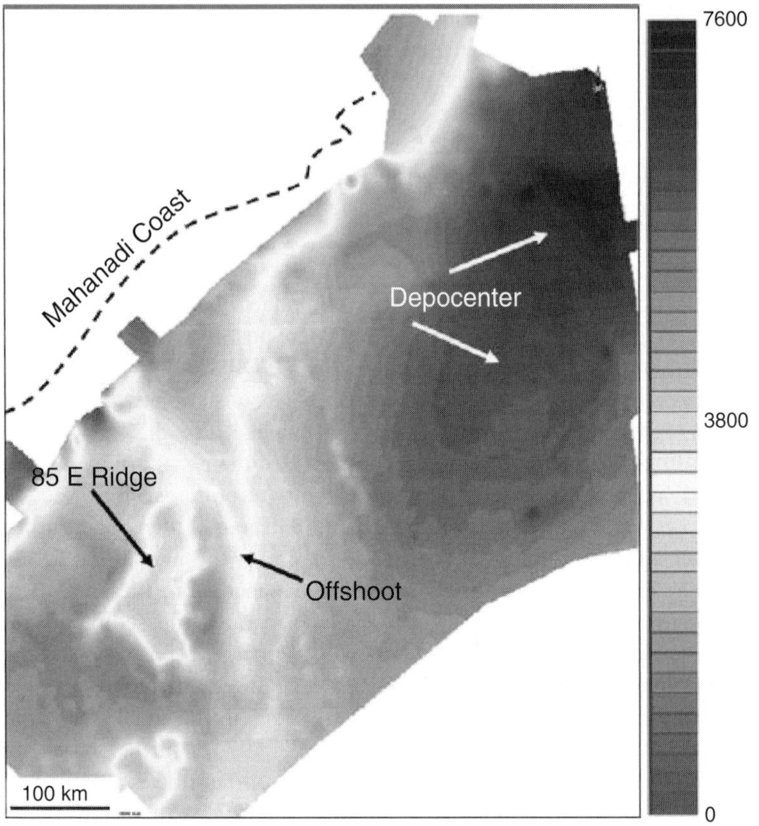

FIGURE 4.18 Sediment isopach (TWT) map of the northern Bengal Fan area showing the total sediment thickness. Increase in thickness towards the north (marked as depocenter) and reduction in thickness over the 85° E ridge can be noticed *(adapted from Bastia et al., 2010c)*. (For color version of this figure, the reader is referred to the online version of this chapter.)

During the precollision time, river system from the west (Mahanadi, Godavari as for example) was mostly active, whereas post-Paleocene onward (following the onset of soft collision), the Ganga–Brahmaputra river system from north became active. The seismic section in Fig. 4.19 depicts the onlapping of Bengal Fan sediments against the Mahanadi slope. However, from a closer view, it appears that the onlap does not follow one single surface, at times (in the post-Eocene), the Mahanadi also contributes sediments in the deeper basin. However, it is very clear that the sediment input from north is much higher in comparison to Mahanadi after the hard collision. Apart from collision between Asian and Eurasian plates which led to the uplift of Himalayas, the emplacement of 85°E ridge had also controlled the sediment dispersal pattern.

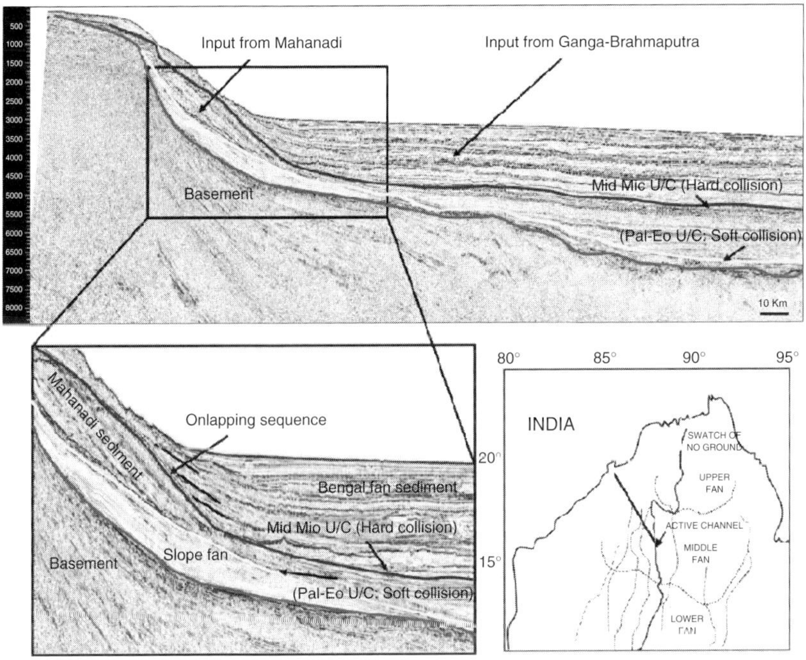

FIGURE 4.19 Seismic section showing the depositional sequences from the Mahanadi river (from west) as well as Ganga–Brahmaputra river system (from north) that were active in different geological times *(adapted from Bastia et al., 2010c)*. (For color version of this figure, the reader is referred to the online version of this chapter.)

In view of this, the sediment dispersal pattern is discussed under three broad divisions (after Bastia et al., 2010c).

4.4.5.1 Precollision (>59 My)

Prior to soft collision (>59 My), the sediment supply to the basin was mostly from west and no active depositional system was reported from north. Both Mahanadi and Godavari rivers are the major sediment source in the precollision pre-Bengal fan time; however, emplacement of the 85°E ridge (Fig. 4.20) during this period affected the sediment distribution pattern. Its presence as a basement high significantly controlled the depositional fairway for the Paleogene sediments. As the ridge was not present as a continuous structural high, minor sediment accumulation east of the ridge within the sub-basin was observed.

4.4.5.2 Between Soft and Hard Collision (59–15 My)

During this period, the Bengal Fan was initiated and the sediment accumulation was mostly confined in the northern part. The time slices of semblance volume (Fig. 4.21) reveals the development of channel system from north which supplied the sediments into the deeper part of the basin.

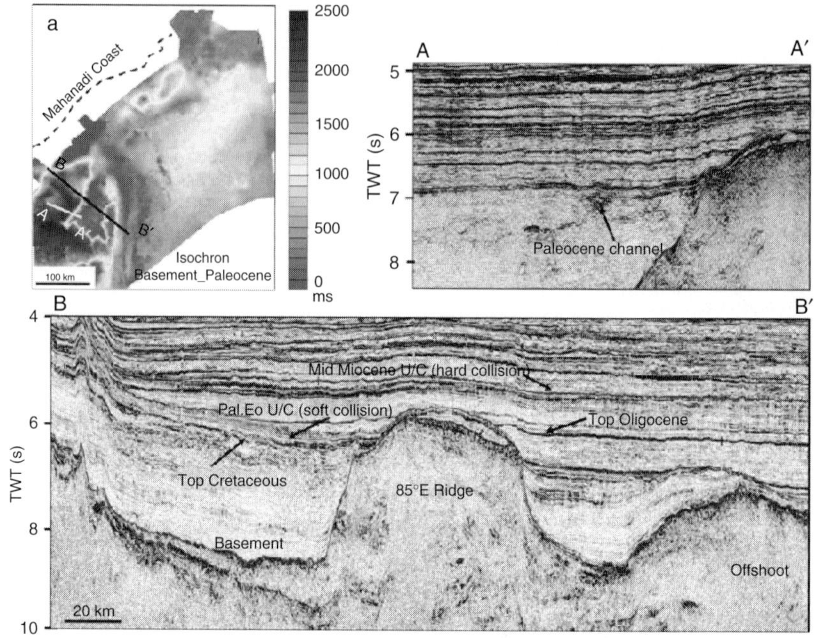

FIGURE 4.20 Isopach map (in time) showing the thickness of sediments in the pre-Eocene time. Note the Cretaceous depocenter west of the 85° E ridge and the profiles AA′ and BB′ showing prominent high-amplitude channel toward western basin of 85° E ridge which wedges out against the ridge *(adapted from Bastia et al., 2010c)*. (For color version of this figure, the reader is referred to the online version of this chapter.)

4.4.5.3 Post-hard Collision (<15 My)

Due to the hard collision and the rapid rise of Himalaya, the sediment supply increased manifold from this time onward and as a result, the depocenter has been shifted southward (Fig. 4.22). Many channel–levee systems have been developed and these helped in transporting the sediments further into the abyssal plain forming splay deposits beyond the toe of the slope. In the proximal part of the fan, the typical sequence is characterized by vertical building of mud flow-splay-channel–levee deposits (Fig. 4.22). Erosion of interchannel lows with formation of HARP (High Amplitude Reflection Pattern) deposits (Schwenk et al., 2005; Weber et al., 1997) gave rise to the formation of active channel–levee system. The levee is formed by overspilling of mud, and HARP is also observed at the base of the channel and referred as splay/frontal lobes in the area (Bastia et al., 2010c).

4.4.6 Seismic Facies Analysis and Evolution of the Channel–Levee System

Five major seismic facies identified from the study area are shown in Fig. 4.23 along with a representative seismic section which covers most of the seismic facies discussed below. Similar facies analysis was made by group of workers in the Amazon fan (Flood and Piper, 1997; Flood et al., 1995). On the basis

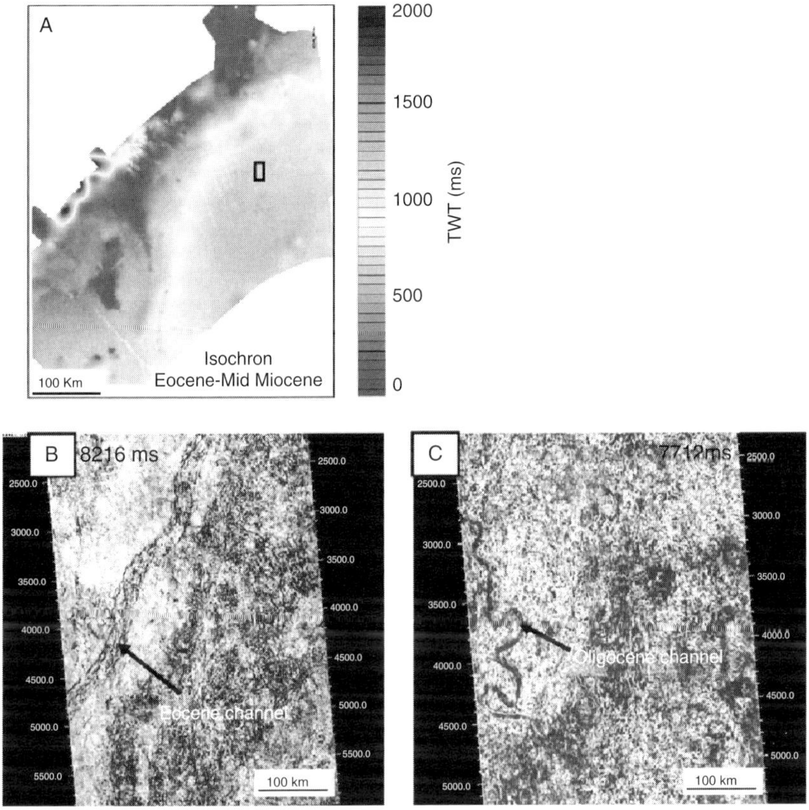

FIGURE 4.21 Isopach (in time) map of Post-Paleocene to Mid-Miocene sediment thickness in the northern part of Bengal Fan. Time slice equivalent to (B) Eocene and (C) Oligocene level showing presence of N–S meandering channel obtained from high-resolution 3D seismic coherency volume (area shown in A) further indicates that Bengal fan system was active as early as Eocene; however, channel activity was less compared to post hard collision growth of the fan *(adapted from Bastia et al., 2010c)*. (For color version of this figure, the reader is referred to the online version of this chapter.)

of seismic calibration in the Amazon fan, Schwenk et al. (2005) carried out detailed facies analysis in the Bengal Fan. In addition to their classification, another commonly observed seismic facies in the area is the one formed by mass transfer or gravity flow, characterized by chaotic HARP. The base of the deposit forms erosional relation with the underlying units formed due to gravity induced mass flow termed as turbidites. According to Schwenk et al. (2005), the channel facies is formed due to filling of the abandoned channel. The upper part of the channel might have been filled by the overspilled deposits from the adjacent channel. The HARP facies is mainly observed at the base of the channel and probably results in the frequent levee avulsion followed by unconfined flow in intrachannel lows (Schwenk et al., 2005). However, in the distal part, this can also be interpreted as part of frontal lobes. The Chaotic High Amplitude reflection (CHARS) pattern at the base of channel axis as well as the

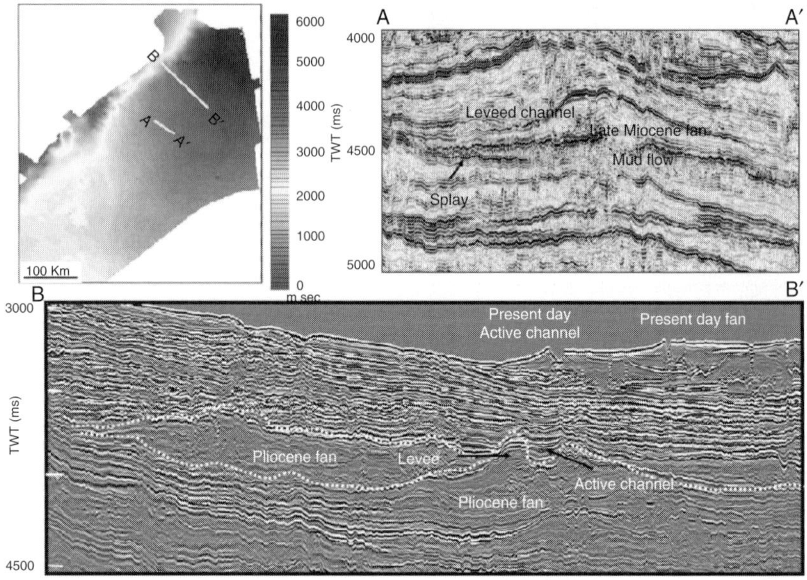

FIGURE 4.22 Isopach map (in time) showing the post-Mid Miocene sediments in the northern Bengal Fan. Note the shifting of depocenter toward north. Characteristic seismic facies during Miocene (A–A′) and Pliocene (B–B′) showing deep water sequence buildup involving mud flow followed by splay and subsequently by channel–levee facies. The present-day active channel is shown in profile B–B′ *(adapted from Bastia et al., 2010c)*. (For color version of this figure, the reader is referred to the online version of this chapter.)

HARS might have been formed due to vertical stacking of blocks. However, lateral migration and aggradation of channel is reported offshore Angola by Kolla et al. (2001).

Similar to other large submarine fans like Amazon, Congo, Indus, and Mississippi fans (Schwenk et al., 2005), in the Bengal Fan also channel–levee systems are the main architectural elements. Like many other fans, it had evolved by stacking and overlapping of channel–levee systems due to frequent avulsions on the upper fan. The levee is formed by overspilling of mud. The active channel–levee system involves filled channels, levee deposit, CHARS/HARS, and HARP (Fig. 4.24). While the coarse-grained material is deposited at the center of the channel, fine-grained material is spilled over to form the levee. Frequent channel avulsion in the proximal fan causes complex depositional pattern. Due to channel avulsion, the channel takes new path forming intrachannel lows which are later filled up by unchannelized flow with formation of the HARP. During the initial stage, vertical aggradation (Pirmez and Flood, 1995) of channel gives rise to CHARS/HARS at the base of channel axis; subsequently, lateral migration becomes more common (Kolla et al., 2001). Channel–levee system was active from Early Eocene time in the northern part of the Bengal Fan (Curray, 1994). From Mid-Miocene onward, it became more prominent in the southern part of the fan which is

Seismic facies	Example	Facies prediction
Low amplitude chaotic		Mass flow
Parallel high amplitude	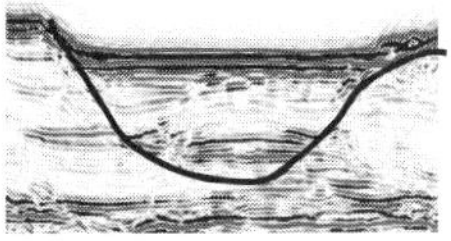	Channel fill
Low amplitude convergent		Levee
Parallel high amplitude continuous		HARP
Chaotic high amplitude	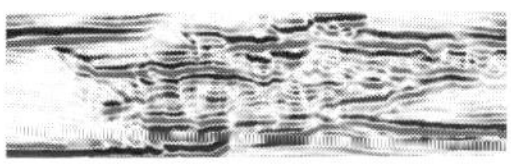	CHARS

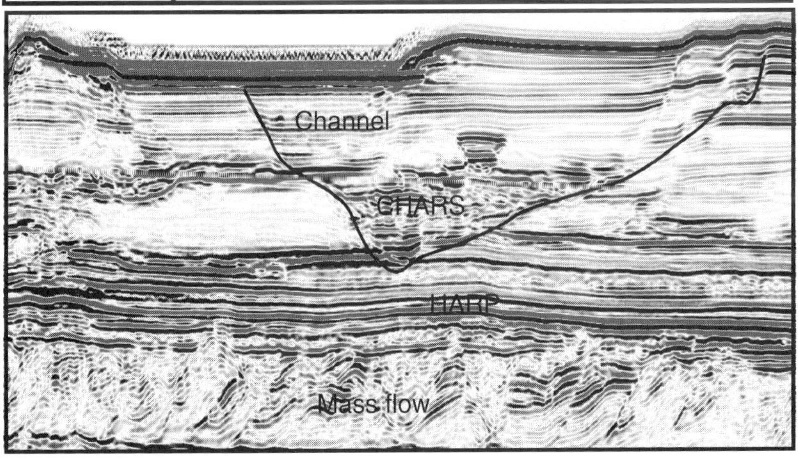

FIGURE 4.23 Representative seismic section (bottom panel) showing important characteristics that are useful in the identification of various seismic facies (top panel) in the Bengal Fan. Some of these characteristics were used earlier by Schwenk et al. (2005) for facies analysis in the Bengal Fan. (For color version of this figure, the reader is referred to the online version of this chapter.)

correlated with increasing rate of erosion after the hard collision. Due to continuous progradation, Plio-Pleistocene channel–levee complex can be observed farther southward compared to Miocene channel complex. Weber et al. (1997) pointed out that the Bengal Fan shows active growth during the most recent sea-level rise as well as the recent sea-level highstand.

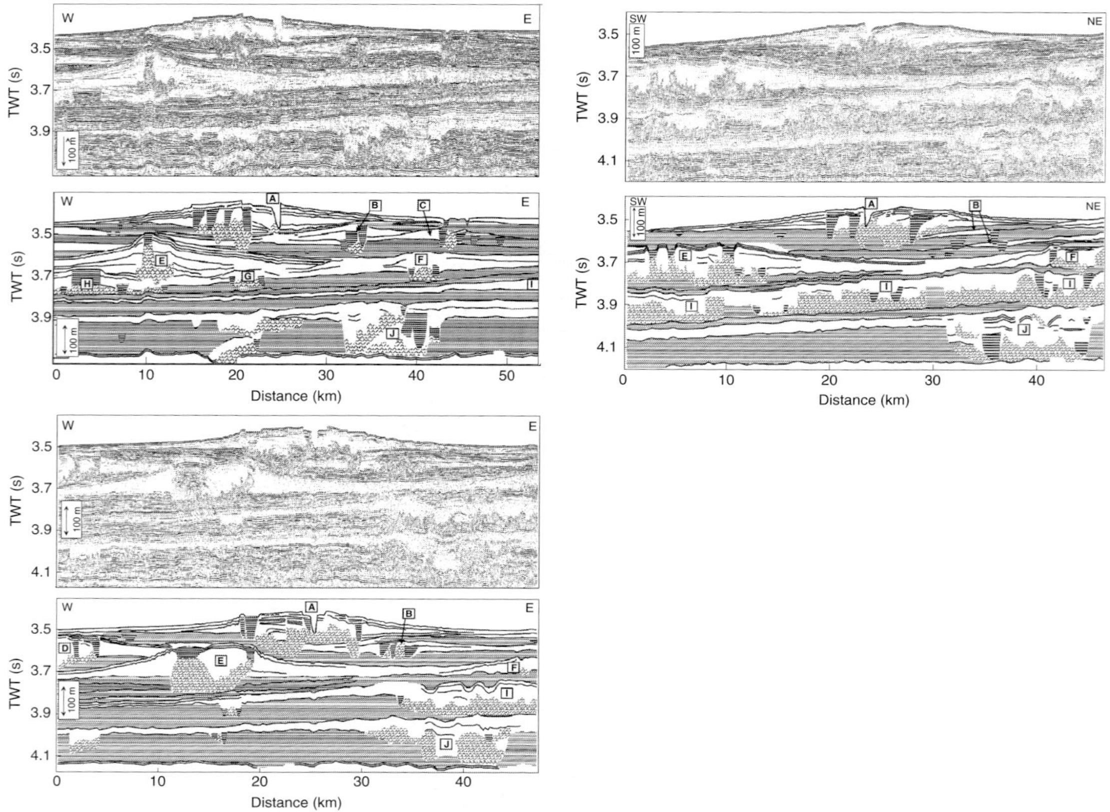

FIGURE 4.24 Seismic sections and the corresponding line drawings along three seismic profiles located in the present-day middle fan showing the interpreted seismic facies in the area *(Reproduced from Schwenk et al., 2005)*. Based on this data, they identified four main seismic facies such as levee deposits, HARPS,

4.4.7 Progradation of the Bengal Fan with Time

Analysis of several of 2D seismic sections in the proximal, middle, and distal fan (Bastia et al., 2010c) revealed continuous progradation of the Bengal Fan with time. In the present-day upper fan, three seismic lines (Fig. 4.25) provide the strike view of deepwater Bengal Fan sediments. The Neogene section shows a vertical change from unconfined distal lobe, and the fan fringe deposits are progressively overlain by channelized units capped by a succession of large channel–levee type systems.

The seismic lines (Fig. 4.26) falling in the present day middle fan provide a strike-section of western part of the deep-water Bengal Fan sediments and record the progradation of the Bengal Fan from lower to middle fan environments during the Neogene. The majority of the section corresponds to the lower fan environment, however, compared to the more proximal sections described above, the channel–levee height has been reduced considerably. The Paleogene section consists of variable amplitude, continuous seismic facies with increasing truncation of reflectors into the Pliocene. The seismic lines from the present-day distal fan records only lower fan lobes (Fig. 4.27) and the parallel high-amplitude reflection is interpreted as occurrence of sheet sands in the frontal lobe.

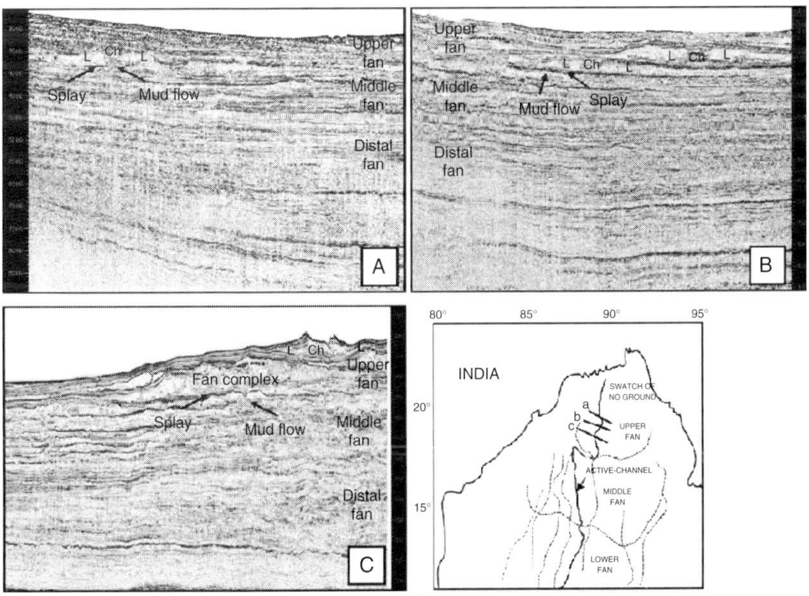

FIGURE 4.25 Seismic sections (A–C) across the present-day upper fan, which indicate shifting of distal fan system to younger proximal fan and continuous progradation of Bengal fan with time *(adapted from Bastia et al., 2010c)*. (For color version of this figure, the reader is referred to the online version of this chapter.)

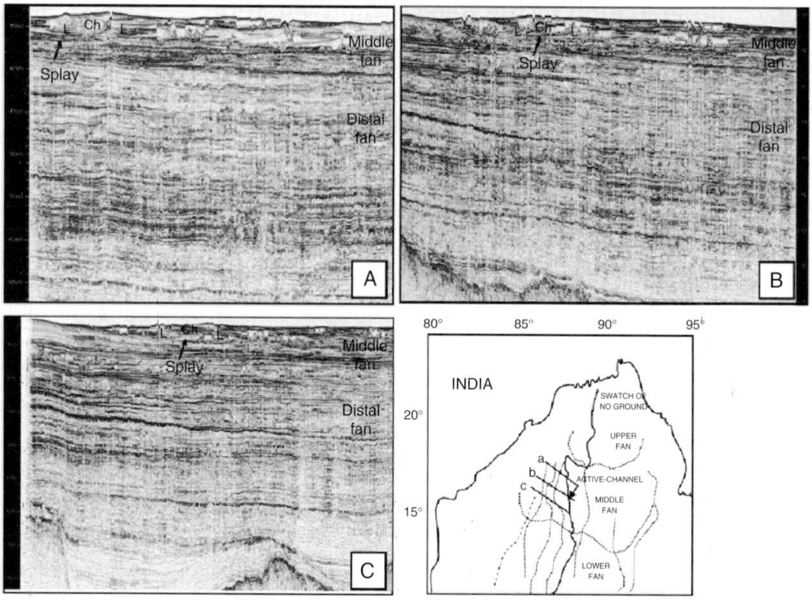

FIGURE 4.26 Seismic sections (A–C) across the present-day middle fan characterized by channel–levee and splays. Notice that the deeper section mostly comprises channel–levee facies. Absence of mud flow in the middle fan can also be noted *(adapted from Bastia et al., 2010c)*. (For color version of this figure, the reader is referred to the online version of this chapter.)

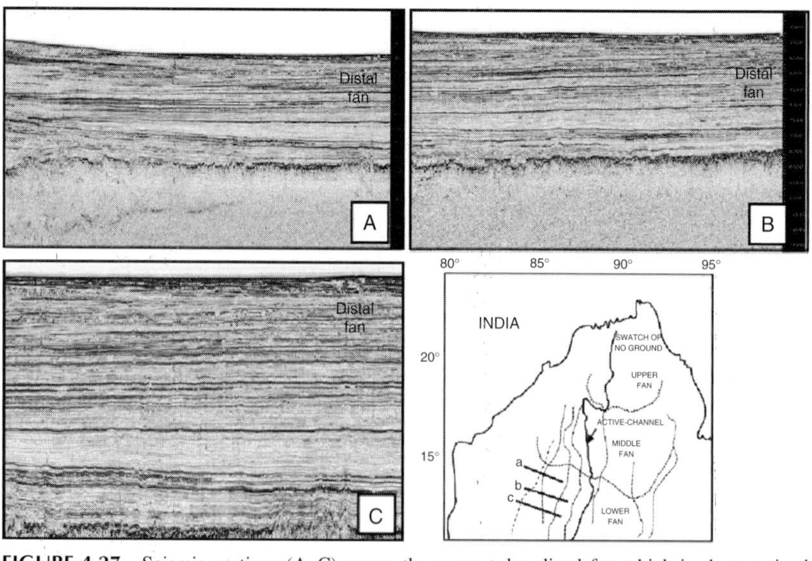

FIGURE 4.27 Seismic sections (A–C) across the present-day distal fan which is characterized by channel system and absence of mud flow and splays. Note that height of the levee reduces considerably in this region to almost absent *(adapted from Bastia et al., 2010c)*. (For color version of this figure, the reader is referred to the online version of this chapter.)

Chapter 5

Sedimentary Basins Along the East Coast of India: Subsurface Geology, Depositional History, and Petroleum Systems

5.1. INTRODUCTION

The East Coast of India is represented by the rifted passive margin that had evolved in response to the process of continental rifting and subsequent seafloor spreading between India, Antarctica, and Australia during early Cretaceous breakup of Eastern Gondwanaland (Powell et al., 1988). Geomorphologically and tectonically, the coast is segmented into several pericratonic rift basins that have emerged because of intrinsic distinctions between different subcrustal blocks during rifting (Lal et al., 2009). The eastern Indian shield margin is characterized by several ancient tectonic blocks, such as the Shillong plateau, Chotanagpur plateau, Eastern Ghats, and the Southern Granulite Terrain (SGT), that are bounded by mega tectonic elements (Radhakrishna, 1989). A brief description of regional geology and drainage pattern along with a detailed account of petroleum systems for individual basins that formed at the eastern margin is provided here.

5.2. REGIONAL GEOLOGICAL SETTING OF THE EAST COAST BASINS

The Precambrian structural grain of the Indian shield to a larger extent controlled the continental fragmentation and the subsequent tectonic reactivations at the western as well as the eastern margin. The East Coast of India is dominated by the presence of Proterozoic mobile belts thrust over the Dharwar, Singbhum, and Bastar cratons. The basement rocks along the coast belong to the Eastern Ghat Mobile Belt (EGMB) in the northern part and the SGT in the south. The boundary between the basements of EGMB and the SGT with

Developments in Petroleum Science, Vol. 59. http://dx.doi.org/10.1016/B978-0-444-53604-4.00005-3

the above-mentioned cratons widely varies in geometry along the east coast. Further, these mobile belts are believed to be an Indian counterpart of the mobile belts of the Enderby Land, Antarctica (Biswal and Sinha, 2004; Black et al., 1987), as evidenced by the occurrence of stacked terrains of various ages of metamorphism in the EGMB and SGT of India, and Napier and Rayner complex of Antarctica. During the Phanerozoic era, the development of several Gondwana grabens along the pre-existing weak zones took place at the east coast; the most distinct among them are the Pranahita–Godavari and the Mahanadi grabens (Fig. 3.16). Both these NW–SE trending Gondwana rift basins roughly oriented perpendicular to the coastline contain Upper Carboniferous–Middle Triassic sediments (Lisker and Fachmann, 2001; Rao, 2001).

5.2.1 Drainage Pattern

The East Coast of India is characterized by the presence of several major rivers that have originated in the Western Ghats or Himalayas and flow into the Bay of Bengal (Fig. 4.13). The most remarkable feature of the drainage pattern over the Indian peninsula is the way in which all the major east flowing rivers (e.g., the Godavari, Krishna, and Cauvery) almost originate at the western edge of the Deccan plateau and traverse through the whole width of the plateau (Gunnell and Radhakrishna, 2001). This easterly drainage is ascribed to an easterly tilt of the Peninsular landmass by Radhakrishna (1993). The antecedent character of the drainage of major rivers flowing in the Peninsula was discussed by Ollier and Power (1985). The major rivers are the Cauvery, Krishna, Godavari, Mahanadi, Ganges, and Brahmaputra, apart from many minor rivers such as the Vaigai, Palar, Pennar, and Vamsadhara. It is an emergent coast with well-defined beaches, many lagoonal lakes, sand dunes, and sand pits. The coastal plain all along is characterized by the presence of major riverine deltas such as the Cauvery, Krishna, Godavari, Mahanadi, and Ganges–Brahmaputra through which huge amounts of sediments are transported into the deepwater areas of the eastern margin and the adjoining Bay of Bengal.

5.2.2 General Description of Basins Along the East Coast and the Andaman Region

The East Coast of India has a strike-length of about 2000 km extending from the Sri Lanka shelf in the far south to the Bengal basin in north; it consists of five major sedimentary basins: the Cauvery, Palar, Krishna–Godavari (K–G), Mahanadi, and Bengal basins (Fig. 5.1). This segmentation could be assigned due to the presence of major river system at the coast (Sastri et al., 1973) as well as the intrinsic subcrustal distinction along the east coast (Radhakrishna, 1989). Lal et al. (2009) observed that these basins are separated by major tectonic elements: the Bengal and Mahanadi basins are separated by a major fault, Visakhapatnam high separates the Mahanadi from K–G, Nayudupeta high separates

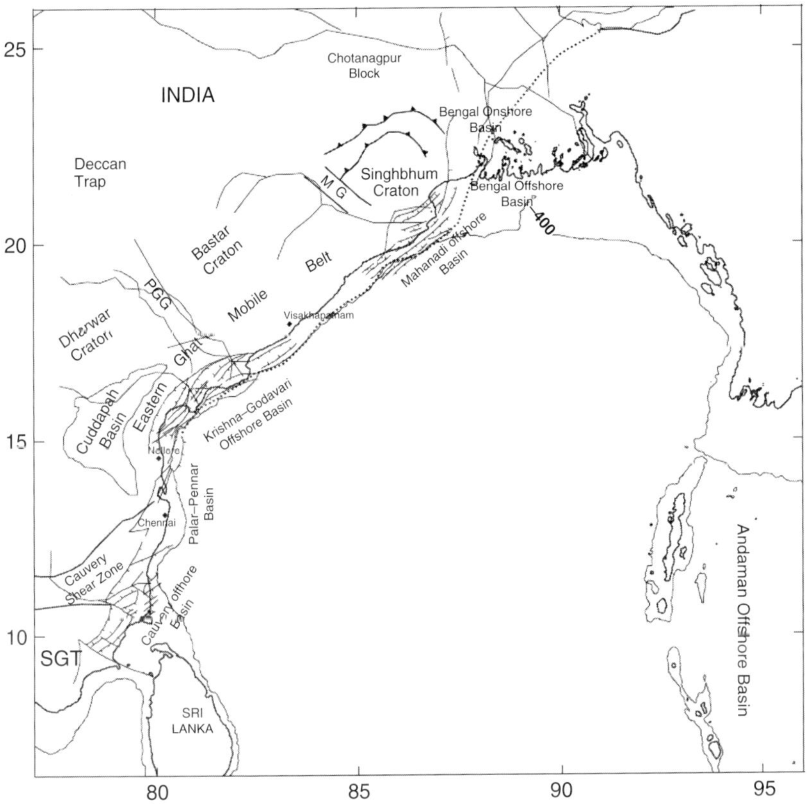

FIGURE 5.1 Map showing geotectonic features and major sedimentary basins along the east coast of India *(modified after Lal et al., 2009).*

K–G from Palar, and Chingleput high separates Palar from the Cauvery basins. On the basis of tectonic style, sedimentation, and basin geometry from the shelf to deepwater areas, Bastia (2006) subdivided the East Coast basins into seven subdivisions. From north to south, these are (i) Ganga–Brahmaputra basin (Bengal offshore), (ii) Mahanadi basin, (iii) Visakhapatnam Bay Basin, (iv) K–G basin, (v) Palar–Pennar basin, (vi) Cauvery basin, and (vii) Gulf of Mannar.

The ONGC and the Geological Survey of India (GSI) have done pioneering geological studies along the east coast for the last several decades, as a result the exposed surfacial sedimentary rock types of east coast basins and the adjoining crystalline rocks have been mapped. Later, ONGC and Oil India Limited (OIL) conducted a systematic exploration work in these basins and acquired a large amount of regional gravity and magnetic data along with detailed 2D multichannel seismic and subsurface lithologic data. As described in Chapter 1, after the NELP era, many other E&P companies started participating in the acquisition and analysis of geophysical data in the basins. Detailed interpretation of geophysical data in these basins (Bastia, 2006; Bastia et al., 2010a–c; Fuloria, 1993; Prabhakar and Zutshi, 1993; Rangaraju et al., 1993; Rao, 2001; Sastri et al., 1973; 1981; Venkatarengan et al., 1993; among many others) revealed NE–SW trending horst–graben structural features which extend into the offshore and merge with the deep-sea areas of the Bay of Bengal. While Sastri et al. (1973) classified the Cauvery, Palar, K–G, and Mahanadi basins as pericratonic, Talukdar (1982) opined that these four basins might have originated as pull-apart basins during the breakup of Gondwanaland. Later, Biswas et al. (1993) classified them as pericratonic rift basins in the onshore part and as Miogeoclinal prism for their offshore counterpart. Chari et al. (1995) presented various geological evidences to suggest that the Cauvery basin is an intracratonic rift basin. As described in Chapter 3, geophysical investigations suggest that the eastern margin can be considered as a composite of two segments: the northern part (north of 16° N) having rifted margin character and the southern part developed as a consequence of shearing in the early stages of continental separation (Chand et al., 2001). Therefore, it can be inferred that the basins in the southern part of the margin might have evolved into the rift basins from the initial pull-apart basins. Further, the K–G and Mahanadi Basins are connected with the Indian shield to the west by the NW–SE trending grabens, called the intracratonic rift basins (Biswas et al., 1993). The Palar and Cauvery basins do not have such a geotectonic feature (Sastri et al., 1973). These major onshore basins located at the east coast extend well into the Bay of Bengal, thereby making the eastern offshore a major deepwater petroliferous province of India (Bastia, 2006).

In addition to the east coast basins and the adjoining offshore, the Andaman offshore basin located in the SE corner of the Indian Territory is another deepwater basin lying along the Andaman trench–arc region, which is an active plate boundary between the Indian and the Eurasian plates (Fig. 3.1). The Andaman–Nicobar Island region belonging to the fore-arc accretionary complex along with the back-arc Andaman basin characterized by the presence of volcanic arc is the only frontier petroleum province of India occurring in active margin tectonic setting.

5.3. BASIN-WISE ANALYSIS OF HYDROCARBON POTENTIAL

According to Magoon and Dow (1994), the essential elements of a petroleum system are the petroleum source rock, reservoir rock, seal rock, and the overburden rock. The processes that enhance the prospectivity are the trap formation, generation, migration, and accumulation of hydrocarbons. Appropriate relative timing of formation of these elements along with the above processes is necessary for hydrocarbons to accumulate and be preserved. Exploration plays and prospects are typically developed in those basins or regions in which a complete petroleum system has some likelihood of existing.

The eastern Indian shelf is a narrow linear belt between West Bengal in the north and Tamil Nadu in the south. As the basins along the East Coast of India form a geographically continuous unit and share a common geological history, the tectonostratigraphic history of the basins has a natural genetic linkage to petroleum systems as well as the distribution of hydrocarbons at the margin (Husain et al., 2000). For example, Pandey and Dave (1998) stated that the West Bengal–Mahanadi shelf constitutes a single stratigraphic unit where Cenozoic sediments overlie the Precambrian or Gondwana rocks or a thin veneer of Cretaceous sediments. In this chapter, a detailed account of hydrocarbon potential of the East Coast basins of India including the deep offshore areas of the margin is presented by regrouping the individual basins into five broad divisions for ease of presentation as well as on the basis of the extent of information available in both onshore and offshore, their potential for hydrocarbon occurrence, and keeping in mind the significant breakthrough in deepwater hydrocarbon exploration in the Indian offshore region in recent times. Therefore, the regrouped basins reported for the present purpose are (i) Cauvery–Palar Basin, which includes the region from Gulf of Mannar to Palar basin, (ii) K–G Basin including the Pennar basin in the south and part of Visakhapatnam Bay basin in the north, (iii) the onshore part of Mahanadi and Bengal basins and the adjoining shelf areas of the North East Coast (NEC) are called the Mahanadi-NEC Basin, (iv) the deepwater petroleum province along the eastern offshore including the 85° E Ridge, and (v) the Andaman offshore Basin. It is relevant to state that exploration activity in terms of geophysical data acquisition and subsurface drilling information drastically varies among these different basins, so we made a sincere effort to provide in this book as much regional information as possible from all the basins but at the same time, wherever possible, provide detailed information available for some of the basins.

5.3.1 Cauvery–Palar Basin

The Cauvery–Palar basin is situated on the southern part of eastern flank of peninsular India covering the coast between Ramnathapuram near the Palk Strait and the region north of Madras along the Coromandel Coast (Fig. 5.2) encompassing an area of about 35,000 km^2 on the land and about 33,000 km^2 of offshore area extending up to 200 m isobath (Rangaraju et al., 1993). In addition, the basin contains more than 95,000 km^2 of deepwater offshore area. The basin is classified as the pericratonic rift basin (Biswas et al., 1993) characterized by the NE–SW horst–graben basement morphology, which came into existence as a result of fragmentation of the eastern Gondwanaland since the late Jurassic. Though exploration for hydrocarbons in the basin was initiated by ONGC as early as 1958, with the first well drilled over Karaikal high in 1964 (Source: DGH) gave indications of active petroleum system in the basin. Subsequently, intense and systematic acquisition and interpretation of geoscientific data both onland and offshore yielded commercially successful oil and gas discoveries after the second phase of exploration program in 1984.

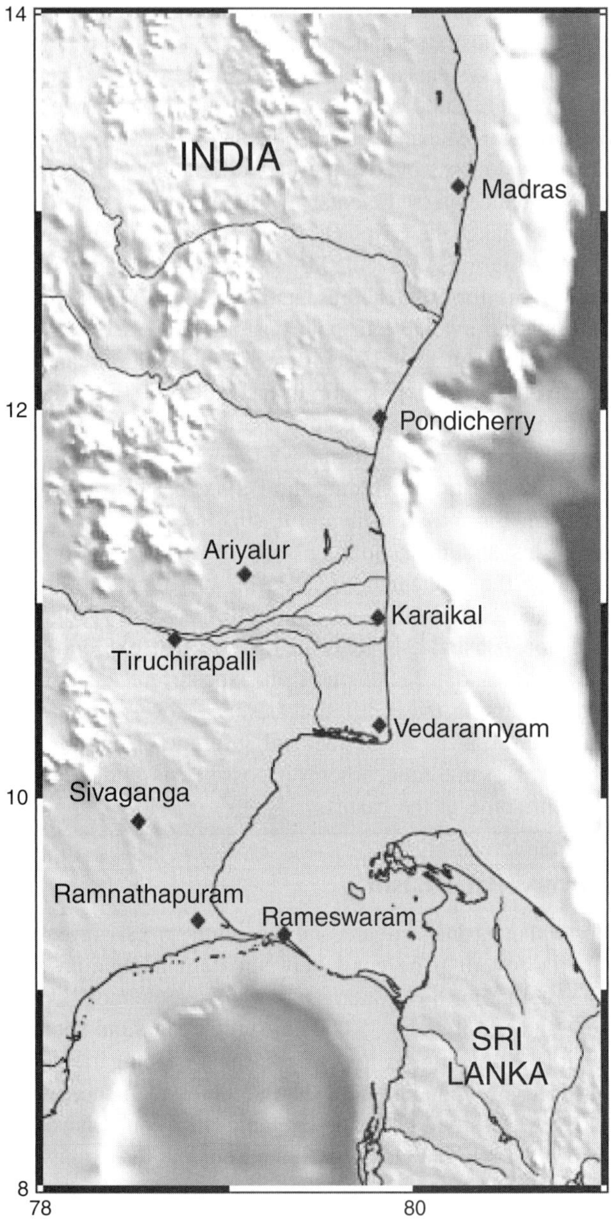

FIGURE 5.2 Morphology of the Cauvery–Palar basin showing river pattern and prominent locations along the SE continental margin of India.

5.3.1.1 Structure and Tectonics

The Cauvery–Palar Basin is a block-faulted pericratonic rift basin comprising horst–graben basement architecture. The structural grain of the basin is dominated by the Jurassic–Early Cretaceous pattern of normal faults that control the system of NE–SW trending horsts and grabens characteristic of rift extension (Prabhakar and Zutshi, 1993). This trend is in conformity with the adjoining Precambrian Eastern Ghat trend and several NW–SE oriented cross-faults cut across this trend. The basement rocks supporting the basin belong to the Eastern Ghat group of rocks mainly composed of charnockites, quartzites, granitic gneiss, and schists. Drilling at many places through the basement revealed fractured and weathered nature of the basement. While the outcropping marine Permian sediments in the Palar basin are significant from the point of view of paleogeography (Sastri et al., 1973), the western margin of the Cauvery basin is demarcated by steep down to basin faults separating the shield crust from the Cretaceous and Tertiary sediments.

The subsurface division of the basin into major depressions separated by several ridges is known for long, initially based on gravity and magnetic studies and later substantiated from seismic data and exploratory drilling. The subsurface structure and tectonostratigraphic evolution of the basin was described in detail by different workers (e.g., Biswas et al., 1993; Chandra et al., 1993; Kumar, 1983; Lal et al., 2009; Prabhakar and Zutshi, 1993; Rangaraju et al., 1993; Sastri et al., 1973; among many others). The major structural and tectonic elements compiled from these studies are (i) Pulicat depression defined by Nayudupeta high in the north and Chingleput high in the south in the Palar basin, (ii) Ariyalur–Pondicherry and Thanjavur–Tranquebar depressions separated by Kumbakonam–Madanam ridge, (iii) Nagapattinam depression limited by Karaikal ridge in the north and Vedaranniyam terrace in the south, (iv) Palk Bay and Thanjavur depressions separated by Mannargudi ridge, (v) Pamban depression separated by Mandapam–Delft ridge from Palk Bay, and (vi) Mannar depression further west in the Gulf of Mannar. As can be seen from Fig. 5.3, the Pondicherry, Tranquebar, Nagapattinam, and the Mannar depressions continue into the deep oceanic regions, while the Palk bay is bounded by Indian and Sri Lanka massifs. Further, the structural features were dislocated by two prominent cross-trends such as the Vedaranniyam–Tiruchirapalli and Madurai–Rameswaram (Prabhakar and Zutshi, 1993) trends. The Kumbakonam and Madanam ridges as well as the Karaikal and Mannargudi ridges are disjointed by Mayavaram and Nannilam saddles, respectively. On the basis of interpretation of IRS data, Mitra and Agarwal (1991) identified several NW–SE cross-trends that are younger than NE–SW main structural features and inferred that they might have played an important role in migration and entrapment of hydrocarbons.

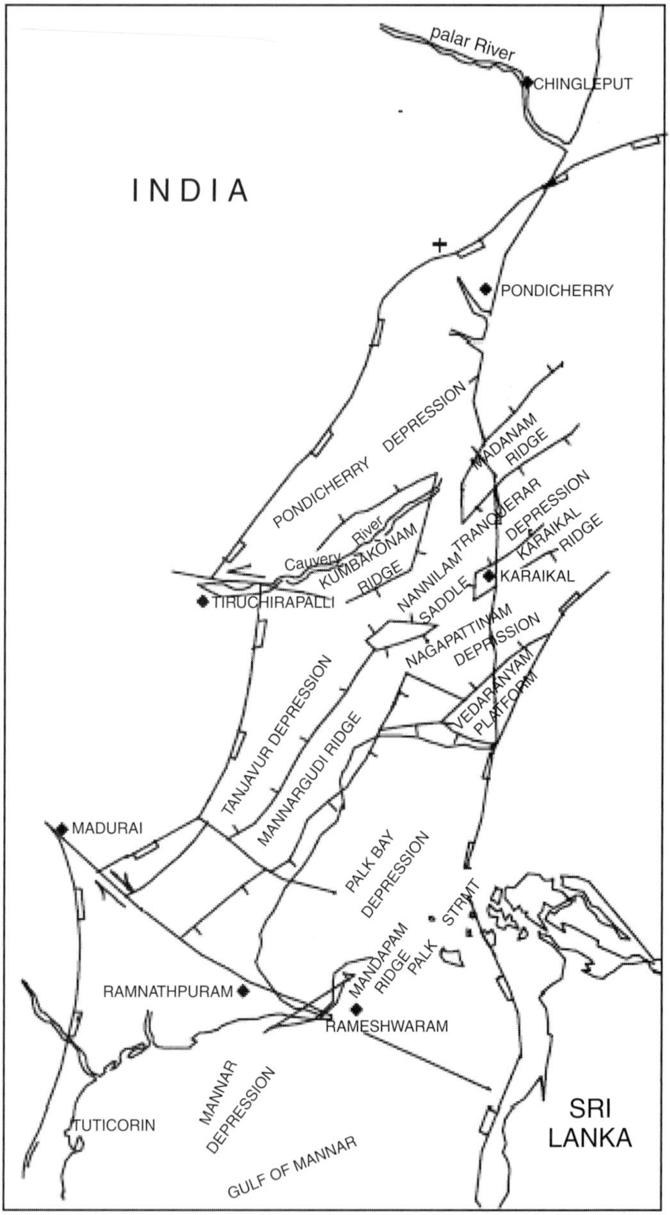

FIGURE 5.3 Tectonic map showing various ridge–depression basement structural configuration in the Cauvery–Palar basin *(adapted from Prabhakar and Zutshi, 1993).*

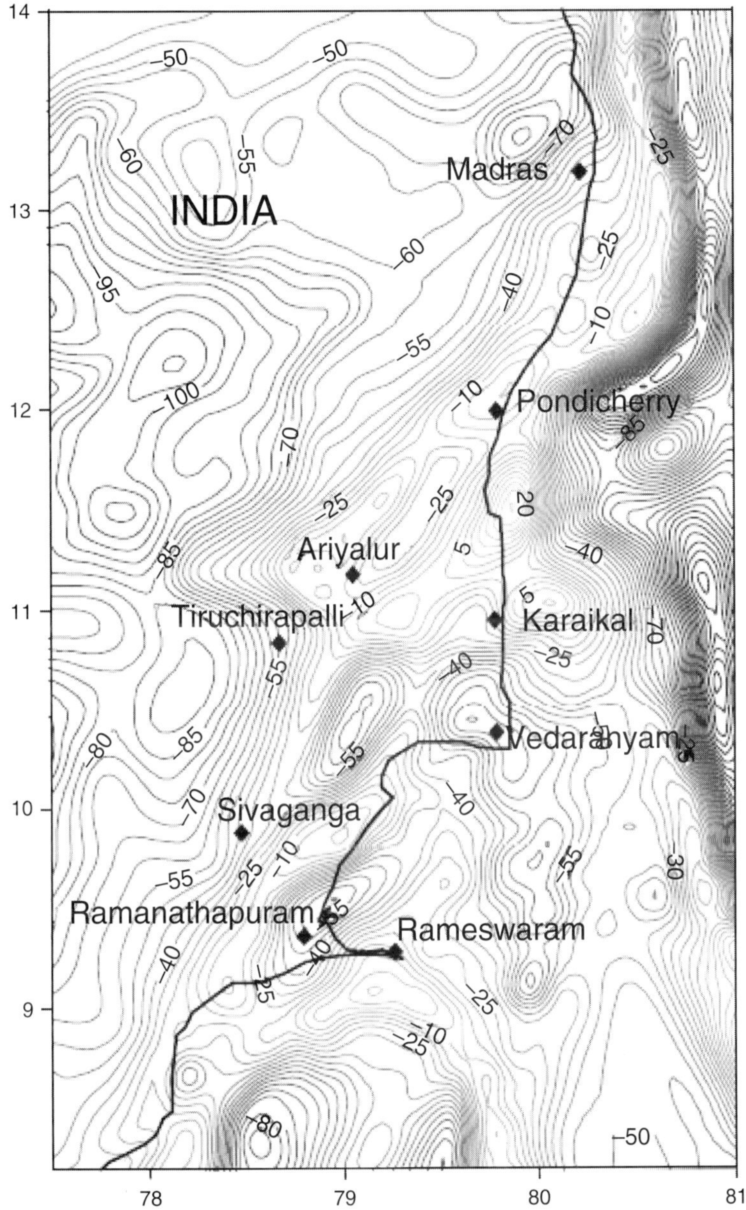

FIGURE 5.4 Gravity anomaly map of the Cauvery–Palar basin using Bouguer anomalies onshore (Verma et al., 1993) and free-air anomalies offshore (Andersen and Knudsen, 2008). Note the correlation of basement structure (Fig. 5.3) with the anomaly trends.

5.3.1.1.1 Gravity Anomaly Map

The gravity anomaly map of the Cauvery basin prepared using the Bouguer anomalies onshore (Verma et al., 1993) and the satellite-derived free-air anomalies offshore is presented in Fig. 5.4. As can be seen, the map shows excellent correlation of various gravity highs and lows of the order of 30–40 mGal with the ridge-depression features mapped from the seismic data. In addition, the offshore region shows continuation of some of these features. In the shelf region, the shelf-edge gravity anomaly can be clearly noticed. Large negative Bouguer anomalies are noticed in the crystalline part of the Indian shield crust in the western part of the map.

5.3.1.1.2 Basement Configuration

The total sediment thickness map (Fig. 5.5) prepared from the seismic data reveals various ridges and depressions discussed above (Rangaraju et al., 1993) and shows that the sediment fill in the basin is highly differential with thickness varying from 1 to 6 km. The thickness of sediments may exceed 6 km in the offshore area east of Pondicherry, where the basin opens up into deep oceanic region. However, presence of higher thickness of sediments in the offshore part can not be ruled out. A brief description of subsurface geometry of various ridges and depressions (Figs. 5.6 and 5.7) compiled from several publications (Chandra et al., 1991; Prabhakar and Zutshi, 1993; Rangaraju et al., 1993; Sahu, 2008; Sastri et al., 1973) is provided here.

5.3.1.1.2.1 Pulicat Depression This depression covers the central part of Palar basin and is separated from the Nayudupeta high by a NE–SW trending fault. It shallows toward southwest flank and appears to be less intensely faulted than the northwest flank. It consists of nearly 3.0 km of sediments belonging to Early Permian and younger.

5.3.1.1.2.2 Ariyalur–Pondicherry Depression This NE–SW trending basin is bounded in the west and northwest by the granitic outcrop of the Indian shield and by Kumbakonam–Madanam ridge toward southeast. The basement is steeper in the northern side and more than 6 km of sediments ranging in age from Late Jurassic to recent are present in the basin. The subsurface lithology suggests that the basin was exposed to open marine conditions for a substantial period of time during its evolution.

5.3.1.1.2.3 Kumbakonam–Madanam Ridge This ridge is the subsurface extension of Precambrian crystalline rocks exposed in the SW of Ariyalur–Pondicherry depression with 1.5–2.0 km of sediments overlying the ridge. Northwestward extension of this ridge is a large domal feature located in the offshore called the Porto-Novo ridge.

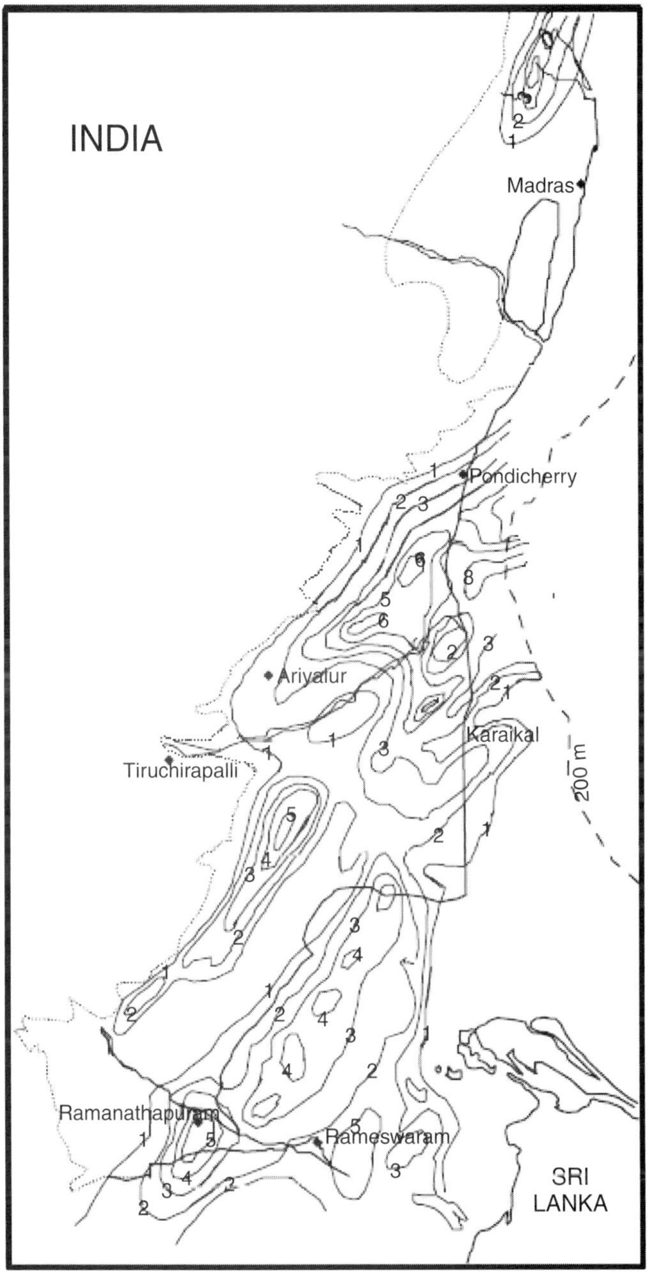

FIGURE 5.5 Sediment fill map (in km) of the Cauvery–Palar basin *(adapted from Rangaraju et al., 1993).*

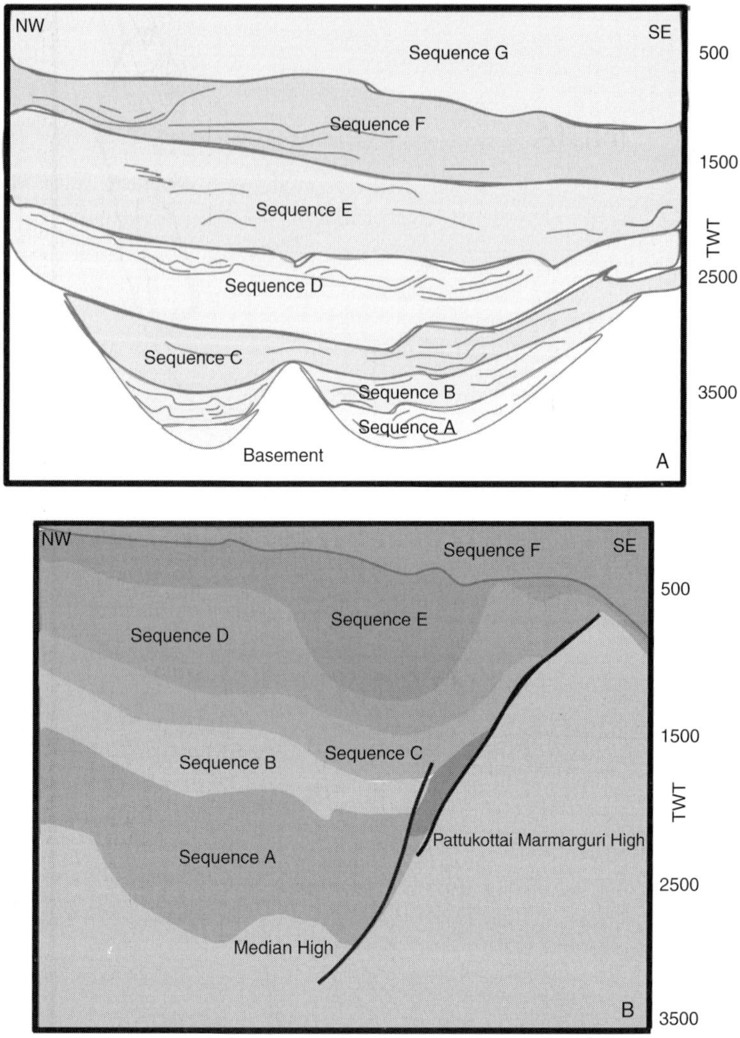

FIGURE 5.6 Seismogeological sections showing various depositional sequences across different ridges and depressions in the Cauvery–Palar basin *(adapted from Sahu, 2008)*. (For color version of this figure, the reader is referred to the online version of this chapter.)

5.3.1.1.2.4 Thanjavur Depression The Thanjavur depression delimited by the Precambrian granitic rocks in the west represents the NE–SW trending half-graben. The basin contains sediments characterized by six depositional sequences ranging in age from Pre-Albian to Recent with lowermost sequences faulted against the crystalline basement. The geometry of the basin reverses toward the Tranquebar depression, which is due to the presence of N–S trending Mayavaram–Nannilam saddles between them.

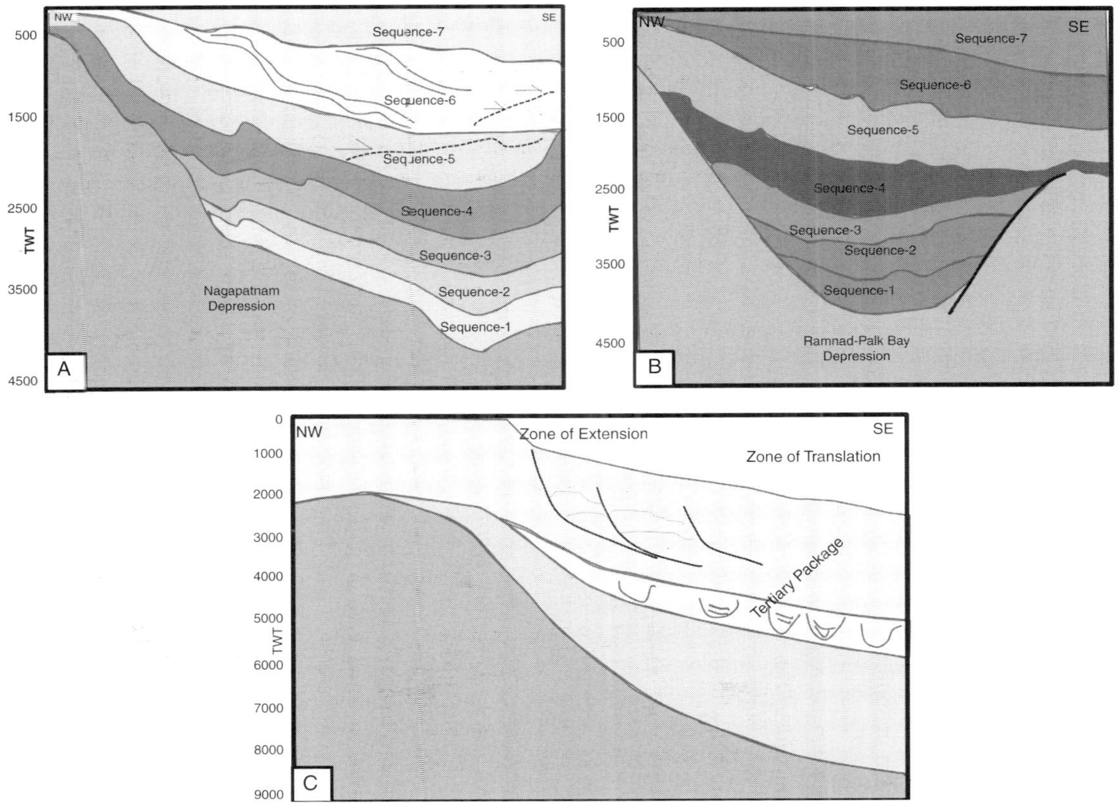

FIGURE 5.7 Seismogeological sections showing various depositional sequences across different ridges and depressions in the Cauvery–Palar basin *(adapted from Sahu, 2008)*. (For color version of this figure, the reader is referred to the online version of this chapter.)

5.3.1.1.2.5 Tranquebar Depression This NE–SW trending depression, flanked by Kumbakonam–Madanam ridge in the northwest and Mannargudi–Karaikal ridge in the southwest, covers both onshore and offshore and constitutes sedimentary sequences ranging in age from Pre-Albian to Recent. More than 6 km of sediments are present in this basin, of which the Tertiary sediments are about half. From southwest to northeast, the basin changes polarity (Sahu, 1995) with the presence of a median ridge as an intra-grabenal high.

5.3.1.1.2.6 Karaikal Ridge This small linear ridge lies in both onshore and offshore. Depth to the basement ranges from 1.0 km in northeast to about 2.5 km in southwest.

5.3.1.1.2.7 Nagapattinam Depression This basin extending offshore is delimited to the west by Mannargudi ridge, in the north by Karaikal ridge, and in southeast by the Vedaranniyam ridge. Six seismic sequences are observed with the maximum depth to the basement of about 5 km.

5.3.1.1.2.8 Mannargudi Ridge This NE–SW trending ridge running along the Palk Bay coast separates the Thanjavur and Palk Bay depressions. The ridge is overlain by sediments mostly of Tertiary age and depth to the basement increases gradually toward northeast up to more than 1.0 km.

5.3.1.1.2.9 Palk Bay Depression Palk Bay is a shallow water sub-basin with the maximum water depth of 20 m in the southern part of the Cauvery basin occupying an area of 7600 km^2 extending from Vedaranniyam in northeast to near Ramnad in southeast. The western side of the Bay comprises alluvial covered coastal part of Tamil Nadu bordered by Miocene, Early Cretaceous/Jurassic, and Precambrian exposures. Further, it is delimited in the northwest by Mannargudi ridge, southeast by the Mandapam–Delft ridge, and northeast by the Vedaranniyam terrace. The basin comprises nearly 5-km thick sediments characterized by seven seismic sequences, out of which nearly 2 km sediments are of Tertiary age. Chandra et al. (1991) identified several basement lows in this sub-basin, and the most prominent among them are Devipattinam and Uppurchattram in the south, Mimisal in the central, and Mullipalam in the northeast Palk Bay. The Pamban depression located in Sri Lankan waters is separated from Palk Bay by Mandapam–Delft ridge.

5.3.1.1.2.10 Vedaranniyam Ridge It is located toward southeast of Nagapattinam depression mostly in offshore. This terrace-like feature with nearly 2.5 km of Tertiary sediments is the subsurface extension of Sri Lankan basement massif toward north.

5.3.1.1.2.11 Mannar Depression This is an offshore basin lying in the Gulf of Mannar and opens into the Indian Ocean in the south. The Madurai–Rameswaram cross-trend delimits the basin from the Palk Bay in northeast. A total of six sedimentary sequences have been identified and a drilled well in this basin encountered basaltic rocks below Upper Cretaceous sediments.

5.3.1.2 *Depositional History*

In order to understand the basin development as well as for constructing the depositional history, all available stratigraphic information, such as the lithology, thickness of seismic sequences, log characteristics, and depositional environment, obtained from seismic and well data in various sub-basins described above are considered. The tectonics operative in different times in the geological past in different parts of the basin should also be considered as an important factor, as it is known to be influencing the spatiotemporal variation of stratigraphic units and stratigraphic evolution (Allen and Allen, 1991; Dickinson, 1974). Another important factor in basin evolution is the position of paleoshelf edge in different geological times related to various transgressive and regressive phases. Figure 5.8 shows the paleoshelf edges identified for the Cauvery basin by previous workers (Kalyanasundaram and Vijayalakshmi, 1991; Venkatarengan, 1987).

5.3.1.2.1 Generalized Stratigraphy

The surface geological mapping and subsurface information obtained from the seismic as well as lithologic data suggest that sedimentary formations belonging to Permian to Holocene interrupted by several major hiatuses are present in the basin. Though some of the older sequences are exposed at places along the shoulders of these basins, the subsurface sedimentary sections are not completely documented in the surface exposures. The generalized stratigraphy of the Cauvery and Palar basins is given in Tables 5.1 and 5.2 (after Rangaraju et al., 1993). However, unlike the Palar basin, about which only limited subsurface information (shallow borehole and seismic) is available, huge amount of seismic and well data acquired by ONGC as well as many other private companies and the Directorate General of Hydrocarbons (DGHs) in the Cauvery basin (both onshore and offshore) made the detailed description of lithostratigraphy of this basin possible (after Raju et al., 2005; Sahu, 2008; Venkatarengan et al., 1993). Some salient details from Sahu (2008) are briefly presented below.

5.3.1.2.1.1 Cauvery Basin

- **Precambrian Crystalline Complex**
  Eastern Ghat group of rocks comprising charnockites, quartzites, granitic gneisses, and schists form the basement supporting overlying sediments in the Cauvery basin. The drilling through the crystalline rocks at many places confirmed weathered and fractured nature of the basement.

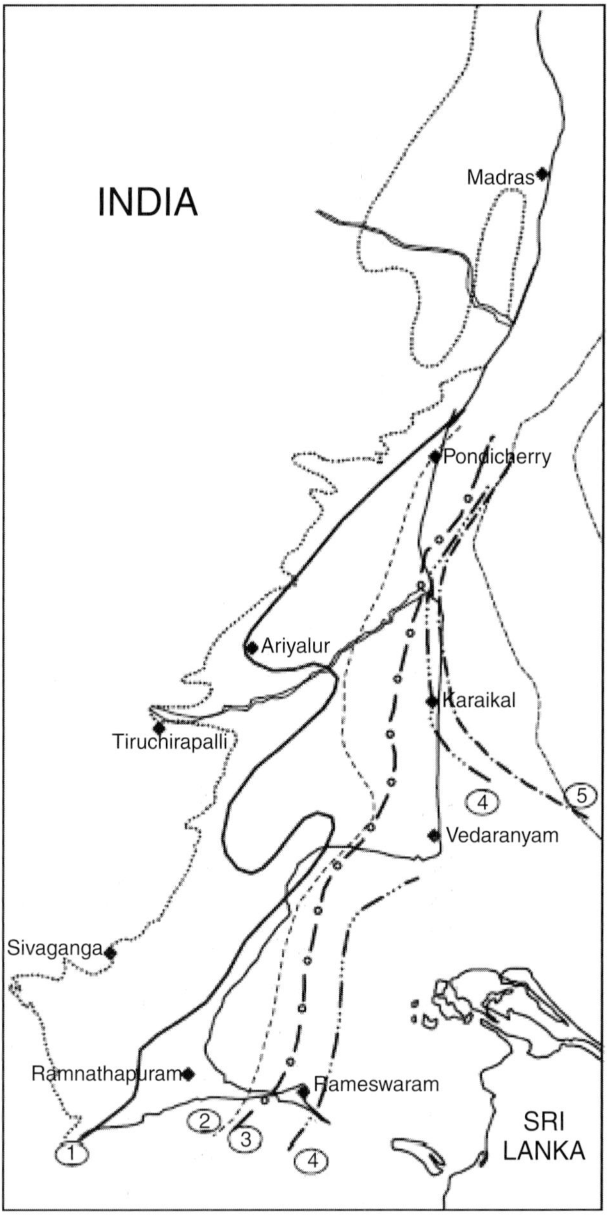

FIGURE 5.8 Map showing the Paleoshelf edges in the Cauvery basin *(adapted from Rangaraju et al., 1993).*

TABLE 5.1 Generalized Stratigraphic Chart for the Cauvery Basin (adapted from Rangaraju et al., 1993).

Age	Outcropping Formation	Gross Lithology and Facies (and Max. Thickness)	
		Outcrops	*Subsurface*
Pleistocene–Recent	Cuddalore Formation	Alluvial sand, clays, and ferruginous grits (100 m)	
Pliocene–Middle Miocene		Deltaic sandstone and kaolinitic claystones with abundant plant fossils in outcrops. Sandstone strata; pebbly/conglomeratic and ferruginous, with thick lignite beds (600 m)	Sandstones, shales, and minor limestone: deltaic to marine (1400 m)
Early Miocene–Middle Eocene	Pondicherry Formation	Not exposed	Sandstones: deltaic and deepwater fans, shales: deltaic and pelagic, and sheital limestone (2200 m)
		Variegated sandy clays and fossiliferous limestone: shallow marine (100 m)	Sandstone and shales: mainly deepwater, occasionally shallow/marginal marines
Early Eocene–Paleocene	Niniyur Formation	Predominantly algal lst. (boundstone) and minor sandy palocypod packstone–wackestone: shallow marine (200 m)	Limestone in limited areas of shell and slope
Maastrichtian–Campanian	Ariyalur Formation	Kallamedu sandstone (Maastritchian): predominantly sandstone with ferruginous claystone bed at the top. Underlain by Kallankunchchi limestone (Masstritchian): grainstone to wackestone, with Orbitoides Beds: lower part aranaceous Sillakudi sandstone (Campanian–Early Maastritchian): sandstone of various composition, from arkose to Orthoquartzite, with occasional bands of claystone and shell limestone (1100 m)	Shale and sandstone, predominantly deepwater. Shallow marine sediments in the western part of the basin (1500 m)

Continued

TABLE 5.1 Generalized Stratigraphic Chart for the Cauvery Basin (adapted from Rangaraju et al., 1993).—Cont'd

Age	Outcropping Formation	Gross Lithology and Facies (and Max. Thickness)	
		Outcrops	*Subsurface*
Santonian–Late Coniacian	Trichinopoly Formation	Basal calc. conglomerate with clumps of corals, ferruginous oolites, and sandy stromatolites, overlain by alternating beds of claystone and cals. pebbly sandstone. Marlstone bands with ammonite concretions are also present (1300 m)	
Turonian–Late Aptian	Uttattur Formation	Paravay sandstone (Turonian/Early Coniacian) occurs predominantly as calc. sandstones with bands of variegated claystone and molluscan limestone. Underlain by Karai shale [Aptian–Early Turonian]: basal/interfingering coralline and reefoidal lst. [Dalmiapuram/Kallakudi limestone: mid-Albian] overlain by gray. Calcareous, gypsiferous, and pyritiferous shales with phosphatic nodules. Uppermost part siltstone and sandstone (1100 m)	Sandstone, shale, conglomerate, and occasional limestone: marginal marine to neritic (1500 m)
Early Cretaceous [Early Aptian–Late Jurassic]	Sivaganga/Therani Formation (syn-rift sequence)	Gritty and pebbly sst., shales. Conglomerate with plant fossils: nonmarine (380 m) Aptain/Hauterivian in outcrops	Pebbly sst., conglomerate and shale: nonmarine to marginal marine (>1000 m)
Precambrian	Dharwarian shield	Crystalline rocks	Crystalline rocks

TABLE 5.2 Generalized Stratigraphic Chart for the Palar Basin (adapted from Rangaraju et al., 1993).

Age	Formation	Gross Lithology	Environment
Pleistocene–Recent	Alluvium	Alluvial sands, silts, and silty clays	Continental
	Laterite	Laterite	
	Conjeevaram gravels (50–100 m)	Pebbly sandstone and gravel	Continental
Mio-Pliocene	Cuddalore Formation (100–150 m)	Pebbly sandstone with abundant plant fossils in outcrops	Continental to marine
Late Cretaceous–Paleogene	Undefined Paleogene (400 m)	Black claystone and bands of thin sandstone (on the basis of borehole data)	Marine
Postrift sequence			
	Late Cretaceous: Unknown, requires confirmation in the subsurface by deep drilling. Likely to be present, analogous to the Krishna–Godavari and Cauvery basins		Marine
Early Cretaceous (Aptain)	Satyavedu Formation (300–2000 m)	Sandstone, ferruginous, purple with plant fossils	Continental
Early Cretaceous (Neocomian)	Sriperumbudur Formation (600–1100 m)	Shale and feldspathic sandstone, with plant fossils and coal streaks or lenses	Lacustrine with marine intercalations
Syn-rift sequence			
Late Jurassic	Likely to be present. Requires confirmation in the subsurface by deep drilling. Analogous to the Krishna–Godavari and Cauvery basins		Continental
Early Permian	Ongur Formation (equivalent of Talchir Fm) (Lower Gondwana (300 m)	Boulder beds and splintery gray/greenish gray shales	Fluvioglacial
Precambrian	Eastern Ghats	Crystalline rocks	

- **Late Jurassic–Early Cretaceous (Upper Gondwana Group)**
  These rocks of Upper Gondwana Group overlie the crystalline basement and are exposed along the margin of the basin. This unit of rocks constituting of medium to coarse grained quartzose and feldspathic sandstone, at places conglomeratic and overlain by white to grayish shales, is represented by Therani Formation in the northern part and Sivaganga Formation in the southern part of the basin. Fossil assemblages have been reported from these rocks (Datta and Bedi, 1968), and in particular, the Therani Formation contains index Gondwana plant fossils. These have been deposited under fluvial-deltaic to brackish water greatly influenced by influx of freshwater (Datta and Bedi, 1968).

- **Early Cretaceous (Uttattur Group)**
  These formations overlie the older Gondwana rocks and basement granites and gneisses. In the outcrops, this group of rocks, dominantly carbonate in the north becoming arenaceous toward the south, is made up of Kalakundi, Karai Shale, and Maruvathur Clay formations. In the subsurface, these rocks are represented by the Andimadam, Sattapadi, and Bhuvanagiri formations.

5.3.1.2.1.1.1 Andimadam Formation The rocks comprise pale gray, fine to coarse grained, micaceous sandstone and micaceous silty shale and at places conglomeratic. The formation is developed in most of the depressions, such as the Ariyalur–Pondicherry, Tranquebar, Thanjavur, and the Ramnad, having the lower boundary marked by the basement rocks and the upper boundary defined by an argillaceous section. On the basis of lithology and fossil assemblages, these are interpreted to have been deposited under fluviatile to lacustrine to shallow marine environment.

5.3.1.2.1.1.2 Sattapadi Shale This formation belonging to Albian–Cenomanian age is mainly composed of silty shale and thin calcareous sandstone. This formation is widely distributed in the basin, but absent in the southeastern part. The arenaceous facies of the Bhuvanagiri Formation marks its upper contact. The environment of deposition is inferred to be middle to outer shelf. This shale sequence is one of the important source rock facies for hydrocarbon generation in the basin.

5.3.1.2.1.1.3 Bhuvanagiri Formation The formation is of Cenomanian–Turonian age and predominantly comprises sandstone with minor claystone and shale. It is developed mostly in the northern and central parts of the basin. The equivalent rocks in the outcrops are represented by an argillaceous unit belonging to upper part of Karai shale/Maruvattur clay. It is inferred to have been deposited in middle shelf to upper bathyal environment.

5.3.1.2.1.1.4 Palk Bay Formation The occurrence of this formation is restricted to the Palk Bay. The lithology is dominantly calcareous sandstone with a few bands of sandy claystone. The depositional environment is inferred to be shallow marine in a fan delta setting.

- **Late Cretaceous (Ariyalur Group)**
  The sediments in the outcrops are classified under two groups, namely, the Trichinopoly and Ariyalur groups, and consist of sandstones and limestone formations.

5.3.1.2.1.1.5 Kudavasal Shale Formation It is equivalent to Sillakudi Formation in the outcrop and present in most areas of the basin except in the Nagapattinam depression. The formation consists of shale/calcareous silty shale with occasional calcareous sandstone bands. It is assigned an age of Coniacian to Santonian and interpreted to be deposited under transgressive sedimentary environment.

5.3.1.2.1.1.6 Nannilam Formation This formation ranges in age from Santonian to Lower Maastrichtian and consists of alternations of shale, calcareous silty shales, and occasional calcareous sandstones. It is conformably overlain and underlain by the Porto-Novo and Kudavasal formations, respectively. It is inferred to be deposited under outer shelf to upper bathyal environment.

5.3.1.2.1.1.7 Porto-Novo Shale It is predominantly developed in the northern part of the Ariyalur–Pondicherry depression west of Karaikal Ridge in the Porto-Novo offshore area as well as in Palk Bay depression. It consists of rocks having predominantly argillaceous with minor siltstone. The age of the formation is Campanion to Maastrichtian deposited under outer shelf to upper bathyal environment.

5.3.1.2.1.1.8 Komarakshi Shale The formation has developed toward the eastern part of the basin. It unconformably overlies the Bhuvanagiri or Palk Bay Formations, and underlies the Karaikal Shale or Kamalapuram Formations. The rocks consist mainly of calcareous silty shale. The age of the formation is Coniacian to Maastrichtian, and the formation is deposited under outer shelf to upper bathyal environment.

- **Tertiary (Nagore and Narimanam Groups)**
  A complete sequence of Tertiary sediments is encountered in the subsurface, whereas the exposed rocks are represented by the Niniyur Formation of Paleocene age and the Cuddalore Sandstone of Mio-Pliocene age. The Tertiary section in the subsurface is considerably thick and is subdivided into two groups: the lower part is named as the Nagore Group and the upper part, the Narimanam Group, and both are separated by an unconformity.

5.3.1.2.1.1.9 Nagore Group The formations of this group overlie the Ariyalur Group. The four formations recognized in this group are described below,

5.3.1.2.1.1.10 Kamalapuram Formation This formation ranging in age from Maastrichtian to Middle Eocene is present in almost entire basin particularly with two major depocenters in the Ariyalur–Pondicherry depression and Karaikal area and consists of shaly sandstones and shales. The Porto-Novo–Komarakshi Shale unconformably underlies the formation, whereas the overlying Karaikal Shale has conformable contact.

5.3.1.2.1.1.11 Karaikal Shale This formation comprises shales that are occasionally calcareous/pyritic. It conformably overlies the Kamalapuram

Formation. It is predominantly present north of Karaikal and near Mandapam extending toward the Gulf of Mannar. The age of the formation ranges from Paleocene to Eocene, and the formation is deposited in middle to outer shelf.

5.3.1.2.1.1.12 Pandanallur Formation It has a limited areal extension and is restricted to Bhuvanagiri area. It consists of claystone sandstone, deposited in middle shelf environment. The age of the formation is Lower Eocene.

5.3.1.2.1.1.13 Tiruppundi Formation The formation is present in large portion of the basin with major depocenters: the Pondicherry offshore, Nagapattinam depression, and south of Palk Bay depression. The formation comprises limestone, siltstone, and sandstone ranging in age from Middle Eocene to Early Miocene. They are deposited under inner to middle shelf condition.

5.3.1.2.1.1.14 Narimanam Group It is the youngest sedimentary sequence and lies unconformably over the Nagore group. The rocks comprised sandstone, clay/claystone, and limestone. This group is subdivided into eight formations.

5.3.1.2.1.1.15 Niravi Formation It is confined to the eastern part of the basin mainly covering Ariyalur–Pondicherry, Tranquebar, western part of Nagapattinam depressions, and southern flank of Karaikal ridge. The formation unconformably overlies the Tiruppundi Formation/Karaikal Shale. The rocks consist of gray colored, fine- to medium-grained, calcareous sandstone unit with intermittent shale beds and range in age from Late Eocene to Oligocene.

5.3.1.2.1.1.16 Kovilkalappal Formation It occurs mainly in Thanjavur and Nagapattinam depressions and overlies the Niravi Formation. It is of Upper Eocene age and argillaceous in nature with significant occurrence of limestone.

5.3.1.2.1.1.17 Shiyali Claystone Formation It is observed to occur in Madanam and Karaikal area. The age of the formation ranges from Oligocene to Lower Miocene, and the formation comprises claystone with minor occurrence of siltstone and sandstone deposited under inner shelf environment.

5.3.1.2.1.1.18 Vanjiyur Sandstone Formation The formation has limited areal extent, which occurs over Karaikal ridge and in Tranquebar depression. It is predominantly arenaceous in character and comprises dark gray, calcareous sandstone and siltstone deposited under inner to middle shelf environment. The age of this formation ranges between Upper Oligocene and Lower Miocene.

5.3.1.2.1.1.19 Tirutaraipundi Sandstone Formation The formation is present in the southern part of the Nagapattinam depression and extends toward the Palk Bay. It comprises mainly sandstones with minor limestone layers in between and deposited in inner to middle shelf environment. The age of this formation ranges between Oligocene and Early Miocene.

5.3.1.2.1.1.20 Madanam Limestone Formation The formation is unconformably underlain by the Tiruturaipundi Sandstone and at places by Vanjiyur Sandstone. It comprises mainly limestone with minor silty clays deposited in shallow inner neritic conditions. This formation is assigned Early Miocene age.

5.3.1.2.1.1.21 Vedaranniyam Limestone Formation The formation occurs only in the southeastern part of the basin. It consists of predominantly coral limestone and minor grainstone deposited in shallow inner shelf regime. The age of this formation is Middle Eocene to Lower Miocene.

5.3.1.2.1.1.22 Tittacheri Formation The formation is present in a large part of the basin. It grades into the Cuddalore Sandstone Formation near the outcrops. It consists of unconsolidated gravelly sandstones and earthy clays and is inferred to be deposited in shallow inner shelf conditions in the onshore. The age of the formation is Lower Miocene to Pliocene.

- **Major Unconformity Surfaces**

  For a proper reconstruction of geological history of any basin, the knowledge of gaps in stratigraphic record is very important. According to Vail et al. (1977), a hiatus is the total interval of geological time that is not represented by strata at a specific position along a stratigraphic surface. The stratigraphic succession of the basin described above is interrupted or separated by major hiatuses and reported by several previous workers mostly on the basis of subsurface sections (Basavaraju and Govindan, 1997; Govindan and Ravindran, 1996; Raju et al., 1994). A description of these hiatuses is summarized in Raju and Ramesh (2005) and Sahu (2008) and a brief excerpt is provided below:

  > Basavaraju and Govindan (1997) reported Early Permian sediments in the Thanjavur basin, where Triassic and Jurassic sediments were not found. This is referred to as Permo-Triassic–Jurassic hiatus. During Turonian, a major hiatus in the subsurface is observed in different wells (Govindan and Ravindran, 1996), and in the outcrops, a minor hiatus equivalent to late Turonian is observed in the basin. At the Cretaceous Tertiary Boundary, a major hiatus of varying time span of <1 to >30 Ma is documented in several parts of the Cauvery basin by Raju et al. (1994). The top of Ypresian (Early Eocene) is characterized by a hiatus having considerable duration in parts of the basin mainly over Mannargudi and Pattukottai structures. Towards the top of Thirupundian (Middle Miocene) another hiatus of varying magnitude is recognized in Tranquebar, Nagapattinam depressions, and Karaikal ridge.

5.3.1.2.1.2 Palar Basin The NE–SW trending basement structure is composed of Archean igneous and metamorphic complex. The poorly exposed sedimentary rocks are mostly seen in outcrops toward the northeastern margin of the basin. The oldest sedimentary rocks consist of boulder beds and greenish shales, which are fluvioglacial deposits of Early Permian age (Venkatachala and Rawat, 1972) and are correlative of Lower Gondwana Talchir Formation (Murthy and Ahmad, 1970; Sastri et al., 1973). According to Sastri et al. (1973), the Early Cretaceous Sriperumbudur beds include a group nearly 600-m thick of splintery, gray and greenish shales containing dark gray gypseous clay and interbedded sandstone, ironstone, and limestone deposited in shallow, brackish water, probably close to the shoreline, and the Satyavedu formation having nearly 2000-m thickness transgressively overlaps the former. The rocks consist of boulder conglomerates interbedded with sandstones deposited under littoral to nearshore fluviatile conditions. The Satyavedu formation is unconformably overlain by a 400-m sequence of black claystone and thin sandstone bands and inferred to be of Paleogene. These Paleogene rocks are overlain by the Miocene–Pliocene Cuddalore Sandstone consisting of conglomeratic and clayey sandstones. They are overlain by the Kanjeevaram Gravel and Pleistocene laterites and conglomerates, which are interpreted to be fluviatile deposits. The overlying Holocene alluvial sands and clays are 45–50 m thick. On the basis of available surface and subsurface information, the Triassic sediments appear to be absent in the Palar basin (Rangaraju et al., 1993).

5.3.1.2.2 Tectonostratigraphy and Structural Evolution of Cauvery–Palar Basin

Detailed seismic data coupled with subsurface lithological information and unconformity surfaces obtained from well data in various ridges and depressions were used by Sahu (2008) to identify several seismic sequences as well as interpret the individual sedimentary packages in terms of their age and depositional environment. We summarize here some salient features of these depositional sequences from the individual depressions in Tables 5.3–5.7. On the basis of this vast compilation, Sahu (2008) defined a total of seven tectonostratigraphic units such as (i) Andimadam, (ii) Sattapadi, (iii) Bhuvanagiri, (iv) Nannilam, (v) Kamalapuram, (vi) Niravi, and (vii) Cuddalore. Detailed description and analysis of these identified units are beyond the scope of this book and can be found in Sahu (2008). We present here some typical seismogeological sections (Figs. 5.9 and 5.10) depicting drastic vertical as well as horizontal lithofacies variations in the basin (Venkatarengan et al., 1993) and control of major transgressive and regressive cycles on the depositional sequences in the Cauvery basin (Sahu, 2008). Rangaraju et al. (1993) presented a detailed synthesis of sedimentation and stratigraphy of Cauvery and Palar basins together and proposed that the entire sedimentary succession called the "East Coast Supergroup" can be subdivided into two broader tectonostratigraphic sequences: (i) the Late Jurassic–Early Cretaceous syn-rift mega sequence comprising of nonmarine units and (ii) Early Cretaceous–Cenozoic postrift mega sequence, comprising marine sediments. While both these sequences directly rest on crystalline basement in the Cauvery basin, in Palar basin, they are underlain by the Lower Permian fluvioglacial formations. On the basis of evidences on subsurface paleontological data and stratigraphic position of volcanic flows encountered in the Gulf of Mannar (Kalyanasundaram and Vijayalakshmi, 1991), Rangaraju et al. (1993) identified the breakup unconformity at Early Aptian age. The present structural styles in the Cauvery and Palar basins discussed earlier came into existence during rifting. One

TABLE 5.3 Tanjore Depression (Six Depositional Sequences) (after Sahu, 2008).

	Age	Thickness (m)	Lithology	Seismic Character	Depositional Environment
Sequence-1	Pre-Albian	>2000	Predominantly arenaceous at the flanks and argillaceous toward the deeper part	Weak and discontinuous reflection patterns	Alluvial fans/cones
Sequence-2	Albian	>600	Calcareous sandstone, shale, and minor limestone/dolostone	Moderate to strong and continuous	Fluvial and/or lacustrinal
Sequence-3	Albian to Cenomanian	>500	Calcareous sandstone, shale, and silty shale		Fluvial and deltaic mode under marine embayment condition
Sequence-4	Turonian to Campanian and at places to Maastrichtian	>1000	Claystone, silty shale, and sandstone	Moderate to good	Deltaic to open marine
Sequence-5	Paleocene to Eocene	<300	Sandstone, shale, siltstone, and silty sandstone	Strong, divergent, sub-parallel and discontinuous	Deltaic and offshore longitudinal bars
Sequence-6	Eocene to Recent	>200	Sandstone, coal, claystone, and limestone	Strong, parallel to sub-parallel, continuous and discontinuous at places	Inner shelf deltaic, continental to lagoonal

TABLE 5.4 Tranquebar Depression (Seven Depositional Sequences) (after Sahu, 2008).

	Age	Thickness (m)	Lithology	Seismic Character	Depositional Environment
Sequence-1	Albian to Pre-Albian	>1500	Sandstone and shale	Weak and discontinuous reflection patterns	Alluvial fans/cones
Sequence-2	Albian to Cenomanian	>800	Dominantly shale with subordinate sandstone	High-amplitude, strong, and continuous reflection pattern	Lacustrinal to marginal marine setup
Sequence-3	Turonian	<600	Coarse to medium-grained sandstone, pebbly at places, silty shale, and shale	Parallel and continuous at the upper part, chaotic at the lower part	The regressive regime being confined to the incised valley
Sequence-4	Campanian to Maastrichtian	>1000	Shale, silty shale, sandstone, and limestone	Strong, divergent and parallel at lower part and upper part, respectively	Transgressive phase followed by regression toward the top
Sequence-5	Paleocene to Eocene	>600	Sandstone, shale, and limestone	Continuous and parallel at the upper part, and chaotic and discontinuous at the lower part	Topographically rugged terrain lower part
Sequence-6	Eocene to Oligocene	>500	Sandstone, shale, claystone, and limestone		
Sequence-7	Miocene to Recent	>1200	Sands, clay, and limestones	Changes from strong, parallel at places to opaque and poor	Canyons and stacking channels in offshore area

TABLE 5.5 Nagapattinam Depression (Seven Depositional Sequences) (after Sahu, 2008).

	Age	Thickness (m)	Lithology	Seismic Character	Depositional Environment
Sequence-1	Albian	<2000	Coarse detritus of rift fill		Predominantly nonmarine and may include fan delta/alluvial fan
Sequence-2	Albian to Cenomanian	1200	Micaceous feldspathic coarse grained, occasionally pebbly sandstone with carbonaceous shale and minor coal	Parallel and continuous reflection with variable amplitude and abruptly cut at the upper part	Lacustrinal to marine
Sequence-3	Cenomanian to Turonian	600	Gray micaceous shale siltstone and sandstone	Discontinuous, divergent, and chaotic nature at places	Mostly nonmarine
Sequence-4	Coniacian to Maastrichtian	1200	Shale, sandstone, and limestone	Continuous, parallel and divergent at places	Transgressive setup that dwindles toward the top
Sequence-5	Paleocene to Eocene		Sandstone, shale, and limestone	Discontinuous and chaotic in nature at lower part whereas parallel and concordant at upper part, also show abrupt terminations	Shallow marine environment with deepwater clastic
Sequence-6	Eocene to Oligocene	1000	Sandstone, shale, claystone, and limestone	Divergent, continuous and variable amplitude at places	Fluvio-deltaic to shallow marine to barrier reef shoal
Sequence-7	Miocene to Recent	>700	Sands, clay, and limestone	Changes from strong, parallel at places to opaque and poor	

TABLE 5.6 Ramnad–Palk Depression (Seven Depositional Sequences) (after Sahu, 2008).

	Age	Thickness (m)	Lithology	Seismic Character	Depositional Environment
Sequence-1		>600	Pebbly coarse to medium-grained sandstone and subordinate shale	Discontinuous and chaotic	Alluvial fan/ cones
Sequence-2	Albian			Downlap relation at the upper surface of the sequence-1 and the conformable sequence in the overlying sequence	Occasional marine transgression
Sequence-3	Cenomanian to Turonian	200	Sandstone, shale, and limestone	Downlap relation with underlying sequence and the onlap relation with the rising basement. They are discontinuous, moderate to strong with variable amplitudes	Shallow marine
Sequence-4	Turonian to Maastrichtian	500 (may exceed 800)	Dominantly sandstone and shale	Continuous and parallel attitude	Marine environment in shelfal setup
Sequence-5	Paleocene to Eocene	<400	Sandstone, shale, and limestone	Downlapping the upper surface of sequence-4	
Sequence-6	Eocene to Oligocene	1000	Sandstone, claystone, shale, and limestone	Good, continuous, and divergent	Progradational multiples
Sequence-7	Miocene to Recent	<350	Sands, clay, and limestone	Changes from strong, parallel at places to opaque and poor	

TABLE 5.7 Ariyalur–Pondicherry Depression (Seven Depositional Sequences) (after Sahu, 2008).

	Age	Thickness (m)	Lithology	Seismic Character	Depositional Environment
Sequence-1	Albian to Pre-Albian	500–1500	Dominantly sandstone and shale	Discontinuous and weak, chaotic to reflection-free zones are also observed and at places patchy reflectors	Alluvial fan/cones
Sequence-2	Albian to Ceromanian	>400	Shale with subordinate sandstone	Strong, continuous, and high-amplitude parallel reflection	Lacustrinal to marginal marine setup
Sequence-3	Turonian	>800	Coarse to medium-grained sandstone, pebbly at places, shale and silty shale	Chaotic at the lower part and continuous and parallel at the upper part	
Sequence-4	Coniacian to Maastrichtian	<1500	Shale, silty shale, sandstone, and limestone	Strong and divergent at the lower part, parallel at the upper part	Transgressive phase followed by regression
Sequence-5	Paleocene to Eocene	<600	Sandstone, shale, and limestone	Continuous and parallel at the upper part whereas discontinuous and chaotic at the lower part	
Sequence-6	Eocene to Oligocene	<500	Sandstone, claystone, shale, and limestone	Good, continuous, and divergent	Progradational multiple units toward southeast
Sequence-7	Miocene to Recent	700, 1700 (at deepwater area)	Sands, clay, and at places limestone especially on the flanks	Changes from strong, parallel at places to opaque and poor	

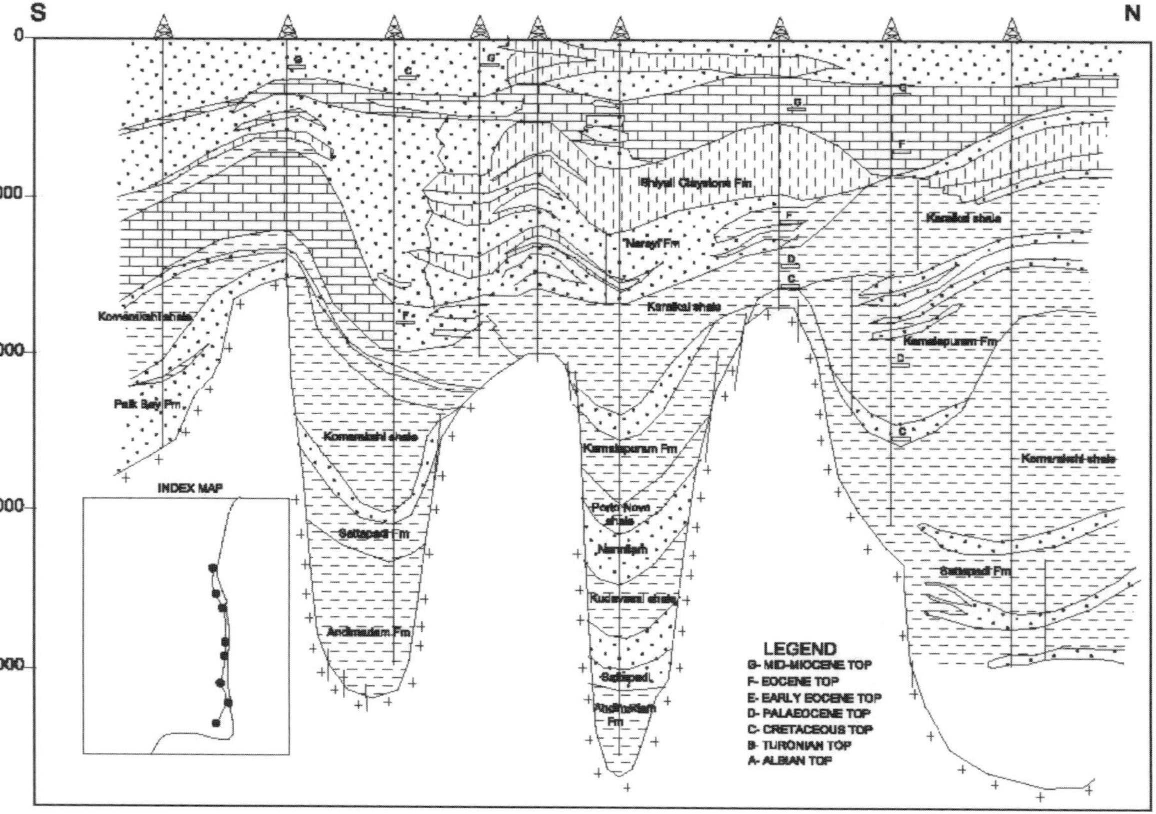

FIGURE 5.9 Cont'd

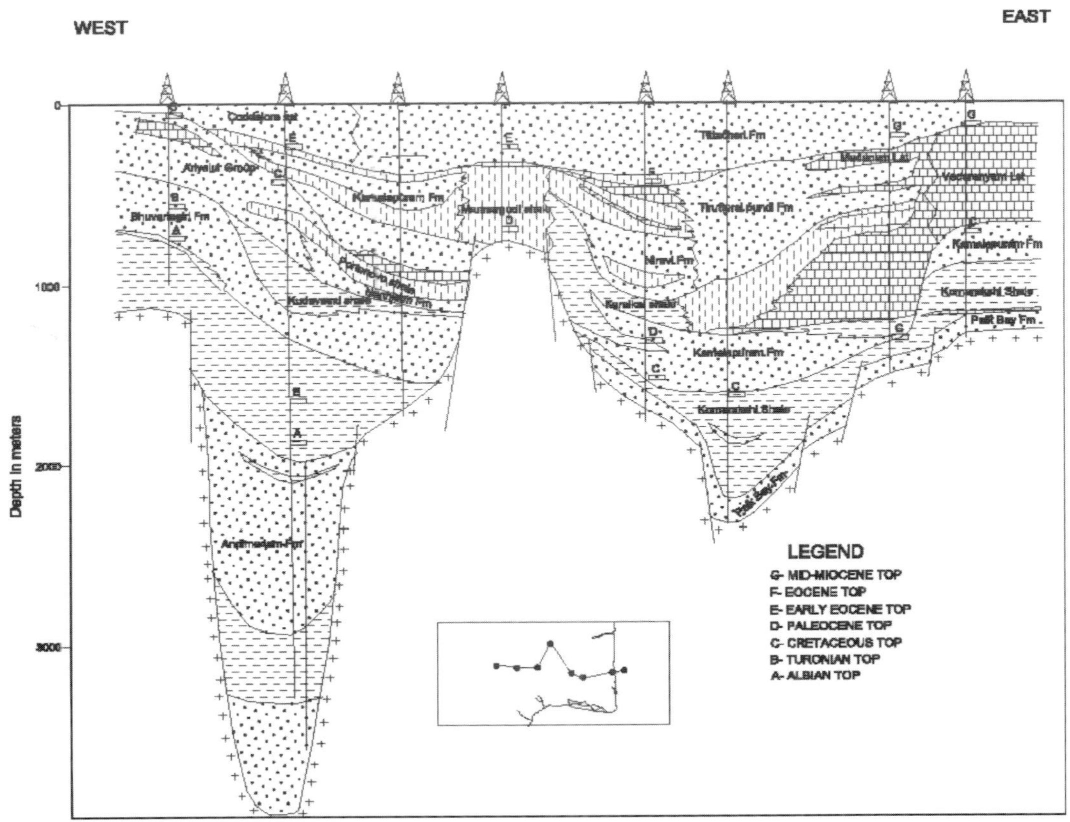

FIGURE 5.9 Lithostratigraphy sections (after Venkatarengan et al., 1993) along and across the Cauvery–Palar basin *(adapted from Sahu, 2008)*.

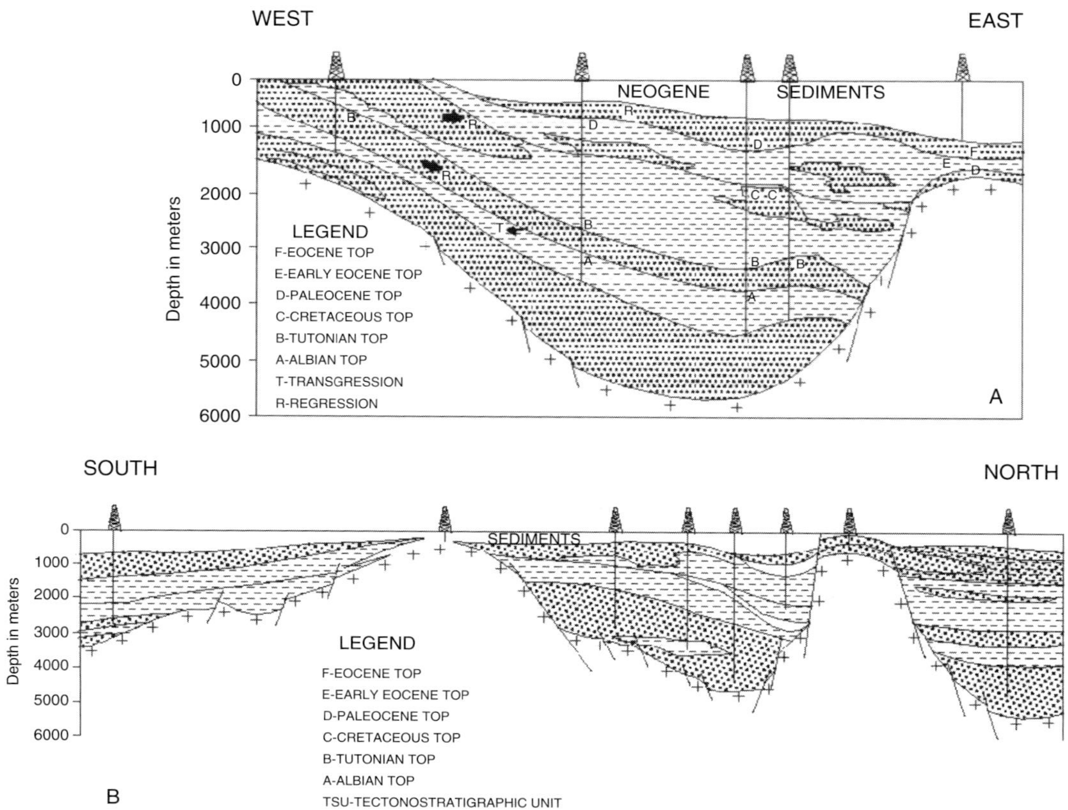

FIGURE 5.10 Tectonostratigraphy profiles showing horst–graben morphology as obtained from the well information (after Venkatarengan et al., 1993). (A) East–West; (B) North–South *(adapted from Sahu, 2008).*

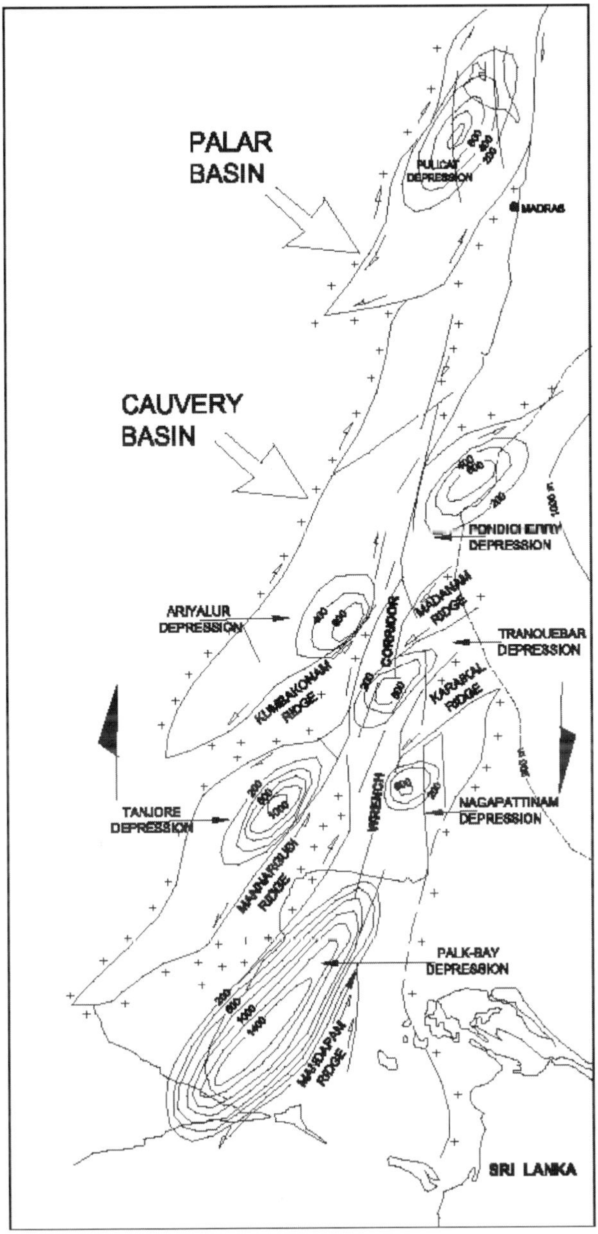

FIGURE 5.11 Schematic diagram showing the tectonic differentiation in the Cauvery and Palar basins. Contours represent the approximate thickness of syn-rift sediments in meters *(adapted from Rangaraju et al., 1993).*

interesting aspect to be noted at this stage is the development of a "wrench corridor" that correlates with the N–S saddle (Fig. 5.11) during rifting (Rangaraju et al., 1993) discussed previously. As this part of the eastern margin is believed to have evolved because of shearing motions as evidenced from the plate reconstruction models and other geophysical observations (Chand et al., 2001; Powell et al., 1988; Subrahmanyam et al., 1999), the observed structural differentiation in the Cauvery and Palar basins further corroborates such shear movements during rifting. However, in the southern part of the basin, the rifting failed to rupture the continental crust between India and Sri Lanka. The thickness of syn-rift sediments is controlled by the NE–SW trending structural geometry of ridges and depressions. The early post-rift sedimentation is controlled by the thermal sag phase dominantly from rift shoulders. During this phase, restricted marine source rocks are deposited. The postrift sediments show an overall thickening offshore. As the syn-rift sediment supply was not adequate to fill the depressions, the topography of antecedent rift controlled the sedimentation even after the extension ceased till the end of Cretaceous (Chaudhari et al., 2009). In fact, the syn-rift Andimadam and Palk Bay formations were terrestrial fans and coarse sand eroded from the higher elevated ridges got deposited in the intervening lows (Husain et al., 2000). As stated above, the rift–drift transition is marked by volcanic activity in the Gulf of Mannar. The basin fill map up to the top of sediments (Basement to Late Albian), thicker sediments are noticed in the Thanjavur, followed by Nagapattinam, Tranquebar, Ramnad, and Ariyalur–Pondicherry depressions (Chaudhari et al., 2009). The early thermal subsidence stage during Cenomanian to Maastrichtian caused basin-wide invasion of sea in many depressions and gave rise to the late Cretaceous sediments such as the Sattapadi and Porto-Novo, Kudavasal shales, and Bhuvanagiri and Nannilam Formations (Husain et al., 2000).

5.3.1.3 Petroleum System

The essential elements of a petroleum system include a petroleum source rock, reservoir rock, seal rock, and overburden rock, while the processes that are required for hydrocarbon occurrence are trap formation and generation, migration, and accumulation of hydrocarbons (Magoon and Dow, 1994). For a better understanding of the generation and migration of petroleum, it is necessary to integrate various aspects, such as sedimentation, burial history, and maturation, which occurred in the basin over geological time.

5.3.1.3.1 Source Rocks and Maturation

The source rock potential of the Cauvery basin has been evaluated from several drilled rock samples related to different sedimentary sequences by various workers (Balan et al., 1993; Chandra et al., 1991; Ramani et al., 2000; Thomas and Sharma, 1993; and many others). Though the proven hydrocarbon fields in the basin indicate the presence of a working hydrocarbon system, detailed analysis reveals that the source rock potential is different in the individual sub-basins. It is known that biological, tectonic, and physicochemical factors influence the deposition of organic matter, and based on the published information and the authors knowledge gathered through recent drilling campaign on source rock in Cauvery basin, the indication is that there is very good correlation of source rock potential with the tectonic evolution of the basins. Cauvery basin was formed during the late Jurassic–Early Cretaceous period with the rifting of the Precambrian basement into a series of horsts and grabens in response to crustal tensional tectonics. Three main tectonostratigraphic packages distinguished with respect to the source rock deposition in the Cauvery basin are as follows: (i) Within the syn-rift, deposition cycles are largely fluvial and are essentially controlled by episodic tectonic creation of accommodation, by extension, and consequent subsidence. In these systems, source rocks are deposited in the fluvio-lacustrine condition. Though very few wells have been drilled up to this section, oil biomarker studies indicate majority of the oil is being sourced from this interval. (ii) The syn-rift to post-rift transition represents a gradual change from lacustrine to deep marine facies. Between syn-rift unit and post-rift unit, there is a breakup unconformity and the source facies deposited in this sequence are dominated by fluvio-deltaic mixed with marine Type-II organofacies. (iii) In post-rift, the major accommodation for marine fill is created by the thermal sag and the later isostatic adjustments due to differential densities of the continental and oceanic crust. Basin restriction coupled with deposition during anoxic events has led to the deposition of over 90 m of organic-rich marine black shales within the Karai Clay Formation (Govindan, 1982, 1993; Tewari et al., 1996). The global anoxic event associated with the Albian–Cenomanian marine transgression, like in many other parts of the world, has pervaded the Cauvery Basin and favored development of good source rocks with Type-III kerogens, whereas the Coniacian–Campanian–Maastrichtian transgressive/regressive phase is identified to be relatively of lesser significance for development of good quality source rocks (Chandra et al., 1991). Apart from this, presence of low-oxygen-tolerant microfauna favoring good source rock corresponding to this event is further supported by the presence of source rock in Nagapattinam and Tranquebar depressions (Govindan, 1993).

Time of generation of hydrocarbon is mainly controlled by the paleo-heat flow related to the rifting, and the overburden thickness hence varies widely from basin to basin. The hydrocarbon generation in the Ariyalur and Tranquebar depression is believed to have started during early Paleocene; the same in Nagapattinam and Palk Bay depression started during Mid-Eocene time (Fig. 5.12). Well-developed fault system associated with the ridges can be considered as conduit for hydrocarbon migration.

5.3.1.3.2 Reservoir

The fluvio-deltaic rift sections in Cauvery basin are proven to be rich in reservoir facies. Coarser clastics have been reported both in Early Cretaceous (Palk

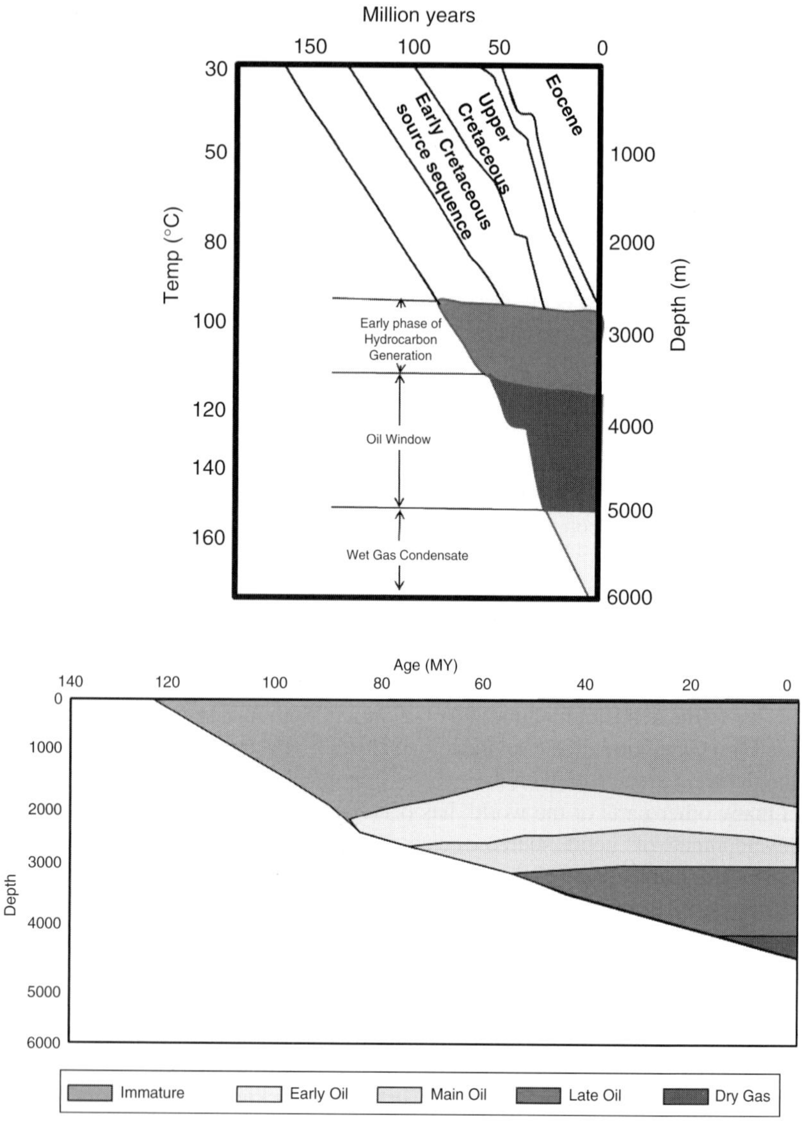

FIGURE 5.12 Hydrocarbon generation history (Kerogen-Type-3) in TKD-1 well, onshore Tranquebar sub-basin (top) and hydrocarbon generation history (Kerogen-Type-3) in PH-11-1 well, offshore Palk Bay sub-basin (bottom) *(adapted from Sahu, 2008)*. (For color version of this figure, the reader is referred to the online version of this chapter.)

Bay) and Late Cretaceous (Bhuvanagiri) with both of these having evidence of hydrocarbon accumulation. In the northern part of the basin, there is a deposition of prograding deltas, and depending on the sea-level variations, carbonate deposition is also reported in the shelf. The post-rift reservoirs are composed of high-density turbidites of Late Cretaceous to Miocene age and relatively smaller areal extent than pre-rift reservoirs. Important reservoir units are present between Kamalapuram shale and Niravi formation of Tertiary age characterized by alteration of sand and shale. The main hydrocarbon pool in post-rift sequence includes Narimanam, Kovikalappal, Vengidengal, Tiruvarur, and Nannilam sequences. Apart from conventional reservoirs, fractured granitic basement also has been proved to be good reservoir in many sub-basins.

5.3.1.3.3 Seal and Entrapment

In the onshore Cauvery basin, important seal facies resulted from the post-breakup are the Kudavasal and Kamalapuram shale and Kovikallapal and Madanam limestone, whereas in the offshore, widespread transgressive shales of Albian and Cenomanian age provide regional seal. For late Cretaceous reservoirs present in both onshore and offshore part of the basin, regional seal is provided by the Paleogene shale. Entrapment conditions vary widely in Cauvery basin. Structural entrapment in the basin is related to the structural closure associated with the basement high. Fault-related structural closure in both hanging wall and footwall side also acts as good entrapment mechanism. Good structural traps are reported from Thanjavur and Tranquebar depressions in the onshore and Ariyalur–Pondicherry, Palk Bay, and Mannar depressions in offshore area. Common stratigraphic trap includes wedgeout sequences against basement high. High-density turbidites, basin floor fan, slope fan, and low-stand wedges, are interesting areas from exploration point of view. In majority of the basin, extensive and thick shale deposits overlying reservoir zones acts as good seal.

5.3.2 K–G Basin

K G basin had developed along the bite of East Coast of peninsular India between Visakhapatnam in the north and Nellore in the south including the Pennar river basin (i.e., between 17°40′ N and 14°30′ N) and covers approximately an area of 28,000 km^2 in the onshore and 24,000–49,000 km^2 in the offshore considering 200 and 2000 m water depths, respectively (Venkatarengan and Ray, 1993). The basin is covered by deltaic and interdeltaic area of two major rivers such as the Krishna and the Godavari (Rao, 2001), which along with their tributaries feed the basin and also discharge into deep offshore regions of Bay of Bengal. This basin, at present classified as a pericratonic rift basin (Biswas et al., 1993), was in fact a major intracratonic rift within the Gondwanaland until the Early Jurassic (Rao, 2001). Archean crystalline basement and Upper Cretaceous sedimentary outcrops demarcate the basin in the northwest part. A significant part of the onshore basin is covered by the Quaternary alluvium.

Following the initial mapping of basin margin outcrops by Blanford et al. (1856) and King (1881), ONGC has been conducting detailed geological mapping and geophysical surveys in the basin since 1959. By the end of

1970, ONGC completed the geological mapping, along with regional coverage of gravity, magnetic, multichannel seismic reflection data, both in onshore and offshore areas of the basin. The GSI and National Institute of Oceanography have also been carrying out regional geological and geophysical investigations in the shelf–slope areas of the basin. These data along with several exploratory wells drilled in the basin gave rise to reasonable understanding of subsurface geometry of the basin. The onshore well (Narsapur-1) drilled in 1978 and the offshore well (G-1-1) drilled in 1980 were the first wells to be drilled, and they have proved the presence of hydrocarbons in the basin. Since then, a number of major oil/gas discoveries have been made in the basin. Deepwater oil and gas discoveries by the Cairn India followed by two giant gas discoveries by the Reliance and GSPC converted the basin into one of India's most important petroliferous basins. The first deepwater oil field MA was discovered by the Reliance.

5.3.2.1 *Regional Geology*

The Archean igneous and metamorphic complex forms the basement and delimits the basin on the west (Fig. 5.13). The Archean rocks exhibit a general ENE–WSW to NE–SW trend, which correlates with the dominant trends of the Eastern Ghats (Rao, 2001). The basement rocks consist of the gneisses, quartzites, charnockite, and khondalite, while the most part of the basin is under the alluvium cover, where the subsurface geology is known from the drilled well data. However, isolated outcrops of Permian, Cretaceous, Paleocene, and Mio-Pliocene rocks are present near the basin edge.

5.3.2.2 *Structure and Tectonics*

The K–G Basin is a pericratonic in character and is perpendicularly juxtaposed to the NW–SE trending Pranahita–Godavari graben (PGG) at the central part of ECMI (Fig. 5.1). It is a polytectonic basin due to the presence of the above two orthogonal structures: one, an older NW–SE trending intracratonic Gondwana rift, the PGG, and other, the younger rift, the NNE–SSW oriented K–G rift basin formed during the rifting of the eastern Gondwanaland and subsequent separation of India from the East Antarctica (Gupta, 2006). The crystalline basement morphology shows that the K–G basin is characterized by the presence of several NE–SW trending horst–graben structural features (Fig. 5.14). Some of the grabens in the basin contain thick sedimentary sequences showing several cycles of deposition ranging in age from late Carboniferous to Holocene (Rao, 2001). Interpretation of gravity data in the onshore part of the basin (Murty and Ramakrishna, 1980) indicated that the basin is divided into three sub-basins: the Krishna, West Godavari, and East Godavari, separated by the Bapatla and Tanuku horsts. Kumar (1983) presented further division of the West Godavari sub-basin into the Gudivada and Bantumilli grabens separated by the Kaza–Kaikalur horst. Subsequently,

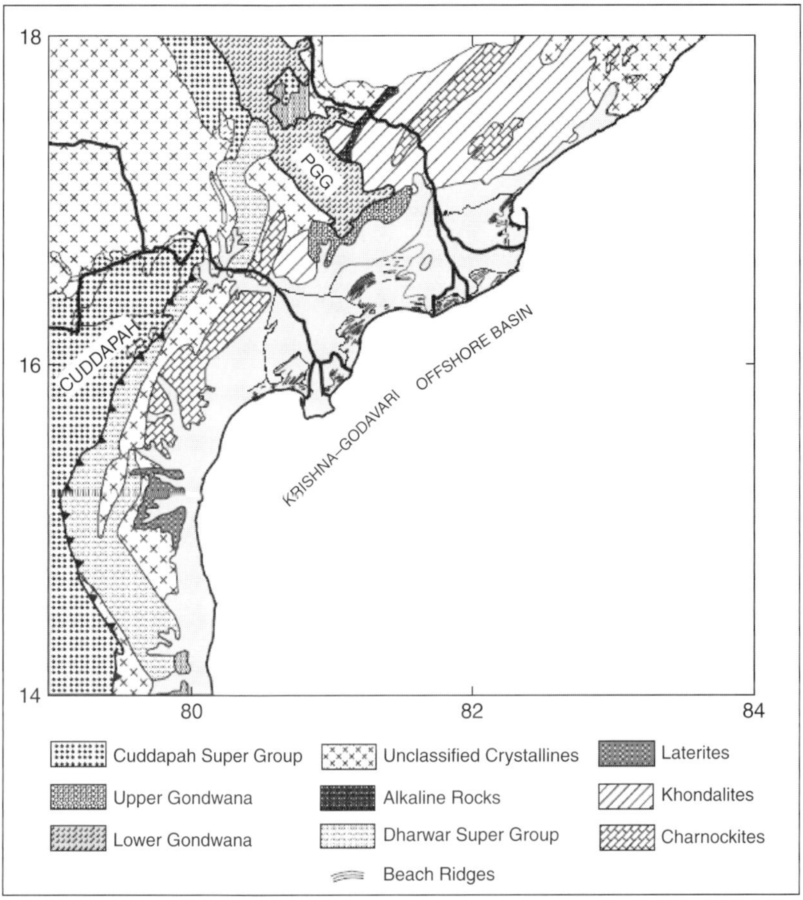

FIGURE 5.13 Simplified geological map of the K–G basin and the adjoining areas. The mapped lithological details are from Ramam and Murthy (1997) and the geomorphological features in the basin are from Biswas (1993).

the seismic as well as the drilled well data in the basin provided detailed basin architecture, subsurface lithology, and basement structure in both onshore and offshore areas (Bastia, 2006; Gupta, 2006; Mohinuddin et al., 1993; Prabhakar and Zutshi, 1993; Rao, 2001; Venkatarengan and Ray, 1993).

5.3.2.2.1 Deep Seismic Sounding Studies

Interpretation of two Deep Seismic Sounding (DSS) profiles in K–G onshore basin (Kaila et al., 1990), the Paloncha–Narsapur profile along the PGG, and the Kallur–Polavaram profile across the graben are discussed in Chapter 3. These provided valuable information on the crustal structure below the basin as well as velocity structure of the shallow sedimentary layers (Fig. 5.15). Subsequently, the velocity structure of the sediments above the crystalline

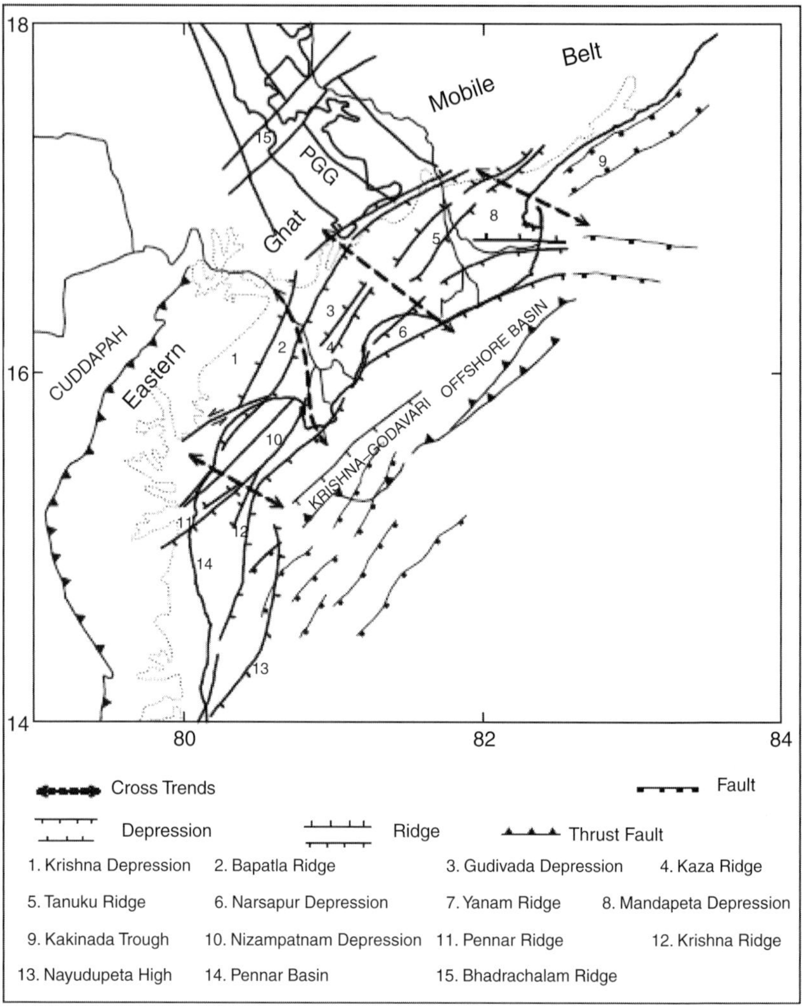

FIGURE 5.14 Map showing various tectonic and structural features in the onshore and the offshore K–G basin compiled from various sources (Choudhuri et al., 2006; Prabhakar and Zutshi, 1993; Venkatarengan and Ray, 1993).

basement for these two profiles is obtained from Ray trace modeling by Reddy et al. (2002). The refraction data along the Narsapur–Paloncha profile indicate that the Bapatla ridge separates Godavari graben from the coastal Godavari basin and also indicate the presence of five layers having velocities 1.8, 2.5, 4.0, 2.8, and 4.7–4.8 km/s in the coastal Godavari basin and two layers with velocities of 2.3 and 3.5 km/s in the Godavari graben area (Reddy et al., 2002). The refraction data integrated with the drilled well information further indicated the subsurface extension of the Deccan volcanism in the coastal areas of the K–G basin.

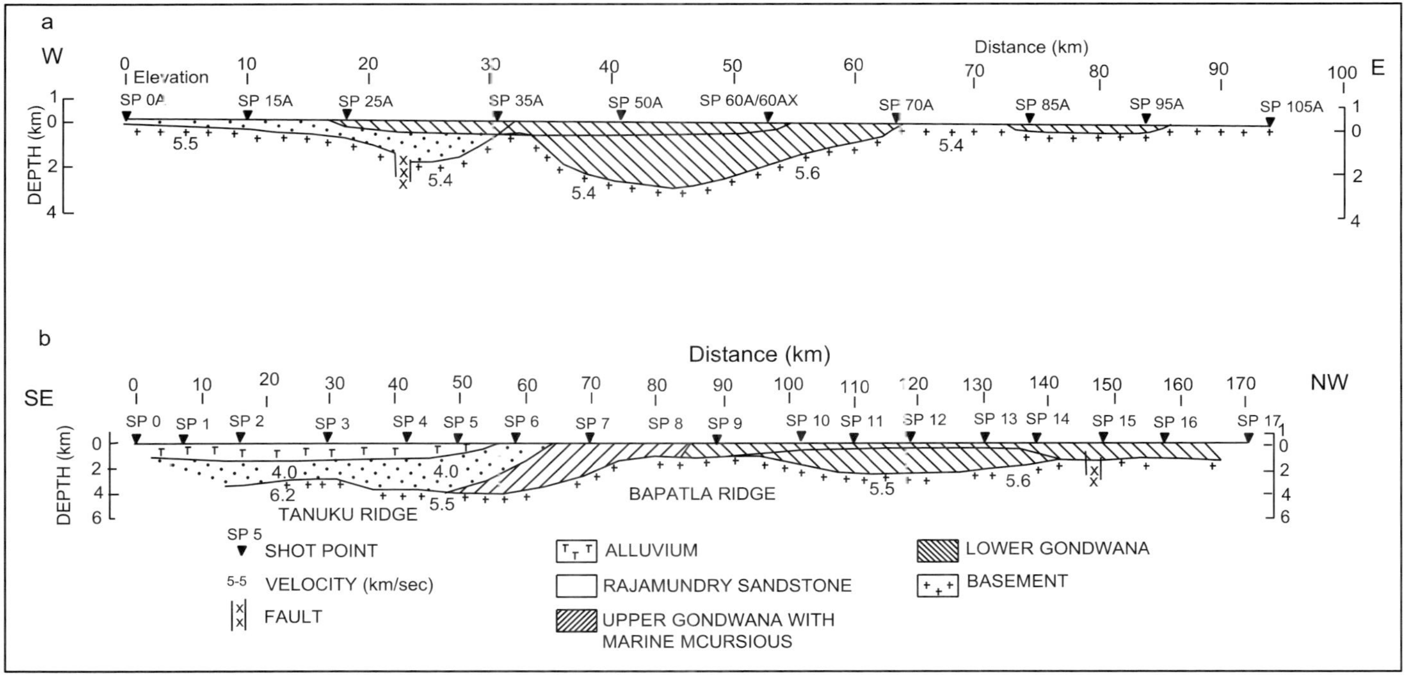

FIGURE 5.15 Crystalline basement configuration and shallow velocity structure along (A) Kallur–Polavaram and (b) Paloncha–Narsapur DSS profiles *(adapted from Kaila et al., 1990a).*

5.3.2.2.2 Gravity Anomalies

Detailed gravity and magnetic investigations in the basin have been carried out by several agencies, mainly ONGC, GSI, NGRI, and IIG, as well as the Andhra and Osmania Universities (Kumar, 1983; Mishra et al., 1989; Murthy and Babu, 2006; Murty and Ramakrishna, 1980; Reddy et al., 1988; among many others). Interpretation of these data along with the seismic information revealed the presence of a number of depressions separated from each other by narrow basement ridges having regional alignment in a NE–SW direction. Simplified gravity anomaly map (onshore Bouguer gravity and offshore free-air gravity) of the K–G basin is presented in Fig. 5.16. As can be seen from the figure, the merged gravity anomalies show excellent continuity of anomaly closures at the coast as well as their trends into the shelf area. The Bouguer anomalies in the onshore clearly reveal the ridge-depression features of the basement, with anomalies ranging between 20 and 30 mGal, based on which Murty and Ramakrishna (1980) and Kumar (1983) delineated several sub-basins (Krishna, East Godavari, West Godavari, and Pranahita–Godavari depressions, etc.) associated with gravity

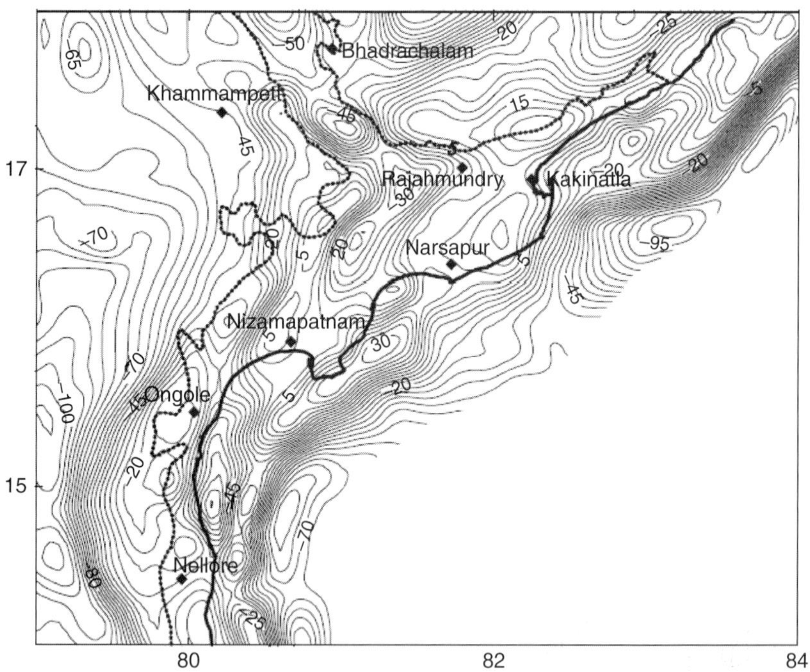

FIGURE 5.16 Gravity anomaly map of the K–G basin with Bouguer anomalies *(adapted from Kumar, 1983)* in the onshore and satellite-derived free-air anomalies (Andersen and Knudsen, 2008) in the offshore.

lows separated by ridges (Tanuku, Bapatla ridges, etc.) associated with gravity highs. The shelf area is also marked by several highs and lows trending along the coast and further displays typical bipolar gravity anomalies in the shelf–slope region. Mishra et al. (1987) have identified the presence of a number of intrusive bodies from gravity and magnetic interpretation and suggested that in the southern part of the basin, these intrusive rocks could be responsible for high heat flow in the basin. Based on the analysis of magnetic anomaly map, Rajaram et al. (2000) identified sources related to the different tectonic phases, the shallow geological formations belonging to the NE–SW trending horst and graben structures, and NNE–SSW trending Eastern Ghat folding superposed on the NW–SE structures that are related to the Godavari graben. From the magnetic anomaly map, the boundaries of the Cuddapah basin and the adjoining crystalline rocks of Dharwar and EGMB with the K–G basin could be clearly delineated. In the K–G basin, the anomaly trends show NW–SE orientation correlating with the Godavari graben as well as the East Godavari sub-basin, and NE–SW trends in the west Godavari and the Krishna sub-basins. These trends are seen to be separated or segmented by the NW–SE trending Pithapuram, Chintalapudi, and Avanigadda cross-trends, indicating basement structure control in defining these anomalies. It is further suggested that the source of magnetic anomalies in the onshore K–G basin could be the surface or subsurface charnockitic rocks, thicker traps below sediments, or supracrustal rocks (Rajaram et al., 2006).

5.3.2.2.3 Basement Configuration

Detailed seismic and drilled well data acquired under the petroleum exploration program of ONGC for the last four decades gave rise to information regarding the subsurface basement configuration in the onshore part as well as the shelf region of the K–G basin (Gupta, 2006, Prabhakar and Zutshi, 1993; Rao, 2001; Sastri et al., 1973; Venkatarengan and Ray, 1993). The basement structure map (Fig. 5.17) reveals several NE–SW trending ridge-depression structural features with sediment fill varying from 2.0 to 4.0 km in most of the basins except in the coastal as well as shelf areas of Godavari basin, where more than 8 km of thick sediments have been noticed. This typical rift basin morphotectonic expression is very well displayed on the gravity anomaly map as discussed previously. From Pennar basin in south to Visakhapatnam in north, a total of 14 major/minor tectonic elements have been mapped from geophysical exploration in the onshore as well as shallow offshore. These are (1) Krishna Depression, (2) Bapatla Ridge, (3) Gudivada Depression, (4) Kaza Ridge, (5) Tanuku Ridge, (6) Narsapur Depression, (7) Yanam Ridge, (8) Mandapeta Depression, (9) Kakinada Trough, (10) Nizampatnam Depression, (11) Pennar Ridge, (12) Krishna Ridge, (13) Nayudupeta High, and (14) Pennar depression. A brief description of subsurface

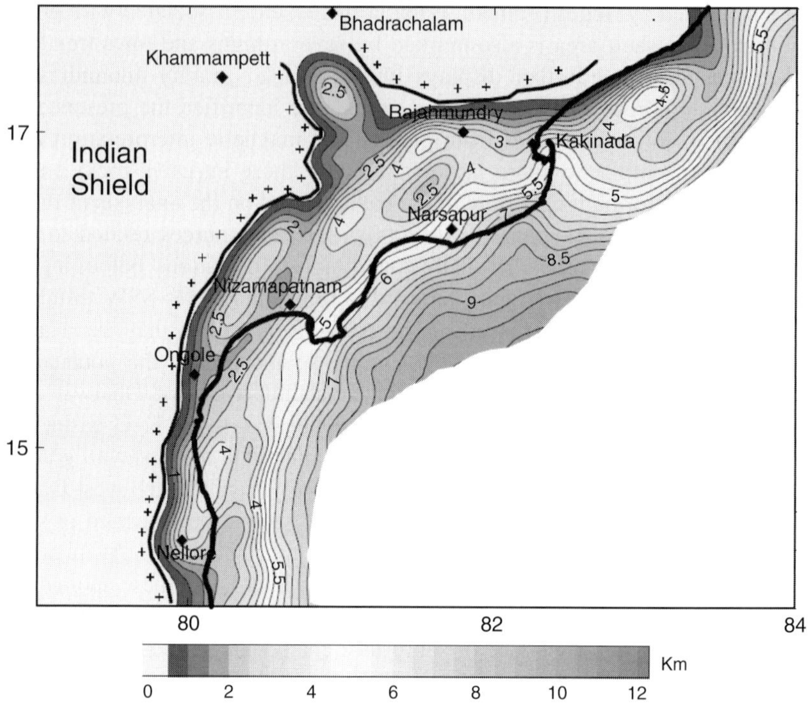

FIGURE 5.17 Basement depth map of the K–G basin onshore and shallow offshore areas *(modified after Prabhakar and Zutshi, 1993)*.

geometry of the major ridges and depressions compiled from several publications (Kumar, 1983; Prabhakar and Zutshi, 1993; Sahu, 2010; Sastri et al., 1973; Venkatarengan and Ray, 1993) is provided here.

5.3.2.2.3.1 Krishna Depression This NE–SW trending sub-basin located in the southwestern part of the basin is delimited by the Indian shield crust in the west and the Bapatla Ridge in the east. The basin contains more than 2500 m of thick sediments having mostly sediments of Mesozoic age, which are unconformably overlain by thin Tertiary and recent formations (Fig. 5.18). The basement structure indicates that it is a half-graben with polarity toward the Bapatla Ridge (Bastia, 2006).

5.3.2.2.3.2 Bapatla Ridge It lies mostly onshore (over a length of 200 km) and extends southeastward into the Nizampatnam Bay characterized by thin sediment cover of less than 500 m. Toward northeast extension of this structural high, Cretaceous to Mio-Pliocene sediments are exposed.

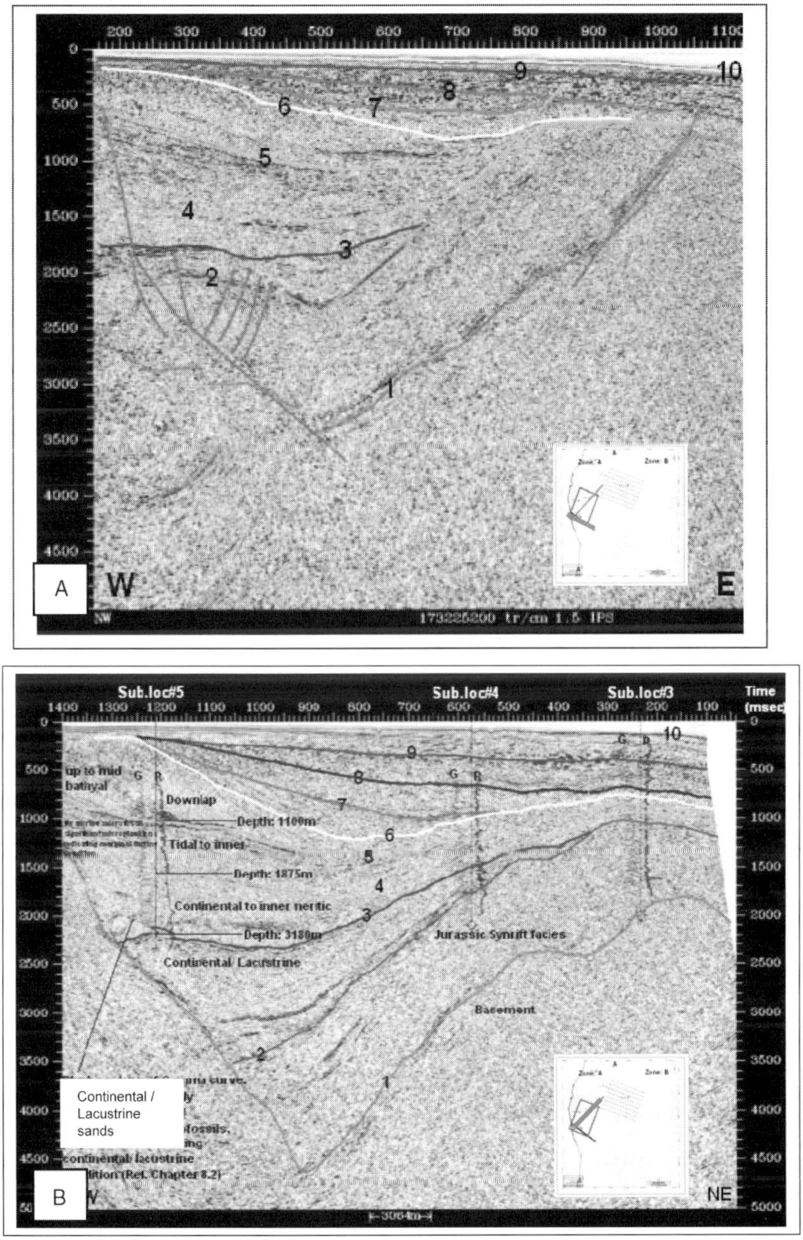

FIGURE 5.18 Seismic sections showing thick Mesozoic sediments in the Krishna depression. Numbers (1–10) indicate different stratigraphic horizons from basement top (1) to the seabed (10) *(adapted with permission from Bastia and Nayak, 2006 © Society of Exploration Geophysicists)*. (For color version of this figure, the reader is referred to the online version of this chapter.)

5.3.2.2.3.3 ***Nizampatnam Depression*** It lies southeast of the Krishna depression covering an area of 7000 km^2 (Sahu, 2010) and is limited by two major cross-trends, the Avanigadda in northeast and the Ongole in southwest. The basin gently dips toward southeast and covers sediments from Jurassic to recent with significant deposition during lower Cretaceous.

5.3.2.2.3.4 ***Tanuku Ridge*** This NE–SW trending ridge separates the Kommugudem and the east Godavari sub-basins. The well drilled in the high showed the presence of late Cretaceous and early Paleocene sediments overlying the basement. In the northern part, volcanic rocks of upper Cretaceous and lower Paleocene are exposed. The ridge continues further southward and is called the Kaza ridge. The Kaza–Kaikalur ridge separates the west Godavari depression into two, the Gudivada depression and the Narsapur depression.

5.3.2.2.3.5 ***West Godavari Depression*** This depression limited by the Bapatla Ridge in the west and the Avanigadda cross-trend in the south is further divided into three sub-basins, namely, the Gudivada, Bhimadole, and Narsapur (Bantumilli) depressions. The Kaza Ridge separates the Gudivada and Bantumilli depressions, while the Chintalapudi cross-trend separated the Bhimadole depression from the above depressions in the area west of Tanuku Ridge. The depressions in the onshore are characterized by nearly 2000–4000-m thick sediments.

5.3.2.2.3.6 ***East Godavari Depression*** It is located in the eastern part of Tanuku Ridge and extends toward the offshore, forming a major sediment depositional site particularly during Miocene to Pliocene times. It appears to be composed of smaller depressions separated by minor structural highs, the Narsapur, Amalapuram, and the Yanam. The major structural features identified in this depression are a series of Paleogene and Neogene growth faults with associated rollover anticlines and shale diapirs.

5.3.2.3 Generalized Stratigraphy

The detailed stratigraphic framework for this basin has been derived essentially from the seismic and subsurface lithologic data and documented by several earlier workers in the onshore part (Kumar, 1983; Rao, 2001; Sastri et al., 1973; Venkatarengan and Ray, 1993) as well as in the offshore (Sahu, 2005). Govindan (1984) made a detailed biostratigraphic zonation on the basis of micropaleontological studies. The Archean basement, pre-Cretaceous red bed, lower Paleocene basalts, and middle Eocene limestone form the distinct litho markers present in the subsurface of this basin. Available literature is extensive and widely varies in terms of formation names as used by different workers. Two seismogeological sections depicting the basin architecture and the various depositional sequences with associated litho units are shown in

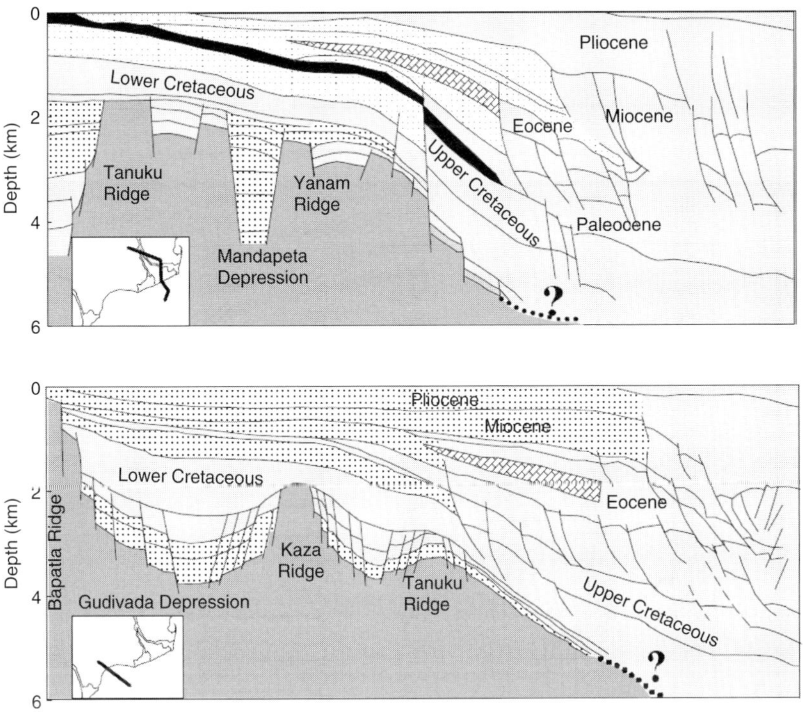

FIGURE 5.19 Seismogeological sections across the K–G basin as revealed from the drilled well data showing the various structural features at different stratigraphic levels *(adapted from Sahu, 2010)*.

Fig. 5.19. The stratigraphic framework (Table 5.8) of this basin is provided in Sahu (2010), a brief description of it is given below.

5.3.2.3.1 Gondwana Stratigraphy

This stratigraphic section is more than 3000-m thick and occurs below PGG. It is composed of three units such as Draksharama Argillite, Kommugudem formation, and Mandapeta Sandstone. This section extends from onshore to offshore.

5.3.2.3.1.1 Draksharama Argillite This formation composed of shale and sandstone is of early Permian age and lies unconformably over the Precambrian basement. The Kommugudem formation unconformably overlies it.

5.3.2.3.1.2 Kommugudem Formation This formation is of Permian age characterized by sand and shale with coal. Presence of spores and pollens of terrestrial plants and the lithological suites indicates fluvial to lagoonal depositional environment. Thickness of this formation is more than 900 m.

TABLE 5.8 Tectono-Stratigraphy of the K-G basin (adapted from Sahu, 2010).

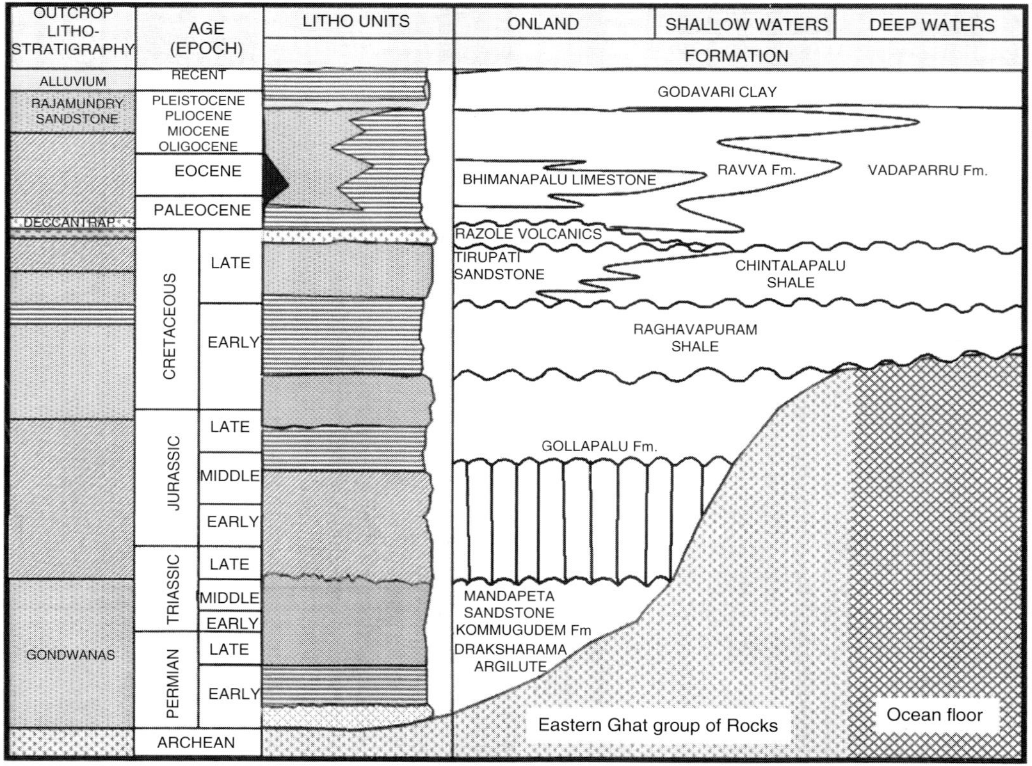

5.3.2.3.1.3 Mandapeta Formation It is composed of dominantly sandstone with minor intercalations of shale having thickness more than 1200 m. This formation deposited under fluvial environment is assigned Permo-Triassic age.

5.3.2.3.2 Jurassic–Cretaceous Stratigraphy

This stratigraphic succession is associated with the rifting and subsequent breakup of India from the eastern Gondwanaland counterparts and gave rise to superposition of NE–SW trending structural grain over the earlier NW–SE trending Gondwana graben. It is represented by four major formations such as the Gollapalli, Raghavapuram, Tirupati, and Chintalapalli.

5.3.2.3.2.1 Gollapalli Sandstone Formation This formation comprises syn-rift sediments in different grabens that were deposited under continental to lacustrinal environment during early subsidence accentuated by tectonics. The lower part of this formation is pebbly sandstone followed by ferruginous claystone, the middle part is sandstone with silts, and the upper part consists of medium- to fine-grained sandstone. This formation is more than 1000-m thick and belongs to late Jurassic to early Cretaceous age.

5.3.2.3.2.2 Raghavapuram Shale Formation This formation lies above the Gollapalli sandstone and is underlain by Tirupati sandstone or Raghavapuram shale characterized by unconformities above and below. Drilled wells in Gudivada and Mandapeta depressions indicate that lower part of it is highly carbonaceous shale. Thickness of this formation is maximum around 2000 m near the coastal area. It is deposited under inner to middle shelf marine depositional environments during Barremian to Campanion age. The presence of organic-rich shales in this formation makes it an important source rock especially in the graben areas.

5.3.2.3.2.3 Tirupati Sandstone Formation The basinward subsidence of crustal rocks resulted in the tilting of the basin toward SW or SE, thereby causing major marine conditions and deposition of thick upper Cretaceous sediments. The lower part of this section is argillaceous while the upper part is arenaceous consisting of medium to coarse grained and ferruginous sands. It is of upper Cretaceous age and consists of thickness more than 1000 m deposited under transitional to marginal marine conditions.

5.3.2.3.2.4 Chintalapalli Shale Formation It is essentially a shale section with fine-grained sands in the lower part as the Tirupati sandstone grades basinward into this formation. It is of upper Cretaceous age having maximum thickness of around 2000 m near the coast as evidenced from seismic data. This formation is inferred to be deposited under inner shelf environment.

5.3.2.3.2.5 Razole Volcanics Razole traps are outcropped near Rajahmundry, and in the subcrop, nearly 15,000 km^2 of area is covered by the basaltic trap in the basin. Four trap flows separated by marine intertrappean sediments comprising clay, sandstone, and limestone have been identified in the east Godavari sub-basin by Govindan (1984). A schematic cross-section of the west and east Godavari sub-basins shows the subsurface distribution of trap (Fig. 5.19). This formation is assigned an age late Maastrichtian to early Danian based on the palynofloral assemblage.

5.3.2.3.3 Tertiary Stratigraphy

This section is composed of Vadaparru shale, Ravva Formation, and Bhimanapalli limestone Formation.

5.3.2.3.3.1 Vadaparru Shale This major argillaceous facies formation overlies Razole trap in the onland area and Chintalapudi shale in the offshore. It is deposited following Razole volcanism and unconformably lies above the trap. The shale is rich in organic matter and occasionally contains thin beds of sandstones and siltstones. It forms an important source rock having thickness ranging from 1500 to 2200 m. The age of this formation ranges from Paleocene to Pliocene.

5.3.2.3.3.2 Bhimanapalli Limestone Formation This formation consists mainly of fossiliferous limestone at places with sand and shale alternations. Maximum thickness of this formation is 600 m and it grades toward basin to Vadaparru shale. It is deposited under marginal marine to shallow marine conditions during the middle Eocene.

5.3.2.3.3.3 Ravva Formation This formation represents alternate bands of sand and shale, at places siltstone, and their thickness varies significantly within the formation. Age of this formation ranges from Miocene to Pliocene and thickness varies between 1000 and 3000 m. Basinward, it changes to Vadaparru shale. The sands in this formation are hydrocarbon bearing and resulted in several important discoveries.

5.3.2.3.4 Quaternary Stratigraphy

This clay sequence overlying Ravva formation is called the Godavari clay and is the youngest stratigraphic unit in the basin. Owing to increased sedimentation in post-Miocene period, this formation became very thick, attaining a maximum thickness of around 2500 m. Sands within this formation in the offshore are proved to be hydrocarbon bearing; sometimes, they are grouped under Ravva formation.

5.3.2.4 *Tectonostratigraphy and Sedimentation*

A tectonostratigraphic chart describing various sequences from onland to deepwater parts of the basin is shown in Fig. 5.20. The stratigraphic framework of the K–G basin detailed above reveals that the age of the formations ranges from Permian to Recent. The basin extends from onshore to deepwater offshore areas having maximum sediment fill exceeding 9 km. However, this huge sedimentary package is divided by several wide unconformities in the basin, which can be correlated with several regional geodynamic events, sea-level changes, or provenance delivery system (Sahu, 2010). The sedimentary sub-packages, marked by the respective unconformities, have formed into different tectonostratigraphic units that can be ultimately linked to the basin evolution. The major geodynamic events are the breakup of India from the eastern Gondwanaland during Jurassic/early Cretaceous, separation of Madagascar from India in mid-Cretaceous, Deccan volcanism in late Cretaceous, and collision events (soft and hard) between India and Eurasia during Eocene (Oligocene?) and Miocene. Apart from these events, the changes in sea level also affected the basin. The following are the unconformity surfaces observed in the basin (after Sahu, 2010): Mio-Pliocene, intra-Oligocene, K/T boundary, Turonian top, Aptian to Albian, Triassic–middle Jurassic, and the base of Permian. The Mio-Pliocene unconformity is a major hiatus recognized in both onshore and offshore and spreads between middle Miocene and early Pliocene. Raju et al. (2005) reported from the paleontological data that the magnitude of this hiatus varies from 5 to 18 Ma. The intra-Oligocene unconformity had formed in the onshore part of the basin because of the global sea-level low stand, which caused the nondeposition or smaller deposition in the shelfal part, and sedimentation in slope and basin floor area of the basin. Throughout the east coast basins, a major hiatus spanning from 5 to 22 Ma (Raju et al., 2005) has been observed in the sedimentary record between Cretaceous and Tertiary sequences and is famously known as the K/T boundary. This boundary is marked by volcanic eruptions related to the Deccan trap volcanics. The separation of Madagascar from India gave rise to an unconformity of smaller magnitude in the K–G basin during intra-Turonian, which is prominently observed in the Cauvery basin. The Albian–Aptian unconformity is well documented in the basin and formed in response to the waning of rifting and concomitant marine transgression into the basin. The Triassic to late Jurassic unconformity is known as the breakup unconformity and marked by Red beds in the basin. The basin has evolved through different Tectonostratigraphic Units (TTU), and these acted as the main building blocks. A total of five TTUs are identified and a detailed description of them is out of scope of this book, but can be referred to in Sahu (2010). These are (i) pre-rift, (ii) syn-rift, (iii) rift–drift transition, (iv) early post-rift, and (v) late post-rift.

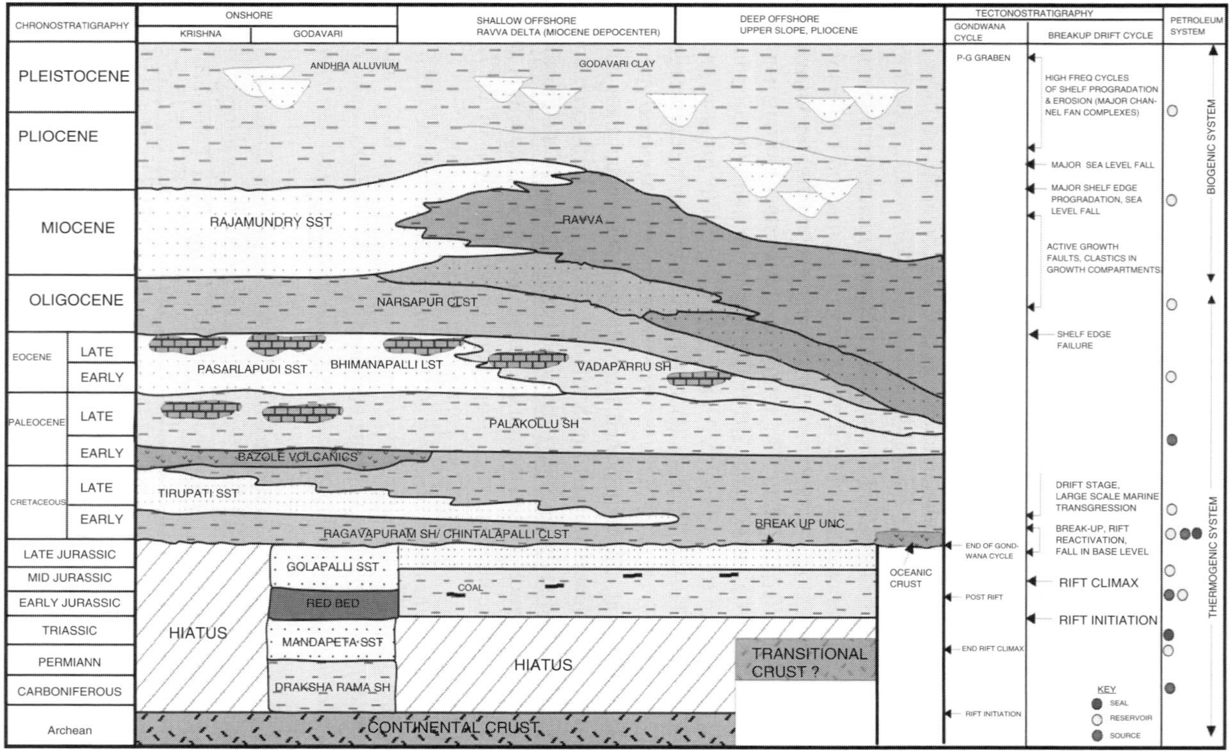

FIGURE 5.20 Generalized stratigraphy of the K–G basin from onshore to deep offshore (modified after Bastia et al., 2007). (For color version of this figure, the reader is referred to the online version of this chapter.)

5.3.2.5 Petroleum System

Krishna–Godavari basin is a proven petroliferous basin with commercial hydrocarbon accumulations in both onshore as well as deepwater offshore.

5.3.2.5.1 Source Rocks and Maturation

One of the most important key factors to exploration strategy of any area is the presence of hydrocarbon source having strong linkage with the tectonics, paleogeography, paleoclimate, source rock distribution, and their effectiveness. Correlation between preservation of source rock, its type, and quality with the tectonic evolution of the K–G basin has provided valuable information on the development of petroleum system in the basin. Compilation of large volume of published literature and from the authors' inferences, the general petroleum system of the basin is described below:

- *Early rift sequence:* These formations were deposited in oxidized areal-subaerial condition with coarser clastics. The sections are characterized by low shale–sand ratio, and the preservation of organic matter is poor due to oxidizing environment. Therefore, it is devoid of any source rock interval; however, these sections have very good reservoir potential.
- *Peak rift sequence:* These sections were deposited in fluvio-lacustrine reducing environment. The shale–sand ratio shows a relative increase from older sequences, and the preservation of organic matter in this sequence is very high (mainly type-I and type-III organofacies). Age of this sequence is Late Jurassic, and it has a good correlation with world analogue basins (North Sea) having the same peak rift sequence.
- *Post-rift sequence:* This sequence deposited under shallow to deep, restricted marine condition has relatively higher shale content compared to its older sections. Preservation of organic matter is relatively high and major organofacies types are of Type-II and III.
- *Open marine sequence (drift sequence):* The postrift sequence is overlain by the open marine sequence which is dominated by shale, and the preservation of organofacies is very poor and of mainly Type-III. The geologic age corresponds to Albian–Maastrichtian.

Compilation of source rock data in different stratigraphic sections of the basin indicates:

- Older formations (Oxfordian and older sections) encountered in the onland and shallow water area have moderate to good source rock potential.
- Almost all drilled wells in offshore basin correspond to Kimmeridgian–Tithonian (Late Jurassic) age and have very good presence of source rock

interval. These are deposited mainly in fluvio-lacustrine conditions and the major organofacies are of Type-I and III.

- The fluvio-lacustrine source rock of Late Jurassic age gradually changed to restricted shallow to deep marine Type-II and Type-III organofacies in the Beriassian–Barremian (Early Cretaceous).
- Late Cretaceous section represents more of open marine condition and the major organofacies are type-III terrestrial. These sections show poor source rock quality except the global anoxic event associated with the Albian–Cenomanian marine transgression, like in many other parts of the world, which favors development of good source rocks with Type-II kerogens.
- Throughout the basin, paleogene shale associated with the thermal maxima shows good to very good preservation of organic matter; however, this interval is mostly immature because of limited overburden thickness. The limited area where the Miocene–Pliocene loading is significant (Ravva field), this source plays an important role in charging the Miocene reservoirs.

The results of source rock modeling in different parts of the K–G basin (after Sahu, 2010) are presented in Fig. 5.21.

5.3.2.5.2 Reservoirs

Krishna–Godavari basin is composed of both carbonate and clastic reservoir facies of different ages. The fluvio-deltaic rift sections in this basin are proven to be rich in reservoir facies. Cretaceous Carbonate units are one of the important reservoirs in this basin (Kaza and Tanuku Ridges). The carbonate draping over these basement highs show development of porosity due to leaching and dolomitization. During Eocene, the development of shelfal carbonate shows varied reservoir quality with some of the terraces; the carbonate development is maximum. The drilling in the offshore shows clastics of Cretaceous having very good reservoir potential. Excellent reservoir quality has been reported from early Miocene Ravva formation. In Razole and Naraspur structures, fractured volcanics of the Razole formation (Upper Cretaceous to Lower Paleocene) has produced gas.

5.3.2.5.3 Entrapment Condition

A number of hydrocarbon plays have been recognized in the K–G basin (Fig. 5.21). Faulted anticlines in the Gondwana and Late Cretaceous sediments are proved to be hydrocarbon bearing. A number of anticlinal and fault closures are identified in the paleogene sediments.

Entrapment condition varies widely in the K–G basin. Structural entrapment in the basin is related to horst blocks which are formed due to rifting (Fig. 5.22). Anticlinal folds related to compression in the toe thrust region are another important trap in this basin. Similarly, growth faults/tilted fault blocks related to

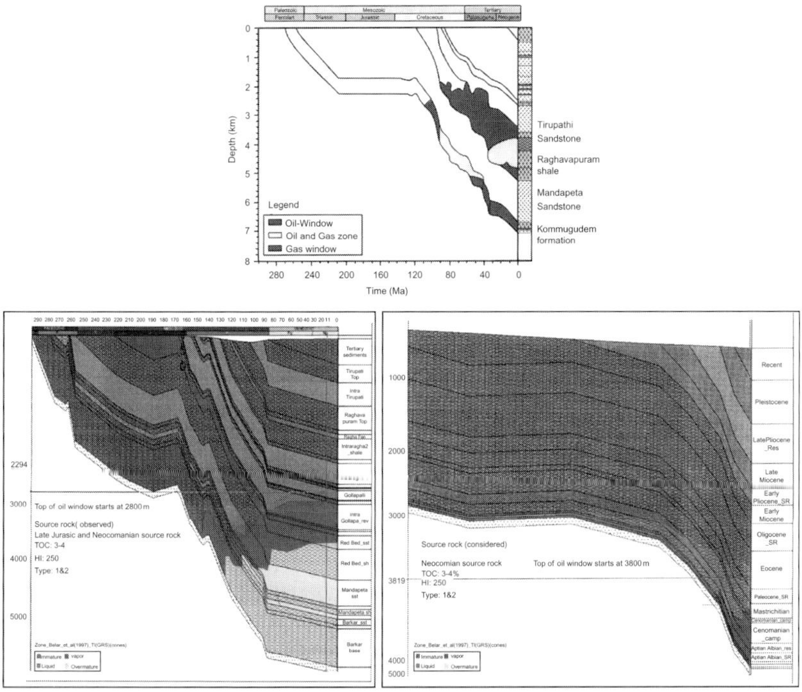

FIGURE 5.21 Results of source rock modeling in the K–G basin onshore (above) and shallow water offshore (below) (adapted with permission from Sahu, 2010 © Technology Publications, Dehradun). (For color version of this figure, the reader is referred to the online version of this chapter.)

syn-sedimentary deformation can be considered common entrapment (Fig. 5.22A) as observed in the Ravva field. Erosional remnants related to unconformity also act as good stratigraphic trap in this basin (Fig. 5.22A and B).

5.3.3 Mahanadi-NEC Basin

5.3.3.1 Mahanadi Basin

The Mahanadi basin lying in the northern part of the east coast of India is another major onshore–offshore basin covering a vast stretch of the continental shelf as well as the deepwater areas of the eastern offshore. The basin covers a total area of 55,000 km^2, out of which about 14,000 km^2 lies in the shallow offshore area with shelf continuing up to 150-m isobath (Source: DGH). Including the deep offshore area, it encompasses an area of about 200,000 km^2. The basin is fed by the major deltaic river system of

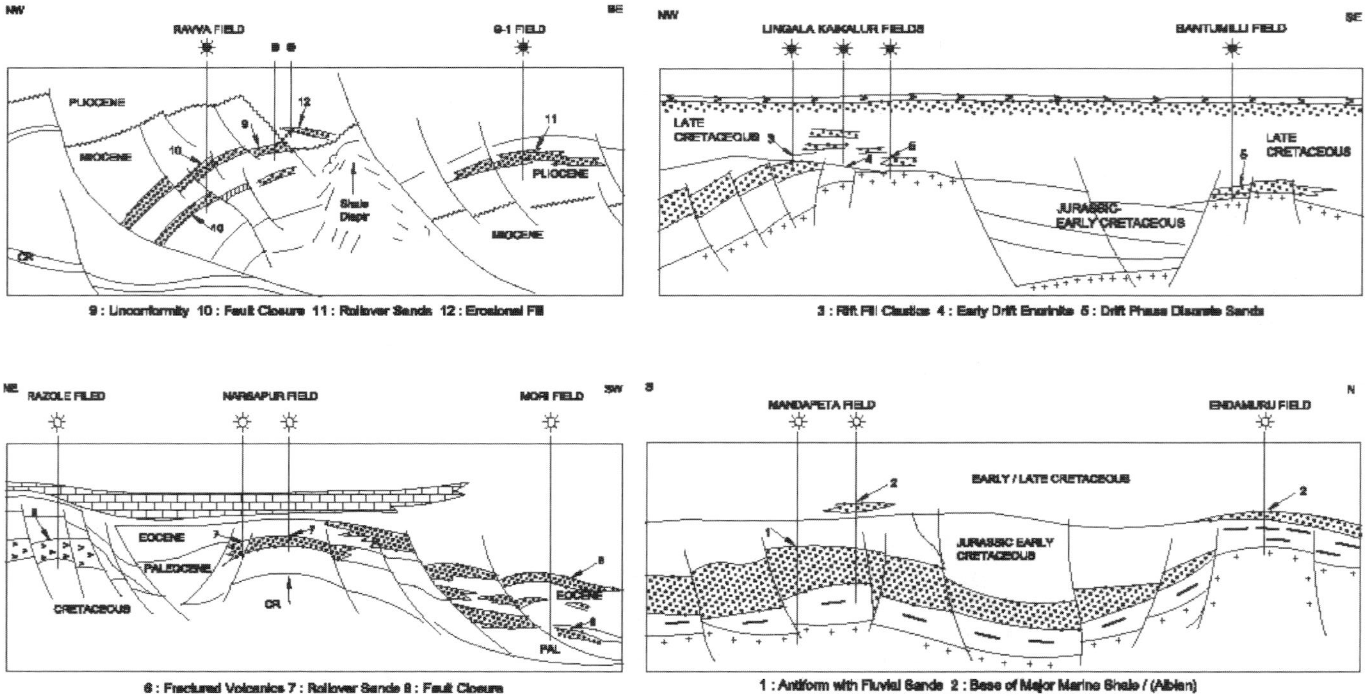

FIGURE 5.22 Seismogeological sections showing different structural/stratigraphic play types and prominent oil/gas fields in the K–G basin *(adapted from Venkatarengan and Ray, 1993).*

the Mahanadi, Baitarani, and Brahmani (Mahalik, 1995), additionally, the Ganges–Brahmaputra system, and Subarnarekha river also drains sediments into this basin. The onland part of the basin is limited to the northwest and west by the Precambrian outcrops belonging to the Indian Shield crust, and toward northeast, the basin merges into onshore Bengal basin. The 85° E Ridge in the region south of the Chilka Lake in offshore forms the approximate southwestern limit of the basin. The Mahanadi-NEC basin was influenced by the huge deltaic sedimentation subsequent to the major collision events during late Oligocene/Miocene due to which the boundary with the Bengal basin became obscured (Fuloria, 1993). A major part of the onshore Mahanadi Basin (Fig. 5.23) is covered with alluvium with various rock exposures of upper Gondwanas, khondalites, granites, gneisses, and laterites (Subrahmanyam et al., 2008b). Sedimentary sequences lying over the metamorphic basement range in age from the Cretaceous to Holocene.

The hydrocarbon exploration activity has been initiated in the basin since 1975 involving several national and multinational oil companies. The first hydrocarbon discovery in the basin was made with the gas strike in 2004. Subsequently, many drilled wells in the onshore as well as the shallow offshore gave encouraging results with hydrocarbon finds in the Tertiary sequence.

5.3.3.1.1 Structure and Tectonics

The Mahanadi basin came into existence during the Jurassic as a result of rifting and breakup of Gondwanaland (Fuloria, 1993). Since the Cretaceous, a number of transgression and regression events have given rise to the development of several elongated depressions and intervening ridges (Fig. 5.24), both in onshore as well as offshore (Fuloria et al., 1992). The regional structural trends in this basin are almost parallel to the well-developed NE–SW trend of the east coast of India. The NW–SE trending Mahanadi Graben orthogonally meets the Mahanadi basin, similar to the K–G basin, and extends into the offshore through the Mahanadi region. The continental sediments of upper Paleozoic and Mesozoic had been predominantly deposited and preserved in the Mahanadi Graben, while rifting and syn-rift infill was continuing in Jurassic and early Cretaceous in the graben.

5.3.3.1.2 DSS Studies

The DSS investigations have been made in the basin by NGRI in collaboration with OIL with an objective of delineating basement structure, thickness, velocity, and densities of various sedimentary layers as well as understanding the deeper structure and tectonics (Kaila et al., 1987). Refraction and wide-angle reflection studies were carried out along three profiles, which include (i) Konark–Cuttack–Mukundpur, (ii) Baliamba–Kendrapara, and (iii) Paradip–Kabatbandha in the basin. The basement configuration as well as the velocities of sediments and the crystalline basement along these

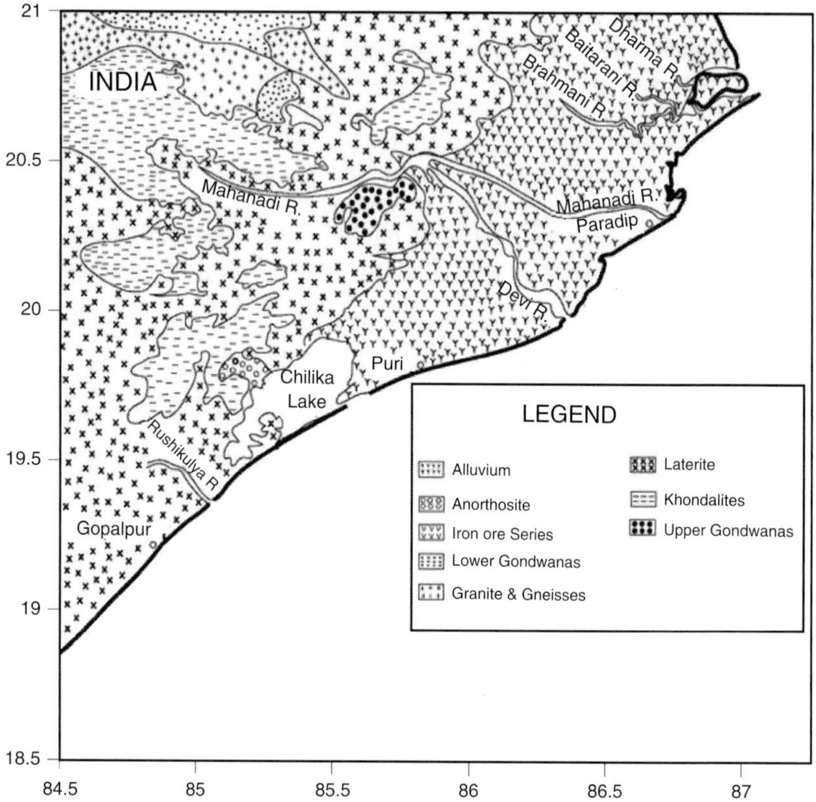

FIGURE 5.23 Geology of the Mahanadi onshore basin *(Redrawn from Subrahmanyam et al., 2008b).*

profiles is shown in Fig. 5.25. The basement in the basin shows velocities in the range of 5.6–6.0 km/s overlain by two or three layers of sediments having variable thickness and separated by faults.

5.3.3.1.3 Gravity Anomaly Map

The Bouguer gravity anomalies (NGRI, 1978) in the onshore for the Mahanadi and Bengal basins together along with the satellite-derived free-air anomalies in the adjoining offshore have been considered in order to compile a composite gravity anomaly map (Fig. 5.26) of the region. The map clearly indicates the horst–graben structural trends of the Mahanadi basin.

5.3.3.1.4 Basement Configuration

The general basement configuration in both onshore and offshore parts of the basin has been obtained from the interpretation of aeromagnetic anomaly map (Baburao et al., 1982). Several exploratory drilled wells as well as the seismic

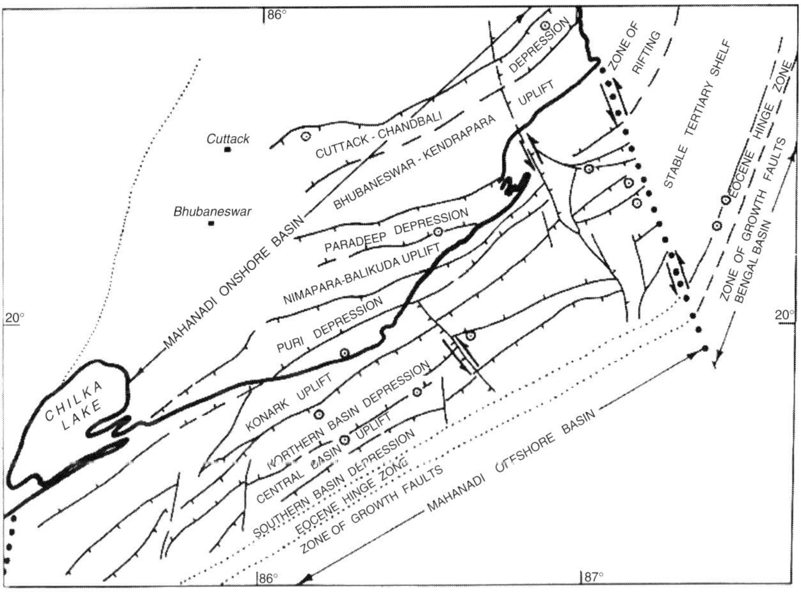

FIGURE 5.24 Tectonic map of the Mahanadi basin *(adapted from Fuloria, 1993)* showing various faults, depressions, and ridges.

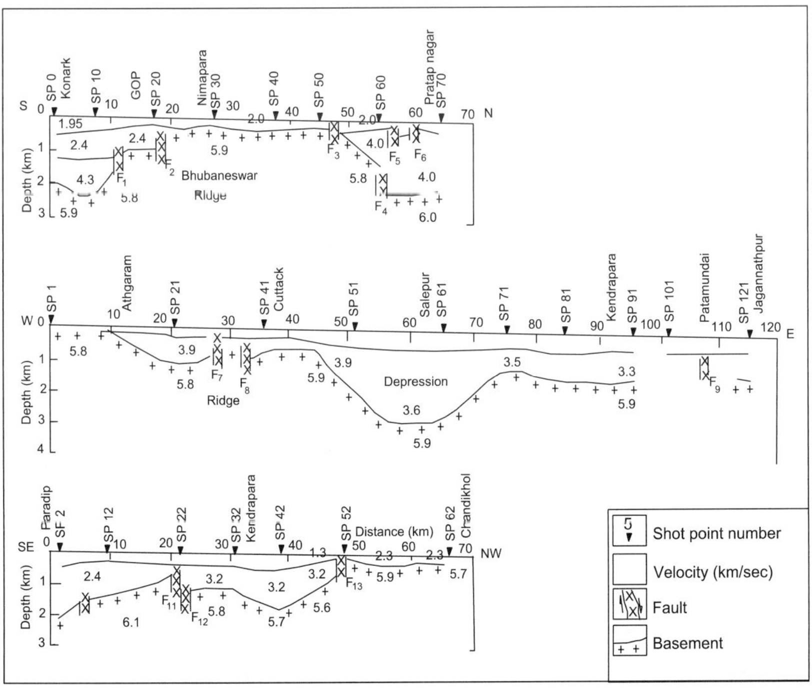

FIGURE 5.25 Basement configuration and shallow sedimentary velocities along (A) Konark–Mukundpur DSS profile, (B) Baliamba–Jagannathpur DSS profile, and (c) Paradip–Kabatabandha DSS profile *(Reproduced with permission from Kaila et al., 1987 © NGRI).*

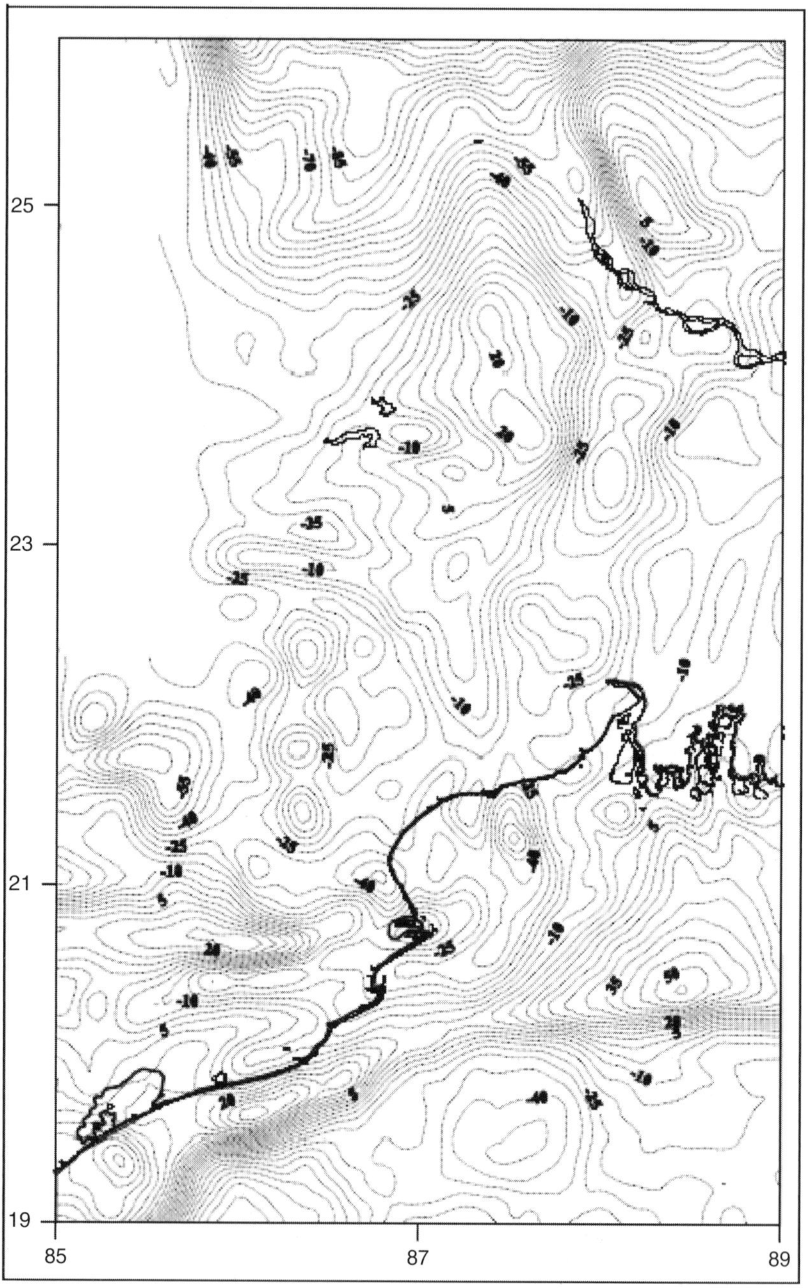

FIGURE 5.26 Map showing the gravity anomaly map of Mahanadi and the Bengal basin region. Bouguer anomalies in the onshore (NGRI, 1978) and satellite-derived free-air anomalies (Andersen and Knudsen, 2008) in the offshore are used to prepare this map.

data subsequently provided valuable information on the basement rocks and the detailed lithostratigraphy characteristics of the overlying sedimentary formations (Fuloria et al., 1992; Jagannathan et al., 1983). On the basis of these data, time structure map of the basement (Fig. 5.27) as well as tectonic map (Fig. 5.24) of the basin onland and shelfal part is discussed by the earlier workers. The continental basement is characterized by a number of major faults: out of these, the ENE–WSW trending faults influenced the basin formation during the Cretaceous period (Fuloria et al., 1992). While the structural configuration at the base of Cretaceous is of horst–graben type, the younger stratigraphic levels have resulted from the tectonic movements along ENE–WSW as well as the transverse faults across them. One such major transverse fault, Dhamara offshore lineament, separates the Mahanadi onshore as well as offshore from the Bengal basin in the east (Fuloria et al., 1992; Subrahmanyam et al., 2008b). Available data indicate that the offshore area represents a pull-apart type of basin, which generally parallels the coast having sedimentary succession rapidly thickening seaward (Jagannathan et al., 1983).

5.3.3.1.5 Description of Sub-basins

Geophysical surveys in the basin indicate that the basin comprises several depressions and ridges, such as Cuttack depression and Bhubaneswar Ridge, as well as the coastal Konark, Paradip, and Chandbali depressions, and the basement depth

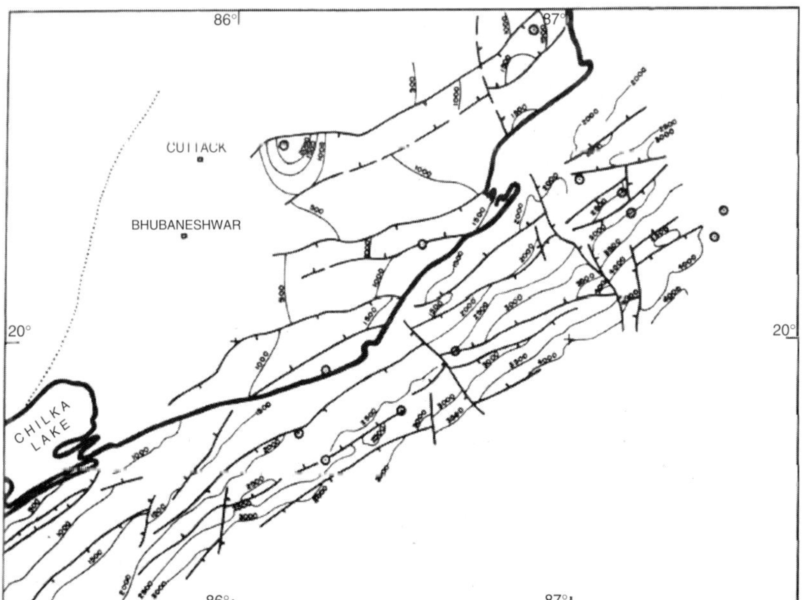

FIGURE 5.27 Time structure map of basement in the Mahanadi basin (onshore and offshore) region *(adapted Fuloria, 1993).*

varies from 200 m near Bhubaneswar ridge to a maximum of 900 m below Cuttack–Chandbali depression (Murthy et al., 1973). According to Sastri et al. (1981), the Cuttack–Chandbali depression is a long, narrow graben, which is aligned ENE–WSW near Cuttack and swings northward to orient in NE–SW near Chandbali. It contains nearly 2000–2500-m thick sediments having mostly late Mesozoic sequence, and it extends into the offshore areas. The Paradip depression is mostly an offshore feature trending NE–SW in the south and almost N–S in the north, likely to contain sediments of the order of 2500 m. The Bhubaneswar Ridge trends almost E–W and has an eastward plunge. The ENE–WSW trending Puri depression is delimited by the E–W trending Bhubaneswar Ridge in the NNW and extends into the offshore toward SSE. The thickness of sediments ranges from 1500 to 3000 m in the depression. A shallow basement region near Bhadrak covers a large part of the basin in the north. The NE–SW trending Rajnagar ridge swings toward N–S and extends further into the offshore.

5.3.3.1.6 Generalized Stratigraphy

The generalized stratigraphy and lithostratigraphic information obtained from different wells that were drilled in the onshore and shallow offshore part of the Mahanadi basin (Fuloria, 1993) are shown in Fig. 5.28. Two transects showing basin architecture and various sequences identified in the basin are presented in Fig. 5.29.

The Talchir boulder bed of Gondwana age is the oldest exposed sediments overlying the Precambrian basement. The upper Gondwana Athagarh sandstone is characterized by plant fossils deposited in the fluvial, fluvio-deltaic conditions. Most of the drilled wells in the onshore have encountered volcanics of Rajmahal age followed by Albian–Cenomanian clastic deposits. These clastics form the major source potential for this basin. During Eocene, major platformal carbonate development is observed in the onshore and shallow offshore part. The soft collision during Oligocene resulted in clastic deposition in the offshore area, while the onland and shallow offshore areas remain starved. Miocene shows the maximum thickness of sediments fed dominantly through Ganges with episodic input from the Mahanadi and Subarnarekha river system.

5.3.3.1.7 Petroleum System

5.3.3.1.7.1 Source Rocks and Maturation The drilled well data in the Mahanadi basin show three major source rock units, such as (i) Early Cretaceous deposited on the sag fill basin, (ii) Eocene–Oligocene related to thermal maxima, and (iii) Miocene clastics-related Himalayan uplift. The Early Cretaceous source rock is of predominantly Type-III with minor Type-II organofacies deposited in an anoxic condition similar to the K–G and Cauvery basins already described. In the Eocene–Oligocene, the source rock interval in the

AGE	LITHOLOGY	DEPOSITIONAL ENVIRONMENT	THICKNESS (m)
Pleist-Recent	Sands, clays, and silts	Deltaic to shallow shelf	200–600
	Contact gradational to unconformable		
Pliocene	Sands and clays	Prodelta to marine	200–700
	Contact gradational to unconformable		
Miocene	Claystones, siltstones and sandstones, fossiliferous patchy limestones in the lower part	Deltaic to open marine	600–1900
	Contact unconformable		
E-M Eocene	Fossiliferous limestones, carbonaceous shales, siltstones and sandstones	Shallow marine (Inner shelf)	200–400
	Contact conformable		
Paleoc.	Argillaceous limestones, shales, siltstones and sandstones	Deltaic to shallow marine	50–600
	Contact unconformable		
L. Cret.	Mainly sandstones with minor shales and limestones	Shallow marine (shelf)	0–500
	Contact unconformable		
E. Crotaceous	Basalts, tuffs and intertrappeans, shales/claystones	Subaerial and subaqueous	25–850
	Contact unconformable		
Pre-camb.	Granites and gneisses (Basement Complex)		

FIGURE 5.28 Generalized stratigraphy and lithostratigraphy information from different wells drilled in the onshore and shallow offshore part of the Mahanadi basin *(adapted from Fuloria, 1993)*.

drilled wells shows similarity with other East Coast basins. However, lack of overburden in the onland and shallow offshore region hinders maturity of the source rock in this basin. On the other hand, in the offshore area, due to higher thickness of Mio-Pliocene sediments, the source rock is well within the maturity zone. The discovery of hydrocarbon in Miocene proved the active petroleum system charged from the matured source rock.

5.3.3.1.7.2 Reservoir Rocks Mahanadi basin witnesses good quality reservoir rocks including both clastics and carbonates. Reservoir facies consists of

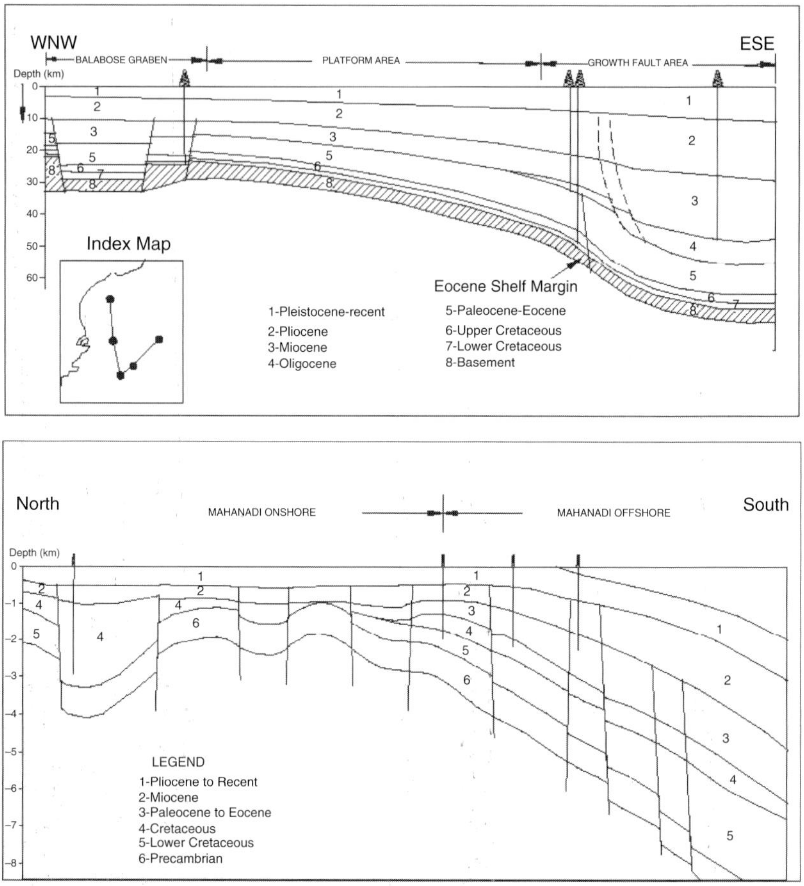

FIGURE 5.29 Seismogeological sections in the Mahanadi offshore basin. Top: adapted with permission from Jagannathan et al. (1983) © Himalayan Time Group and bottom: adapted from Fuloria (1993).

prograding delta, deepwater turbidites, low-stand wedges, and basin floor fan. In general, rocks show good porosity and permeability. The intertrappean sequences with associated fractured and weathered volcanic flows can also act as potential reservoir facies in this basin.

5.3.3.1.7.3 Entrapment Types Similar to the K–G basin, in the Mahanadi basin also, varied entrapment conditions prevail. Structural entrapment in the basin is related to the faulting-related traps as well as ridge-related structural closure in the offshore. Common stratigraphic trap includes wedgeout sequences against the ridge, pinchout sequences, and slope fan. Older turbidite fans cut by younger mud-filled canyons are another important entrapment condition present mostly near the shelf–slope region.

5.3.3.2 *Bengal Basin*

The Bengal basin is a major pericratonic basin situated in the northeastern part of Peninsular India and extends into the Bangladesh covering 56,000 km^2 of area onland. The basin is bounded in the west by the Precambrian basement rocks, north by Shillong massif, east by the Arakan Yoma belt, and toward south, it opens into the Bay of Bengal (Das Sharma et al., 1993). The basin contains thick sedimentary succession of continental to marine clastic and carbonate sediments ranging in age from early Cretaceous to Holocene (Banerji, 1981).

5.3.3.2.1 Structure and Tectonics

Except few patches of Gondwana and post-Gondwana sediments, the entire basin is covered by the Alluvium with Ganges and its tributaries forming the main drainage of the basin. The basin is divided into four structural elements (Fig. 5.30) depending on the tectonic evolution (after Das Sharma et al., 1993): (i) NE–SW trending fault zone demarcates the western margin of shelf with the crystalline basement, (ii) the western foreland shelf constituting the area to the west of Sylhet slope break, (iii) the hinge zone separating the western foreland shelf and frontal flank of the subsiding trough, and (iv) the basin deep experiencing rapid subsidence to the east of Eocene slope break. Alam et al. (2003) described the basin as composed of three geotectonic provinces, such as (i) the stable shelf, (ii) the central deep basin, extending from Sylhet trough in northeast to Hatia trough in south, and (iii) Chittagong–Tripura fold belt. A geological cross-section through the stable shelf and the central deep basin provinces is shown in Fig. 5.31. Banerji (1981) synthesized the complex geological history of this basin and suggested that the basin had evolved in two phases: the first, a marine transgressive phase ending after the late Eocene, and second, a regressive phase that had resulted in continental deltaic deposition. The first phase starts soon after the Gondwana tectonism with the stage of initial subsidence in early Cretaceous. Figure 5.32 shows the genetic relationship between the Gondwana basin and the marine Cretaceous–Paleogene basin (Banerji, 1981).

5.3.3.2.2 Generalized Stratigraphy

Except few outcrops bordering the eastern Indian shield, stratigraphy of the Bengal basin is obtained from a large number of exploratory drilled wells in the basin. The generalized subsurface stratigraphy of the basin is presented in Fig. 5.33 (after Berger et al., 1983) which shows the stratigraphic sequences ranging in age from lower Gondwana to Recent.

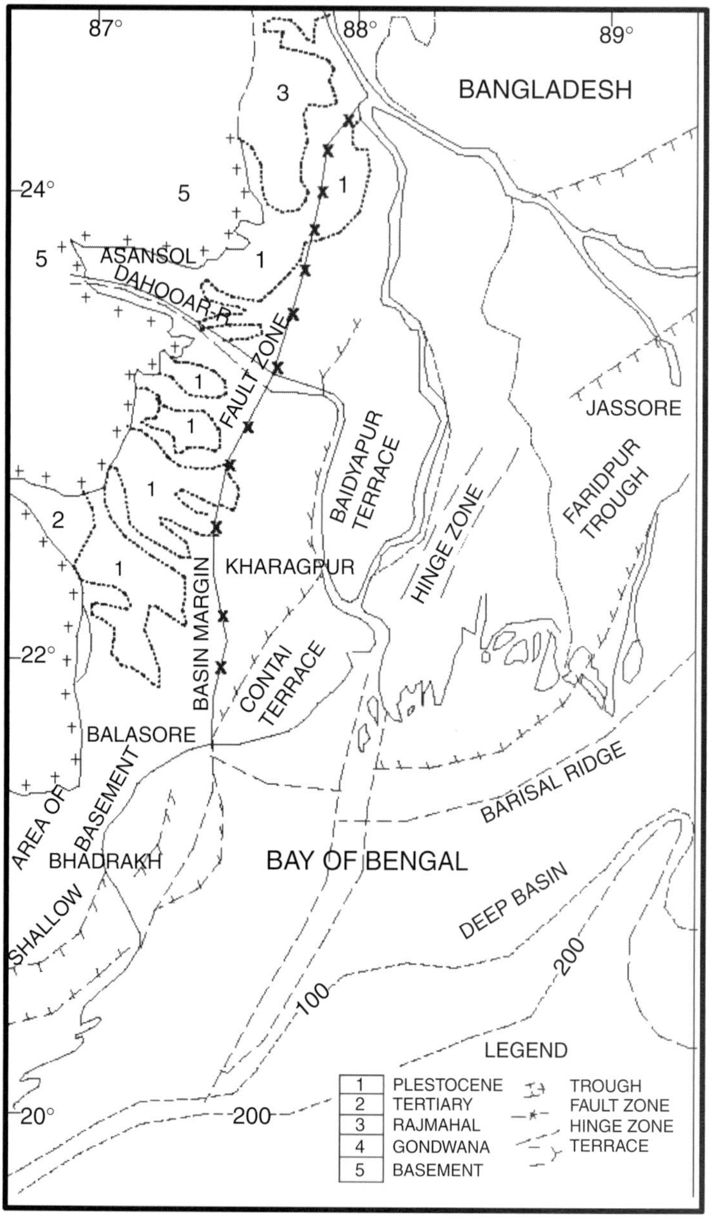

FIGURE 5.30 Simplified tectonic map of the Bengal Basin showing different geological provinces and major tectonic elements *(adapted from Das Sharma et al., 1993)*.

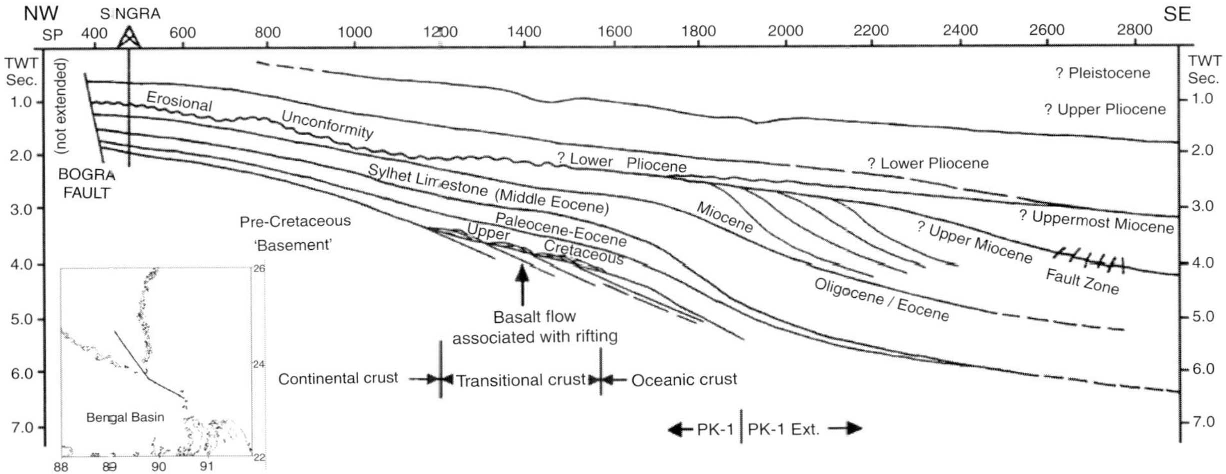

FIGURE 5.31 Geological cross-section through the Bengal Basin. Significant thickening of the Tertiary sedimentary succession from northwest toward south-east can be observed *(adapted from Alam et al., 2003)*.

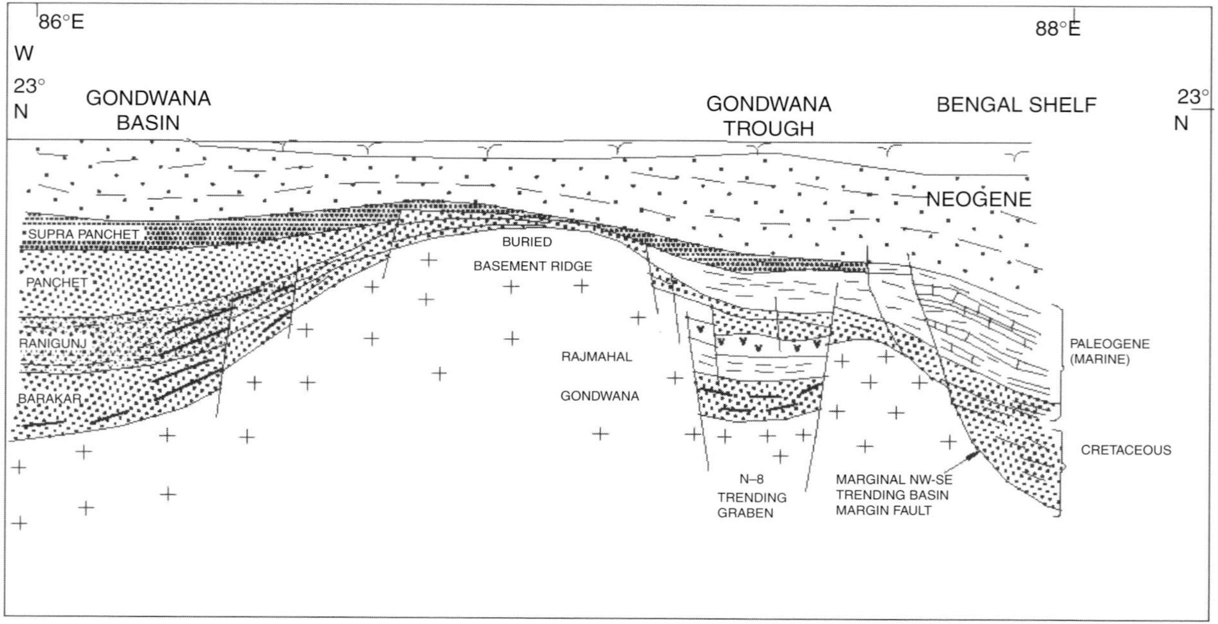

FIGURE 5.32 Schematic section in the western part of the West Bengal shelf. The genetic relationship between the Gondwana basin and the marine Cretaceous–Paleogene basin can be observed *(adapted from Banerji, 1981)*.

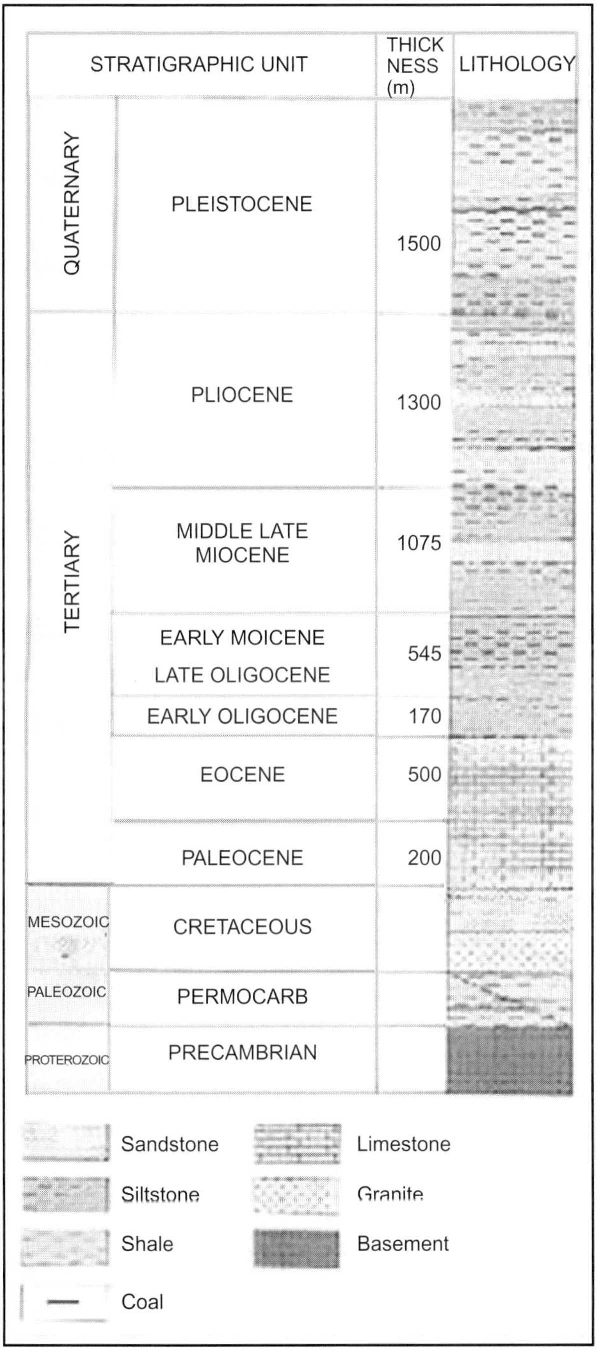

FIGURE 5.33 Generalized stratigraphy of the Bengal basin (Reproduced with permission from Berger et al. (1983) © Schlumberger Ltd). (For color version of this figure, the reader is referred to the online version of this chapter.)

5.3.3.2.3 Petroleum System

- **Source rock**

  Source rock facies are well established in Permian coals and shales in the grabens and half-grabens. Younger Tertiary shales in the deltaic environment also contain source rock facies. Tertiary sources with good organic matter found to become oil mature during Oligocene. The restricted basin setup is the ideal source rock locii which are expected to contain terrestrial hydrocarbon facies in this basin.

5.3.4 Deepwater Petroleum Prospects Along the Eastern Offshore of India

5.3.4.1 Importance of Deepwater System

The term deepwater geologically refers to the environment on the seaward side of the continental shelf break. The hydrocarbon exploration and production activities in the deepwater (400–2000 m) and ultra-deepwater (>2000 m) basins have significantly expanded worldwide since 1990. As the deepwater reservoirs cannot be reached easily, clear understanding of deepwater systems depends mainly through remote observations using geological and geophysical techniques and success in their exploration mostly depends on how well we will be able to integrate these datasets. The study of deepwater deposits makes use of data from detailed examination of outcrops, from 2D and 3D reflection seismic surveys, and from cores, log suites, and biostratigraphy. Reflection seismic analysis can be used in various ways to produce images of significant depositional or structural features. While the deepwater deposits located at shallow depths can be very well imaged by high-resolution 3D seismic techniques, these provide useful analogues for the more deeply buried deposits. From the drilling point of view, technological revolution has expanded the options for exploration of deepwater deposits, as well as shelf and slope deposits, which were previously beyond the reach of the drill bit in the deepwater settings. Although a number of prominent oil and gas discoveries have been made in some of the basins around the world, most of the E&P activity is concentrated in mainly three areas of the world: the northern Gulf of Mexico, Brazil, and West Africa and a part of Asia (Nilsen et al., 2007). Globally, deepwater remains an immature frontier with many deepwater sedimentary basins being poorly explored.

The term "deepwater" is informally used by the petroleum industry in two ways (Shanmugam, 2006). First, it refers to the sediments deposited at water depths considered to be deep and located somewhere in the upper to middle slope region to the floor of a basin (Fig. 5.34), and second, it indicates the present-day water depths more than 400 m (Shanmugam, 2006). While the geoscientists use the term to convey the deepwater depositional origin of a buried reservoir, the drilling engineers would use the term to denote the

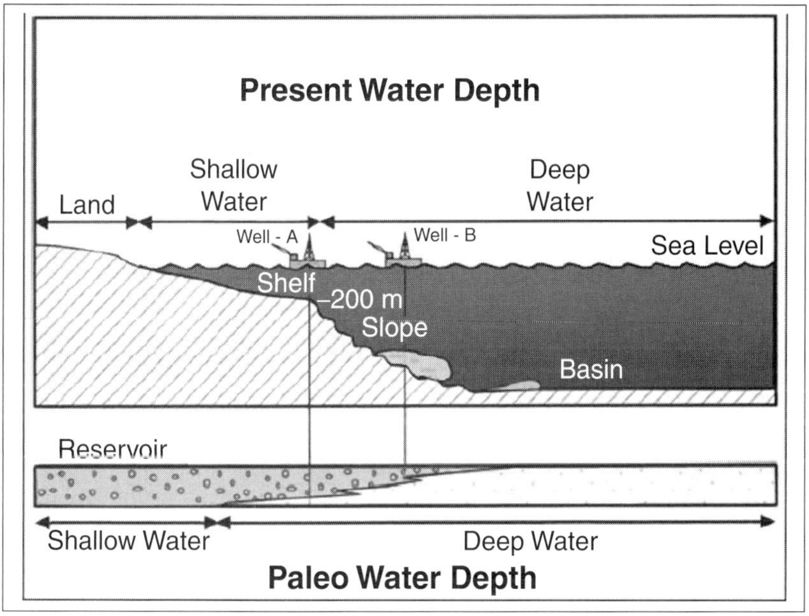

FIGURE 5.34 Diagram illustrating the deepwater (>400 m) terminology *(adapted from Shanmugam, 2006)*. (For color version of this figure, the reader is referred to the online version of this chapter.)

deepwater drilling depths. In this book, the deepwater is referred to the present offshore water depths from 400 to 2000 m, and ultra-deepwater is the area lying deeper than 2000 m.

5.3.4.2 Deepwater Area of the Eastern Offshore

Large sedimentary basins have formed at the deltas of the major rivers in the east coast of India which are potentially petroliferous and attract increasing levels of exploration activity (Bastia, 2007). Large volumes of seismic, gravity, and magnetic, multibeam surveys, and seabed logs acquired along the eastern Indian offshore brought out considerable knowledge of the individual basins. In a broader sense, the deepwater eastern offshore area of India is discussed here under the following two types of basins based on the tectonic regime: (i) the divergent margin basins of the east coast and (ii) the convergent margin basins along the offshore Andaman and Nicobar islands.

The discovery of giant sized oil and gas fields in the offshore K–G basin has shifted the focus toward the vast unexplored deepwater sediments of the Bay of Bengal. Sedimentary accumulations of the Indian East Coast range from rifted Mesozoic to Recent sediments. These were deposited when the continental rifting and subsequent drifting of India away from Antarctica ushered in the accumulation of marine sediments along the east coast.

5.3.4.2.1 Connecting Shelf to the Deepwater: Evidences from the East Coast of India

In the case of continental margin, the shelf region acts as the main staging area for the sediments to disperse into the deepwater areas. Due to tectonic movements and continuous changes in sea level at the coast, topographic changes takes place in the shelf as well as slope areas. These changes affect the sediment pathways leading to deepwater basins and therefore influence the deposition of deepwater systems. As many canyons observed in the shelf and slope regions act as conduit for sediment transportation, connecting shelf to deepwater is an important step in understanding the deepwater depositional systems. Identification of different stratal terminations from seismic data and interpretation of shoreline shifts are important in the sequence stratigraphic framework. In fact, the sequence stratigraphy is extremely helpful in determining the facies associated with shelf, slope, or basinal part (Catuneanu, 2005). A shoreline shift can be transgressive or regressive, which can help in characterizing various offshore reservoirs. For example, during transgressive phase, sediments transported to the deepwater are mostly fine grained, while during regressive phase, the channel system will be active. The continental shelf along the east coast of India is narrower when compared to the west coast. A number of canyons have been observed in the slope area and the presence of these canyons at the shelf–slope break (Fig. 5.35) acts as inlet for sedimentation to deep basin. The shelf canyons can be connected to the deepwater channel system and a variety of shelf margin and slope processes influence the sediment pathways into the deepwater basins (Figs. 5.36 and 5.37).

5.3.4.3 Deepwater Depositional Elements

The mechanisms and physics of the flows which produced sedimentary deposits in the submarine slope, submarine canyons, submarine fans, and the submarine basin plain are known as deepwater processes (Shanmugam, 2006). Sediments come from erosion of land and also from biological processes. In deepwater environments, they can be deposited by various processes of erosion, transportation, and deposition. These are classified into the following broad categories (i) Pelagic and hemipelagic settling, (ii) Mass transport processes, (iii) Sediment gravity flows, and (iv) deepwater bottom current processes (after Shanmugam, 2003). To properly interpret reservoir characteristics such as sand content, sand trend, continuity, connectivity, and reservoir quality, we must be aware of the processes by which the deposit formed. Several comprehensive summary books and papers have addressed the major controls on the deposition of deepwater systems (Bouma et al., 1985; Nelson and Nilsen, 1984; Pickering et al., 1989; Richards et al., 1998; Shanmugam, 2003, 2006; Stow et al., 1985). A review

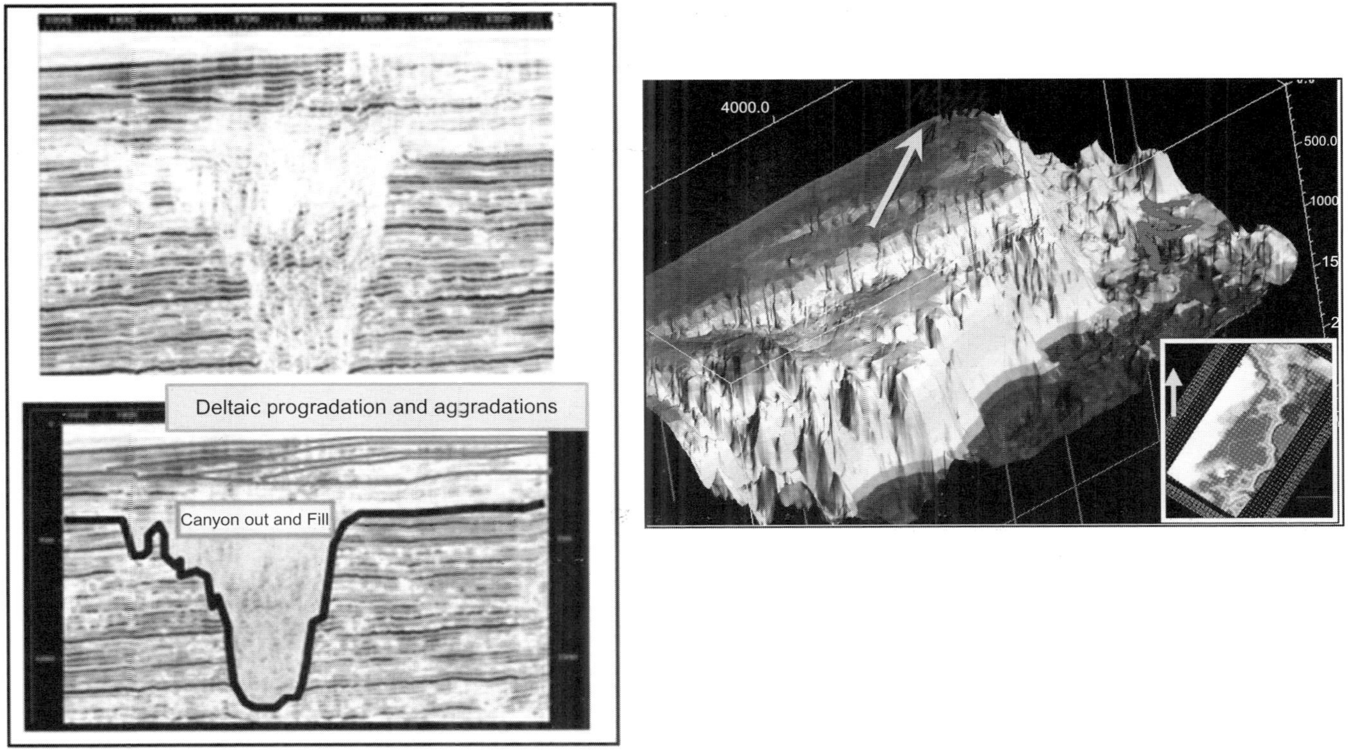

FIGURE 5.35 Uninterpreted and interpreted seismic sections (left) showing deep shelfal canyon cut above which progradational and aggradational delta pack was deposited during high stand. A massive shelfal canyon cut of Pleistocene age as mapped within the seismic volume (adapted with permission from Bastia, 2007 © Technology Publications, Dehradun). (For color version of this figure, the reader is referred to the online version of this chapter.)

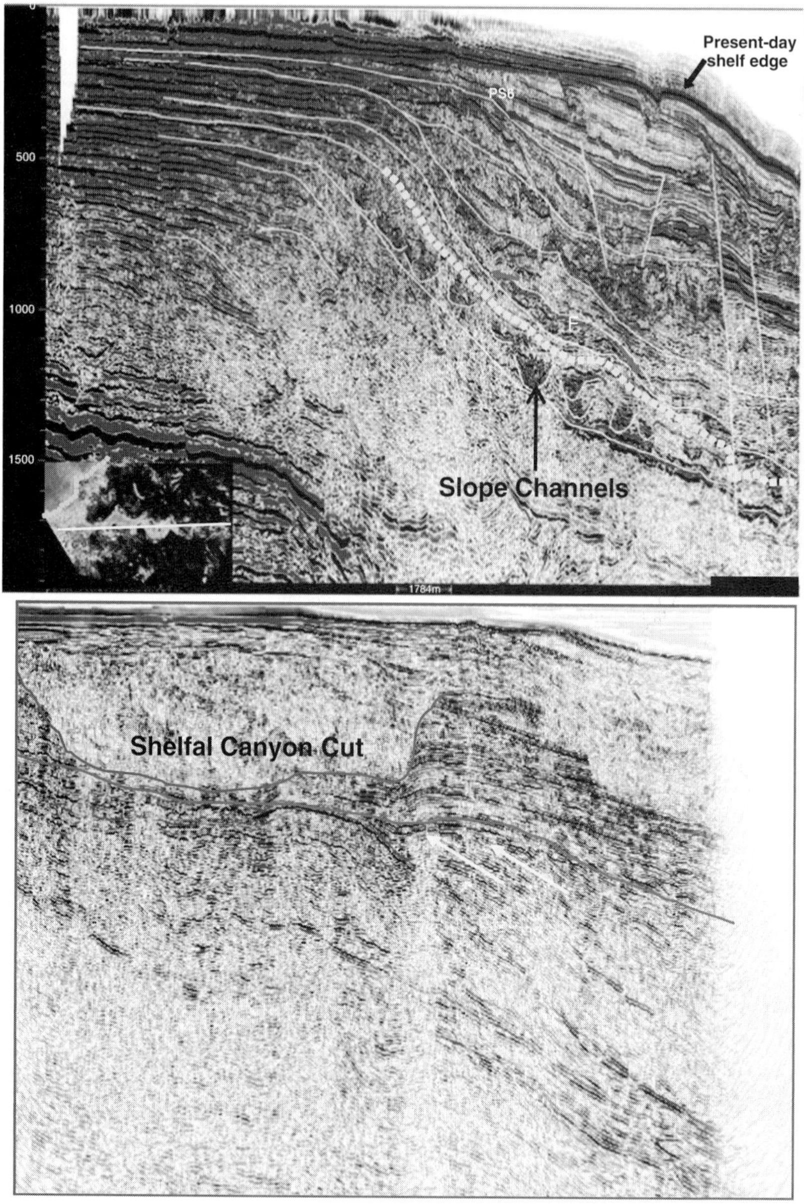

FIGURE 5.36 Seismic sections depicting the slope channels and shelfal canyon cut (adapted with permission from Bastia, 2007 © Technology Publications, Dehradun). (For color version of this figure, the reader is referred to the online version of this chapter.)

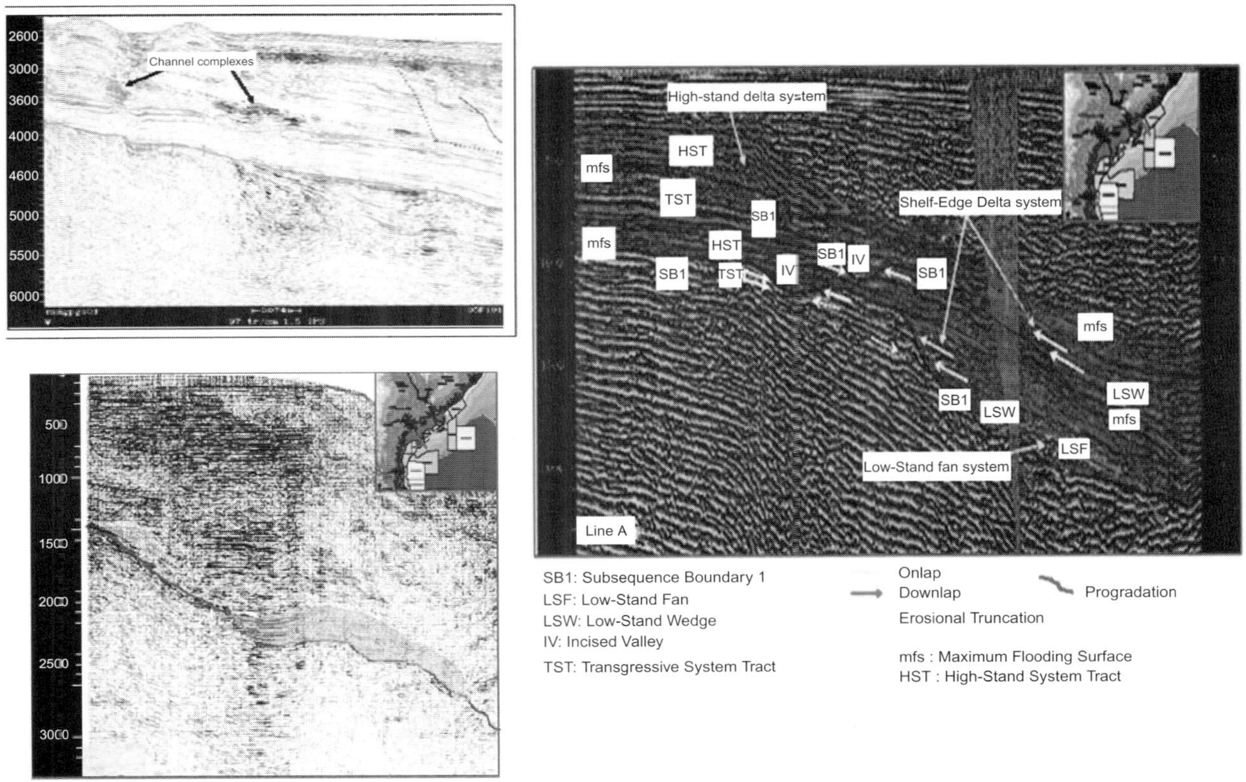

FIGURE 5.37 Seismic section showing stacked channels (above left), seismic phase section showing major sequence stratigraphic elements connecting high-stand prograding delta on the shelf, low-stand wedge and slope/basin floor fan (above right), and seismic section showing shelf-edge delta progradation and low-stand fan complex (adapted with permission from Bastia, 2007 © Technology Publications, Dehradun). (For color version of this figure, the reader is referred to the online version of this chapter.)

of these aspects is beyond the scope of this book; however, more details can be found in Bastia (2007).

The main depositional elements observed in the deep-sea basins of the east coast of India can be broadly subdivided into (i) Tertiary formations consisting of delta–distributary channel complex, shelfal canyon cut-and-fill sequences, and deepwater channel complex (Figs. 5.35 and 5.36) and (ii) Mesozoic rift-related structural plays. We present here seven representative seismic sections or seismogeological sections prepared from the high-resolution MCS lines (Bastia, 2010; Bastia et al., 2010a) covering the shelf to deepwater eastern offshore (location in Fig. 5.38) to understand the deepwater subsurface geology along the east coast of India. As these profiles lie across the margin in the offshore continuation of the major onshore basins of the east coast, they are described from the point of view of the respective basins. Some salient aspects of each profile are described below.

5.3.4.3.1 Bengal Offshore

This north–south trending section passes through the Ganga–Brahmaputra Basin in the offshore with water depth varying from 70 to 1900 m. The basin is fed by two major river systems of the Ganges and Brahmaputra and extends in the east into Bangladesh and, in the south, into the deepwaters of the Bay of Bengal (Curray and Moore, 1974). The depositional elements in the offshore part of the basin are typically represented by numerous repeated canyon cuts caused by mass wasting, with intervening fills in the form of deepwater to subaerial channels and progradational deltaic deposits. The section (Fig. 5.39) shows the shelf–slope system with prograding shelf edge with time and importantly images the morphology of present-day channel–levee overbank system, which are known to be the prominent play type in the deepwater basins around the world. The most interesting seismic signatures identified in the section are the prograding clinoforms, which shows the basinward progradation of shelf edge in the Tertiary section (Fig. 5.40). Nearly 5000 m of Tertiary sediments are observed, which is caused by the mighty Ganges and Brahmaputra rivers. A number of canyon cuts are identified at the shelf and are formed during the low stand when the shelf has been exposed.

5.3.4.3.2 Mahanadi Offshore

This NW–SE trending section passes through the Mahanadi basin at the water depth from 80 to 2600 m (Fig. 5.41) and shows the shelf and slope fan extending to basin floor. The depositional elements are the channel–levee complex identified as bright amplitude package. The section also displays a slope fan lying at the slope and part of basinal side. The younger sediments are cut by numerous vertical to subvertical polygonal fault systems associated with differential

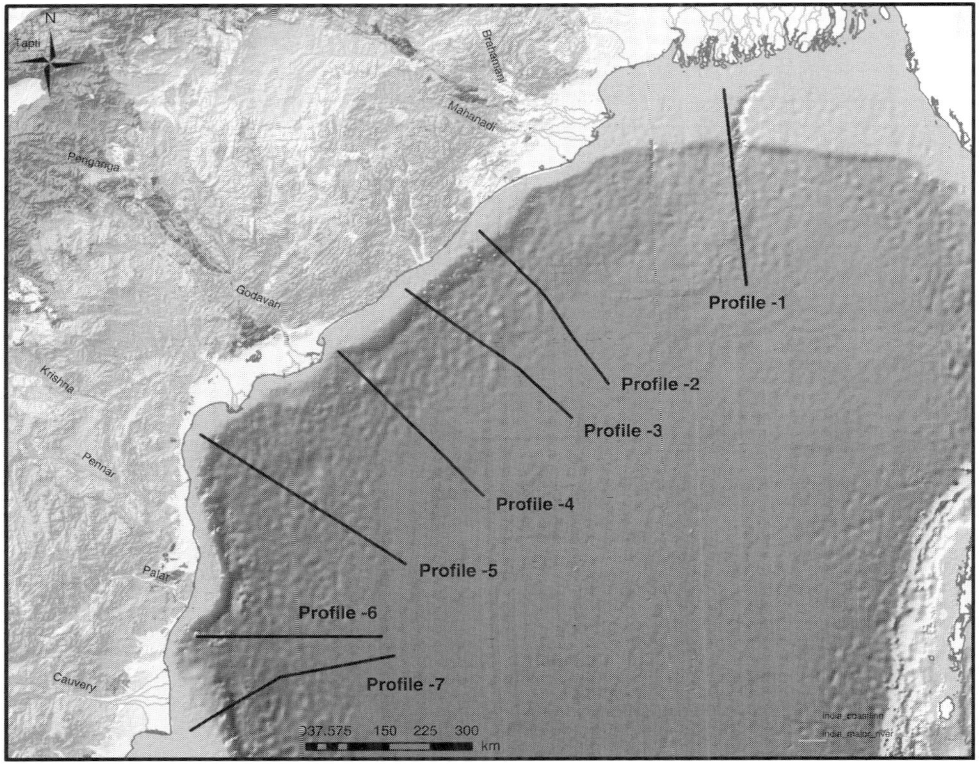

FIGURE 5.38 Topographic relief map of the eastern offshore showing the location of MCS lines used to describe various deposition elements along the margin. Interpreted sections are from Reprinted with permission from Bastia, 2010 © Society of Petroleum Geophysicists, India and seismic sections are from Bastia et al. (2010a&c). (For color version of this figure, the reader is referred to the online version of this chapter.)

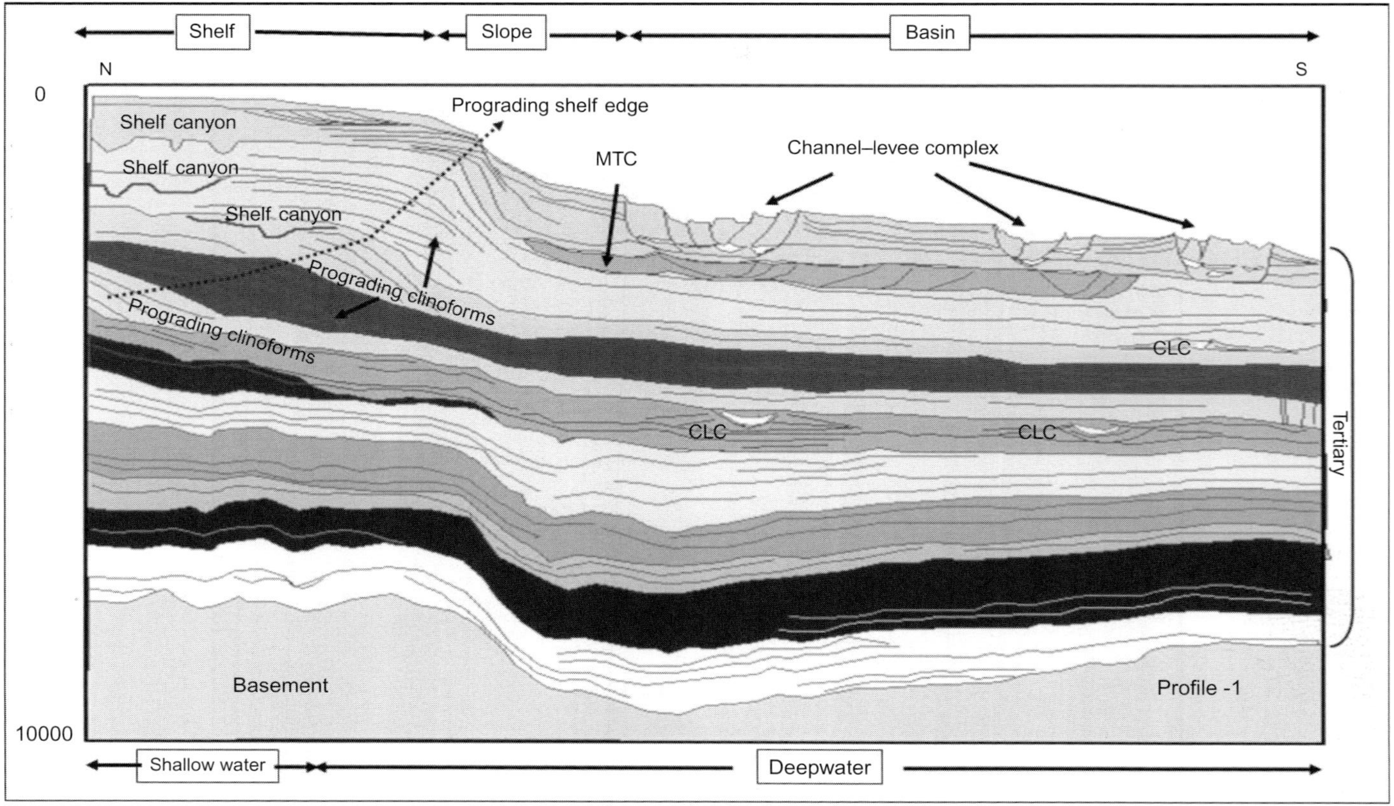

FIGURE 5.39 Interpreted seismogeological section representing shelf–slope system and depositional elements of Ganga–Brahmaputra (Bengal) basin. Various features identified on the section are prograding shelf edge through geological time and morphology of present-day channel–levee complex, shelf canyons, and mass transport complex. MTC, mass transport complex; CLC, channel–levee complex (adapted with permission from Bastia, 2010 © Society of Petroleum Geophysicists, India).

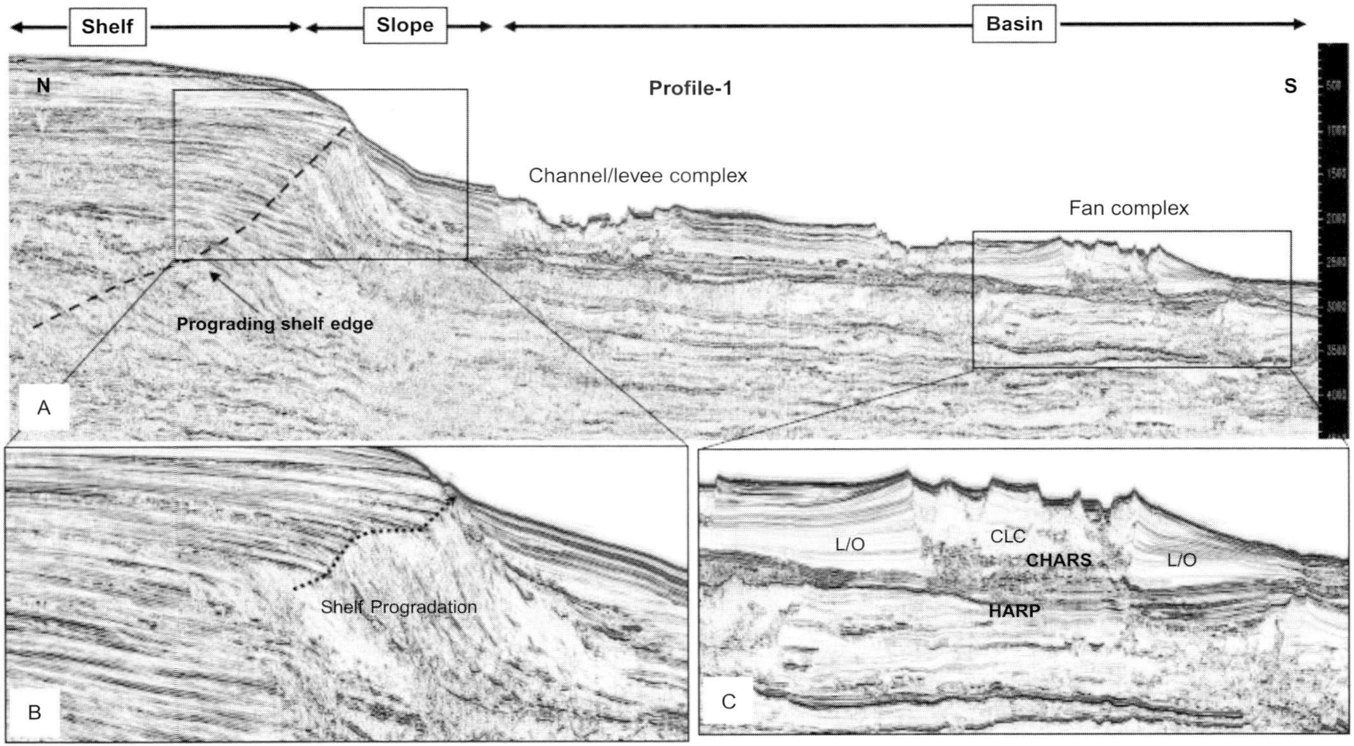

FIGURE 5.40 Seismic section showing (A) shelf-slope system and depositional elements in the Bengal offshore, (B) prograding shelf edge in the younger Tertiary time, and (C) morphology of present-day channel–levee complex. L/O, levee/overbank; CLC, channel–levee complex; CHARS, chaotic high-amplitude reflectors; HARP, high-amplitude reflection package (adapted with permission from Bastia et al., 2010a © Elsevier). (For color version of this figure, the reader is referred to the online version of this chapter.)

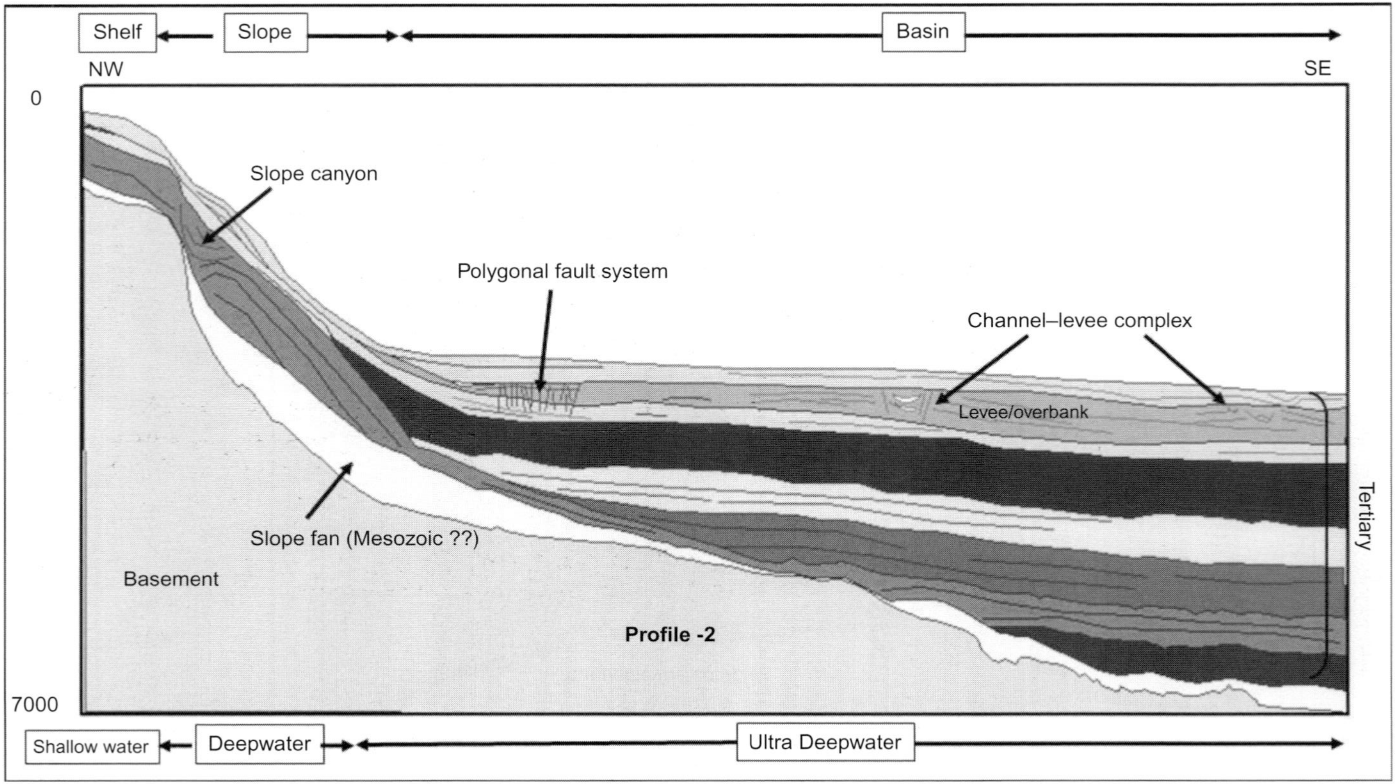

FIGURE 5.41 Interpreted seismogeological section in the Mahanadi offshore showing thick Tertiary sequence. One prominent slope fan is identified in the section. This fan is characterized by transparent zone in seismic section (Fig. 5.42). A number of channel–levee complexes are identified in the Tertiary section which form a part of basin floor fan. *Adapted with permission from Bastia, 2010 © Society of Petroleum Geophysicists, India.*

compaction and fluid expulsion. Onlapping reflectors suggest the presence of wedgeout plays (Fig. 5.42). High-amplitude bodies within the canyon cuts represent deepwater channel complex, which is another well proven play type around the world.

5.3.4.3.3 85° E Ridge

The 85° E ridge is a major N–S trending subsurface ridge passing through the Bay of Bengal and meets the east coast in the Mahanadi offshore (Bastia et al., 2010b). The ridge played a significant role through geologic times in confining and distributing the sediments from the coast to deep sea. Part of the ridge was aerially exposed during early Paleogene period and acted as local provenance for sediment (Dangwal et al., 2008). The seismic section (Fig. 5.43) shows the presence of two local lows on the flanks of the ridge. These isolated lows will behave as sub-basins within which the sediment dispersal pattern could vary. A number of channel–levee complexes are also identified.

5.3.4.3.4 K–G Offshore

This NW–SE trending section passes through the deepwater area of the K–G basin. The K–G basin has evolved as a composite rifted basin characterized by horst–graben structural features since the late Jurassic period. The basin is fed by two major river systems, namely, Krishna and Godavari. The initial rifting–drifting phase had initiated a number of fluvio-lacustrine sediments throughout the basin. The regional seismogeological section (Fig. 5.44) depicts thick pile of Mesozoic and Tertiary sedimentary sequences. The Mesozoic sequence represents rifted geometry, whereas the Tertiary sequence exhibits very bright amplitude package suggesting numerous vertically stacked sinuous deepwater channel–levee complexes (Bastia, 2006). Drill well results show that the sand to shale ratio within these channels is very high, providing an excellent reservoir facies for petroleum accumulation. As can be seen, there are a number of mini basins created during the rift process. However, from the seismic data alone, it is difficult to differentiate between the pre-rift and syn-rift sediments. Prominent structuration in the Tertiary sequence is due to the shale movement.

5.3.4.3.5 Pennar–Krishna Offshore

The section is located in the offshore area north of Pennar basin and southern part of the K–G basin. The section (Fig. 5.45) depicts Mesozoic rifts with associated half grabens and high-amplitude onlapping Tertiary sequences. Numerous polygonal faults are also noticed in the younger Tertiary section. Another important event in the younger Tertiary section is the presence of thick mass transport complex, and an understanding of these mega features would immensely help in describing the basin evolution.

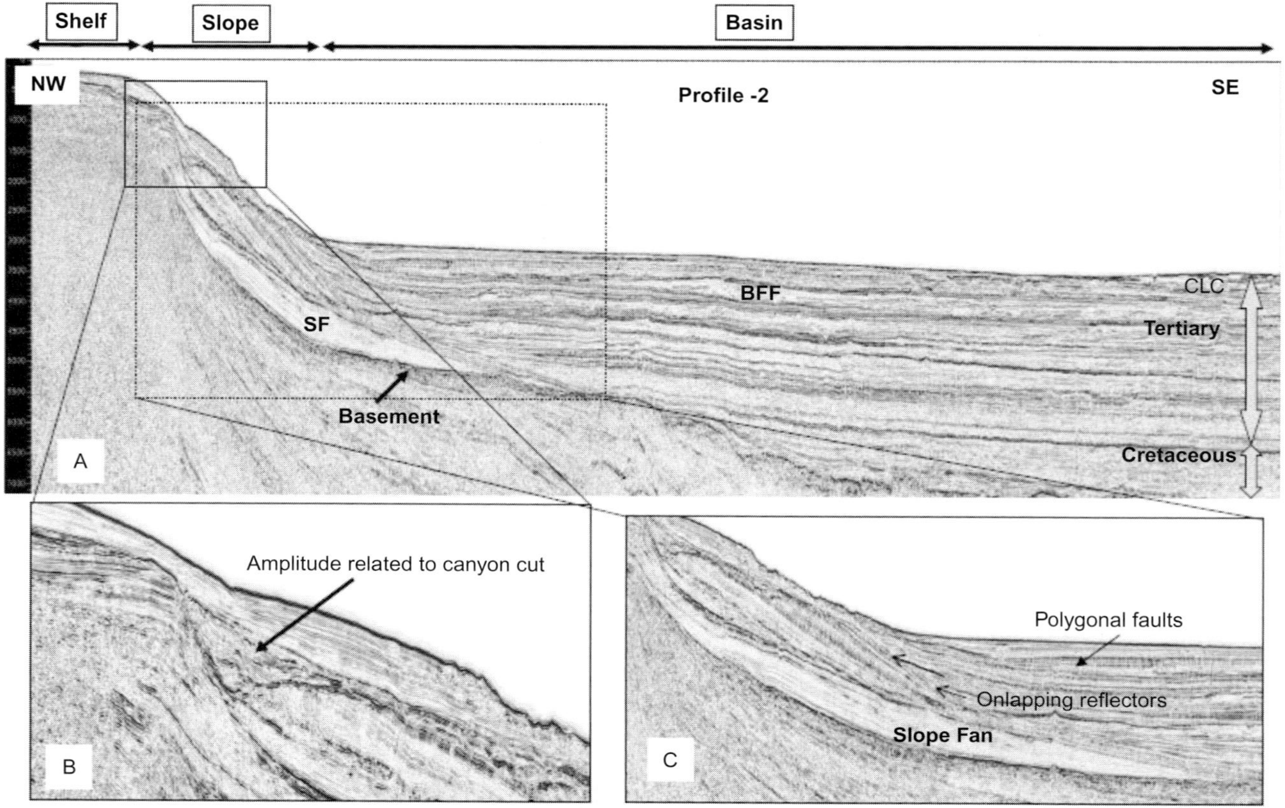

FIGURE 5.42 (A) Seismic section through the Mahanadi basin. SF, slope fan; BFF, basin floor fan; CLC, channel–levee complex. (B) High amplitude associated with canyon cut. (C) Transparent zone representing the slope fan. Also shown are the polygonal faults and onlapping reflectors (adapted with permission from Bastia et al., 2010a © Elsevier). (For color version of this figure, the reader is referred to the online version of this chapter.)

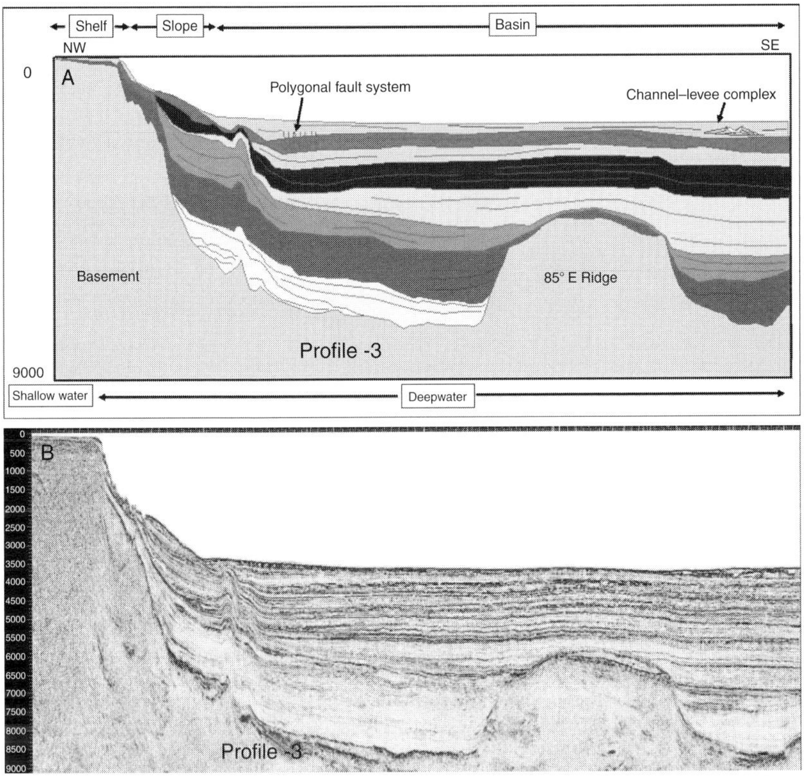

FIGURE 5.43 MCS section (below; after Bastia et al., 2010a) and the interpreted seismogeological section (adapted from Bastia, 2010) depicting the expression of 85° E ridge. Two lows flanking the ridge can be a good place for source rock generation. Channel–levee complex and polygonal faults are also seen. *Reprinted with permission from Bastia, 2010 © Society of Petroleum Geophysicists, India.* (For color version of this figure, the reader is referred to the online version of this chapter.)

5.3.4.3.6 Cauvery Offshore

This section lying in the southern part of east coast of India between few meters and 3800 m bathymetry helps to discuss the deepwater subsurface in the Cauvery offshore. This basin represents a series of horst–graben features that are oblique to the coast (Balakrishna and Sarma, 1981). The grabens are filled with Mesozoic sediments. Thick Tertiary sediment pack mostly contributed by the Cauvery river system is accumulated as a thick wedge on the uniformly eastward sloping platform. The section shows the prominent basement highs and the Mesozoic sediments abutting against these highs (Fig. 5.46). Presence of a steep slope allows bypassing of the shelf sediments to the deepwater (Fig. 5.47); as a result, the slope region is almost starved and sediments of huge thickness are deposited in the

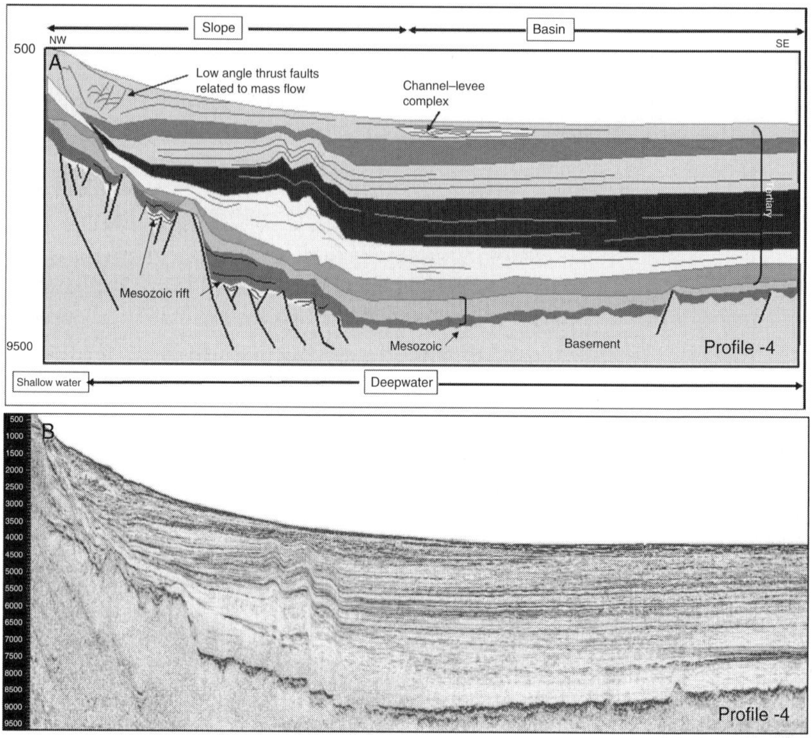

FIGURE 5.44 MCS section (below; after Bastia et al., 2010a) and the interpreted seismogeological section (above; after Bastia, 2010) across the Krishna–Godavari basin depicting Mesozoic and Tertiary sedimentary sequences. Interpreted faults in the Mesozoic section form well-defined half-grabens. Tertiary section is characterized by channel–levee complex and mass transport complex. *Reprinted with permission from Bastia, 2010 © Society of Petroleum Geophysicists, India.* (For color version of this figure, the reader is referred to the online version of this chapter.)

basinal part. Closely spaced polygonal faults are also observed in the section. The onlapping sequences in the older sections could be good plays to follow.

5.3.4.4 Basin Architecture and Associated Play Types

Extensive coverage of 2D offshore seismic data along the entire east coast of India provided an excellent framework for describing the continental-scale sedimentary basin evolution and associated play types in the deepwater eastern offshore (Bastia, 2006; Bastia et al., 2010a). The map showing the basement architecture and the location of the seismic profiles (AA′ to GG′) utilized here (after Bastia, 2006) pertaining to individual basins is presented in Fig. 5.48.

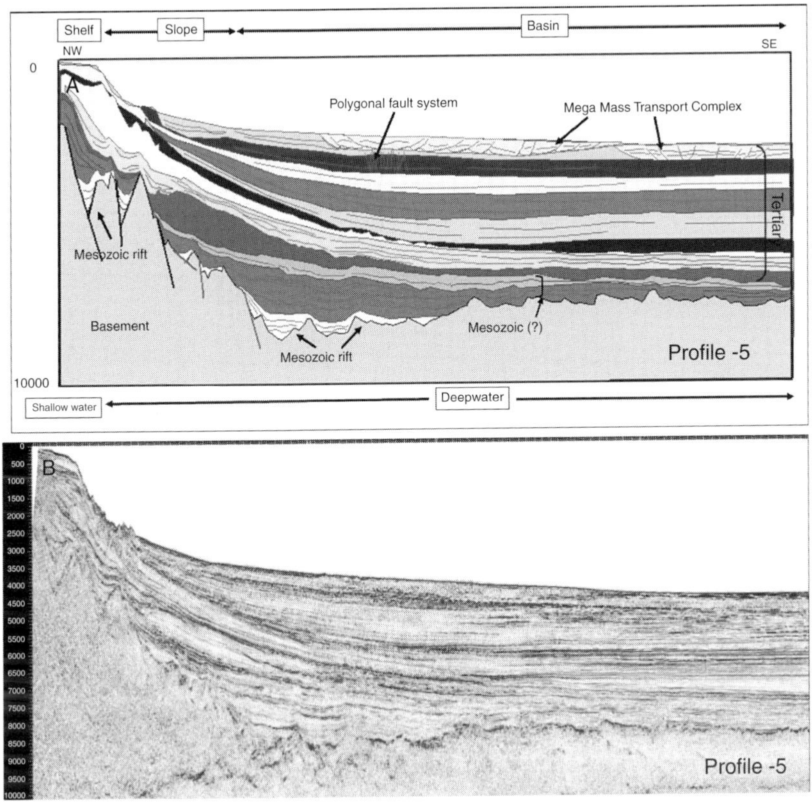

FIGURE 5.45 MCS section (below; after Bastia et al., 2010a) and the interpreted seismogeological section (above; after Bastia, 2010) connecting the shelf to abyssal plane across the Pennar–Palar basin. Mesozoic rift fills and a thick Tertiary sequence can be identified. Two huge mass transport complexes are seen in the younger Tertiary section. A number of closely placed polygonal faults are also seen. *Reprinted with permission from Bastia, 2010 © Society of Petroleum Geophysicists, India.* (For color version of this figure, the reader is referred to the online version of this chapter.)

5.3.4.4.1 Cauvery Basin

The basin contains a series of horst–graben features associated with rifting episodes. The grabens are primarily filled with syn-rift Mesozoic rocks followed by a Tertiary sediment fill as thick wedge.

Wedgeout and pinchout against basement highs (Fig. 5.49) are quite prevalent in the basin. They are recognized as the potential hydrocarbon traps in the basin. The Paleocene–Eocene channel and fan system (Fig. 5.50) may contain a good reservoir facies which are connected to the underlying source by the major fault system. These are considered to be favorable hydrocarbon accumulation traps.

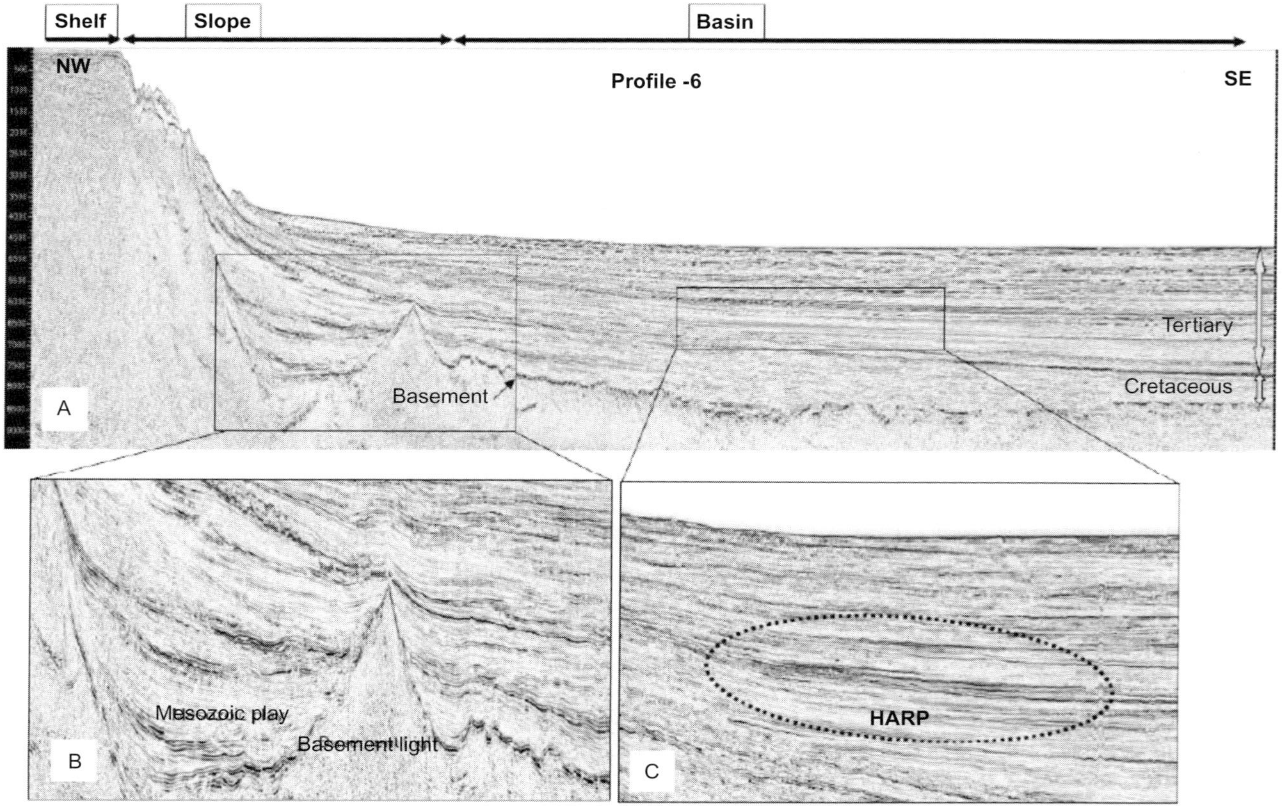

FIGURE 5.46 (A) Seismic section across the Cauvery basin showing prominent basement highs and associated Mesozoic–Tertiary sequences, (C) Mesozoic play abutting the basement high, and (C) Tertiary high-amplitude reflection package (HARP) representing channel fills *(adapted with permission from Bastia et al., 2010a © Elsevier)*. (For color version of this figure, the reader is referred to the online version of this chapter.)

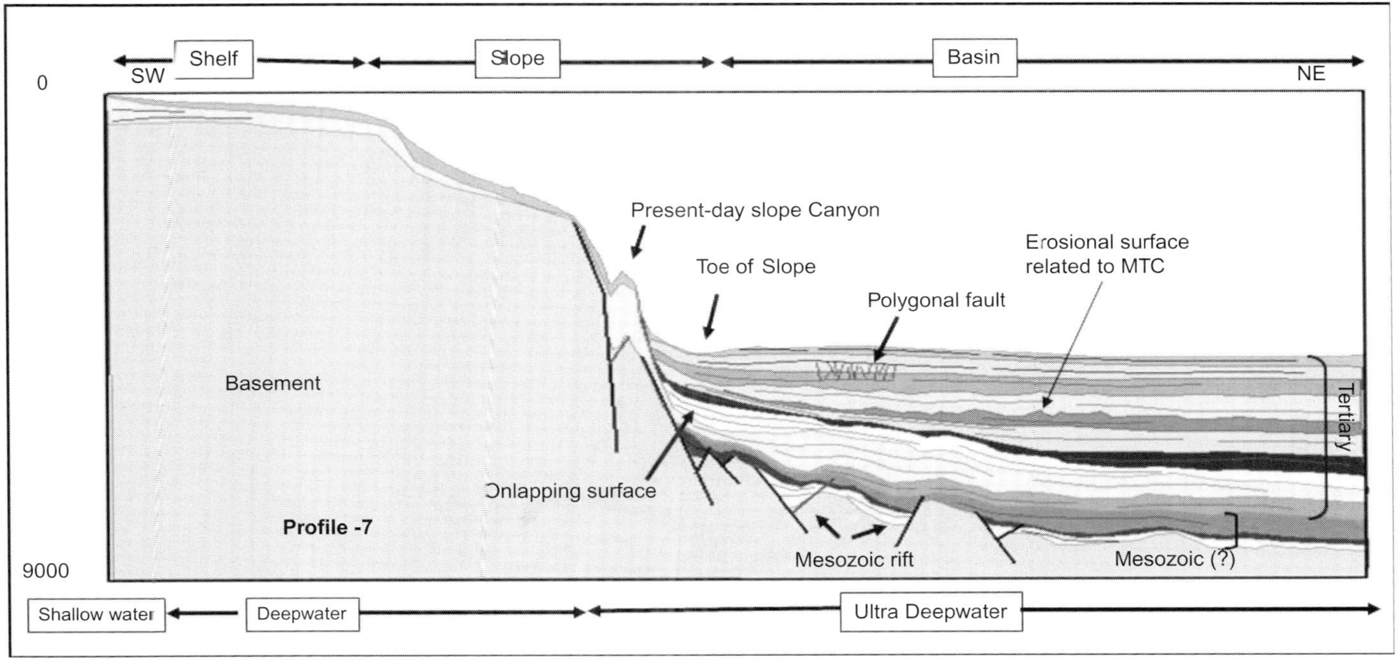

FIGURE 5.47 Interpreted seismogeological section in the Cauvery basin showing prominent basement highs and associated Mesozoic–Tertiary sequences. Sediments abutting against the basement can be a good target for hydrocarbon exploration (adapted with permission from Bastia, 2010 © Society of Petroleum Geophysicists, India).

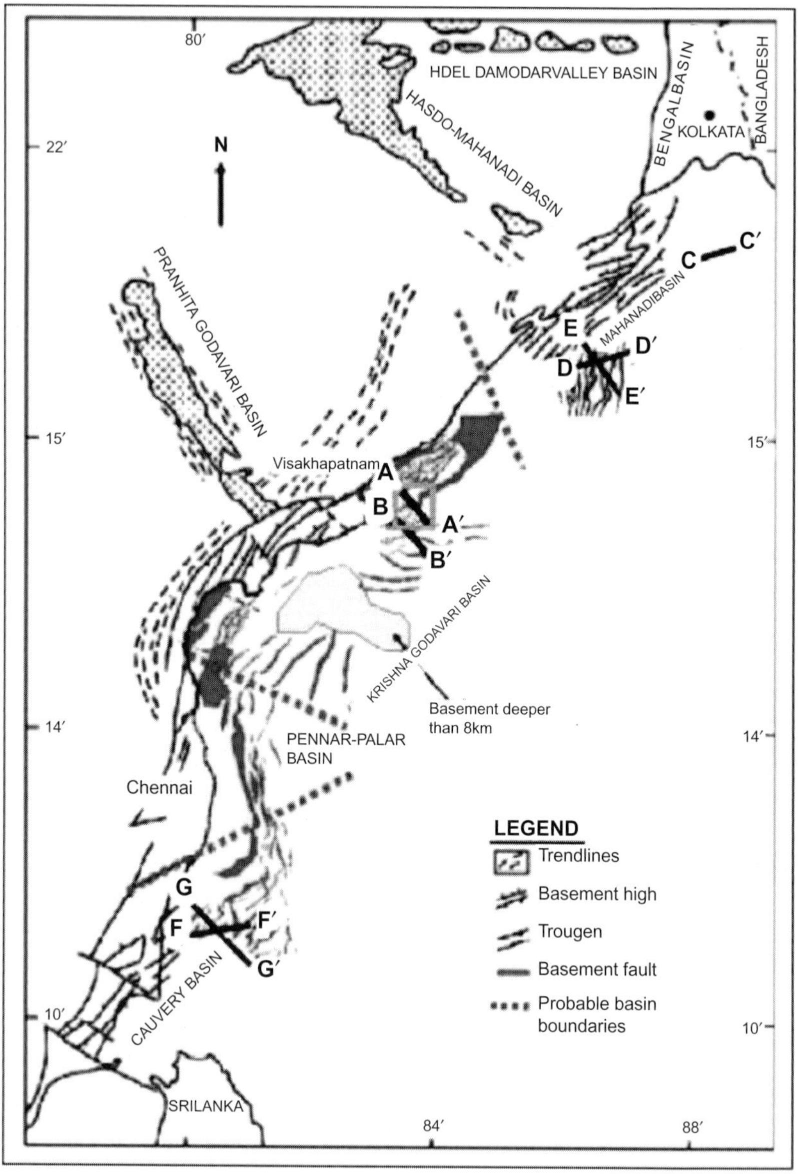

FIGURE 5.48 Basement structural trends along the eastern offshore area showing the location of seismic sections (AA′–GG′) utilized to present various deepwater depositional elements that are recognized as potential hydrocarbon traps (Reproduced with permission from Bastia, 2006 © Society of Exploration Geophysicists). (For interpretation of the references to color in this figure legend, the reader is referred to the online version of this chapter.)

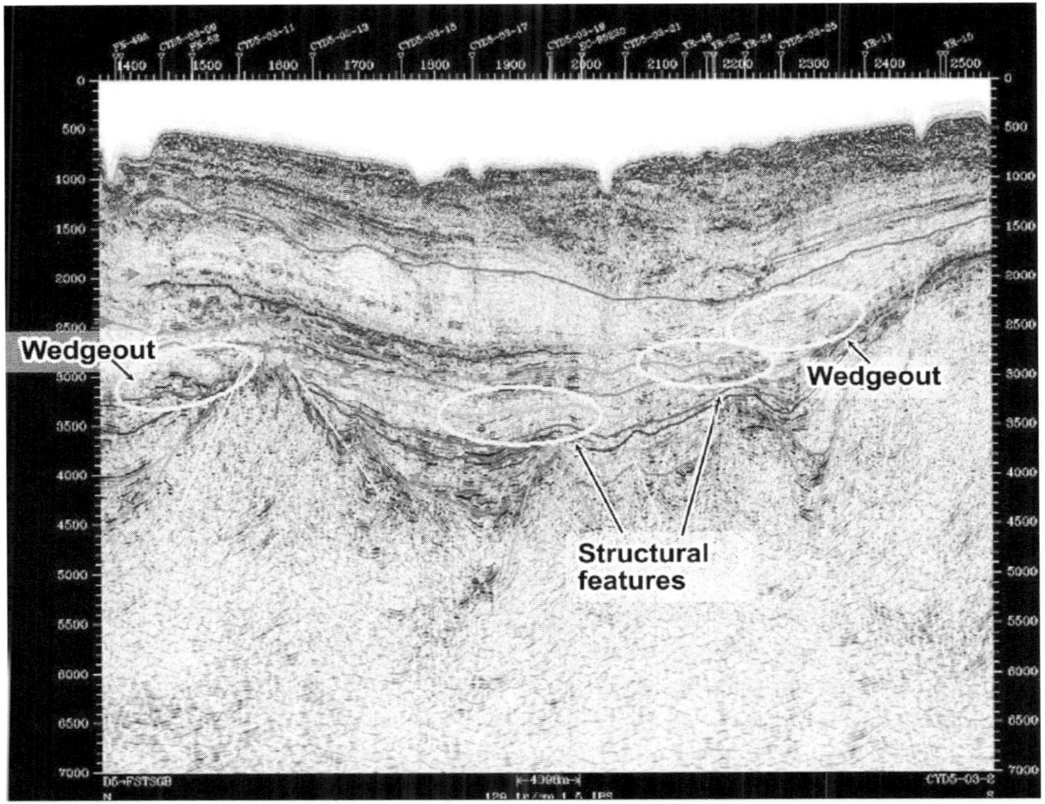

FIGURE 5.49 Seismic section (FF′) showing the rift-related horst and graben formations in the Cauvery offshore. The wedgeouts and pinchouts are observed against the basement highs (Reproduced with permission from Bastia, 2006 © Society of Exploration Geophysicists). (For color version of this figure, the reader is referred to the online version of this chapter.)

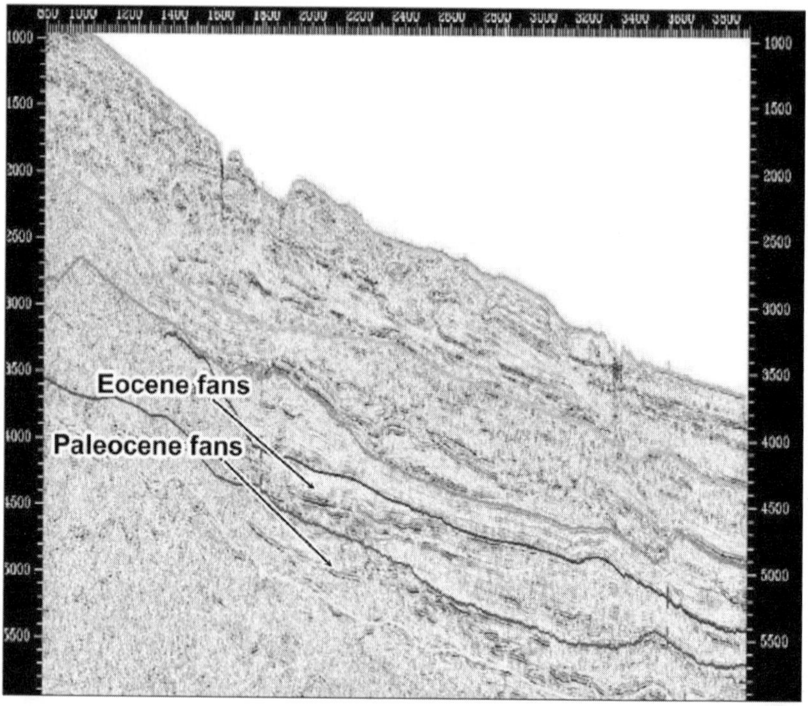

FIGURE 5.50 Seismic section (GG′) showing the Paleocene–Eocene fan system in the Cauvery offshore (Reproduced with permission from Bastia, 2006 © Society of Exploration Geophysicists). (For color version of this figure, the reader is referred to the online version of this chapter.)

5.3.4.4.2 Krishna–Godavari Basin

Study of available seismic data of the deepwater areas reveals that this basin is characterized by multiple play types. The early syn-rift infill character is recognized as stratal thickening into the bounding fault, with onlap thinning to the east onto the rotated fault block complex. The syn-rift sequence of upper Jurassic–lower Cretaceous and late post-rift fill of Tertiary age hold the major petroleum accumulation in this basin. The play types discovered so far could be divided into two broader types, namely, the structural and stratigraphic plays as follows:

Structural play

- *Rift-related structures*
- *Large anticlinal features*
- *Roll over anticlines*
- *Toe thrust plays related to growth faults*

Stratigraphic play

- *Wedgeouts against basement highs*

The distribution of different unconformity surfaces in K–G Basin which holds the petroleum prospectivity has been described by Sahu (2010) as given in Fig. 5.51.

The syn-rift package with lacustrine shale in the axial part of the graben followed by interbedded sandstone and shale unit is shown in Fig. 5.52. The subsequent deposition was marginal marine with alluvial fan to deltaic fan setting. Late post-rift depositional systems are demonstrated with several identifiable seismic sequences (Fig. 5.53). The amplitude extracted slice (Fig. 5.54) between early Pliocene and late Pliocene horizons clearly brings out the sinuous channel–levee lobes. The geometry and architecture of the canyons and channel forms have been predominantly controlled by the basin forming tectonic trends and relief of the continental margin. The fill of the channels suggests episodic sand and mud-rich flows. The sinuosity of the channel is directly correlatable with the slope gradients, more sinuous channels follow lesser gradient outside the canyon. Vertical stacking of the channels along with the lateral shift of the axis in both space and time is shown in Fig. 5.55. The aggradational deepwater channels are associated with levee–overbank deposits (Fig. 5.56). The levees are often either of the same width or even wider than the associated channels. The sinuous channels result in terminal lobes toward the deeper part of the slope with huge width (Fig. 5.54) and are predominantly filled with silt and clay.

5.3.4.4.3 NEC-Mahanadi Basin

The thick pile of Tertiary (more than 6 km) and Mesozoic sediments holds considerable sediments for hydrocarbon generation in this basin. The Paleocene–Eocene marine shales which were deposited in the lows of the basement are expected to contain major source rock facies. The presence of basement ridges (viz., 85° E Ridge) along the eastern coast of the Indian offshore restricted the ocean bottom currents and hence created an euxinic environment during the Cretaceous and early Tertiary.

Major play types observed in the basin can be categorized as under

- *Structural closure against major fault system*
- *Canyon and channel fills*
- *Fan deposits*
- *Wedgeout prospects*

The submarine canyon and turbidite fan system (Fig. 5.57) are described which are predominantly mud-rich. However, the proximal end of the fan system may have a silt and mud facies together. Syn-tectonic rift sedimentation (Fig. 5.58) is quite evident in the basin with localized thickening of sediments. Channel fan complex (Fig. 5.59) may contain coarser clastics in deepwater as compared to the submarine fan deposits.

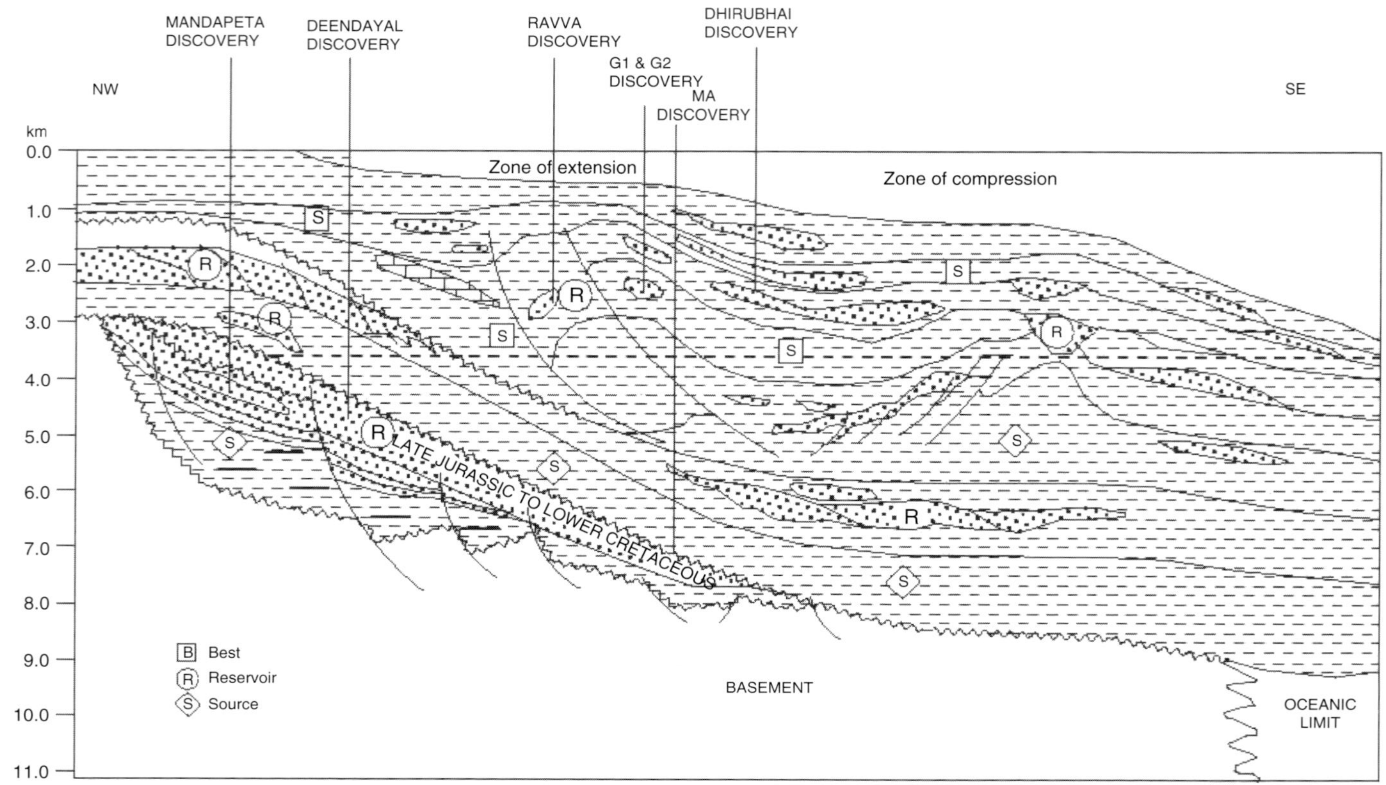

FIGURE 5.51 Distribution of different unconformity surfaces in the K–G basin *(Reproduced with permission from Sahu, 2010 © Technology Publications, Dehradun).*

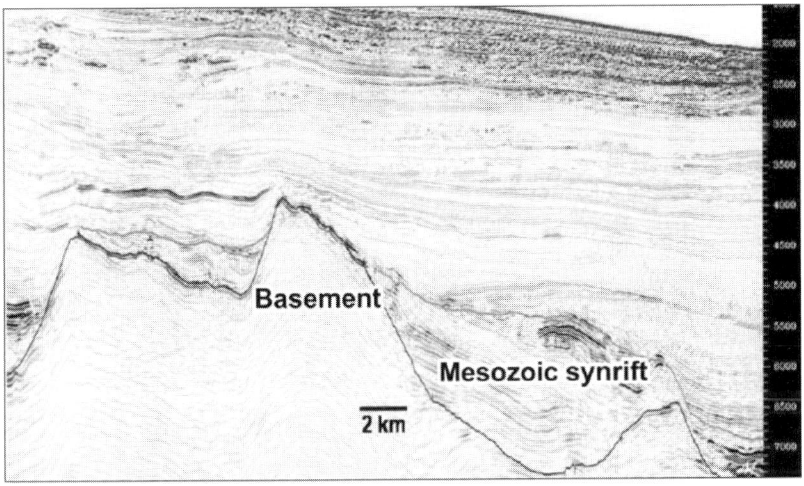

FIGURE 5.52 Seismic section (AA′) showing the Mesozoic syn-rift package in the K–G offshore (Reproduced with permission from Bastia, 2006 © Society of Exploration Geophysicists). (For color version of this figure, the reader is referred to the online version of this chapter.)

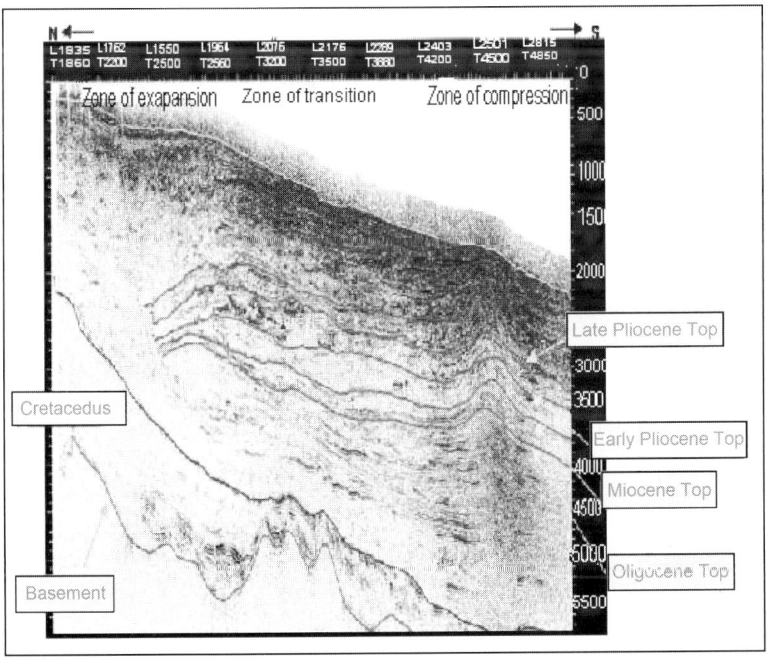

FIGURE 5.53 Seismic section (BB′) in the K–G offshore showing regional sequence stratigraphic correlation. Late post-rift sequences in this basin demonstrated classical deepwater depositional system (Reproduced with permission from Bastia, 2006 © Society of Exploration Geophysicists). (For color version of this figure, the reader is referred to the online version of this chapter.)

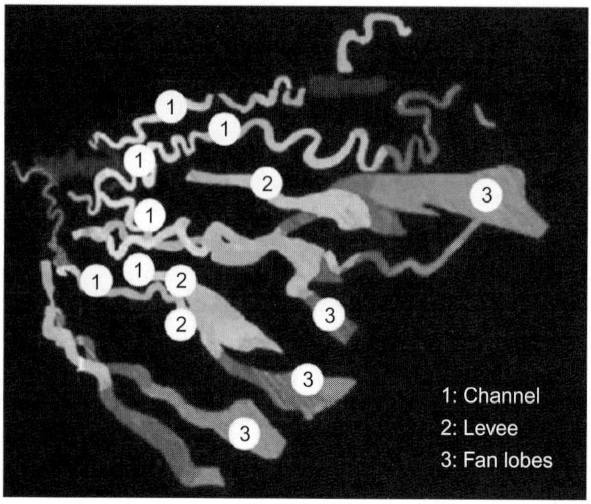

FIGURE 5.54 Stratal amplitude slice extracted from an area in the K–G offshore (marked green in Fig. 5.48) shows sinuous channel–levee lobes in various sequences in Neogene (Reproduced with permission from Bastia, 2006 © Society of Exploration Geophysicists). (For color version of this figure, the reader is referred to the online version of this chapter.)

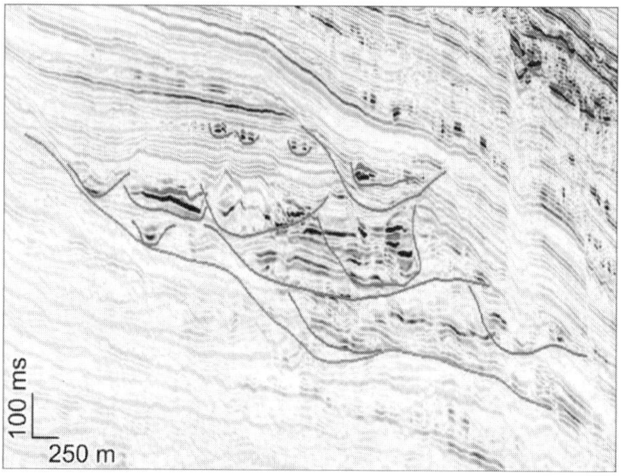

FIGURE 5.55 Vertical stack of channel fill within the channel complex (Reproduced with permission from Bastia, 2006 © Society of Exploration Geophysicists). (For color version of this figure, the reader is referred to the online version of this chapter.)

5.3.4.4.3.1 85° E Ridge The 85°E and the 90°E ridges are the two important N–S trending aseismic ridges in the BOB. As discussed in Chapters 3 and 4, the ridge is an enigmatic structural feature due to its positive basement relief, its characteristic gravity low as well as the complex magnetic signature associated with it. The ridge north of 5°N is deeply buried below thick BOB sediments; however, toward south, it is intermittently exposed on the seafloor

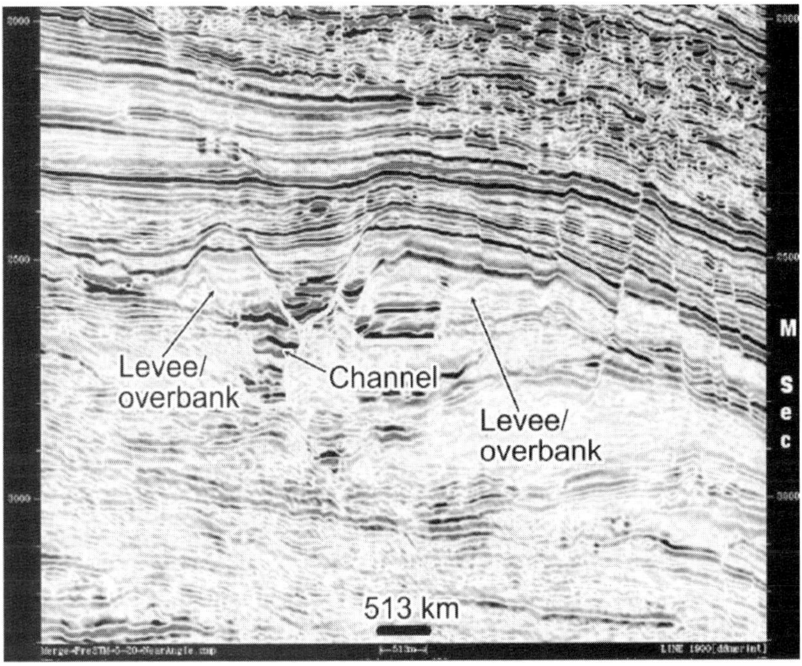

FIGURE 5.56 Section across the channel showing vertical channel stacking and development of levee/overbank (Reproduced with permission from Bastia, 2006 © Society of Exploration Geophysicists). (For color version of this figure, the reader is referred to the online version of this chapter.)

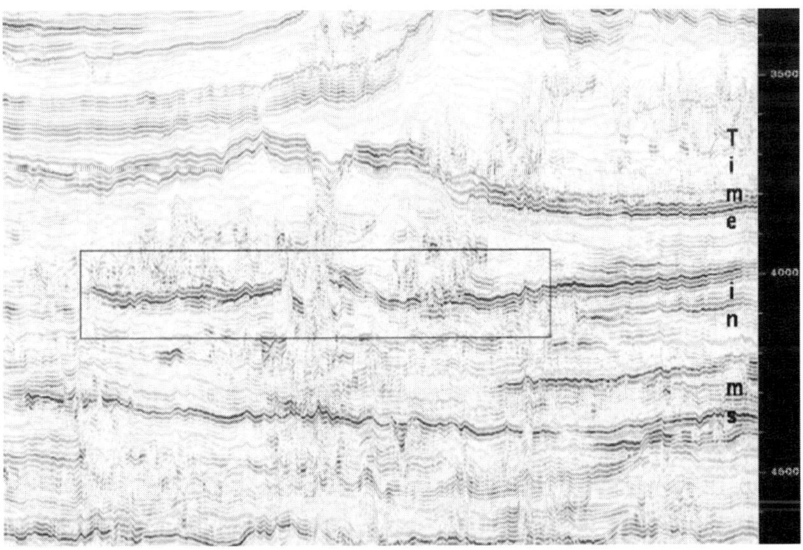

FIGURE 5.57 Seismic section (CC′) showing the deepwater turbidite system in the Bengal offshore (Reproduced with permission from Bastia, 2006 © Society of Exploration Geophysicists). (For color version of this figure, the reader is referred to the online version of this chapter.)

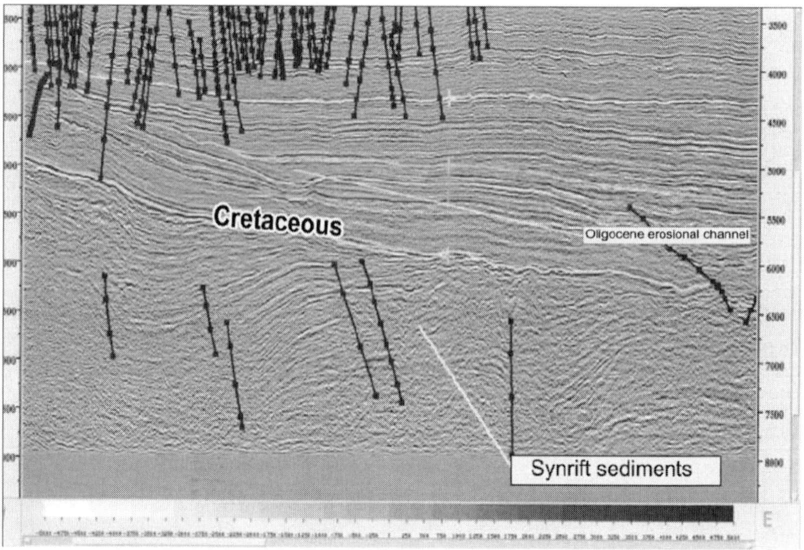

FIGURE 5.58 Seismic section (DD′) showing the syn-rift sediment package in the Mahanadi offshore (Reproduced with permission from Bastia, 2006 © Society of Exploration Geophysicists). (For color version of this figure, the reader is referred to the online version of this chapter.)

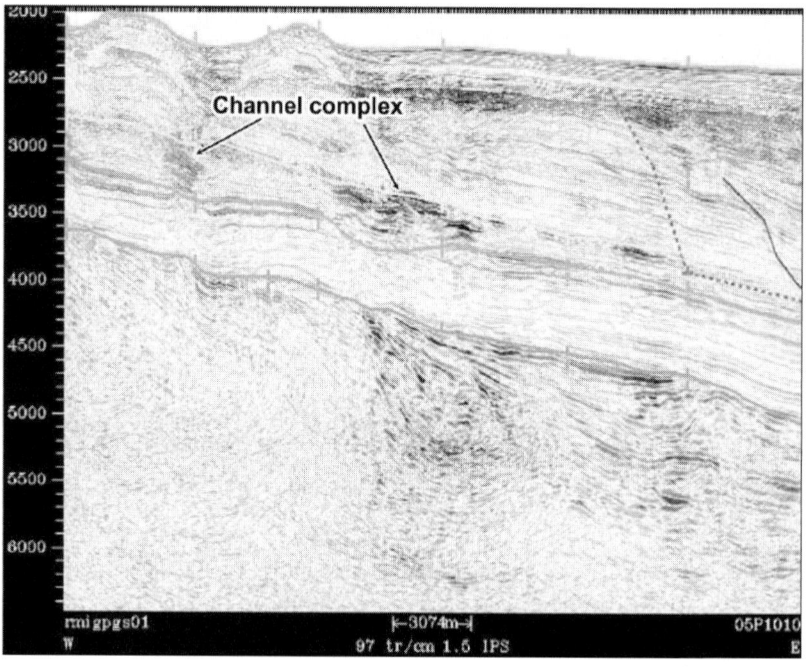

FIGURE 5.59 Seismic section (EE′) showing stack channels in the Mahanadi offshore (Reproduced with permission from Bastia, 2006 © Society of Exploration Geophysicists). (For color version of this figure, the reader is referred to the online version of this chapter.)

up to the Afanasy Nikitin Seamount region. Several hypotheses have been proposed on the genesis of the ridge. While the plume-related origin (Curray and Munasinghe, 1991) is the most widely accepted, other important alternate model is that the ridge is made up of continental crust (Sar et al., 2009).

5.3.4.4.3.1.1 Age of Ridge The age as well as the mode of emplacement of the ridge is purely based on geophysical models in the absence of drilled well information. The accurate age information is critical as it can provide the insight about the oldest sediments present in the region which in turn may indirectly throw light on the source rock potential of the basin (Nayak et al., 2011). While Curray and Munasinghe (1991) propose the formation of ridge during 117–184 Ma, other workers infer the age of emplacement during 105–170 Ma (Radhakrishna et al., 2012; Subrahmanyam et al., 1999) and post-85 Ma (Krishna, 2003). It is interesting to note that the above authors support plume-related volcanic emplacement of the ridge, though the source of the plume is still not clear. However, one common aspect that emerges out of these investigations is that the ridge must be younger to the BOB crust.

5.3.4.4.3.1.2 Source Potential Around 85° E Ridge The ridge is emplaced over the preexisting BOB oceanic crust in the deep offshore areas and gave rise to few structural highs and lows. The basement time-structure map indicates that the ridge is flanked by well developed lows and some of these structural features are separated by the NW–SE trending fracture zones of the oceanic crust. Further, closer investigation of the map reveals the presence of well-developed and restricted lows between the western and eastern mounds of the ridge (Nayak et al., 2011). Such confined lows can be inferred to act as a good source kitchen area for the generation of hydrocarbons.

The isopach of the total Cretaceous sediment thickness map reveals that the depression is filled with thicker sedimentary sequences characterized by more than 1.5 km of sediments within the lows. Parallel seismic reflectors within this sequence indicate the presence of low-energy sediments, which can be good source rock (Nayak et al., 2011). It can be therefore concluded that the lows flanking the ridge can be considered as having good source rock potential around the ridge.

Thick pile of sediments belonging to the Tertiary and Mesozoic holds considerable source for hydrocarbon generation in this basin. Marine shales of Paleocene–Eocene deposited in the basement lows are expected to be rich in the source rock for the generation of hydrocarbon. The shales in the distal part of the Bengal Fan and Mahanadi Delta are also considered to be another potential kitchen for the hydrocarbon generation (Nayak et al., 2011).

- **Structures associated with 85° E ridge**
  Sediments overlying the 85° east ridge (mostly Tertiary age) show well-developed structural closures (Fig. 5.60). Well-developed channels indicating reservoir conduits could also be identified to be associated with the ridge (Fig. 5.61). These could be good locale for hydrocarbon accumulation.

- **Carbonate buildups on the ridge**
  The presence of lava delta within the 85° E ridge suggests that the ridge must have been formed in a shallow water condition (Nayak et al., 2012). This is a favorable environment for carbonate buildup (Fig. 5.60). Probing these carbonate buildups on the ridge could give some exciting results.

5.3.4.5 Andaman Offshore Basin

5.3.4.5.1 Structure and Tectonics

The regional geotectonic history of the Andaman arc-sea region was discussed in detail in Chapter 3; however, we discuss here the structure and tectonics of the Andaman offshore basin with regard to understanding it from the point of view of hydrocarbon exploration. The region surrounding Andaman–Nicobar Islands and the adjoining back-arc Andaman Sea together forms the Andaman offshore basin and is an important transitional tectonic link between the Burmese arc in the north and the Indonesian arc in the south (Hamilton, 1979). Situated in southeastern part of the Bay of Bengal, this basin is associated with the convergent plate boundary between the Indian and Eurasian plates (Curray et al., 1982), and occupies 47,000 km^2 of area including the deepwater region. Formation of this basin is genetically linked to the lithospheric evolution in the Bay of Bengal (Curray et al., 1979). The basin comprises of major morphotectonic units; from west to east, these are the Andaman trench, the accretionary prism, outer structural high formed along the Andaman–Nicobar Islands, fore-arc basin, volcanic arc, and the back-arc basin (Fig. 5.62). An E–W section along the 11° latitude across the Andaman basin (Fig. 5.63) shows the tectonic relationship between the individual tectonic elements in an island-arc setting (Roy and Das Sharma, 1993). The outer high is formed by a series of islands belonging to the Andaman–Nicobar group of islands between 7° and 14° N and consists of partially metamorphosed Cretaceous sedimentary rocks, ophiolite suites, and melanges, characterized by narrow parallel folds, faults, and eastward dipping thrusts (Roy, 1983). The fore-arc basin lying between the outer high in the west and the volcanic arc in the east contains more than 5000 m of thick sediments ranging in age from late Cretaceous to Recent, and is characterized by the presence of mapped anticlinal structures (Dangwal et al., 2008; Roy and Das Sharma, 1993). A prominent unconformity is observed in this basin between the Neogene and Paleogene sediments (Roy, 1983). A detailed tectonic framework of this basin

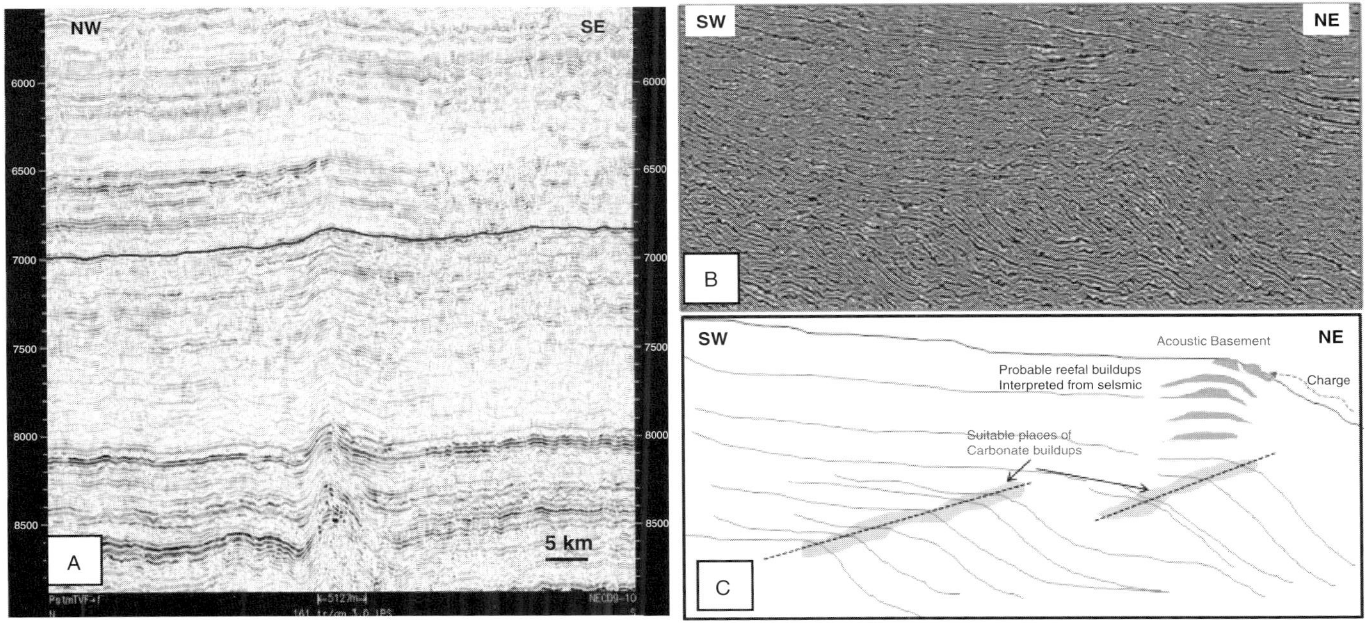

FIGURE 5.60 A) Seismic section showing structure associated with 85° E Ridge, B) Instantaneous phase section and inferred location of carbonate buildup (in C). (adapted with permission from Nayak et al., 2012). (For color version of this figure, the reader is referred to the online version of this chapter.)

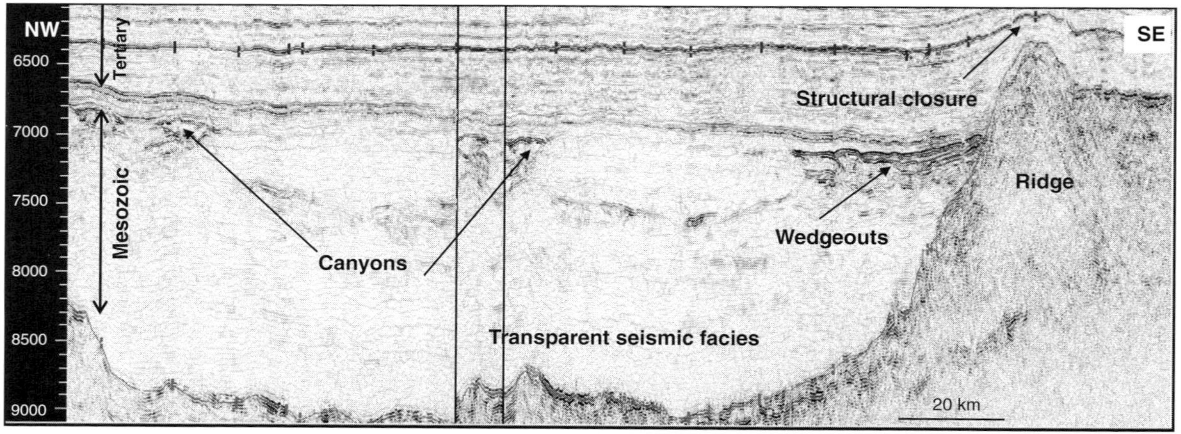

FIGURE 5.61 Seismic section showing 85° East ridge morphology. Deepwater channels and splay complex abutting against ridge are good play to focus. Tertiary structures associated with ridge form good traps (adapted with permission from Nayak et al., 2011). (For color version of this figure, the reader is referred to the online version of this chapter.)

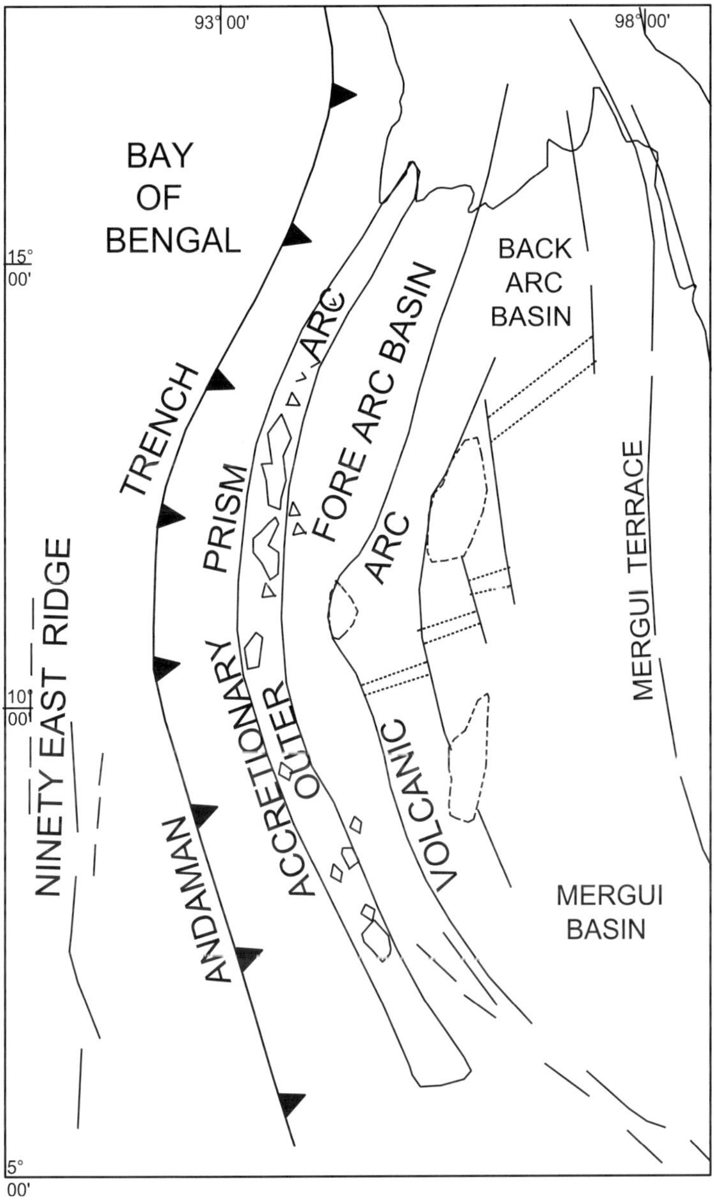

FIGURE 5.62 Map showing broad morphotectonic units of the Andaman Offshore Basin (source: DGH). (For color version of this figure, the reader is referred to the online version of this chapter.)

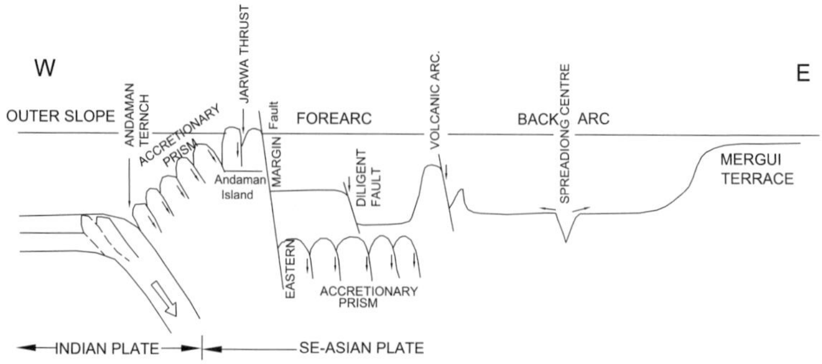

FIGURE 5.63 Structural cross-section depicting various arc-related features across Andaman Sea at 11° N latitude *(adapted from Roy and Das Sharma, 1993).*

is given in Roy and Das Sharma (1993) and is presented in Fig. 5.64. Further east, the volcanic arc represented partly by the active Barren and Narcondam volcanic islands is characterized by the Cretaceous ultrabasic outcrops, Eocene tuffs, and early and middle Miocene volcanic ash beds. The back-arc basin further east of the volcanic arc is a region of high heat flow due to the presence of active Andaman spreading center. Several normal faults separate the back-arc sub-basin from the Mergui terrace of the Malayan Peninsula.

From the hydrocarbon exploration point of view, the entire fore-arc basin and several sub-basins in the volcanic arc and back-arc basin areas became the main target zones of interest. Though exploration activity has been initiated in this basin as early as 1959 by ONGC, the drilling could be started only in 1980. The first exploratory well drilled in this basin resulted in discovery of gas from the middle Miocene limestone formations; subsequently, 14 wells were drilled in this basin, which showed indications of gas from Miocene sediments.

5.3.4.5.2 Generalized Stratigraphy

The stratigraphy of Andaman Basin is constructed based on surface geology over Andaman–Nicobar islands, Richie Archipelago located in the fore-arc, and the subsurface geology obtained from the data of nearly 14 drilled wells in the region. Available information from the DGH as well as other compilations (after Berger et al., 1983) facilitated the description of Paleogene and Neogene sections (Fig. 5.65) as given below:

i. *The Port Meadow Formation (Cretaceous).* It is the oldest sedimentary section of the basin which was not reached by any offshore wells. This formation dominantly consists of cherts, jaspers, slates, cherty limestone, marble, quartzites, and phyllites.

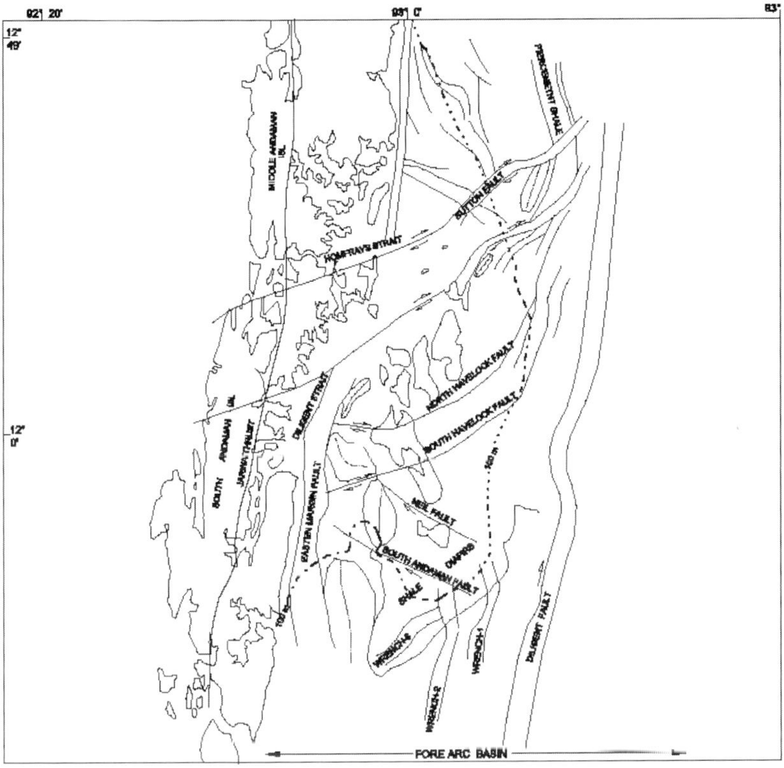

FIGURE 5.64 Tectonic map of the Andaman fore-arc basin *(adapted from Roy and Das Sharma, 1993).*

ii. *The Bartang Formation (Late Cretaceous–Eocene).* It overlies the Port Meadow Formation and consists of predominantly of argillaceous shale, interbedded limestone, siltstone, and shale with minor sandstones and limestone.

iii. *The Port Blair Formation (Oligocene).* It is best observed in the outcrop near the Port Blair town. Lithologically, it comprises alternating sequence of sand and shale, and at its contact with the Bartang Formation, the sequence comprises conglomerate and grit beds.

iv. *The Strait Formation.* It is underlain unconformably by the Port Blair formation and comprises the lower part of the Round Formation. It consists of moderately hard sandstone of gray and cream color, sandy limestone with hard calcareous bands, and white to grayish white tuff and ash layers with siliceous chalk and silt in the upper part. The Long-Guitar formation lies above the Strait formation and predominantly comprises calcareous, gray to bluish clays, silty clays, and hard to calcareous bands in the lower part.

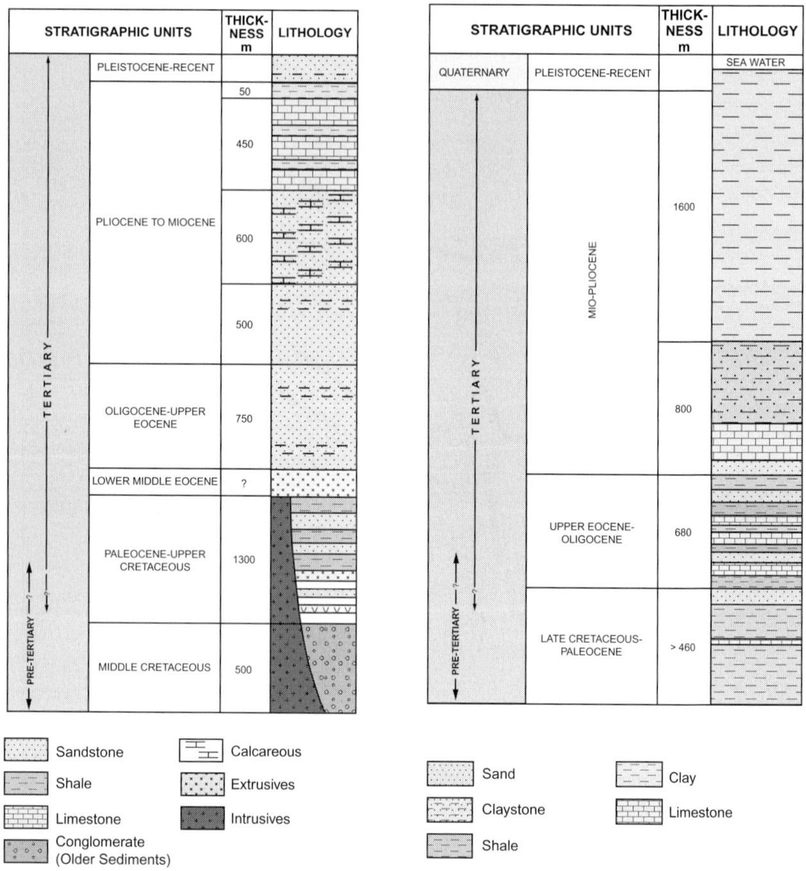

FIGURE 5.65 Generalized stratigraphy of the Andaman Basin as revealed from drilled well data (Reproduced with permission from Berger et al., 1983 © Schlumberger Ltd). (For color version of this figure, the reader is referred to the online version of this chapter.)

5.3.4.5.3 Petroleum System

Though petroleum system of the Andaman basin is not yet well understood, information obtained from the drilled well data (source: DGH) in the region indicates that the thick pile of sediments pertaining to the Tertiary (more than 5 km) and the Cretaceous holds considerable source potential for hydrocarbon generation in the basin. Reservoirs include sands of fluvial and shallow marine environments and reefoidal features within the Neogene. Further, the Paleogene turbidite deposits in the inner slope region have rich potential for organic matter.

Source rock

The thick pile of sediments from Cretaceous to Recent of different depositional environments, viz., fluvial (Irrawaddy), shallow, and deep marine

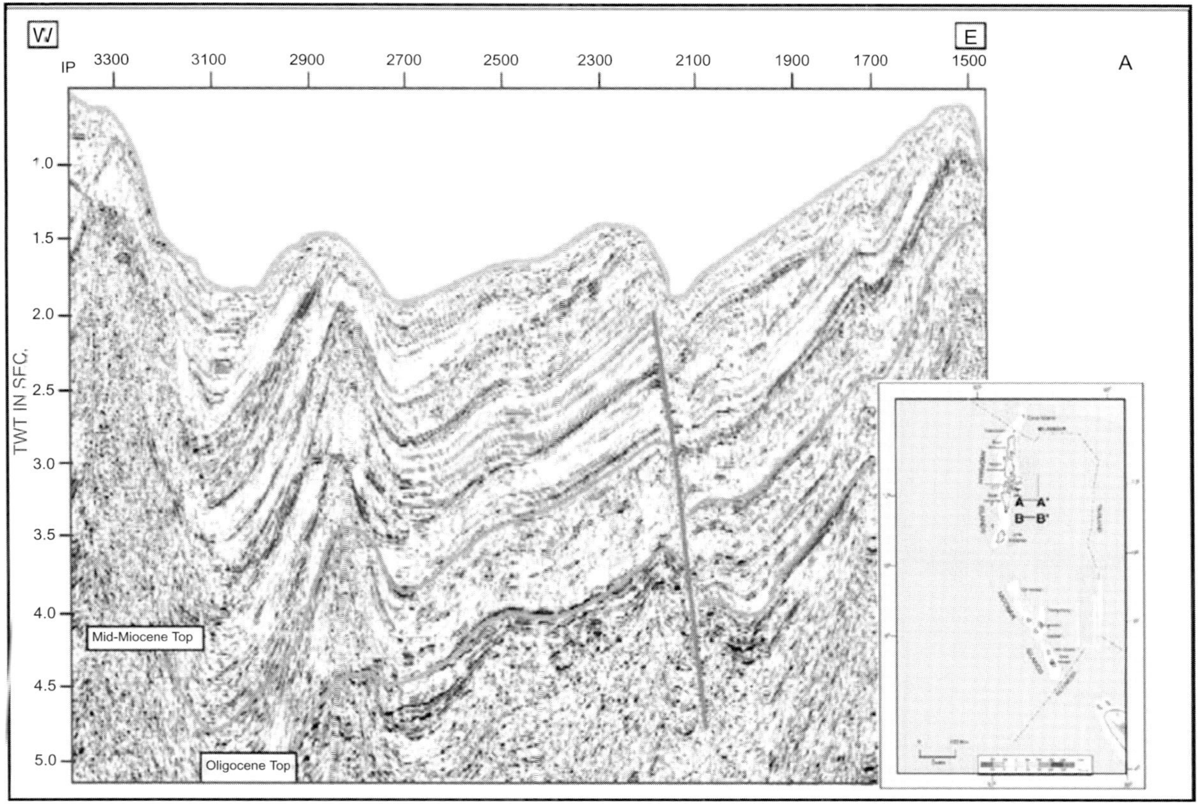

FIGURE 5.66 Cont'd

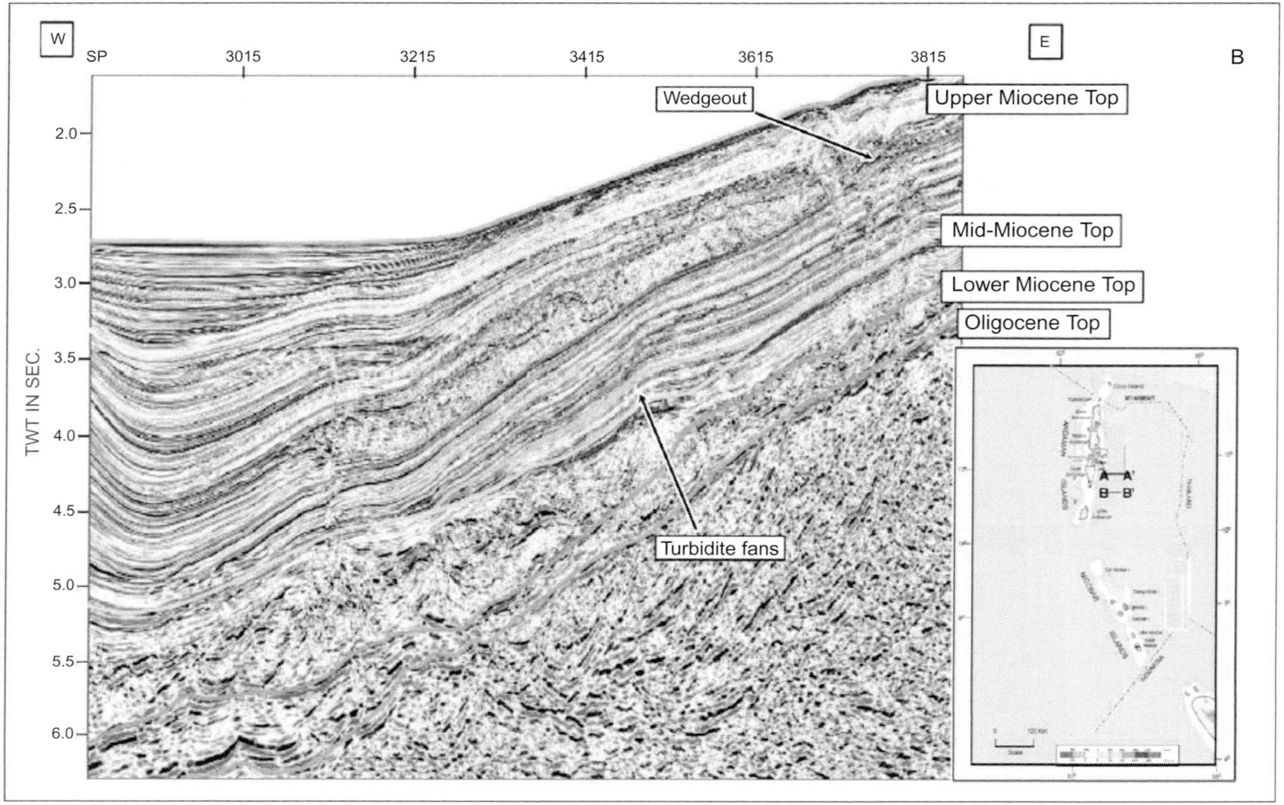

FIGURE 5.66 Seismic sections in the Andaman fore-arc basin. (A) Structural traps (A–A′ in inset) and (B) turbidite fans (B–B′ in inset) (Reproduced with permission from Bastia, 2006 © Society of Exploration Geophysicists). (For color version of this figure, the reader is referred to the online version of this chapter.)

deposits, and the proximity to spreading center is likely to provide optimum heat flow for hydrocarbon generation. In the North Sumatra basin, the Bampo Shale of Oligocene age is a proven source rock.

Reservoir rock

Potential reservoir rocks include Oligocene Parapat Quartzose Sandstone of Oligocene; Belumai Sandstone, Arun Limestone, and Telga Limestone of Early–Middle Miocene; and Keutapang Sandstone of Pliocene. Abundance of shales and claystones in Miocene and Pliocene sequence is expected to provide effective seal. Possibility of the deltas stretching onto shelf edge and beyond during the low-stand times of Mid-Oligocene and Mid-Miocene suggests the presence of reservoir bodies with seal facies in the basin.

Entrapment style

Drilled well data confirm that effective cap rock is in the form of claystones and shales within different stratigraphic horizons such as Bampo shale (Early Miocene) and Baong shale (Lower Miocene–Pliocene). The Miocene section has fault closures and likelihood of stratistructural turbidite prospects. In the fore-arc basin, main types of plays are structural features (Fig. 5.66A) and turbidite fans (Fig. 5.66B).

Chapter 6

Subsurface Geology, Depositional History, and Petroleum Systems Along the Western Offshore Basins of India

6.1. INTRODUCTION

The Western Continental Margin of India (WCMI) in contrast to the ECMI discussed in Chapter 5 has a different evolutionary history. Large onshore basins as well as thick offshore sedimentary depocenters observed along the ECMI are typically absent along the WCMI. The margin consists of five major offshore sedimentary basins (Fig. 6.1) such as the Kutch, Saurashtra, Mumbai, Konkan and Kerala basins separated by transverse basement arches (Biswas and Singh, 1988; Singh and Lal, 1993). Further, the eastern Arabian Sea and the adjoining West Coast of India contain several structural features which have evolved mostly as a consequence of rifting and seafloor spreading between India, Madagascar, and Seychelles (Biswas, 1987; McKenzie and Scalter, 1971; Naini and Talwani, 1982). While the northern part of the West Coast is geologically prominent due to the presence of extensive Deccan flood basalt province which masks the preexisting Precambrian geology of the region, the southern part is composed of Archaean charnockites, gneisses, and granites with limited extent of the Tertiary sediments along the coast of Kerala in the far south. It is very well known that the dominant structural and tectonic trends along the West Coast (Fig. 3.5), such as, the Dharwar trend (NW–SE to NNW–SSE), the Aravalli trend (NE–SW), and the Satpura trend (ENE–WSW to E–W) have a bearing on the north to south sequential rifting of the Indian Subcontinent during the breakup (Biswas, 1982). In this chapter, a brief description of regional geology, subsurface lithostratigraphic history of the offshore sedimentary basins along the WCMI, and the development of petroleum systems for the individual basins is provided.

Developments in Petroleum Science, Vol. 59. http://dx.doi.org/10.1016/B978-0-444-53604-4.00006-5

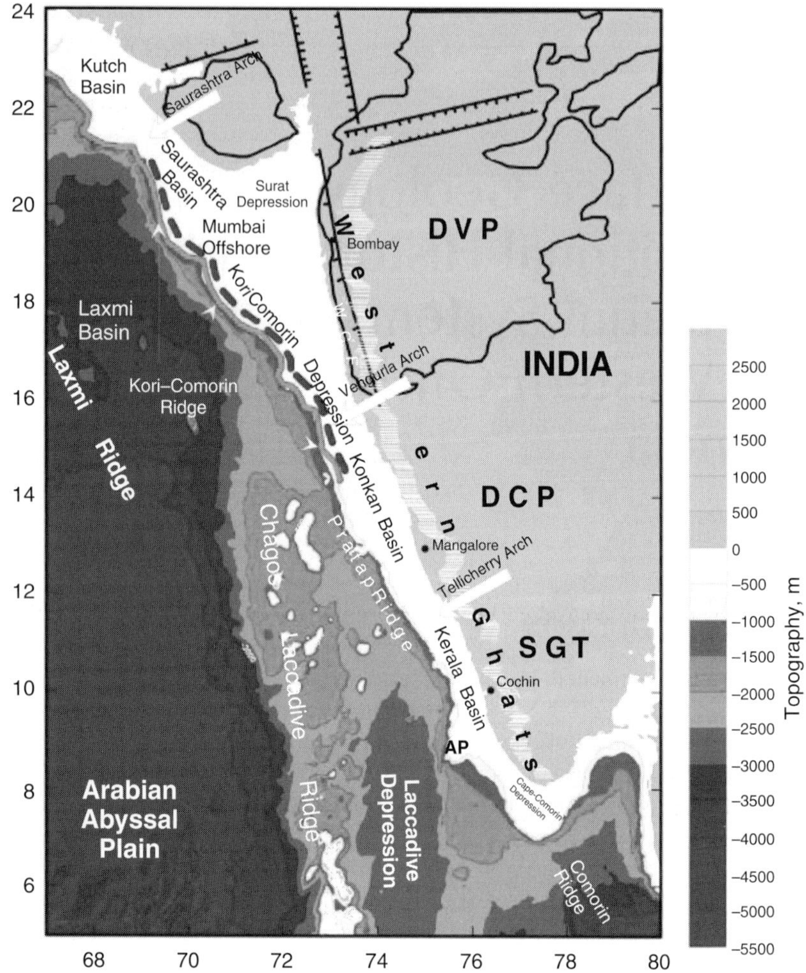

FIGURE 6.1 Bathymetric map showing major structural and tectonic features along the WCMI. WCF, West Coast Fault. Thick marking onland parallel to the coast indicates the location of Western Ghat scarp. SGT, Southern granulite terrain; DCP, Dharwar Craton Province; DVP, Deccan Volcanic Province *(adapted from Singh et al., 2007)*. (For color version of this figure, the reader is referred to the online version of this chapter.)

6.2. REGIONAL GEOLOGICAL SETTING OF THE WEST COAST BASINS

The WCMI had evolved after the breakup of Madagascar in the mid-Cretaceous (~88 Ma) and the Seychelles micro-continent in late Cretaceous (~65 Ma) from India (Norton and Sclater, 1979; Royer et al., 1989; Storey et al., 1995). The later breakup event is also known to have associated with a widespread volcanism which led to the formation of Deccan Continental

Flood Basalt (CFB) Province (Subba Rao and Sukheswala, 1981). A causal link between the breakup and CFB formation to the Reunion plume is supported by many workers (e.g., Courtillot et al., 1986). Further, as India was moving northward, the plume had emplaced the Chagos–Laccadive Ridge and also caused the reorganization of nearby spreading centers. This widespread volcanism observed below the Tertiary sedimentary strata almost throughout the western offshore inhibits our understanding of earlier rift configuration.

The WCMI is characterized by wider continental shelf which increases from around 50 km off Cochin in south to more than 300 km off Saurashtra–Mumbai coast in north. On the other hand, the continental slope is narrower in the north and widens toward south (Biswas, 1987). Further, the margin hosts a number of basement ridges and depressions, and other structural features such as Laccadive, Pratap, Laxmi, and Panikkar ridges (Fig. 6.1). Biswas and Singh (1988) identified six tectonic elements in the offshore running N–S in the shelf and deep oceanic areas of the Arabian Sea (Fig. 6.2), and these are shelfal horst–graben complex, Kori–Comorin depression, Kori–Comorin ridge, Laxmi–Laccadive depression, Laxmi–Laccadive ridge, and the Arabian abyssal plain. According to Kolla and Coumes (1990), some of these offshore features have been controlled by the Precambrian structural grain along the coast.

6.2.1 Drainage Pattern

The significant characteristic of the drainage pattern of the Indian Peninsula is that except the Narmada and the Tapti rivers, all the other major rivers, such as the Godavari, Krishna, and the Cauvery take their origin on the crest of the Western Ghats and flow eastward all the way across the Peninsula and join in the Bay of Bengal (Fig. 6.3). While major rivers Sabarmati, Narmada, and Tapti discharge the sediments into the Gulf of Khambhat, the Luni river joins the Arabian Sea at the Gulf of Kutch. The easterly drainage of the Indian Peninsula appears to be a post-Deccan trap feature caused by the cymatogenic upwarp of the crust along an axis west of the present-day Western Ghat scarp (Radhakrishna, 1993). Some of the westerly flowing rivers (e.g., Sharavati, Kali Nadi) which are shorter and flow rapidly across the narrow coastal tract of Konkan plains between the Ghats and the West Coast have captured the easterly drainage over the Ghats and join the Arabian Sea (Ramakrishnan and Vaidyanathan, 2008). Further south, because of the steepness of the Western Ghats flanks on either side, the rivers Noyil and Vaigai flow eastward, whereas the Ponnani, Periyar, Pamba, and Achankovil flow westward to discharge sediments into the Arabian Sea (Soman, 1997; Valdiya, 2010). The above drainage distribution along the West Coast is relevant in the present-day context. However, with regard to the sedimentation along the western offshore, it is important to understand the spatiotemporal variations of the

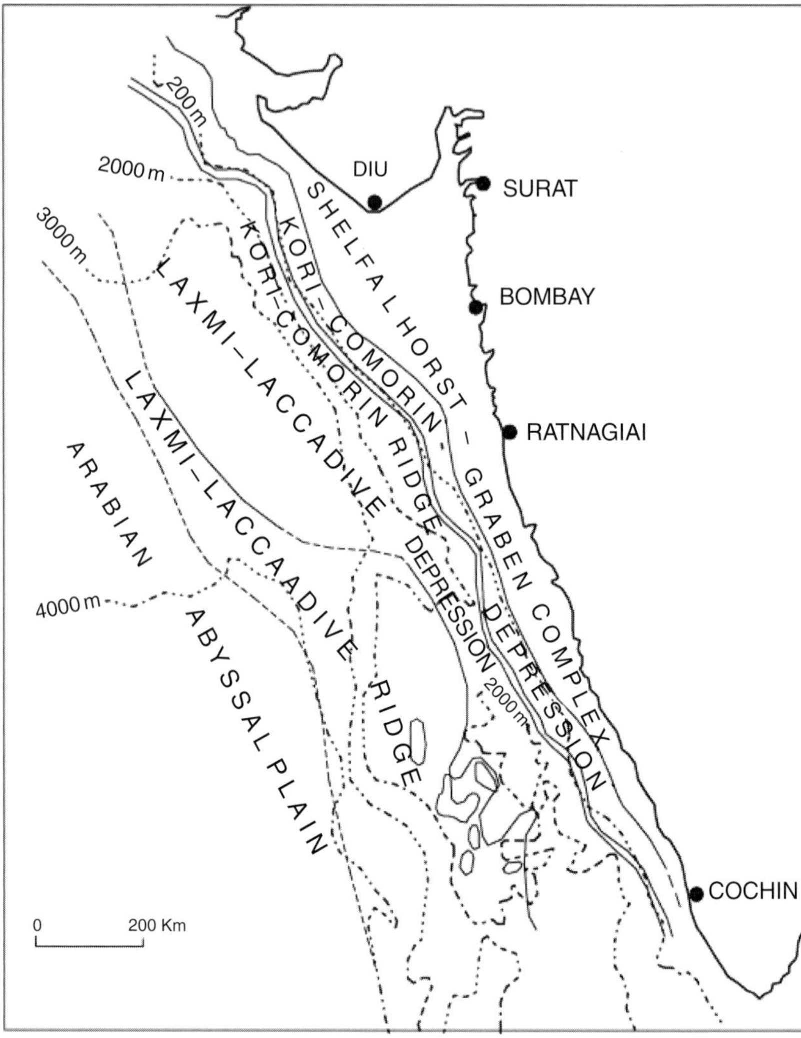

FIGURE 6.2 Major tectonic elements in the shelf and deep oceanic areas of the Western offshore *(adapted from Singh and Lal, 1993).*

drainage system in the geological past which normally would have been controlled by the tectonic–climate interactions and other geological factors (Gunnell and Radhakrishna, 2001) that affected the morphology of the region.

6.2.2 General Description of the Basins

As stated earlier, several basement Arches trending perpendicular to the West Coast divide the shelf region into various thick sedimentary basins. These

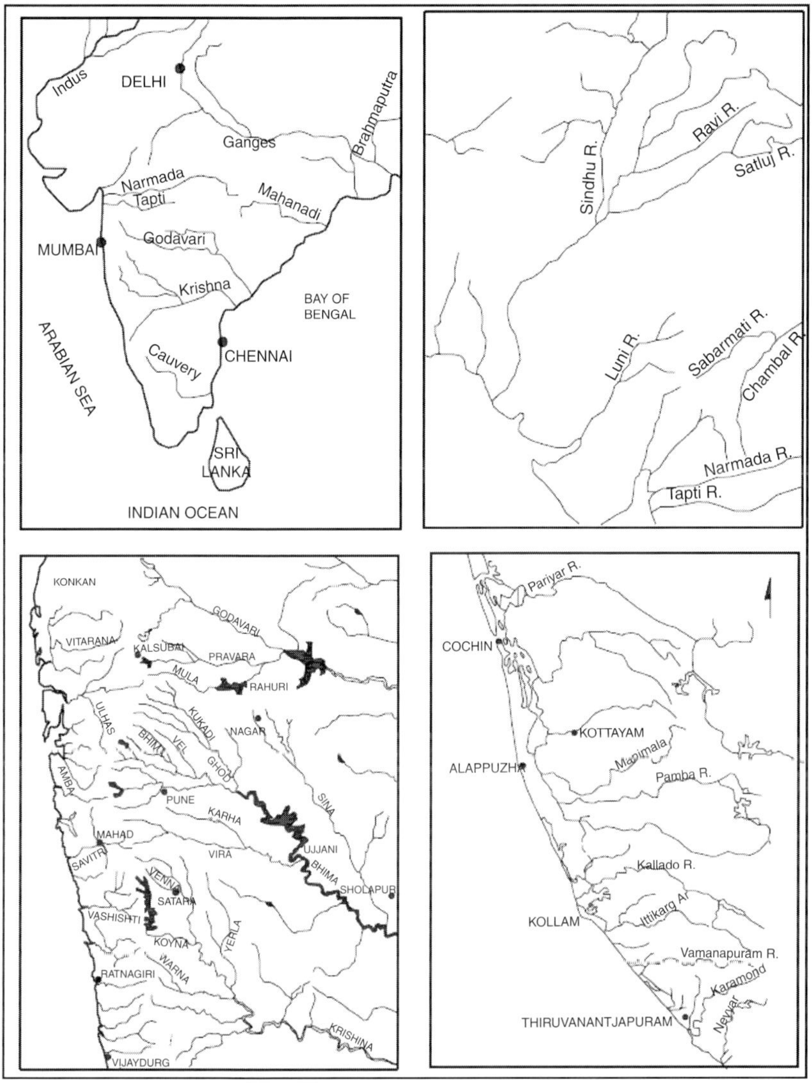

FIGURE 6.3 Regional drainage pattern map of the Peninsular India showing dominantly east flowing river system as well as rivers flowing westward into the Arabian Sea (source: Ramakrishnan and Vaidyanathan, 2008; Soman, 1997; Valdiya, 2010).

include from north to south, the Kutch, Saurashtra, Mumbai, Konkan, and Kerala basins. In the northern most part, the basins of Kutch and Saurashtra are separated by a southwesterly plunging basement high (Fig. 6.1) called the Saurashtra Arch (Biswas, 1987). The Mumbai and Konkan basins are separated by the southwesterly plunging Vengurla Arch in the central part, and the Konkan and Kerala basins are demarcated by the Tellicherry Arch further south

(Mathur and Nair, 1993). Major rivers such as the Indus, Narmada, and Tapti are draining sediments into the Arabian Sea and as a result produced a very thick shelf sedimentary column along the northwest margin. The major conjugate rift systems, the Narmada and Cambay, cross each other south of Saurashtra peninsula in the Surat offshore and form a deep depression called the Surat depression. It is known well that the margin has evolved through an early rift phase and postrift phase of divergent margin development (Biswas and Singh, 1988). The sedimentary basins along the West Coast of India and the adjoining offshore basins can be grouped into two categories: (i) Mesozoic marginal rift basins and (ii) Tertiary offshore rift basins, where Mesozoic prospects are believed to be existing below thick basaltic traps.

6.2.2.1 *Mesozoic Marginal Rift Basins*

According to Biswas (1987), the northern most part of the West Coast was the first to be subjected to continental rifting since the late Triassic giving rise to the three pericontinental Mesozoic marginal rift basins in the onshore such as the Kutch, Narmada, and Cambay basins at the northwest coast (Fig. 3.5). All the three basins had evolved in different periods in the Mesozoic and contain varying thickness of Mesozoic sediments. The Kutch basin formed during the late Triassic is bordered by the Nagar Parkar ridge in north, the Radhanpur–Barmer arch in the east, and the Kathiawar uplift in the south. The basin is styled by three main uplifts along the three master faults with intersecting half-grabens (Biswas, 1980). The Cambay basin is an intracratonic graben bordered by Radhanpur–Barmer arch in the west and the Aravalli orogenic belt on the east. The basin is bounded by enechelon faults paralleling the Dharwar trend and lies across the Aravalli and Narmada trends. Several transverse faults divide the Cambay basin into six tectonic blocks. Geophysical studies in the basin indicate that at least 1200 m of Mesozoic sediments could be lying below the basaltic traps which form the floor of nearly 5000-m-thick sedimentary sequence (Biswas, 1982). The Narmada basin which formed as a rift basin in the late Cretaceous is the southern most of the three marginal basins. It embayed into a narrow graben along the Narmada–Son lineament. Among these three basins, marine sediments of late Cretaceous occur only in this basin, whereas in the other two basins, this is a period of nondeposition and volcanic activity (Biswas, 1987). A generalized stratigraphy of these three basins (after Biswas, 1987) is given in Table 6.1.

6.2.2.2 *Tertiary Basins Along the Western Offshore*

The rift–drift events related to the separation of Madagascar during the mid Cretaceous and the Seychelles in late Cretaceous from India gave rise to the formation of major offshore rift basins such as the Kutch, Saurashtra, Mumbai (Ratnagiri), Konkan, and Kerala basins along the western offshore of India. While the Kutch and Saurashtra have major onshore counterparts that are

TABLE 6.1 Generalized Stratigraphy of the Marginal Rift Basins along the Northwest Coast of India (adapted from Biswas, 1987)

	Kutch Basin			Cambay Basin			Narmada Basin		
Age	*Lithology*	*Geological Environment*	*Events*	*Lithology*	*Geological Environment*	*Events*	*Lithology*	*Geological Environment*	*Events*
Pliocene	S.st/ conglomerate, minor shale (1000 ft)	Deltaic	Regression, major tectonic movement	Sandstone/shale/ conglomerate (3500 ft)	Fluvio-Deltaic to marine	Regression, major tectonic cycle			
E. Miocene	Shale/L.st/ siltstone (600 ft)	Shallow Marine	Transgression						
Oligocene	L.st/shale (126 ft)	Shallow Marine	Regression, tectonic cycle, transgression	Shale/S.st/L.st (600–900 ft)					
L. Eocene									
M. Eocene	L.st/minor shale (185 ft)	Shallow Marine	Regression, tectonic cycle, transgression	S.st/siltstone/ shale/coal (600–900 ft)	Marine/ Deltaic	Oscillatory, transgression/ regression cycle			
E. Eocene	Shale/L.st (135 ft)	Shallow Marine	Regression, tectonic cycle, transgression	Black shales (1500–4500 ft)	Marine/pro-Deltaic	Transgression		Terrestrial	Regression/ uplift/ volcanicity/ major tectonic cycle
						Graben formation	Deccan traps (>3000 ft)		

Continued

TABLE 6.1 Generalized Stratigraphy of the Marginal Rift Basins along the Northwest Coast of India (adapted from Biswas, 1987)—Cont'd

	Kutch Basin			Cambay Basin			Narmada Basin		
Age	*Lithology*	*Geological Environment*	*Events*	*Lithology*	*Geological Environment*	*Events*	*Lithology*	*Geological Environment*	*Events*
Paleocene	Laterite/tuff/Fe clay/S.st (163 ft)	Continental	Erosional/ peneplanation	Volcanic conglomerate/ trap wash/Fe Clay stone (60–4000 ft)	Fluvial/ Lacustrine	Uplift/ erosional/ volcanicity/ major tectonic cycle	L.st/marl/S.st (250 ft)	Marine	Transgression
L. Cretaceous	Deccan trap; basalt flows (1500 ft)	Terrestrial	Major tectonic movement and volcanicity	Deccan traps (3000 ft)	Terrestrial	Rifting	S.st/shale (300 ft)	Deltaic/ Continental	Rifting
E. Cretaceous	S.st/shale/iron stone (3500 ft)	Deltaic	Regression	Feldspathic S.st (300 ft)	Fluvial/Deltaic				
U. Jurassic	S.st/shale (200 ft)	Shallow Marine							
E–M Jurassic	L.st/shale (2500 ft)	Shallow Marine	Regression transgression						Uplift/ erosion
L. Triassic	Arkosic S.st (150 ft)	Continental	Rifting						
Pre-Mesozoic							S.st/shale/L.st	Marine	Transgression
Precambrian basement	Granite/ metamorphics			Granite			Metamorphics/ granite		

related to the earlier Mesozoic rifting event discussed above, the Mumbai, Konkan, and Kerala basins are essentially offshore basins with an exception of Kerala basin having a minor onshore part (Singh and Lal, 1993). From north to south, the offshore basins occupy an area of 250,000 km^2 and including the deepwater area, the total area is around 4,71,000 km^2 up to the Exclusive Economic Zone (Thakur et al., 1999). Vast amount of seismic data as well as drilled well information defined the basin architecture and indicates that Deccan volcanics form the floor for the Tertiary sediments in most parts of the basins (Singh et al., 2007). Further, the subsurface geology and depositional and tectonic history of the Konkan-Kerala (KK) offshore have a genetic linkage to the denudational and uplift history of the Western Ghats (Chand and Subrahmanyam, 2003; Campanile et al., 2007; Gunnell and Radhakrishna, 2001).

6.3. BASIN-WISE ANALYSIS OF HYDROCARBON POTENTIAL

The hydrocarbon potentiality of any basin mainly depends on those controlling factors by which a complete petroleum system has a likelihood of existing. The WCMI underwent continental rifting initially at the northwest coast which started during the late Triassic period and gave rise to Mesozoic marginal rift basins such as the Kutch, Cambay, and Narmada basins (Biswas, 1987). The WCMI also contains several offshore sedimentary basins extending from Kutch in the north to Cape Comorin in the south which had formed at the continental margin since the breakup of Madagascar around 88 Ma (Storey et al., 1995). In the onshore part of the northwest coast, the Cambay basin is one of the most prolific oil and gas producing basins of India; however, as it is completely an onshore rift basin, it is out of the scope of this book to provide details of basin development and its petroleum prospectivity and hence avoided. Along the western margin, the Mumbai offshore basin is a major producer of hydrocarbons, while the Kutch and the KK offshore basins have given indications of hydrocarbons in the shallow wells within the Tertiary formations. It is believed that both Kutch and KK offshore might be containing thick Mesozoic sedimentary succession below the Deccan basalts which increases the petroleum potential of the region. In this chapter, we present a detailed account of the hydrocarbon potential of the western offshore basins such as the Kutch, Saurashtra, Mumbai, Konkan, and Kerala basins. As stated earlier, except the Kutch and Saurashtra, all the other basins are essentially limited to offshore areas only. Therefore, depending on the information available, extent of exploration activity, and their potential, the basins are regrouped into three broad divisions for ease of presentation. These are (i) Kutch–Saurashtra Basin which includes the onshore part of Kutch as well as the Saurashtra basins, (ii) Mumbai offshore Basin, and (iii) KK offshore Basin.

6.3.1 Kutch–Saurashtra Basin

The Kutch–Saurashtra basin is a part of the three marginal rift basins that have formed at different times since the late Triassic at the northwest coast of India (Biswas, 1982). These rift basins are bounded by intersecting sets of faults that follow the Precambrian tectonic trends and the Saurashtra peninsula occurs as a horst block between the rifts (Biswas, 1987). The basin contains substantial offshore part characterized by the presence of shelf around 160 km wide in Kutch offshore and slightly lesser along the Saurashtra offshore with shelf break occurring at 200 m bathymetry. Several structural features observed in onshore area are seen to be extending toward the shelf as well as the deep offshore areas. The entire basin is important from the point of petroleum prospectivity as it is surrounded by the basins having prolific oil and gas discoveries such as the south Indus basin of Pakistan in west, Cambay basin in east, and the Mumbai offshore basin in south (Biswas, 1982).

6.3.1.1 Onshore

The geological and tectonic history of the northwest shield margin is studied in detail by various earlier researchers and with special reference to the tectonic evolution of the Kutch–Saurashtra rift basin. In this regard, the contributions made by Biswas (1982, 1987) and subsequent investigations are noteworthy. Some details on the general geology, structure and tectonic framework of the basin is described below:

6.3.1.1.1 Tectonic Framework

Three Precambrian orogenic trends, the NNW–SSE Dharwar trend, the NE–SW Aravalli trend, and the ENE–WSW Satpura trend dominate the regional structural fabric of the basin and the contiguous western Indian shield (Fig. 3.5). The NE–SW trending Aravalli trend continues across the Cambay graben into Saurashtra as a southwesterly plunging basement arch, while the E–W trending Delhi Fold Belt (the northern component of Aravalli trend) continues into the Kutch basin which has undergone rifting along this trend (Biswas, 1987). The Saurashtra horst within these intersecting rift systems has undergone uplift and is characterized by faulted western margin.

6.3.1.1.2 General Geology and Tectonics

6.3.1.1.2.1 Kutch Basin It is a pericratonic embayment through marginal graben lying between the Nagar Parkar block in north and the Saurashtra block in south (Biswas, 1982). The basin slopes toward WSW and is characterized by numerous late Cretaceous uplifts (highlands) with associated plutonic bodies, and intervening Tertiary basins (plains). Middle Jurassic to lower Cretaceous rocks are exposed on the highlands such as the Kutch mainland, Wagad, Pachham, Khadir, Bela, and Chorar, while the Tertiary sediments occur in the bordering plains (Biswas, 1987) including the plains of Rann (Fig. 6.4A and B). The Deccan traps in the basin are restricted toward the southern part of Kutch

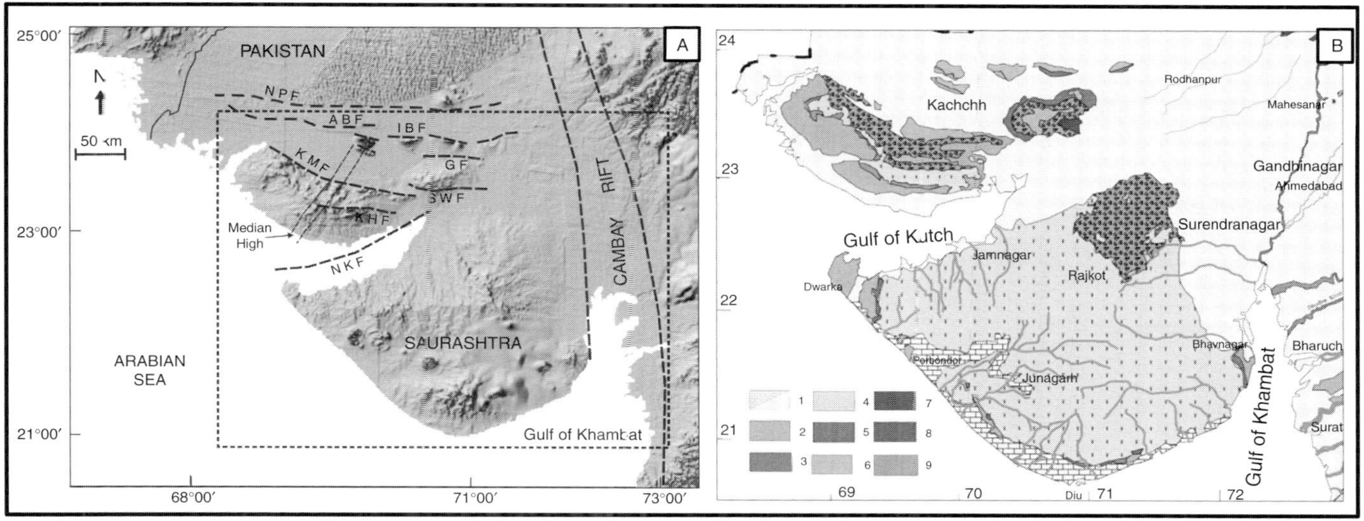

FIGURE 6.4 (A) Shaded relief map showing major faults in the Kutch mainland (after Mallik et al., 2009). NPF, Nagar Parkar Fault; ABF, Allahbund Fault; IBF, Island Belt Fault; GF, Gedi Fault; KMF, Kachchh Mainland Fault; SWF, South Wagad Fault; KHF, Katrol Hill Fault; NKF, North Kathiawar Fault. Rectangle represents the geological map shown in (B). (B) Detailed geological map of Kutch–Saurashtra onshore basin. 1, Quaternary; 2, Tertiary Rocks; 3, Laterite; 4, Deccan Trap; 5, Cretaceous Rock; 6, Jurassic Rocks; 7, Ambaji Granite; 8, Kumbhalgarh Group; 9, Gogunda Group *(adapted from Merh, 1995)*.

mainland between Lakhpat in west and Anjar in east. The traps gradually thin toward north, and according to Merh (1995), the pretrappean topography has controlled the trap distribution in the basin. Differential movements of discrete basement blocks parallel to the Precambrian tectonic trends resulted in uplifts and gave rise to four subparallel ridges or positive topographic elements, such as the Nagar Parkar–Tharad, the Island Belt, Wagad, and the Kutch Mainland along five principal faults (Fig. 6.5): (i) the Nagar Parkar fault, (ii) the Island Belt fault, (iii) the South Wagad fault, (iv) the Kutch Mainland fault, and (v) the North Kathiawar fault (Biswas, 1987; Mathew et al., 2006). A NNE–SSW trending median high occurs along the hinge line of the basin and passes across the E–W trending structural elements (both uplifts and depressions) of the basin (Fig. 6.5). The median high controlled the facies change and sediment thickness, while the basin on its west is deeper with thick sediments showing change of facies from shallow to deeper shelf, in its east, the basement is shallow having facies change from shallow marine to littoral and fluvial (Biswas, 1987).

The Bouguer anomaly map of the Kutch basin (Chandrasekhar and Mishra, 2002) shows several gravity highs and lows representing the local uplifts and depressions of the basement (Fig. 6.6A). The central gravity high over the Kutch mainland uplift is surrounded by strong gradient zones indicating the fault-controlled uplift. Several large amplitude circular gravity highs over Pacham, Khadir, and Bela islands may indicate basement uplifts due to subsurface volcanic plugs.

6.3.1.1.2.2 Saurashtra Basin The most part of the onshore Saurashtra basin is covered by the Deccan traps (Merh, 1995) except in the periphery; the Tertiary sediments are present along the coastal strip with sediments of lower Cretaceous exposed in the northern part (Fig. 6.4B). In the western and the southeastern part of Saurashtra, the exposed Deccan traps consist of several volcanic plugs of alkaline complexes which are pipe-like igneous intrusions having limited horizontal dimensions from few hundred meters to a few kilometers (Chandrasekhar et al., 2002). The region is also characterized by the presence of E–W- and NE–SW-oriented dyke swarms of Deccan volcanic activity (Mishra et al., 2001) and coincides with the dominant structural trends of the region. The Bouguer anomaly map of the region (Fig. 6.6B) reveals several circular gravity highs coinciding with the volcanic plugs, and the overall gravity anomaly trends indicate their association with fracture zones (FZs) or deep-seated faults (Chandrasekhar et al., 2002). The residual Bouguer anomalies computed from this map by Singh and Arora (2008) show a prominent circular gravity high in the northwestern part, a spread out regional low in the east, and further east NE–SW aligned circular highs.

6.3.1.1.3 Stratigraphy

6.3.1.1.3.1 Kutch Basin The generalized stratigraphy of the Kutch basin is summarized along with the Cambay and Narmada marginal rift basins in

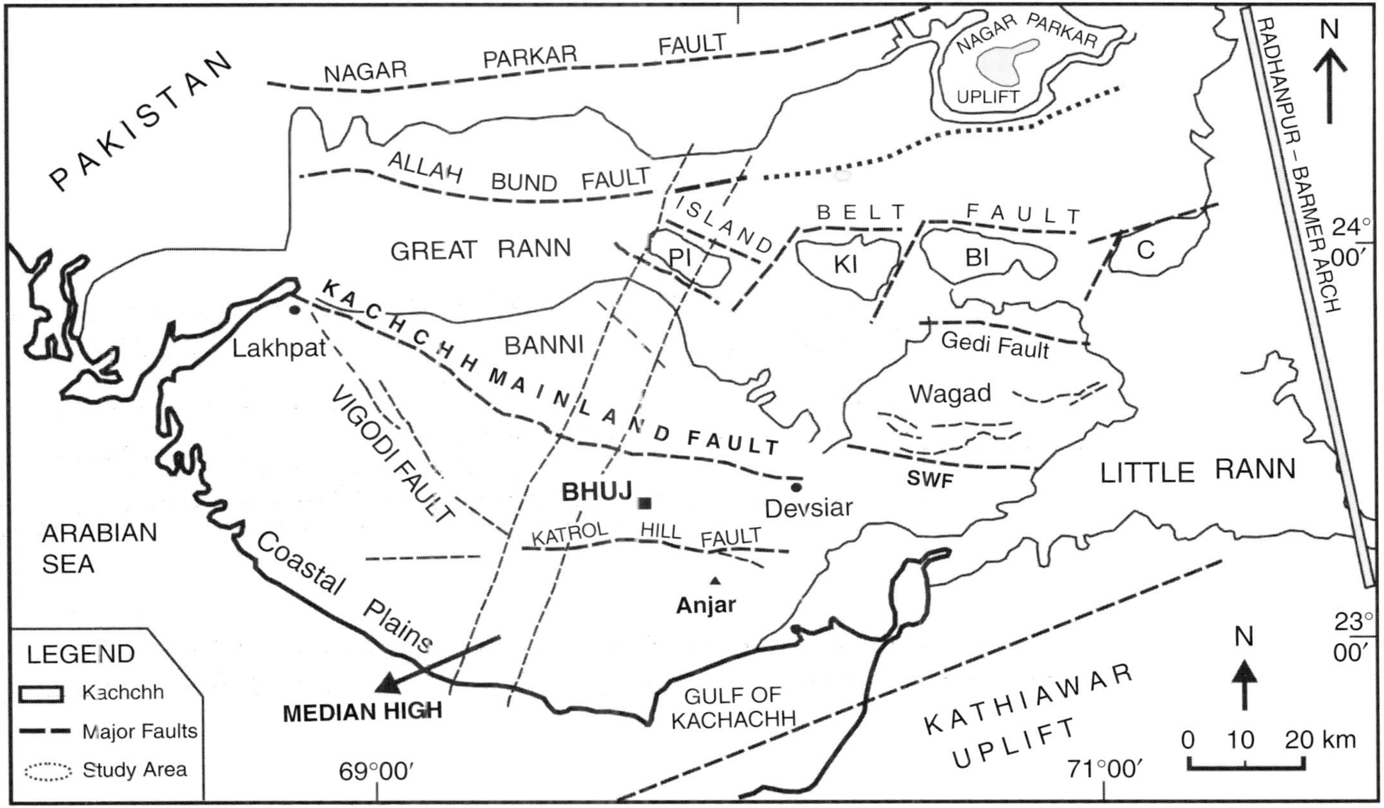

FIGURE 6.5 Detailed tectonic map of the Kutch onshore basin. SWF, South Wagad Fault; PI, Pacham Island; KI, Khadir Island; BI, Bela Island; C, Chorar Island (Reproduced with permission from Biswas, 2005 © Indian Academy of Sciences; Mathew et al., 2006).

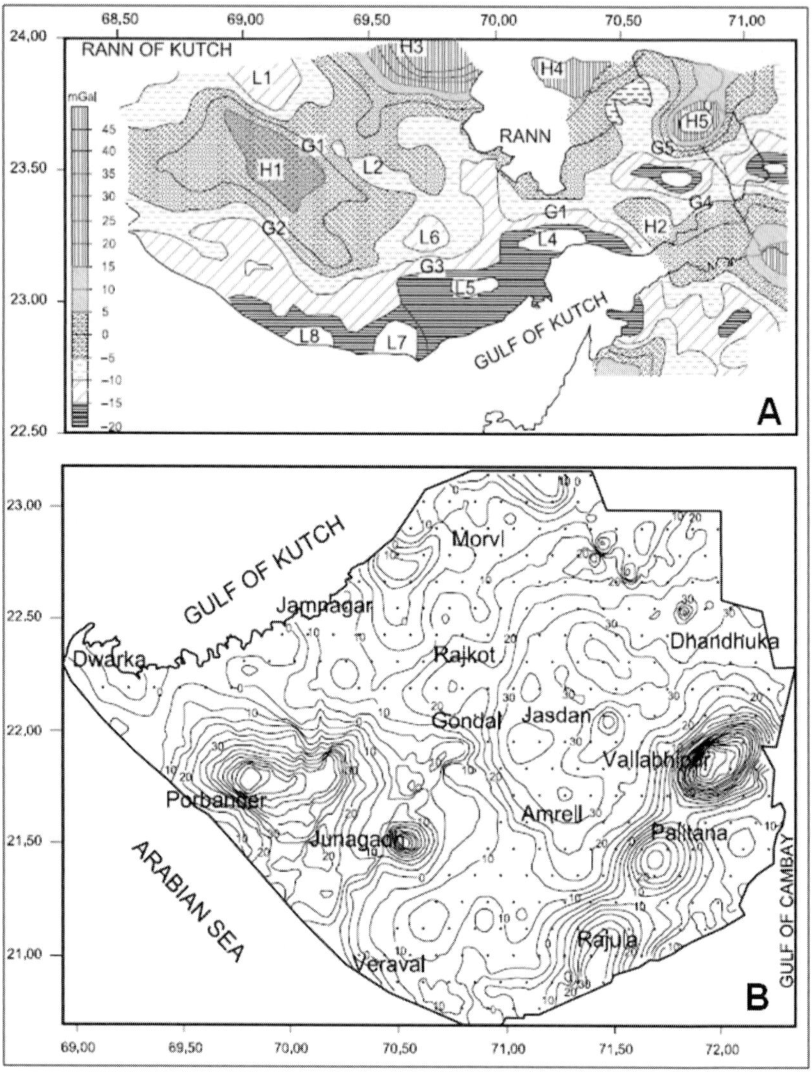

FIGURE 6.6 (A) Bouguer anomaly map of the Kutch onshore basin. Gravity highs (H1–H5), gravity lows (L1–L8), and gradients (G1–G5) are indicated on the map *(adapted from Chandrasekhar and Mishra, 2002)*, (B) Bouguer anomaly map of the Saurashtra onshore *(adapted from Chandrasekhar et al., 2002)*. (For color version of this figure, the reader is referred to the online version of this chapter.)

Table 6.1. The basin has the most complete stratigraphic record from the middle Jurassic to the Holocene and is characterized by thick accumulations of Mesozoic sediments of late Triassic to lower Cretaceous that have been deposited in two major cycles (Biswas, 1981). The middle Jurassic transgressive cycle constitutes carbonates and shales, whereas deltaic clastics are related to the late Jurassic–early Cretaceous regression. During the late Cretaceous–early Paleocene time, these sediments are uplifted, folded, and covered by the Deccan lava flows. The Paleocene is represented by terrestrial volcano-clastic sediments. Tertiary sedimentation had commenced in the basin with the early Eocene transgression and ended with late Miocene–Pliocene regression (Biswas, 1982). These sediments are deposited in the peripheral lows bordering the highs. According to Biswas (1974, 1981), the stratigraphy of Kutch basin is marked by several unconformities and diastems which can be correlated with periods of uplift and erosion. Major stratigraphic gap is during the late Cretaceous to early Paleocene which culminated with the Deccan volcanic activity, and other gaps in stratigraphy include upper Paleocene, post-lower Eocene, upper Eocene, post-Oligocene, and upper Miocene.

6.3.1.1.3.2 Saurashtra Basin The generalized stratigraphy of onshore Saurashtra basin (Merh, 1995) consists of Precambrian basement overlain by Mesozoic sediments, Deccan traps, and thin cover of Neogene as well as Quaternary sediments with exposures of Mesozoic and Cenozoic sedimentary rocks and Deccan traps in the basin (Table 6.2).

6.3.1.1.4 Sub-Basalt Investigations

For the sub-basalt sediment investigations, the subsurface imaging in basalt-covered regions becomes difficult from seismic data alone due to velocity inversion between the basalt-sediment interface; therefore, in such areas, integrated interpretation of complimentary geophysical datasets as applied to

TABLE 6.2 The Generalized Stratigraphy of Saurashtra Onshore Basin (adapted from Merh, 1995)

Coastal deposits, Rann sediments, Alluvium, etc.;	Quaternary
Marine to fluviomarine rocks with associated Aeolian derivatives (miliolite)	
Marine and fluviomarine rocks	Upper Tertiary (Neogene)
Unconformity	
Laterite	Paleocene
Deccan Trap-basalt flows and intrusive bodies	Upper Cretaceous to Lower Eocene
Unconformity	
Marine to fluviomarine (Surendranagar and Wadhwan Formations)	Upper Jurassic to Lower Cretaceous
Unconformity	
Subsurface basement of granite	Proterozoic

complex geological settings have proved to be very successful (Colombo and De Stefano, 2007; Gallardo and Meju, 2007; Pandey et al., 2009). In this context, some excellent case studies are available from the Kutch as well as the Saurashtra region, wherein thick Mesozoic sedimentary sequences have been delineated below the basalts based on integrated geophysical data interpretation (Dixit et al., 2000; Harinarayana, 1999; Kumar et al., 2004; Manglik and Verma, 1998; Pandey et al., 2009; Sain et al., 2002; Satpal et al., 2006; Singh and Arora, 2008). Here, we present two interpreted profiles which reveal thick Mesozoic sediments, one in Kutch basin (Pandey et al., 2009) and the other in Saurashtra basin near Lodhika drilled well (Singh and Arora, 2008). The profile in Kutch basin includes wide-angle seismic and magnetotelluric data and passes along the southern coastal plains between Jakhau and Mundra. The interpreted section (Fig. 6.7A) shows the presence

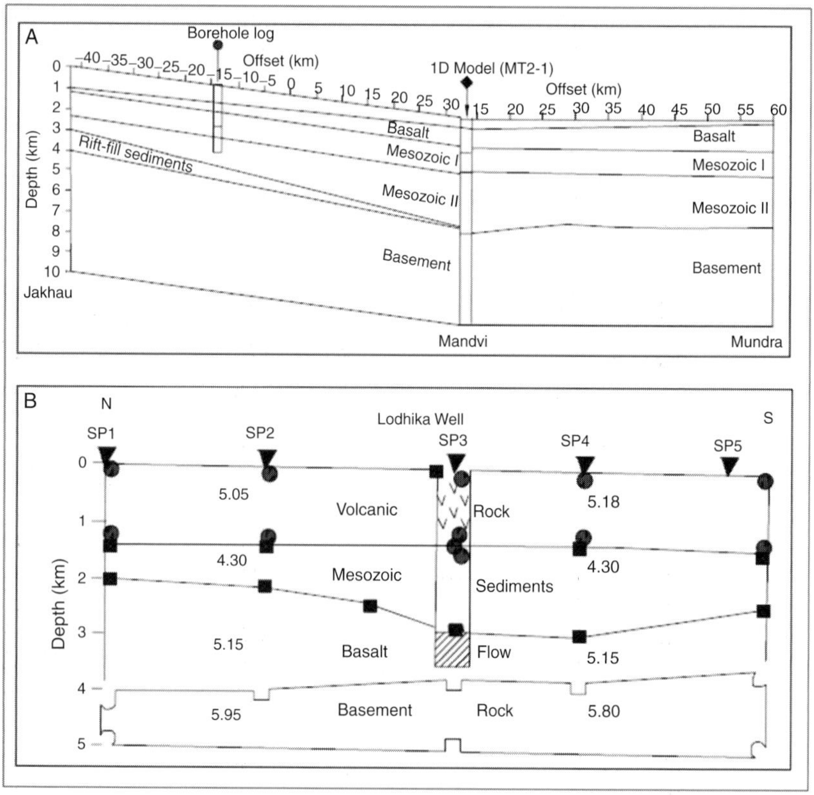

FIGURE 6.7 (A) 2D composite subsurface model interpreted based on seismic and MT studies in the Kutch onshore basin *(adapted from Pandey et al., 2009)*. (B) Interpreted seismic data around Lodhika well in the Saurashtra onshore. SP, shot points. Values are velocities of different layers in km/s *(adapted from Singh and Arora, 2008)*.

of more than 3 km thick Mesozoic sediments having variable thickness of overlying basaltic layer with maximum thickness of around 1.2 km (Pandey et al., 2009). A parametric well has been drilled down to 3500 m in the midst of trap-covered Saurashtra at Lodhika near Rajkot for the purpose of judging the hydrocarbon potential of the Mesozoic sequence in this basin (Singh et al., 1997). Though this well did not indicate any hydrocarbon shows, it provided an opportunity to understand better the subsurface geology, stratigraphy, and for the analysis of source rocks as well as reservoir rocks. In order to understand the areal extent of the Mesozoic sediments, integrated geophysical studies have been carried out by the National Geophysical Research Institute and corroborated with the Lodhika well. We present here a N–S seismic section (Fig. 6.7B) constructed around this well which shows four layers having variable thickness of Mesozoic sediments (Singh and Arora, 2008).

6.3.1.2 Offshore

6.3.1.2.1 Structure and Tectonics

The Kutch–Saurashtra offshore basin lies west of the Kutch and Saurashtra coast at the NW margin approximately between 20° and 23°N latitude. The coastline is characterized by the presence of two indentations, the Gulf of Kutch and Kori creek in Kutch region and the shelf is around 160 km wide. The shelf break is variable and occurs between 130 and 155 m characterized by steeper slope off Kutch and gentler slope off Saurashtra (Chauhan and Almeida, 1993). The factor of regional tectonics can also be noticed in the offshore morphology as evidenced by the changes in shelf break. The offshore continuation of ENE–WSW trending Saurashtra arch into the offshore is seen to be correlating with the significant change in the orientation of shelf break (Chauhan and Almeida, 1993), and this feature divides the offshore basin into two distinctly different sub-basins, the northern Kutch offshore and the southern Saurashtra offshore (Fig. 6.8A). While the ENE–WSW Satpura trend influences the structural grain in the Saurashtra offshore, the E–W structural elements of Kutch mainland continue further west toward offshore (Biswas, 1987). The shallow shelfal part of the basin is characterized by horst–graben structures and detailed geophysical and geological investigations in the deepwater (between 400 and 3000 m water depth) offshore area lying between Indus basin of Pakistan in north and the Laxmi basin in south by Sriram et al. (2006) and Bisen et al. (2010) revealed several structural and tectonic features (Fig. 6.8B). From the hydrocarbon exploration point of view, the deepwater Kutch–Saurashtra can be considered frontier area, therefore, the study by above investigators is very valuable in that it provided a first-order picture of the structural elements for the first time from this basin. Some salient results of their investigations are discussed here. According to Sriram et al. (2006), the Saurashtra Arch which is generally referred as an

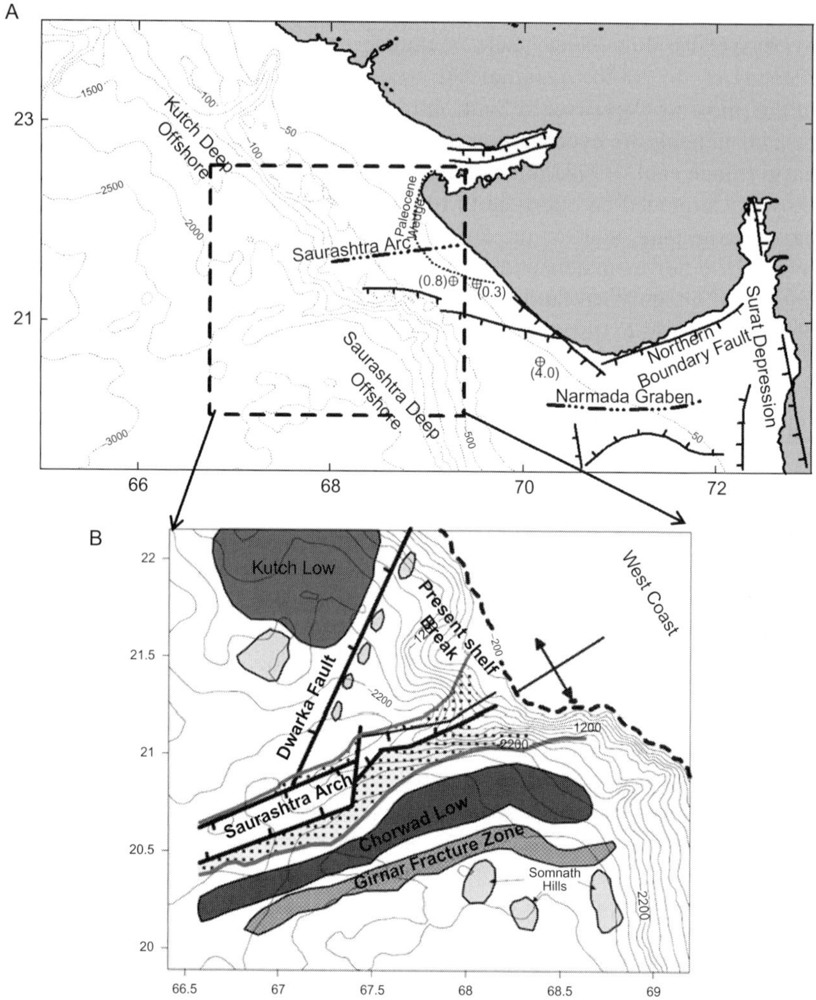

FIGURE 6.8 (A) Map showing major tectonic elements in the Kutch–Saurashtra offshore area. Area marked in thick dashed line is shown in (B). (B) Sketch map showing the structural elements identified in the deepwater Kutch–Saurashtra offshore (modified after Bisen et al., 2010).

anticlinal feature in the shelf region which plunges from the continental hinge into the deepwater area. The Arch is more than 250 km long and 60 km wide characterized by the horst–graben architecture. Apart from this megastructure, other important structural elements (Fig. 6.8B) have been mapped by Bisen et al. (2010) and these are: (i) Girnar FZ, (ii) Dwarka fault, (iii) Somnath hills, (iv) Kutch Low (depression), and (v) Chorwad Low (depression). These are briefly described below.

Saurashtra Arch

As already mentioned, it is the most important structural element oriented in ENE–WSW direction and extends far into the deep offshore areas. It contains horst–graben morphology, divides the Kutch–Saurashtra offshore into two sub-basins, and plays an important role in controlling the Indus Fan sedimentation during the basin evolution. Volcanic origin of this feature is proposed by Sriram et al. (2006) based on the presence of dykes, sills, and hydrothermal vents, and suggested that the Arch had evolved in three different stages such as (i) late Cretaceous volcanic ridge, (ii) early Paleocene crestal collapse and rifting, and (iii) lower to middle Miocene tilting.

Girnar FZ

It is an important ENE–WSW trending feature observed in deepwater Saurashtra offshore south of the Arch. This feature is originally mapped by Malod et al. (1997) in the area north of Laxmi ridge, and the eastward continuation of it is noticed from seismic data in Saurashtra offshore by Bisen et al. (2010) as a 20–30-km-wide fracture characterized by basaltic hills and carbonate growth activity.

Dwarka Fault

It is a 190-km-long NNW–SSE trending basement-controlled normal fault mapped in the area north of the Saurashtra Arch. Bisen et al. (2010) inferred that the fault is active from the Paleocene time, the eastern upthrown side of the fault is characterized by carbonate buildup activity all along the fault, and the southern limit of the fault terminates at the Saurashtra Arch.

Somnath Hills

These are buried subsurface carbonate hills located over the volcanic rocks in the area south of the Girnar FZ. Bisen et al. (2010) suggested that these highs are located at the intersection of Girnar FZ and the NW–SE trending Kori–Comorin ridge located further south.

Kutch Low

It is a deep basement depression (~7.0 s TWT) zone located in the Kutch deepwater offshore area in northwest of Dwarka fault as revealed from the seismic data and part of this low could be located in offshore Pakistan (Bisen et al., 2010).

Chorwad Low

It is an ENE–WSW trending basement depression zone lying between the Saurashtra Arch and the Girnar FZ. At places, the basement reaches to more than 6.0 s TWT.

6.3.1.3 Hydrocarbon Potential of the Basin

6.3.1.3.1 Petroleum System

6.3.1.3.1.1 Kutch Basin

- **Source rocks**

  Compilation of source rock data from outcrops and drilled wells indicate the presence of source potential in the Mid Jurassic to Lower Cretaceous interval, in the western onland part of the basin with adequate maturity. (Biswas, 1982). Further, the Eocene and Oligocene sediments from this area have good source potential. Mature source rocks are observed in the Upper Cretaceous and Paleocene sequences in the offshore. Increased exploration activity in recent times in the offshore Kutch resulted in the discovery of oil and gas in the Late Cretaceous Mundra clastics (Ravindran et al., 2012). According to these authors, in the offshore Kutch, Mesozoic sequence particularly Cretaceous forms significant part of total sedimentary cover. While the Early Cretaceous carbonaceous shales formed in fluvio-deltaic condition contain mature and fairly rich source facies, the Late Cretaceous Mundra clastics lying close to these might have been charged with hydrocarbons.

- **Trap and seal**

  Structural traps in this basin are prevalent mainly due to the superposition of two major rift related tectonic cycles, that is, the early Mesozoic and the Late Cretaceous-Paleocene rifting episodes both giving rise to horst–graben structural formation (Biswas, 1982). The carbonate buildup is a major strati-structural prospect in this basin.

6.3.1.3.1.2 Saurashtra Basin

- **Source rocks**

  In the onshore Saurashtra basin, the Lodhika well which drilled through thick traps has provided first order information regarding the source rock potential of Mesozoic sequences (Singh et al., 1997). The analysis indicates the possibility of hydrocarbon mature Mesozoic sediments from early Cretaceous onward. In the offshore part of the basin, the Late Paleocene to Middle Eocene shale sequence contains fair to good amount of organic matter having marginal hydrocarbon potential. Similar to that observed in the offshore east coast of India, the possibility of shallow biogenic gas generation and accumulation exists in this basin also. The major hydrocarbon migration in this area is possibly through faults from Paleocene matured source to the Oligocene-Miocene reservoir.

- **Reservoirs**

  The expected reservoir rocks in Saurashtra basin are (i) Mesozoic sands, (ii) Intratrappean sequences within Deccan Trap, (iii) Volcanoclastic and sand bodies within the Paleocene to early Eocene units, (iv) Carbonates of Eocene to Middle Miocene, and (v) Delta-related sand bodies of early Eocene to early Oligocene.

- **Cap rocks and entrapment condition**
  The inter-bedded shales and low-porosity limestones of middle Eocene to middle Miocene carbonate sequences may constitute the regional cap rock within the Tertiary formations in offshore Saurashtra Basin, and in the case of Mesozoic sediments, the impervious Deccan Traps act as cap rock. The hydrocarbon accumulation in the basin is primarily controlled by stratigraphic and structural traps as well as their combination.

6.3.2 Mumbai Offshore Basin

The Mumbai offshore basin is the most prolific hydrocarbon-bearing basin having significant discoveries and is famous for the location of "Bombay High" giant oil field discovery in 1974 by the ONGC. From the point of view of occurrence of oil and gas, it is the principal producing basins of India. This completely offshore basin had evolved in a divergent margin setup and covers an area of 1,20,000 km^2 from the coast up to the 200 m bathymetry (Mathur and Nair, 1993). The basin is located on a very large continental terrace characterized by a wide shelf and bounded by the Saurashtra Peninsula and Cambay basins in north, Western Ghats in east, Arabian abyssal plain in west, and the Konkan basin in south (Singh et al., 2007).

6.3.2.1 Structural Framework and Tectonic History

In general description of the structural framework, many earlier workers (Mohan, 1985; Mathur and Nair, 1993) considered the contiguous Saurashtra offshore as part of the Mumbai offshore basin by extending the basin limit up to the Saurashtra arch. As the Saurashtra offshore is separated from the Surat depression by the N–S trending Diu arch, and the former one has already been described as part of the Kutch–Saurashtra basin in the previous section (Section 6.3.1), we avoid further discussion on Saurashtra offshore and limit here to other structural elements of the basin. The Mumbai offshore basin is a composite basin with each of the major structural units having several well-marked subunits (Fig. 6.9A). The basin is characterized by the carbonate-dominated platform structural units surrounded by the clastic-dominated major depressions (Mathur and Nair, 1993). Several NW–SE- to N–S-oriented basement-controlled faults segment the entire shelf between the coast and the shelf margin zone and give rise to many horst–graben features. The central platform structure is flanked by the horst–graben system in north as well as east, and by shelf margin basin in west. Tectonically, the basin can be defined by several structural elements (after Mathur and Nair, 1993), such as (i) Surat

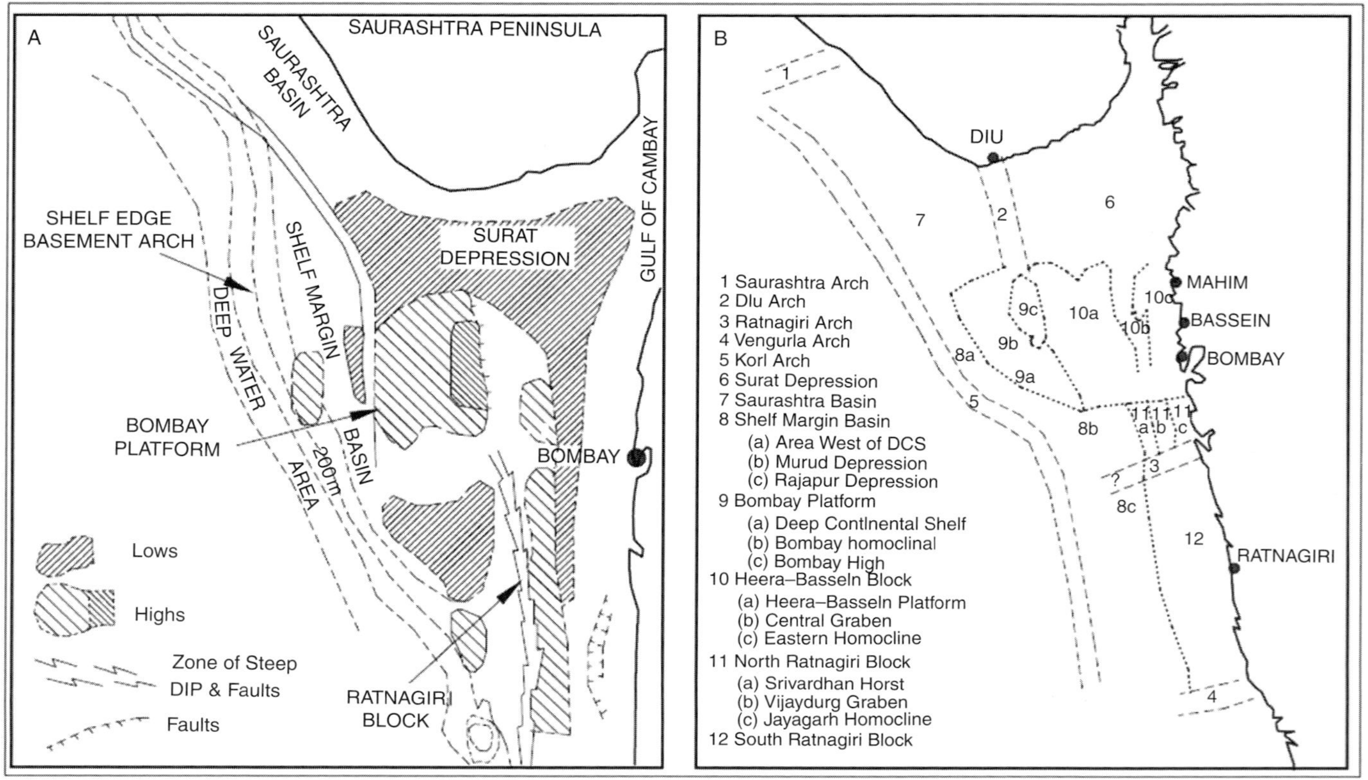

FIGURE 6.9 (A) Structural elements demarcating horst–graben structures in the Mumbai offshore basin *(adapted with permission from Berger et al., 1983 © Schlumberger Ltd)*. (B) Various structural units of the Mumbai offshore basin *(adapted from Mathur and Nair, 1993)*.

depression, (ii) Murud depression, (iii) Rajapur depression, (iv) Bombay platform, (v) Heera–Bassein block, (vi) Ratnagiri block, and (vii) shelf margin basin (Fig. 6.9B), and some of these are separated by various basement arches or major faults. The N–S trending Diu Arch separates the Surat depression from the Saurashtra basin, NE–SW trending Ratnagiri arch separates the Murud and Rajapur depressions as well as the north and south Ratnagiri blocks, Kori arch, a semi-continuous regional arch, separates the shelf margin basin from the deep sea basin, and the Vengurla arch acts as the southernmost limit of the Mumbai offshore basin. Seismic sections shown in Fig. 6.10 display wrench faulting, and the flower structures associated with such faulting have proved to be locations where many prospects have been identified in this basin (Singh et al., 2007). The individual structural elements that include platformal as well as basinal areas are briefly described (after Mathur and Nair, 1993) below:

Surat depression

The N–S trending Diu Arch separates the Surat depression from the Saurashtra basin. This depression forms the southward extension of the Cambay basin and is one of the major clastic-filled basins. It comprises thick sediments of more than 6000 m (Singh et al., 2007) and is characterized by many structures of multiple origins such as basement-controlled anticlines, differential compactions, negative flowers and inversion features, growth faults, and rollovers.

Shelf Margin Basin

The Shelf Margin Basin lies to the west of Bombay platform and Ratnagiri platform. Kori High marks its western limit, while the northern boundary is indistinguishable and may be continuing further along the Saurashtra Basin. It is further divided into three sub-basins such as the area west of deep continental shelf (DCS), Murud depression, and Rajapur depression. While the area west of DCS is a narrow corridor lying between the Kori High and the Bombay platform, the Murud and Rajapur depressions are separated by the NE SW trending Ratnagiri Arch. Both Murud and Rajapur depressions experienced maximum subsidence starting from late Oligocene and middle Miocene times, respectively.

Bombay platform

The Bombay platform is bounded by the Shelf Margin basin in the south as well as the west, by the Saurashtra Basin and Surat depression in north, and by the Heera–Bassein block in the east. It consists of the DCS area and the well-known Bombay High which is an uplifted block, and both are separated by a monoclinal rising area.

Heera–Bassein Block

Located east of the Bombay platform, the block is further subdivided longitudinally into the western platform arch, the central graben, and the eastern homoclinal area.

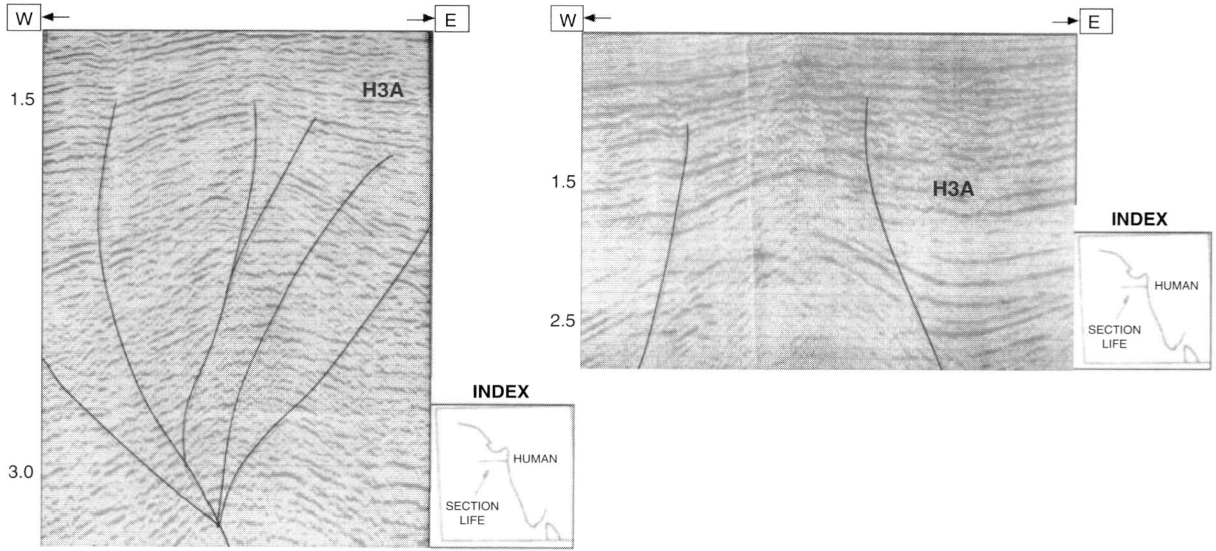

FIGURE 6.10 Seismic sections across the Ratnagiri offshore showing flower structures and wrench faulting *(adapted from Singh et al., 2007)*. (For color version of this figure, the reader is referred to the online version of this chapter.)

Ratnagiri Block

The Ratnagiri block is divided by the NE–SW trending Ratnagiri Arch into the north Ratnagiri and south Ratnagiri blocks. The north Ratnagiri block is a southward continuation of the Heera–Bassein Block, and similar to that it is also divided into three such as Jayagadh homocline, Vijaydurg graben, and Srivardhan horst. It is noticed that the Vijaydurg graben and the central graben in Heera–Bassein block are continuous in the beginning and are connected with the Surat depression; however, subsequent ENE- or NE-strike-slip faulting activity has brought out complexities. These three depressions are known to contain substantial Paleocene–Eocene clastic fill. The south Ratnagiri block is also having similar horst–graben structures with less clastic fill during Paleocene–Eocene.

6.3.2.1.1 Generalized Stratigraphy

The stratigraphy of Mumbai offshore basin is discussed in detail by several earlier workers (Basu et al., 1982; Mathur and Nair, 1993; Nair et al., 1992; Rao and Talukdar, 1980; Singh et al., 2007; among many others). Mathur and Nair (1993) observed that the nomenclature proposed by Basu et al. (1982) based on the detailed stratigraphic scheme is more relevant for the carbonate platform areas. However, the regional chronostratigraphic correlations made by Nair et al. (1992) have been utilized by Mathur and Nair (1993) to describe the distribution of stratigraphic units and their facies content along and across the basin by considering different cross-sections. We present here two sections depicting the depositional sequences within the depressions and over the platform areas cutting across the Heera–Bassein block, Bombay platform, Ratnagiri block, and the shelf margin basin (Fig. 6.11). These sections suggest that during Paleocene to Eocene age, the clastics are primarily deposited in the central graben (Heera–Bassein block) and the Vijaydurg graben (north Ratnagiri block). During this period, the shelf margin basin remained sediment-starved due to the presence of Diu Arch, Bombay high, and the Srivardhan horst. The biostratigraphy, rock stratigraphy, and seismic stratigraphy of the Mumbai offshore basin (after Singh et al., 2007) are presented in Table 6.3.

6.3.2.2 Tectonostratigraphy and Sedimentation

The tectonostratigraphic analysis of the Mumbai offshore basin suggests that the first marine transgression took place toward the close of Paleocene. During this transgression, the Mumbai High as well as parts of Heera and Ratna being the paleohighs remained unaffected; however, from late Oligocene onward, these highs have also come under sedimentation (Singh et al., 2007). The basin configuration as revealed by the sediment fill maps (Fig. 6.12) during the Eocene time as well as in present-day reveals the influence of Dharwar trend as indicated by the Mumbai high, Bassein, Ratnagiri,

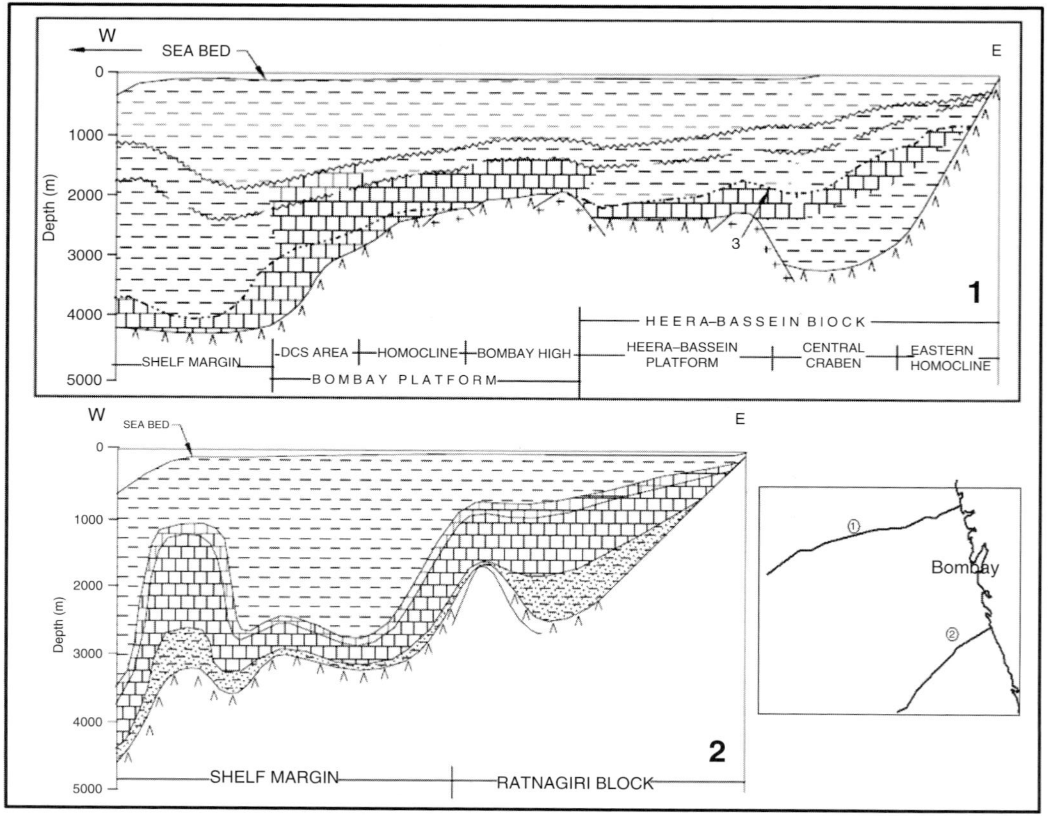

FIGURE 6.11 Seismogeologic sections: (A) section from Mahim to the shelf margin; (B) section across the Ratnagiri block and shelf margin *(adapted from Mathur and Nair, 1993).*

TABLE 6.3 Biostratigraphy, Lithostratigraphy and Seismic Stratigraphy of the Mumbai Offshore Basin (adapted from Singh et al., 2007)

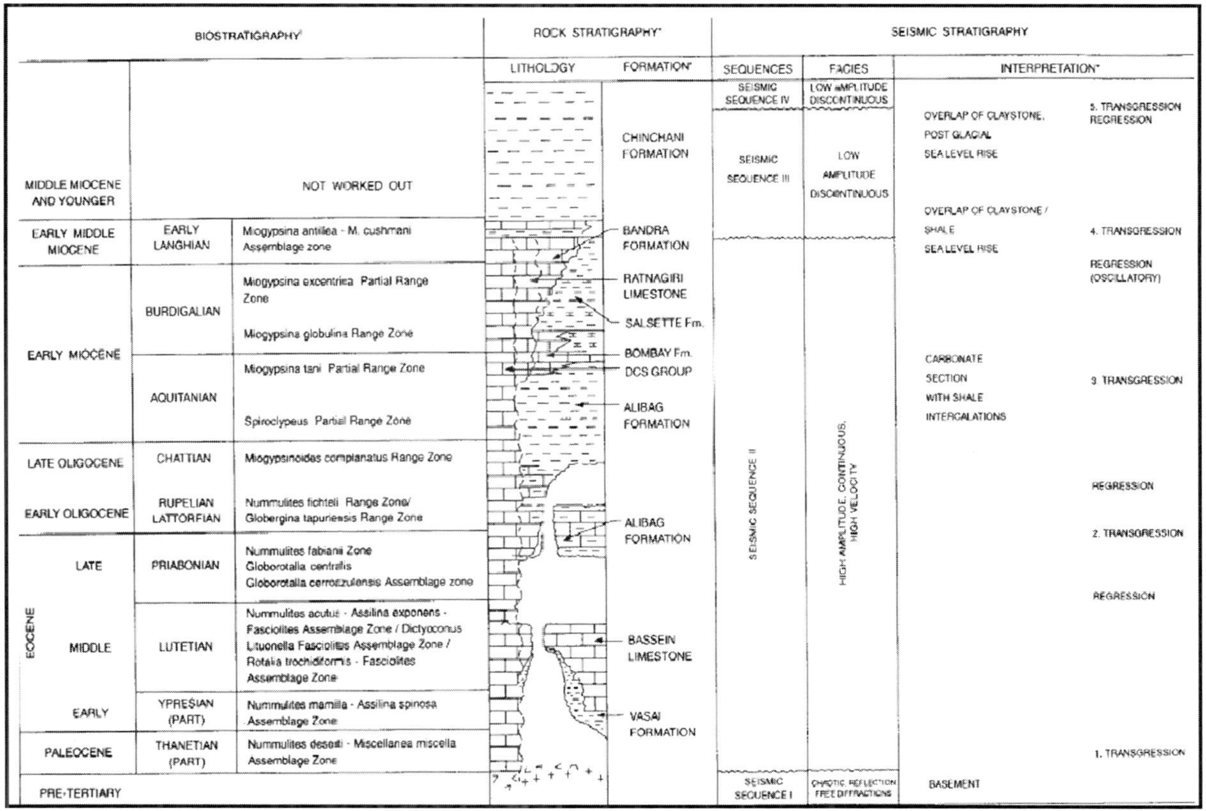

Biostratigraphy*			Rock Stratigraphy* Formation*	Seismic Stratigraphy Sequences	Facies	Interpretation*	
Middle Miocene and younger	Not worked out		Chinchani Formation	Seismic Sequence IV; Seismic Sequence III	Low amplitude discontinuous; Low amplitude discontinuous	Overlap of claystone, post glacial sea level rise	5. Transgression Regression
Early Middle Miocene	Early Langhian	Miogypsina antillea - M. cushmani Assemblage zone	Bandra Formation	Seismic Sequence III	Low amplitude discontinuous	Overlap of claystone / shale sea level rise	4. Transgression
Early Miocene	Burdigalian	Miogypsina excentrica Partial Range Zone; Miogypsina globulina Range Zone	Ratnagiri Limestone; Salsette Fm.; Bombay Fm.	Seismic Sequence II	High amplitude, continuous, high velocity		Regression (Oscillatory)
	Aquitanian	Miogypsina tani Partial Range Zone; Spiroclypeus Partial Range Zone	DCS Group; Alibag Formation			Carbonate section with shale intercalations	3. Transgression
Late Oligocene	Chattian	Miogypsinoides complanatus Range Zone					Regression
Early Oligocene	Rupelian Lattorfian	Nummulites fichteli Range Zone/ Globergina tapuriensis Range Zone	Alibag Formation				2. Transgression
Eocene Late	Priabonian	Nummulites fabianii Zone Globorotalia centralis Globorotalia cerroazulensis Assemblage zone					Regression
Eocene Middle	Lutetian	Nummulites acutus - Assilina exponens - Fasciolites Assemblage Zone / Dictyoconus Lituonella Fasciolites Assemblage Zone / Rotalia trochidiformis - Fasciolites Assemblage Zone	Bassein Limestone				
Eocene Early	Ypresian (Part)	Nummulites mamilla - Assilina spinosa Assemblage Zone	Vasai Formation				
Paleocene	Thanetian (Part)	Nummulites deserti - Miscellanea miscella Assemblage Zone					1. Transgression
Pre-Tertiary				Seismic Sequence I	Chaotic reflection free diffractions	Basement	

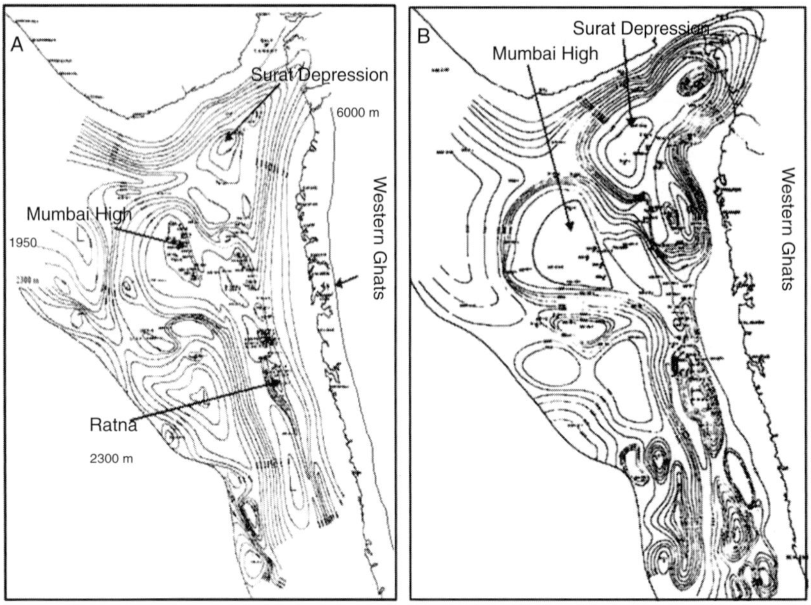

FIGURE 6.12 Sediment isopach maps in the Mumbai offshore basin showing major structural highs and lows: (A) Present; (B) Paleocene–Eocene *(adapted from Singh et al., 2007).*

and other NNW–SSE trending basement features. The basement depth map shows more than 6000 m of sediments in the Surat depression which is the deepest part in the entire basin, about 2300 m of sediments in grabenal area of Ratnagiri block, and around 1950 m over the Mumbai High (Singh et al., 2007). In the basin south of Surat depression, Rao and Talukdar (1980) identified three distinct megasequences with characteristic lithotypes within the Tertiary sediments. These are (i) the lowermost syn-rift or early-rift clastic facies of late Paleocene–early Eocene sequence, (ii) middle Eocene–Oligocene carbonate facies, and (iii) lower and middle Miocene fine clastic facies. However, the lower Miocene sediments are still carbonates over the Mumbai High, and the middle Miocene is characterized by shale, sandstone, and thin limestones. Throughout the basin, the post-Middle Miocene consists of clastics all over the basin. In the Surat Depression, however, Tertiary clastics from Paleocene to recent are present. Detailed facies analysis and paleogeography with reference to major geological events in different time intervals are presented by Mathur and Nair (1993). Some salient aspects of their analysis along with the maps are briefly presented here.

6.3.2.3 Paleogeography and Facies Distribution

Subsequent to the late Cretaceous Deccan volcanic activity, the sediment deposition in the Mumbai offshore basin has commenced since the Paleocene times. The depositional history of the basin has been described between the seven chronostratigraphic units (Mathur and Nair, 1993) that are traced in the entire basin.

Late Paleocene–Early Eocene

During this period, the Surat depression is the main clastic depocenter; however, the central graben and Vijaydurg graben along with the Surat depression have been filled by organic-rich, shallow shelf, marine shale with subordinate sands, and continental to paralic shales, sands, and corals (Fig. 6.13A). The outer shelf and bathyal open-marine claystones have been deposited in the partly starved Murud depression and Saurashtra and Shelf margin basins. The deposition of carbonate facies is mainly restricted to the DCS area. The end of this period witnessed a general regression and development of unconformity.

Middle–Late Eocene

This period witnessed basin-wide transgression after the peneplanation, and except Bombay High, the entire platform area is converted into a

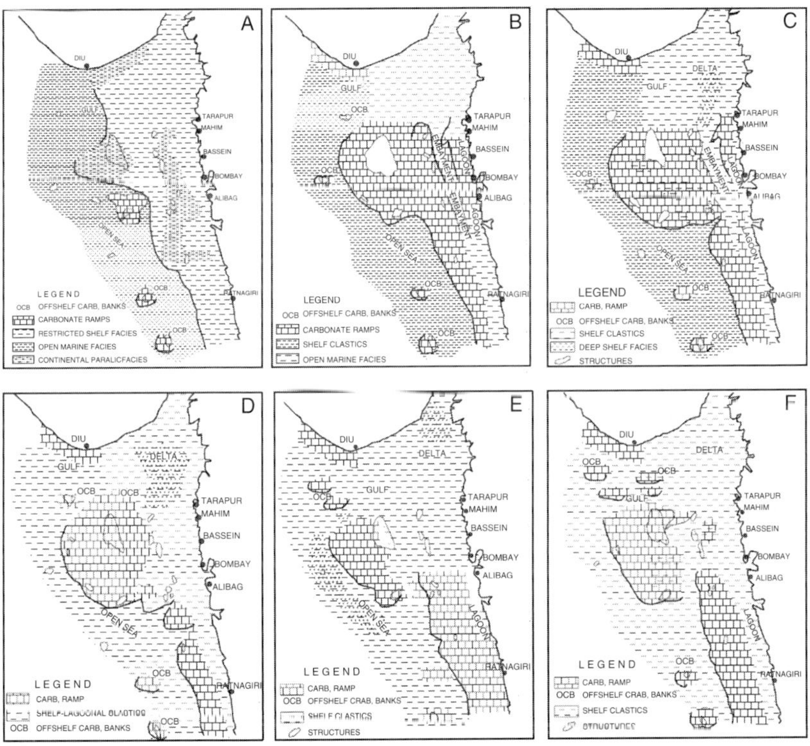

FIGURE 6.13 Paleogeography history and Facies distribution in the Mumbai offshore basin: (A) Paleocene–early Eocene, (B) middle–late Eocene, (C) early Oligocene, (D) late Oligocene, (E) early Miocene, and (F) middle Miocene *(adapted from Mathur and Nair, 1993)*.

homoclinal ramp and gave rise to deposition of carbonates (Fig. 6.13B). The shallow shelf areas of the Bombay platform, Heera–Bassein block, and the Ratnagiri blocks are characterized by extensive carbonate deposition during this period. The end of this period is marked by a major regression and development of regional unconformity.

Early Oligocene

This period witnessed a transgression covering larger areas of the Bombay platform though it did not reach the Eocene depositional limit which was followed by a regression. Most of the depositional aspects remain similar to that observed during the earlier period (Fig. 6.13C). The deposition of carbonates on ramps and off-shelf carbonate banks has continued; however, ramps witnessed shallower bathymetry at places, rate of clastic input has slightly increased, and partial starvation in the Saurashtra basin and shelf margin basin continued.

Late Oligocene

This period is marked by a eustatic rise in sea level which continued until early Miocene with remarkable increase in clastic sediment supply (Fig. 6.13D). The basin experienced westward tilt. Surat depression was subsiding lesser and receding coastline gave rise to delta-related sands. Rapid subsidence of Saurashtra basin and the shelf margin basin gave rise to deposition of clays dispersed from west. The muds dispersed southward have spread over the Heera–Bassein block and to some extent on the north Ratnagiri block and the Bombay platform.

Early Miocene

This period is characterized by general rise in sea level punctuated by brief fall, abundant supply of clastics with reduced rate in comparison to the late Oligocene, northward shifting of depocenter in Surat depression, deposition of large quantity of muds in Saurashtra and the shelf margin basins, and continuation of carbonate deposition on the Bombay platform and Ratnagiri block (Fig. 6.13E).

Middle Miocene

During this period, a brief rise in sea level followed by general fall gave rise to increased supply of clastics, and as a result, large parts of Bombay platform including the Mumbai High came under clastic deposition. The spell of the clastic input came to a halt during later part of mid-Miocene. The carbonate deposition then continued over the Mumbai High, Ratnagiri block, DCS and off-shelf banks (Fig. 6.13F).

Middle Miocene–Holocene

This period is characterized by extensive transgression that covered the entire basin and remarkable increase in clastic sediment supply.

6.3.2.4 Petroleum System

As stated earlier, the Bombay Offshore basin is the most prolific oil and gas producing basin credited with a number of major discoveries that include Bombay High, Bassein, Heera, Panna, Ratna, Neelam, and Mukta fields, and occurrence of oil and gas in these fields is observed at different stratigraphic levels. Most of the hydrocarbon occurrence in this basin is structurally controlled and from the middle Eocene to middle Miocene carbonate reservoirs, though few clastic reservoirs have been observed as stratigraphic plays or combination of both structural and stratigraphic plays. The Eocene, Oligocene, and early Miocene sequences also have proved to be oil/gas bearing in many structures in this basin. Three petroleum systems are present in this basin; while Panna-Bombay is the major petroleum system, other important petroleum systems are the Panna-Bassein and Panna-Mukta (Raju et al., 2005).

- **Source rocks**
  According to Biswas (1991), major source rocks in the Mumbai offshore basin lie in the Paleogene section. Geochemical studies carried out in this basin indicate that the Paleocene–early Eocene Panna formation having TOC in the range between 0.5% and 20% contributed mostly for the generation and migration of hydrocarbons (Chandra et al., 2005). The pre-middle Miocene shales in the Surat depression, shales which alternate with limestones of the Bombay platform, and shales in the shelf margin basin are the most possible source rocks with the presence of more than 0.5% organic content in this basin (Berger et al., 1983). According to Mathur and Nair (1993), the sediments in the regional lows as well as localized depressions within the platforms have undergone thermal maturation under fairly higher geothermal gradients of 4–7 °C/100 m, and the burial history and maturation curves constructed by them for the Surat depression and central graben area (Fig. 6.14) indicate that Paleocene–Eocene sediments reached peak generation potential for Type-III organic matter.

- **Reservoir rocks**
  Four oil/gas bearing zones (L-I, L-II, and L-III are limestones, S-I sand) have been identified within the Miocene sequence (Berger et al., 1983). The lower Miocene (L-III) constitutes the main reservoir in the Bombay High, while the middle Miocene (L-I, L-II) sequences are the youngest reservoir rocks. Other important reservoirs are the middle Eocene–lower Oligocene Bassein limestone and the lower Oligocene Mukta formation (Chandra et al., 2005).

- **Cap rocks**
  The intervening shales mainly act as cap rocks in the reservoirs of this basin. The middle Miocene Bandra formation acted as regional seal.

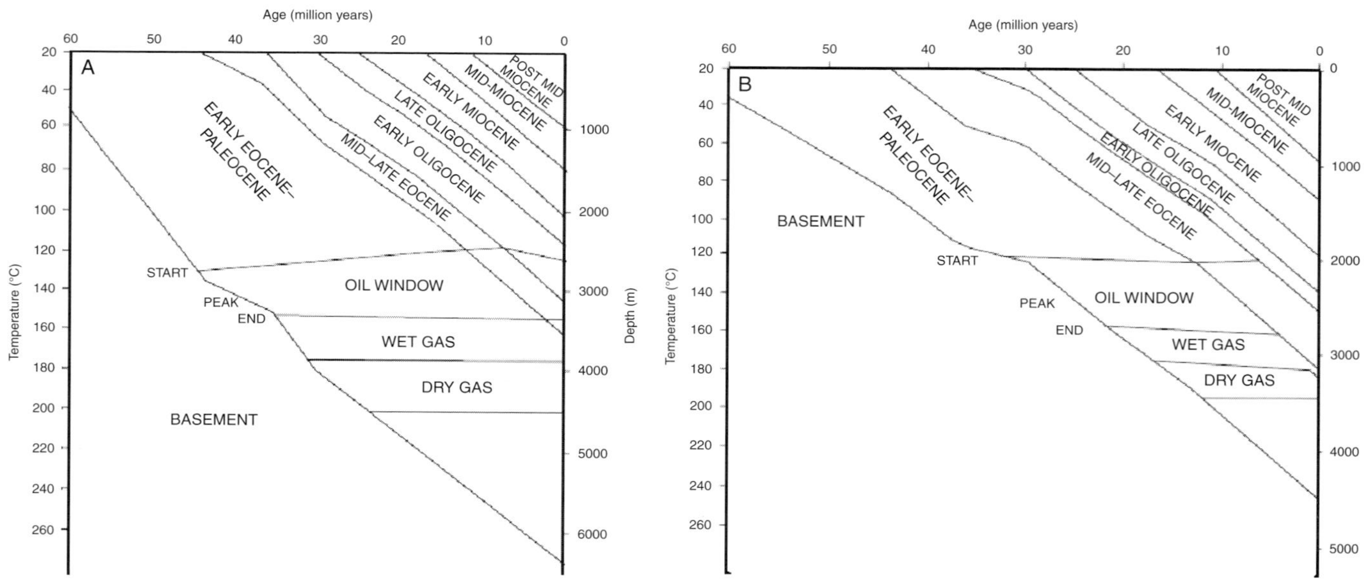

FIGURE 6.14 Subsidence and maturation curves: (A) Surat depression; (B) Mahim depression *(adapted from Mathur and Nair, 1993).*

A number of hydrocarbon-bearing zones from the Basement (deepest fractured basement reservoir) to middle Miocene (shallowest reservoir) occur in the Mumbai offshore basin. Most of the discoveries in this basin are in large structural accumulations, or smaller structures, barring few in the stratigraphic setup. The Paleocene to early Eocene Panna Formation consists of extensive clastic reservoirs, and overlying these clastic reservoirs, a series of carbonate reservoirs ranging in age from Eocene to Miocene have been encountered in the basin. The hydrocarbon play map of the basin is presented in Fig. 6.15 which shows the occurrence of hydrocarbons at six stratigraphic levels (after Mohapatra et al., 1996). Three geologic sections depicting the areas of oil/gas occurrence and potential plays in the basin are shown in Figs. 6.16–6.18. While 60% of the inplace oil/gas discovered comes from the Bombay and Bandra formations, the Bassein limestone, which is the most widespread oil/gas play accounts for nearly 30% of inplace hydrocarbons.

6.3.3 Konkan–Kerala Offshore Basin

The KK offshore basin forms the southern part of the WCMI and extends from Vengurla Arch in the north to the area south of Cape Comorin at the southern tip of India (Singh and Lal, 1993). While the Vengurla Arch which is a NE–SW trending basement arch separates the Mumbai offshore from the Konkan offshore, another basement arch, called the Tellicherry Arch, separates the KK offshore into the northern Konkan basin and the southern Kerala basin (Fig. 6.1). Along the adjoining West Coast of India, the region north of Vengurla Arch is characterized by the extensive Deccan flood basalts, and south of it, the rocks are composed almost entirely of Archaean charnockites, gneisses, and granites with limited extent of the Tertiary sediments (Miocene outcrops) along the coast of Kerala further south. The KK offshore basin contains an area of around 77,000 km^2 up to 200 m bathymetry, and about 92,000 km^2 including the deepwater area up to 2000 m bathymetry (Singh and Lal, 1993). The present understanding of the basin is mainly based on seismic, gravity, and magnetic data as only few wells (around 17) have been drilled till now. While few of the drilled wells penetrated the Paleocene basalts, the oldest sediments encountered in the wells are of Santonian age (Dirghangi et al., 2000; Singh and Lal, 1993).

6.3.3.1 Structure and Tectonics

According to Biswas (1987), the basic framework of the WCMI has been established by the end of late Cretaceous. Unlike the Mumbai offshore basin, available seismic as well as well data information in the Konkan–Kerala (KK) offshore is very sparse. The shelfal horst–graben complex, Kori–Comorin depression, Kori–Comorin ridge, Laxmi–Laccadive depression, Laxmi–Laccadive ridge, and Arabian abyssal plain are the important tectonic elements (Fig. 6.2) identified

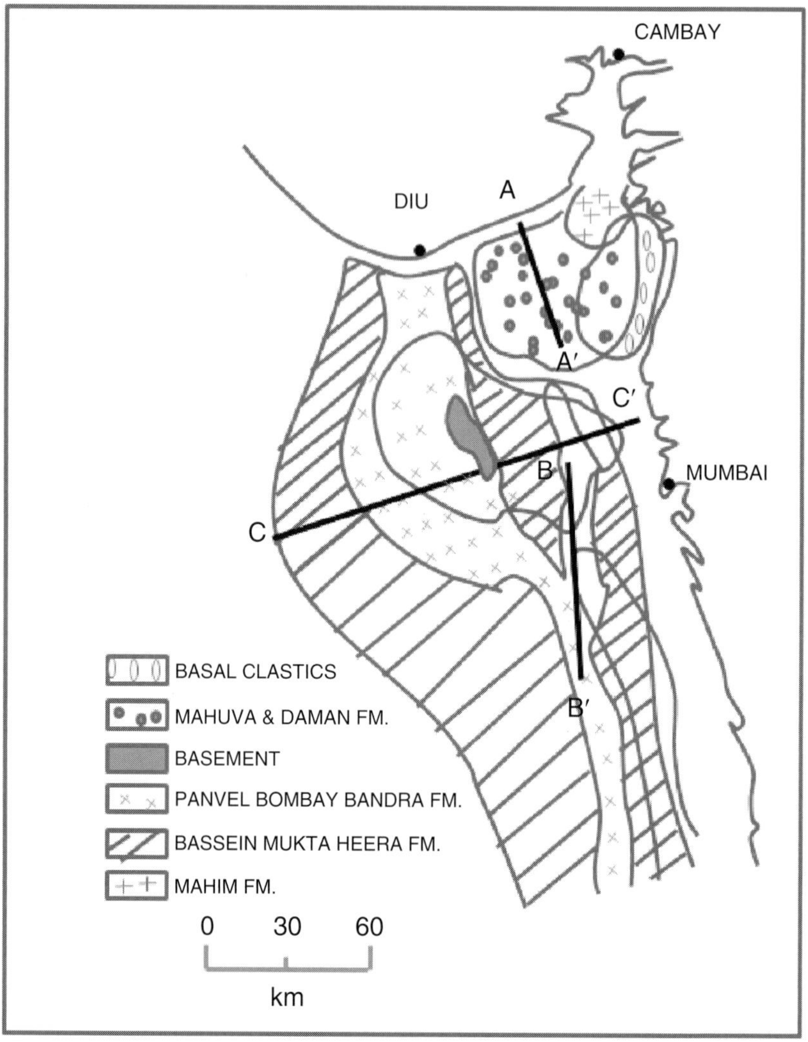

FIGURE 6.15 Hydrocarbon play type distribution map of the Mumbai offshore basin *(adapted from Mohapatra et al., 1996)*. AA′–CC′ are geologic cross-sections, depicting potential play types in Figs. 6.16–6.18. (For color version of this figure, the reader is referred to the online version of this chapter.)

by Biswas and Singh (1988). The Kori–Comorin ridge which is a NW–SE trending linear fault bounded structural high extends up to the Pratap Ridge in the southern part of WCMI (Fig. 6.1). Lying between the Laccadive Ridge in the west and the Pratap as well as Comorin ridges in the east, the Laccadive depression is a prominent subbasin characterized by half-grabens as well as volcanic features (Chaubey et al., 2002a; Thakur et al., 1999). The shelfal horst–graben complex can be further subdivided into three depressions such as the Konkan depression,

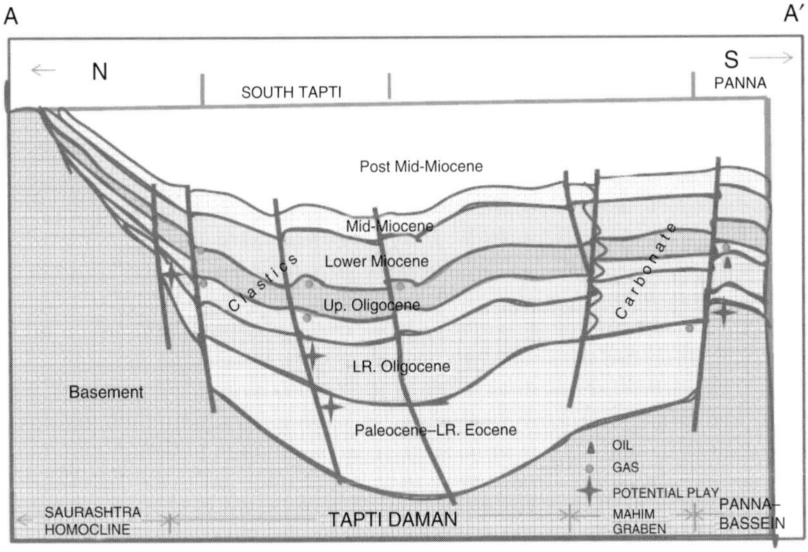

FIGURE 6.16 Schematic geologic section AA´ in the northern Mumbai offshore basin *(adapted from Mohapatra et al., 1996)*. (For color version of this figure, the reader is referred to the online version of this chapter.)

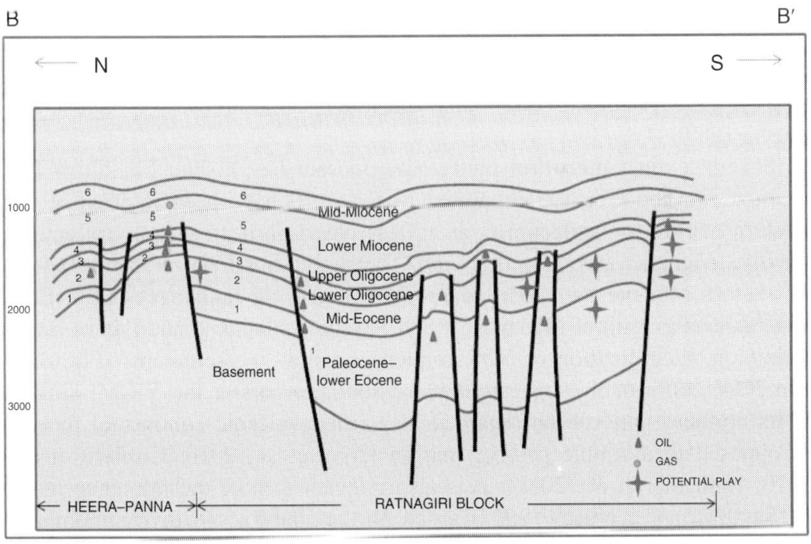

FIGURE 6.17 Schematic geologic section in the south Mumbai offshore basin *(adapted from Mohapatra et al., 1996)*. (For color version of this figure, the reader is referred to the online version of this chapter.)

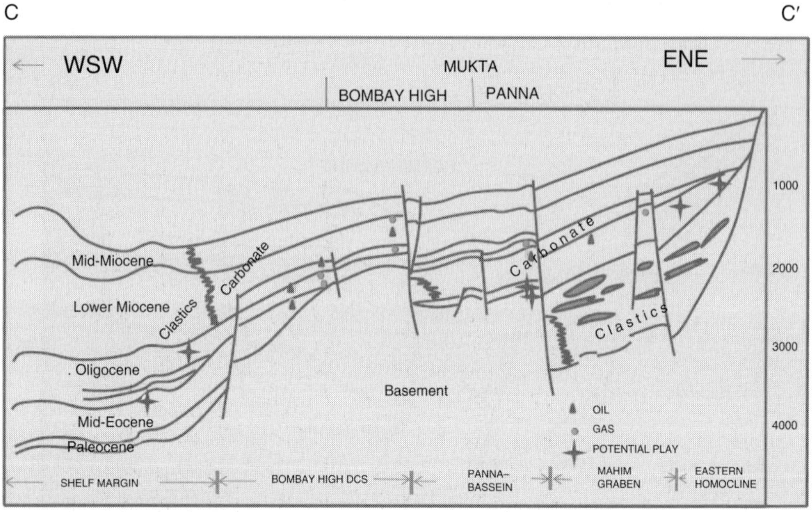

FIGURE 6.18 Schematic geologic section across the Mumbai offshore basin *(adapted from Mohapatra et al., 1996)*. (For color version of this figure, the reader is referred to the online version of this chapter.)

Cochin depression, and the Cape Comorin depression (Singh and Lal, 1993), and the Alleppey platform forms an important tectonic element of this complex (Fig. 6.1). In the Konkan offshore, Berger et al. (1983) delineated several structural features such as the Pratap Ridge complex, the shelf-margin basin, the mid-shelf basement ridge, and inner shelf graben. While the sedimentation is limited to the coast parallel N–S trending grabens separated by local basement highs in the offshore areas, the presence of differential vertical movements characterized by wrench faults, reverse faults, and folds has been observed (Ghosh and Zutshi, 1989).

Several extinct transform faults, viz., Indrani FZ, Vishnu FZ, Shiva FZ, Bramha FZ, and Goa FZ illustrated in Fig. 6.19 have been inferred at the western margin by Nathaniel et al. (2011) based on deep seismic reflection, gravity, magnetic, and refraction data. The shelfal horst and graben province of western offshore are modified by these transform faults (presently, FZs) generated as a result of Madagascar and Seychelles breakup and Laxmi Basin formation. Identification of SDR sequences below the sediments (Fig. 3.13 and 3.14) confirmably suggests that the southern part of the WCMI similar to its northern part can be regarded as a rifted volcanic continental margin as opposed to a simple passive margin (Ajay et al., 2010; Corfield et al., 2010; Nathaniel et al., 2011). It is also relevant to note the inference made by Savostin and Kerusov (2002) based on the analysis of gravity anomalies integrated with seismic data who opined that both Mumbai and the KK offshore areas are not clearly distinct from each other. Based on identified SDRs along with the crustal thickness models, Nathaniel et al. (2011) proposed the

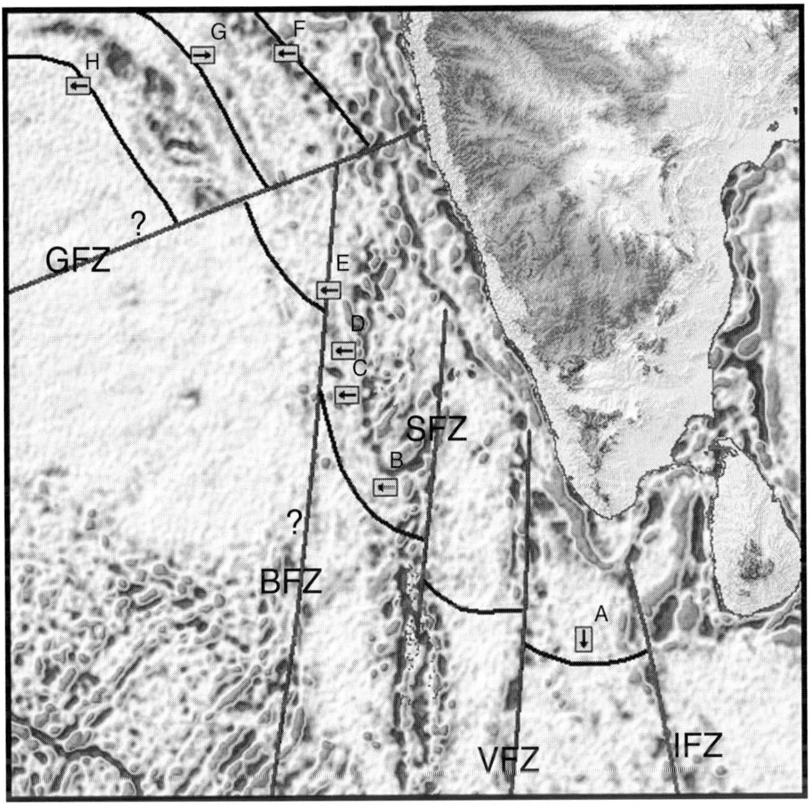

FIGURE 6.19 Map showing locations of SDRs (A–H) identified from the seismic data by various investigators in the western offshore along with proposed Ocean–Continent boundary (after Nathaniel et al., 2011). Arrow marked in A-H indicates the direction of SDR's. GFZ, Girnar FZ; BFZ, Brahma FZ; SFZ, Shiva FZ; VFZ, Vishnu FZ; IFZ, Indira FZ. (For color version of this figure, the reader is referred to the online version of this chapter.)

Continent Ocean Boundary for the entire West Coast of India as a working model for petroleum exploration (Fig. 6.19). From the previous geophysical studies in the region as well as based on the above discussion, we can summarize that the southwest continental margin could be a widely stretched margin, both Laxmi and Laccadive ridges could be continental fragments which were subsequently affected by volcanism, and the Laccadive basin could be underlain by stretched continental crust due to the absence of magnetic anomaly identification. The above points will have relevance and wider implications as far as the hydrocarbon exploration in the western offshore is concerned.

6.3.3.2 *Cenozoic Basin Configuration*

Several ridges separated by depressions, which rapidly subsided, characterized the KK basin development in the early stages, and sediments derived from north as well as west have filled these depressions (Singh et al., 2007). Campanile et al. (2007) compiled the Tertiary sedimentary thickness for the entire WCMI which suggests that the depressions south of Vengurla Arch are devoid of thick sedimentary pile except three localizations of more than 3.0 km sediments. The isopach map of sediments in the KK basin presented by Rao and Srivastava (1984) indicates Cenozoic sediment thicknesses maximum of around 4 km. We present here two seismo-geological sections (Fig. 6.20) across the KK offshore depicting the Tertiary sedimentary sequences overlying the thick Cretaceous volcanic traps. Three distinct tectono-sedimentary stages related to rift–drift episodes such as (i) pre-rift prior to base Tertiary, (ii) syn-rift from Paleocene to Lower Eocene, (iii) post-rift from Middle Eocene (through Oligocene and Middle Miocene) to Recent have been identified in the Konkan offshore by Thakur et al. (1999). On the other hand, the Cenozoic sedimentary package is divided into three regionally correlatable sequences: a Paleocene- late Oligocene sequence, a middle Miocene sequence, and an upper post-Miocene sequence by Campanile et al. (2007) using the published sediment thickness maps of Rao and Srivastava (1984). The isopach maps compiled by these authors for the above three ages are presented in Fig. 6.21. Further, the litho-log data at 11 well locations in the basin allowed Campanile et al. (2007) to further subdivide the sedimentary succession as shown in Fig. 6.22. Based on

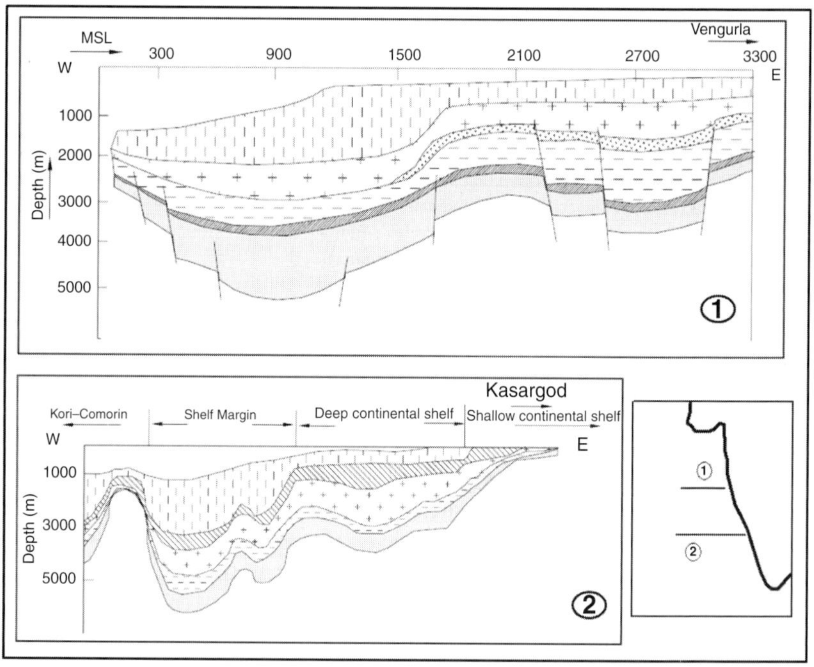

FIGURE 6.20 Seismogeologic sections in the Konkan–Kerala offshore basin *(adapted from Singh et al., 2007).*

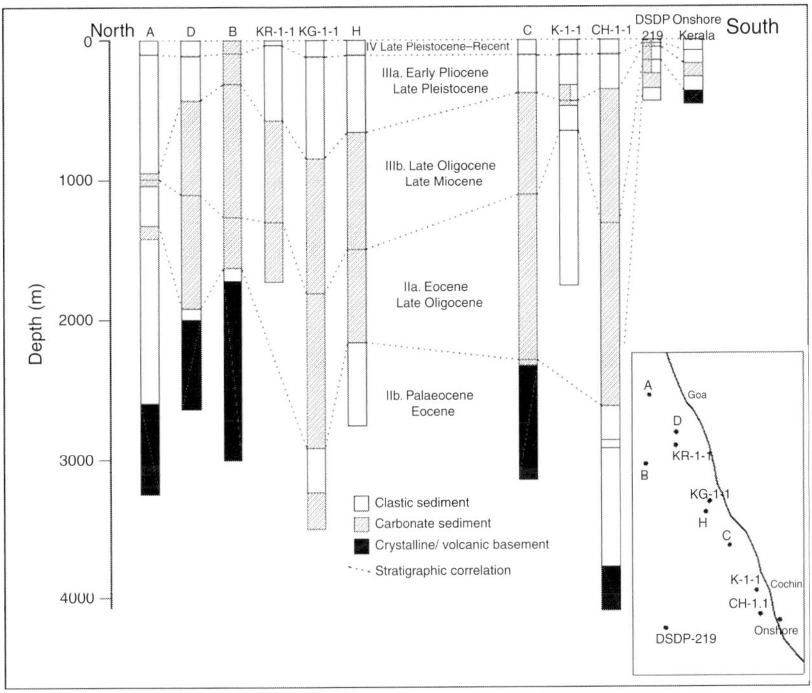

FIGURE 6.21 A N–S section constructed from available wells in the Konkan–Kerala offshore, showing the lithological variation in sedimentary sequences. Dashed lines are stratigraphic correlations *(Reproduced with permission from Campanile et al., 2007 © Wiley-Blackwell).*

the flexural isostatic models, they estimated that the KK basin contains nearly 109,000 km^3 of clastic sediments and suggested that these sediments have been derived through the denudation of elevated rift-flank topography of the Western Ghats.

6.3.3.3 Generalized Stratigraphy

The generalized stratigraphy (Table 6.4) of the basin (after Singh et al., 2007) and the stratigraphic column (Fig. 6.23) as depicted at Karwar-1 (Konkan offshore) and CH-1-1 (Kerala offshore) well locations (after Singh and Lal, 1993) provide valuable information with regard to the sedimentation history in the basin. According to the above data, the sediment deposition is characterized by clastic facies during the late Paleocene–early Eocene syn-rift stage, while Eocene to middle Miocene sedimentary layers consist of limestone, shale, and sandstone predominantly carbonate in the north and clastics in the south. The sediments over the Alleppey Platform consist mainly of the limestone of late Miocene age. However, the post-middle Miocene onward, the sediments are mainly clastics derived from the Western Ghats.

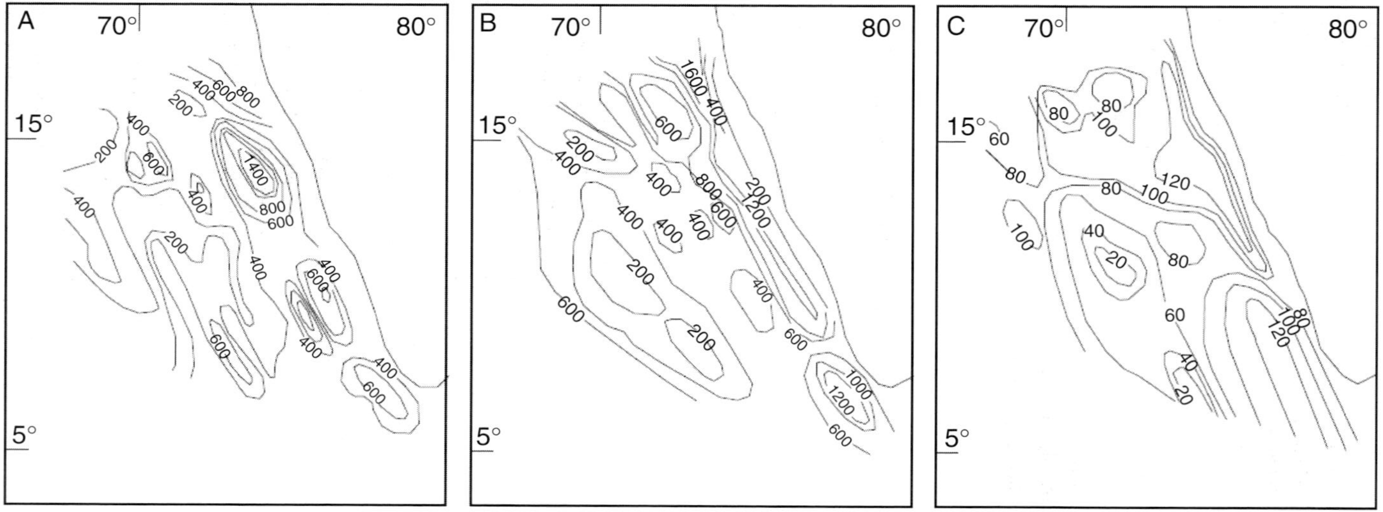

FIGURE 6.22 Sediment thickness maps (TWT) for the three sedimentary subdivisions in the Konkan–Kerala basin. (A) Paleocene-late Oligocene, (B) Miocene, and (C) post-Miocene *(Reproduced with permission from Campanile et al., 2007 © Wiley-Blackwell).*

TABLE 6.4 Lithostratigraphy of the Konkan-Kerala Offshore Basin (adapted from Singh et al., 2007)

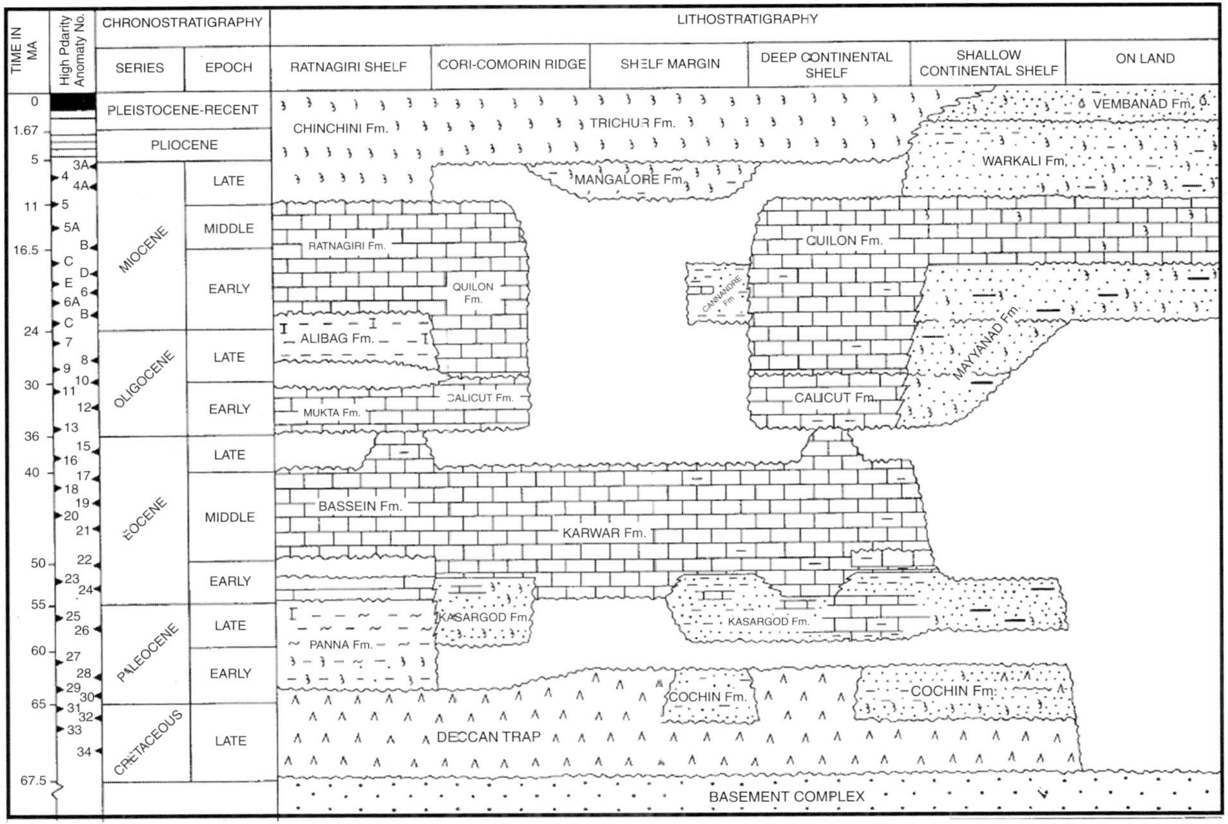

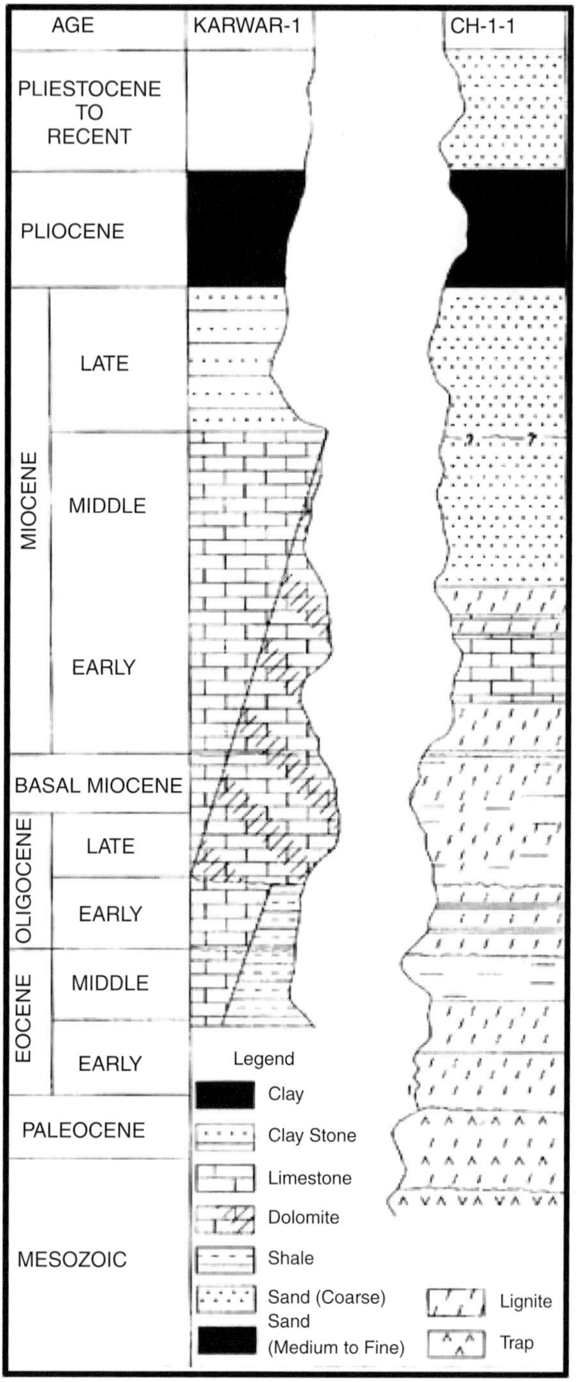

FIGURE 6.23 Generalized stratigraphic column of the Konkan–Kerala offshore basin, showing sequences and source rock character *(adapted from Singh and Lal, 1993).*

6.3.3.4 *Probable Mesozoic Prospectivity of the KK Offshore*

All earlier efforts for hydrocarbon exploration in the KK offshore basin have mainly focused on the Tertiary section. While Singh and Lal (1993) pointed out several well-defined structures and possible source areas in the Tertiary and viewed the basin as prospective for exploration, Gombos et al. (1995), on the contrary, argued the lack of adequate riverine sediment input during the Tertiary as a reason for unsuccessful exploration in this basin. According to Gombos et al. (1995), the close proximity of continental drainage divide to the west coast of India greatly reduced the drainage area available to the small river systems draining into the Arabian Sea due to which the clastic sediment influx to the western offshore basins of India particularly the KK offshore was less compared to that of the east coast India

However, in recent times, the focus had shifted toward understanding the Mesozoic prospectivity of the KK offshore basin based on the premise that significant thickness of sediments belonging to either during or prior to rifting between India and Madagascar is present below the Paleocene trap in the basin (Kenneth et al., 2008; Nathaniel et al., 2008; Radha Krishan et al., 2012). It is pertinent to recall here the inference of Gombos et al. (1995) regarding the existence of drainage pattern from east to west prior to the separation of Madagascar from India in mid-Cretaceous. This older drainage system must have given rise to the deposition of clastic sediments for a prolonged time. Such an inference gained momentum after the successful delineation of Mesozoic sediments below the basalts in well GK-28-2 in Kutch basin (Radha Krishan et al., 2012). While earlier seismic data could map only the Tertiary sediments that overlie the regionally extensive trap cover in the basin, recently acquired high-resolution seismic data with state-of-art processing techniques made the possibility of mapping the sub-basalt sedimentary layers (Kenneth et al., 2008; Fainstein et al., 2012; Radha Krishan et al., 2012).

6.3.3.4.1 Stratigraphy

The stratigraphy of the offshore western margin of India that includes the Mesozoic succession is postulated by Roberts et al. (2008) based on the pre-drift paleogeographic reconstructions evidenced by well and outcrop data from India, Madagascar, and the Seychelles. They highlighted the significance of stratigraphic information from the three offshore wells in the Seychelles, and implications for depositional environments in the adjacent offshore margin of India. The stratigraphic history summarized by Roberts et al. (2008) is briefly presented with particular reference to Indian margin here:

Triassic–Early Jurassic

Karoo-type sediments sourced from the Precambrian uplands of Madagascar, Seychelles, and India.

Middle Jurassic

Marine conditions developed between Madagascar, Seychelles, and India with deposition of shallow marine sediments in the Kutch graben.

Early Cretaceous

Rifting along the Narmada and Cambay fault trends created grabens with alluvial systems. This culminated the delta complexes and fans in the vicinity of the Surat depression.

Late Cretaceous

Restricted marine conditions prevailed in the zone between the Seychelles and India with deltas and fans providing sediments from both landmasses. Shelf carbonates and open-marine conditions developed in south of the Seychelles might characterize the deposition in the Konkan–Kerala region.

Late Cretaceous/Early Tertiary

Significant volcanic activity developed in association with rifting and ocean spreading. Deccan Trap flood basalts covered much of the western India and extended offshore to the Seychelles and beyond. The acquisition of long-offset and their processing and reprocessing of the modern seismic data had revealed significant sedimentary section preserved below the Tertiary sediments. *Reprinted with permission from Roberts et al. (2008) © EAGE Publications.*

6.3.3.4.2 Seismic Interpretation

We present here few characteristic seismic sections (Fig. 6.24 and 6.25) that have mapped the sub-basalt sequences as well as some well-defined structures. The application of state-of-the-art processing has made it possible to visualize the sub-basalt structures. Imaging has improved considerably through reprocessing and has helped in making a better interpretation of the sub-basalt features (Fainstein et al., 2012).

The opening of the Mozambique Proto-ocean initiated the creation of accommodation space between Madagascar and India (Kenneth et al., 2008). This was followed by dextral and then sinistral strike-slip movement of India with respect to Madagascar causing the inversion of existing grabens. The sinistral strike-slip movement of India along the Vishnu FZ during the breakup from Seychelles caused the tightening of the older inversion structures.

The Vishnu FZ divides the Kerala–Konkan Basin into the northern KK basin and the much older Southern Tip of India basin whose architecture and sedimentation history are related to the Mannar Basin (Nathaniel et al., 2008). The Tertiary sediments, as opposed to the Mesozoic, are thickest on the shelf and become very thin beyond the shelf edge. Integration of these ideas and interpretations made Nathaniel et al. (2011) to suggest that the Southern Tip of India and Gulf of Mannar could be a ramp margin that attracted sediments predominantly from the eastern side.

6.3.3.4.3 Hydrocarbon Potential

Traditional seismic diagnostic aids in resolving the reservoirs such as seismic attributes and velocity are incapacitated for this Mesozoic stratigraphy due to

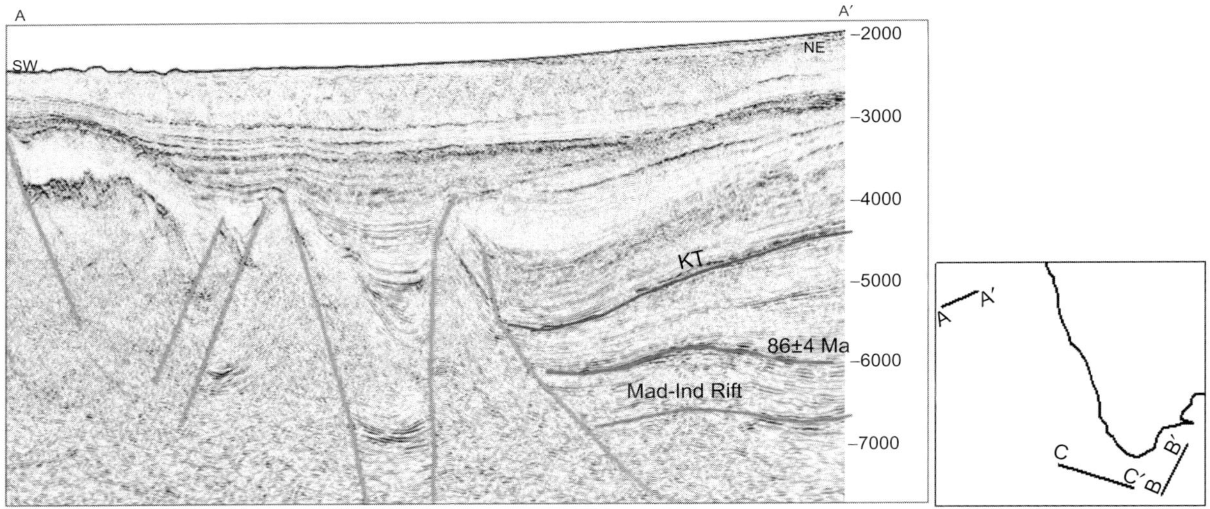

FIGURE 6.24 Seismic section in the Konkan offshore depicting the sedimentary sequences below the K/T boundary (Deccan basalts). Structural inversion trends and faults can be noticed (adapted with permission from Kenneth et al., 2008 © Society of Petroleum Geophysicists, India). (For color version of this figure, the reader is referred to the online version of this chapter.)

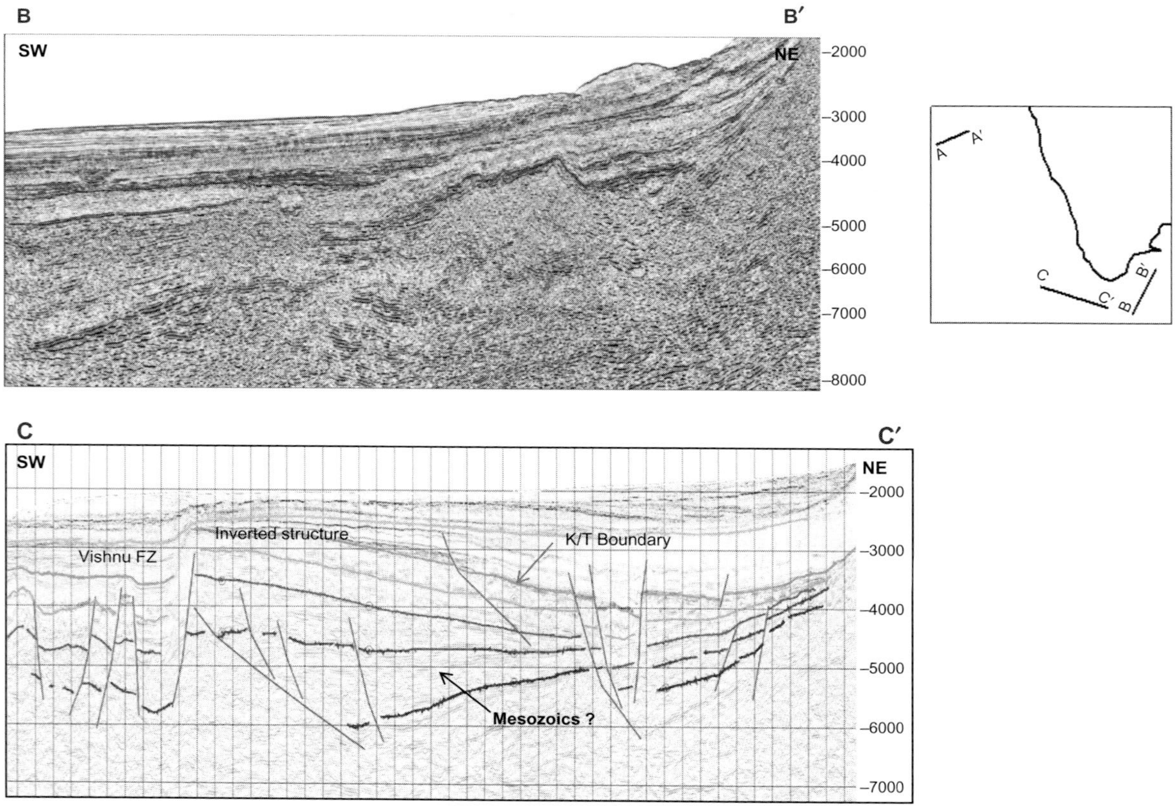

FIGURE 6.25 Seismic section in the Mannar basin (BB′) and interpreted section in the southern part of Kerala basin (CC′) showing deeper sedimentary strata below the basaltic top (adapted with permission from Kenneth et al., 2008 © Society of Petroleum Geophysicists, India). (For color version of this figure, the

overlying volcanic extrusives, which distorted the seismic imagery below the basalt. Therefore, structural play attains immediate exploration attention to avoid risks and uncertainties.

- **Source rocks**
  Mesozoic source rock potential as derived from the Seychelles (Roberts et al., 2008) and the recent hydrocarbon discovery from the Mesozoics in the Gulf of Mannar propose the consequential potential of the Mesozoics in the KK offshore of India (Kenneth et al., 2008; Nathaniel et al., 2008; Radha Krishan et al., 2012).

With the exception of Cochin-1-1 and K-1-1 wherein thick Cretaceous sequences have been drilled in KK basin, the knowledge on the Mesozoic petroleum systems is necessarily conjectural. Based on regional plate tectonic geological considerations, Braillie et al. (2004) suggested several source rock sequences that may exist in the basin, related to the successive rifting and breakup of Eastern Gondwanaland.

A synthesis of the analysis made by Braillie et al. (2004) along with the deeper subsurface imaging data presented here (Fig. 6.24 and Fig. 6.25) can be used to model and infer the potential of hydrocarbons in the KK offshore. New seismic data in the KK offshore indicates the presence of thick sedimentary rocks below the volcanic top. Accordingly, potential source rocks could be from the lower Cretaceous and probably the underlying syn-rift sections (Braillie et al., 2004). Adequately mature source rocks with high hydrocarbon generating potential could be expected. In addition, other source intervals could be within the open marine late Cretaceous sediments (Braillie et al., 2004; Nathaniel et al., 2008).

- **Reservoir rocks**
  The earlier drilled well CH-1-1 in KK offshore had encountered the Cretaceous sandstone of high reservoir quality and is similar to the Pearl-1 drilled well in the Gulf of Mannar. Potential clastic reservoirs could also be present towards the base that are immediately overlying the breakup unconformity. Similar to the Gulf of Mannar (Braillie et al., 2004), the new seismic data in the KK offshore also reveal major fairway of stacked turbidites.

- **Traps**
  Seismic data from the deeper layers suggests that the structural features in the basin could have formed due to differential movements along various transform faults now extinct and identified as FZ's. A number of horsts and tilted fault blocks, compactional drape and the compressionally induced traps may possibly be the potential trapping mechanisms in the basin.

- **Maturation**
  Synthetic burial history modeling (Fig. 6.26) carried out by Somen et al. (2011) for understanding the sub-basalt prospectivity in the basin suggests

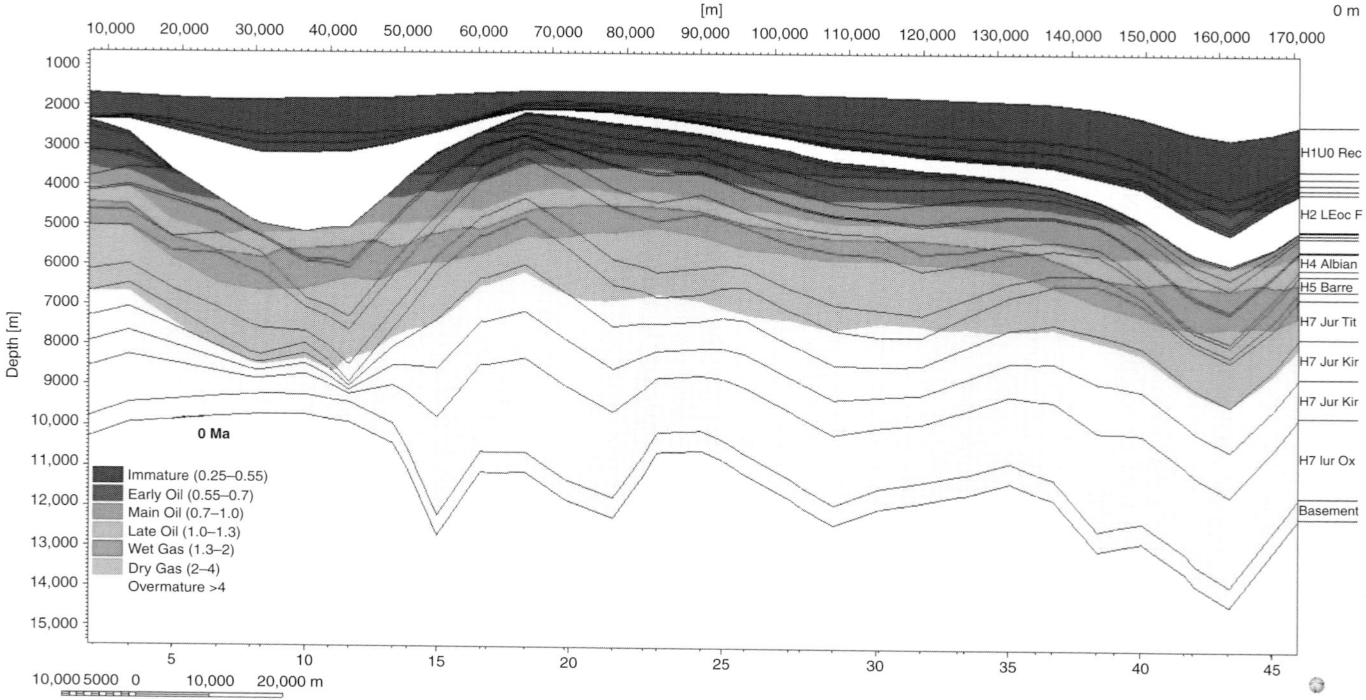

FIGURE 6.26 Thermal maturity results of the 2D model showing oil and gas windows *(adapted with permission from Somen et al., 2011)*. (For color version of this figure, the reader is referred to the online version of this chapter.)

that the deeper sedimentary successions are within the oil and gas maturity windows showing peak generation and expulsion in the late Cretaceous, and minor generation and expulsion in the late Tertiary. Such multiple peaks of generation and expulsion also noticed in the Gulf of Mannar basin by Braillie et al. (2004) has been ascribed to multiple high-temperature events and the deposition of thick post breakup sediments. The results of different simulation types confirm the possibility of gas in the sub-basalt sediments, (Radha Krishan et al., 2012) however, the depth window may vary between different depocenters. Therefore, the sub-basalt Mesozoic sediments in the KK offshore basin may in all possibility hold good hydrocarbon potential, however, given the geological uncertainties and assumptions, such a proposition needs further detailed investigation. Further, it is relevant to state that the presence of a currently active petroleum system was confirmed by satellite seep analysis (Nathaniel et al., 2008; Roberts et al., 2008), which indicates the presence of several seeps over the main syn-rift section in the Konkan and Mannar basin depocenter.

However, deep drilling data is essential in this basin in order to confirm the seismic interpretation of Mesozoic sediments. In case the above inferences are proven true, it will certainly add a new dimension to the hydrocarbon potential of the deepwater realm of the Indian continental margin.

Chapter 7

Exploration in the Indian Offshore Basins—Some Challenging Issues Related to Imaging and Drilling

7.1. SUMMARY OF STRUCTURAL AND PETROLEUM SETUP OF INDIAN OFFSHORE BASINS

The varied tectonic history and the manner in which the development of major petroliferous basins took place in the onshore and offshore realms have given rise to diversity in basins along the East and West Coasts of India. It is clear that both the Kerguelen and the Reunion mantle plumes had variable impact with regard to the volcanism along the eastern and western margins. Similarly, the India–Asia collision history and subsequent development of Fan systems, the drainage pattern of major east flowing river system, resulted in large variations in the sedimentation history and depositional style along both these margins. These variations significantly impacted the development of offshore reservoirs at the margins. While the reservoirs in the western offshore are generally carbonates, in the eastern offshore, these are mainly sands as seen in the comparative stratigraphic summary (Fainstein et al., 2009) of these two areas (Fig. 7.1).

As discussed in Chapter 1, the sedimentary basins along the Indian offshore region can be divided into seven geologically defined segments such as the Kutch–Saurashtra, Cambay/Mumbai Offshore, Konkan-Kerala, Cauvery-Palar, Krishna–Godavari (KG), NEC-Mahanadi, and the Andaman–Nicobar Islands. Such a division will help in planning the exploration programs because of their distinction in terms of the geological history. In order to understand the imaging challenges and drilling hazards in the Indian offshore basins, a summary of the structural and petroleum setup of these basins (derived from Chapters 5 and 6) is very useful. A comparative analysis of the basins will help in the application of modern technologies to the effort of uncovering the exploratory potential of offshore basins in India.

Developments in Petroleum Science, Vol. 59. http://dx.doi.org/10.1016/B978-0-444-53604-4.00007-7

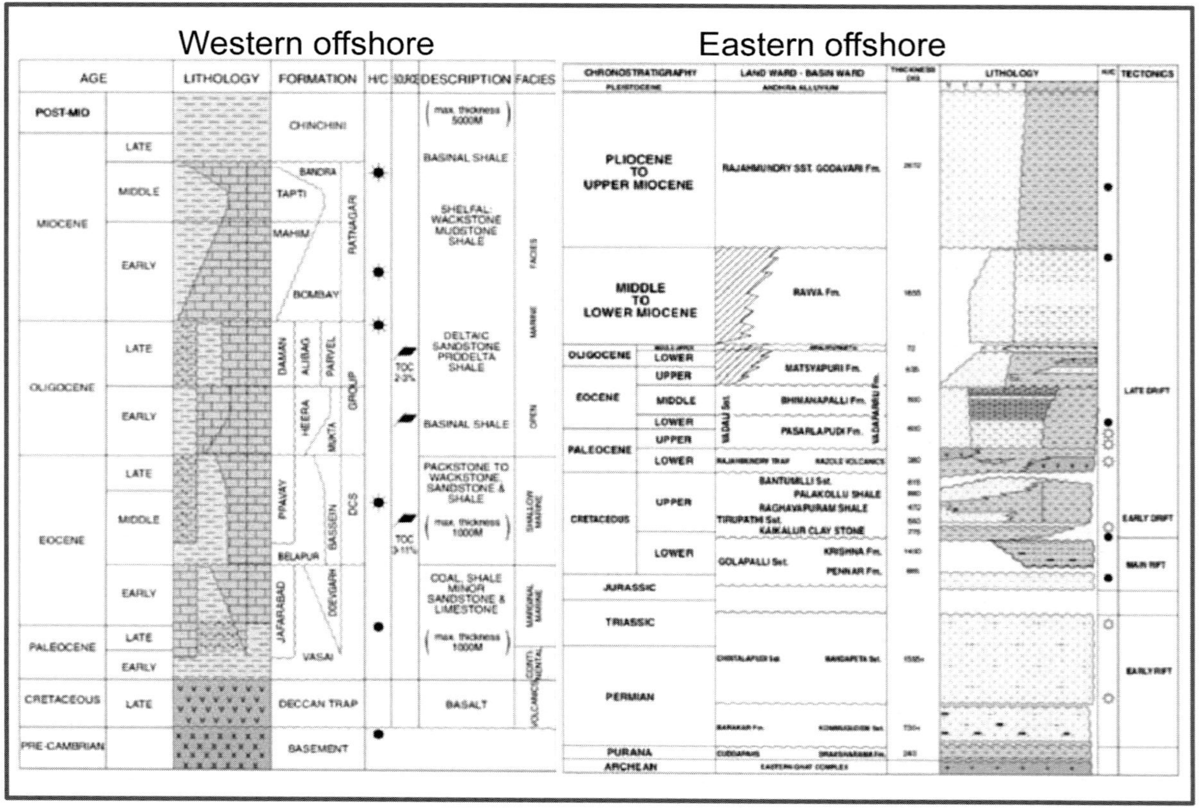

FIGURE 7.1 Comparative stratigraphy of western and eastern offshore of India as revealed from the Mumbai High region in west and the K-G offshore in the east *(adapted with permission from Fainstein et al., 2009 © AAPG conference)* (modified after Jenkins, 1992). (For color version of this figure, the reader is referred to the online version of this chapter.)

The Kutch–Saurashtra basin lying along the west coast of Gujarat adjacent to Pakistan border defines the eastern margin of the Indus Fan. The basin comprises a total sedimentary section of ~7000 m ranging from Mesozoic to Recent with main plays defined by Carbonate Buildups/Platforms, Stratigraphic Pinchouts/Wedgeouts, and Structural Transpressional rollovers (Biswas, 1982; Biswas and Deshpande, 1983; Satpal et al., 2006). The Mumbai offshore basin runs from the area south of Saurashtra and Gulf of Cambay to Goa along the west coast of India and is famously known for the giant Mumbai High field located on a regional high ~200 km from the coast off Mumbai. The Tertiary sedimentary section is ~7000 m and ranges from Paleocene to Recent with play types such as Platform and fractured carbonates, Reefal buildups, Anticlinal/fault-related closures, Wedgeouts against basement highs, and Tilted fault blocks (Biswas, 1982). The Konkan-Kerala basin lying between Goa and south of the southern tip of India is a major frontier deepwater basin. The sedimentary section is estimated to be maximum ~5500 m ranging from Cretaceous to Recent with basalt cover all over and is expected to be underlain by thick Mesozoic prospective section. The major speculative play types are Structural Highs, Carbonate Buildups, Turbidites, and Distal fans (Fainstein and Kalra, 2010). The Cauvery basin is situated along the south east coast of Peninsular India off the coast of Tamil Nadu. The sedimentary section is maximum ~8000 m thick ranging from the Jurassic to Recent with play types having a variety of Stratigraphic traps (pinchouts, Wedgeouts, lenticular sands) and several Structural Type Traps (Rangaraju et al., 1993). The KG basin lying along the coast of Andhra Pradesh is another significant offshore oil-producing basin of India, apart from the Mumbai High, and is the site for several new oil and gas discoveries particularly in the deepwater areas. The estimated sedimentary section is ~7000 m ranging from Proterozoic to Recent with play types such as Structural Highs/Fault Closures, Unconformity-related traps, Wedgeouts, and Channel-Levee systems (Gupta, 2006; Venkatarengan and Ray, 1993). The NEC-Mahanadi basin lies along the east coast north of Visakhapatnam off the Orissa and West Bengal coast and covers the deepwater prospective areas of 85° E Ridge and the Bengal Fan. The sedimentary section in the Mahanadi offshore is of the order of 5000 m ranging from Early Cretaceous to recent with major play types such as Anticlinal/Fault-related, Wedgeouts, Carbonate buildup, and Turbidite submarine fans. In the NEC region, the sedimentary section is estimated to be more than 5000 m with play types as Canyon Cut/Fill, Channel-Levee Complexes, and Turbidite submarine fans similar to the Mahanadi offshore (Dehadrai, 1970; Maksoud, 2007). Finally, the Andaman and Nicobar Islands, lying along an Island Arc system, encompasses an outer-arc accretionary wedge as well as the fore-arc, volcanic-arc, and back-arc basins. The estimated sedimentary section in both fore-arc and back-arc areas exceeds 8000 m. In the fore-arc, the main speculative play types consist of Stratigraphic

Pinchout, Turbidite reservoirs (Bastia, 2006). The deeper lithology play types consists of shallow water-rimmed platform carbonates. In the back-arc basin, there are major inverted anticline structures forming large four-way dip closures and three-way fault closures (source: DGH).

7.2. OFFSHORE EXPLORATION IN INDIA—CHALLENGES AND TECHNOLOGICAL ADVANCEMENTS

Historically, the offshore geophysical data in India is comprised of regional gravity, magnetic, detailed 2D seismic reflection, and refraction data as well as few 3D seismic blocks. As highlighted in Chapter 1, subsequent to the initiation of NELP, several advancements took place practically in all segments of exploration activity and made significant impact with regard to offshore petroleum exploration in India. The role of modern technology is quite impressive in India as the exploration and production (E&P) industry has positively responded to the changed environment after NELP. However, due to large variations in the structural setup and depositional history related to petroleum systems of the individual offshore basins, the E&P industry quite often faces several challenges related to imaging, more challenging in the case of exploration in deepwater areas. For example, the advent of advanced seismic acquisition methods and the consistent utilization of conventional 3D seismic data in the Indian offshore have improved the seismic resolution of thin beds yielding for precise facies evaluation of amalgamated channel reservoirs in the deepwater slope realm (Bastia, 2007). In the areas of complex structures and rapid vertical as well as lateral changes in the velocity field, the Prestack Depth Migration (PSDM) is a very important image enhancement tool in the offshore and future exploration in India shall be highly dependent on such advanced techniques (Fainstein et al., 2009).

7.2.1 Imaging Issues in the Indian Offshore

In the Mumbai offshore basin, the Mumbai High and adjoining shallow water regions present a wide range of geologic and geophysical challenges. The presence of shallow-water multiples, fractured carbonates, shallow refractions, high absorption, and the complex basement structure make the seismic imaging difficult. Multiples represent possibly the largest challenge in this basin with velocity modeling being quite complex. The Konkan-Kerala basin has seen several major attempts in data acquisition to understand the basin through an overall improved resolution, larger bandwidth, and higher S/N ratio; however, the sub-basalt events still show variability in continuity, possibly due to potential depth imaging and illumination issues. Possible solutions to look into this problem could be 3D Wide Azimuth acquisition with long offsets in order to preserve low frequencies for deeper penetration, PSDM using MMT to constrain basalt thickness, and separate up and down going waves to attenuate free surface multiples (e.g., Fainstein et al., 2012).

The perturbations in data quality associated with a nonflat sea surface could possibly be tackled through rough sea deghosting. In the Cauvery basin, major issues seem to be the basement imaging, multiple attenuation, and anisotropic environment. In the KG basin, targets tend to be thin sands leading to generally higher frequency acquisition to ensure vertical resolution. The KG basin exhibits extreme structural complexity and a huge array of fault types making the velocity modeling difficult. Mapping of Tertiary and Upper Cretaceous structures and understanding of channel-sands connectivity through seismic attributes are other major challenges in this basin. The complex geologic structures and depositional regimes in the NEC-Mahanadi basin pose complexities in velocity modeling. Accurate knowledge of pore pressure is important both in exploration as well as in drilling. During the exploration stage, pore pressure prediction can be used to develop fluid migration models so as to rank the prospects (Sayers et al., 2002).

7.2.1.1 Data Processing/Prestack Time Migration

In the KG and Cauvery offshore basins, processing of high-resolution 3D seismic data with Prestack Time Migration (PSTM) has remarkably improved the resolution of deepwater thin sands. Application of multiple attenuation techniques has benefited a lot in many areas. For example, the use of Surface-Related Multiple Attenuation technique (SRME) is demonstrated (after Mishra et al., 2008) for one of the 2D seismic line in Konkan-Kerala offshore (Fig. 7.2). Specialized processing techniques are generally applied to attenuate the water bottom energy and reduce noise after migration. In one of the sub-basalt prospect areas of the Konkan-Kerala basin, Fainstein et al. (2012) recently demonstrated that advanced 3D seismic data acquisition techniques along with state-of-art processing work flow methodology will bring out improved resolution of sub-basalt structural features (Fig. 7.3). Similarly, better processing strategies would result in improved imaging of the subsurface and enhance the signal-to-noise ratio (Fig. 7.4).

7.2.1.2 Depth Imaging

As the standard PSTM would not be able to handle significant lateral change in velocities, the depth migration is performed. Depth migrated images can significantly change our understanding of the deposition and structure which used to be solely based on the time processing results. In the depth migrated imaging, the positioning of the flanks may change or apex of the structure could move laterally and hence is critical in the making or breaking of a potential prospect discovery (Fainstein et al., 2009). Thus, significant investment could be saved due to an improved positioning of the exploratory well. For example, the results from a depth migrated section show marked improvement over a prestack time migrated section (Fig. 7.5) for imaging the sub-basalt structures (after Fainstein et al., 2009).

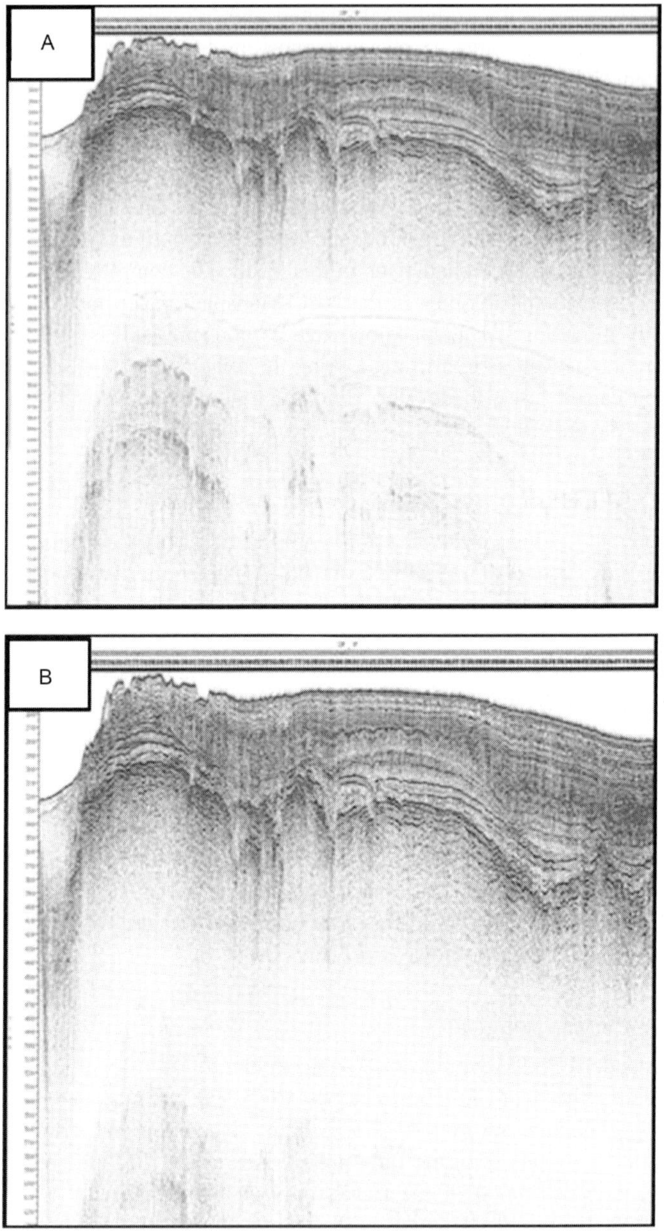

FIGURE 7.2 Stacked sections from KK offshore basin showing marked improvement in resolution: (A) before the application of SRME technique and (B) after SRME application *(adapted with permission from Mishra et al., 2008 © Society of Petroleum Geophysicists, India)*. (For color version of this figure, the reader is referred to the online version of this chapter.)

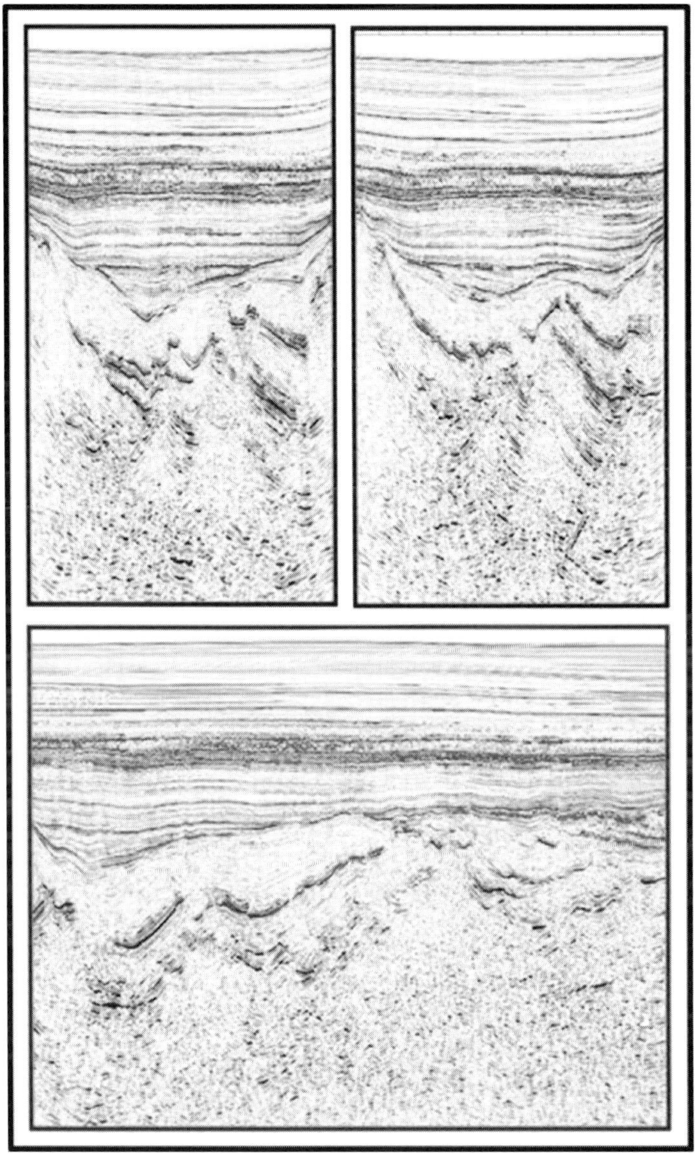

FIGURE 7.3 Results of 3D over–under seismic survey and special processing workflow for imaging the sub basalt structures in the KK offshore basin *(adapted with permission from Fainstein et al., 2012 © Society of Petroleum Geophysicists, India)*. Two successive cross-line sections (top panel) and one in-line section (bottom panel) show several features such as sag basin, faults, anticline structures, etc. below the basaltic top. (For color version of this figure, the reader is referred to the online version of this chapter.)

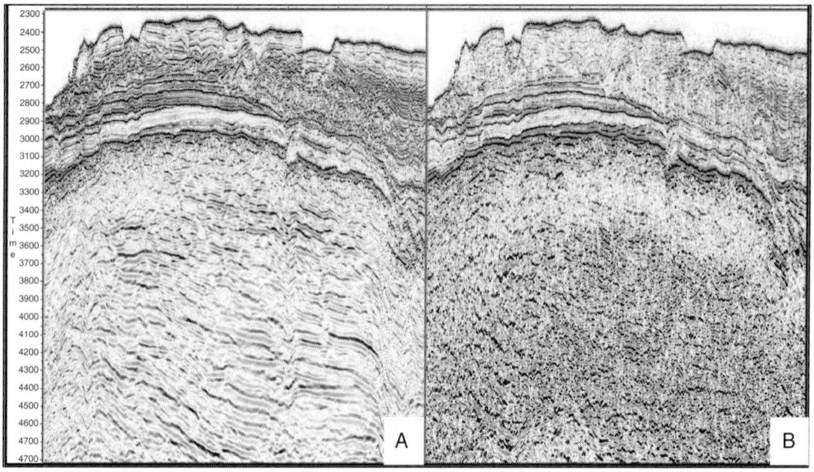

FIGURE 7.4 A representative seismic section from Konkan–Kerala basin showing improvement in resolution: (A) advanced processing and (B) conventional processing *(adapted with permission from Bastia, 2010 © Society of Petroleum Geophysicists, India)*. (For color version of this figure, the reader is referred to the online version of this chapter.)

7.2.1.3 Seismic Interpretation

The improved seismic resolution through application of standard PSTM and PSDM algorithms yields a regional view of maps and prospective oil structures which greatly helps the modern seismic interpretation (Roberts et al., 2008). The regional mapping of source patterns and migration pathways from source rocks to reservoirs has been significantly improved. The offshore datasets are interpreted for key seismic horizons and the structural maps are produced as well as prospects outlined. Attribute analysis of the amplitude, frequency, and phase is carried out along selected 2D lines or time-slices within volume for the purpose of stratigraphic lateral facies distribution and reservoir facies characterization (Fig. 7.6). Further, 3D seismic data provide distribution of faults within the volume and with the help of isochron and isopach maps, the basin reconstruction is subsequently attempted. Real-time visualization of mapping and seismic attributes are now a days integrated with modern deepwater drilling, completion, and production technology. Petrophysics and Formation Micro Imager (FMI) analyses became extremely useful for understanding the matrix porosity and permeability of carbonate reservoirs in the Mumbai High region of western offshore (Rai et al., 2004) and for the facies analysis and distribution of amalgamated deepwater turbidite sand reservoirs in the eastern offshore (Thakur et al., 2010).

7.2.1.4 New Technique Synergies—Electromagnetic, Magnetotelluric, and Ocean Bottom Cable 4C Methods

Conventional seismic and seismic reprocessing of vintage data has shown some limitations in illuminating what lies beneath the basalts especially in the areas

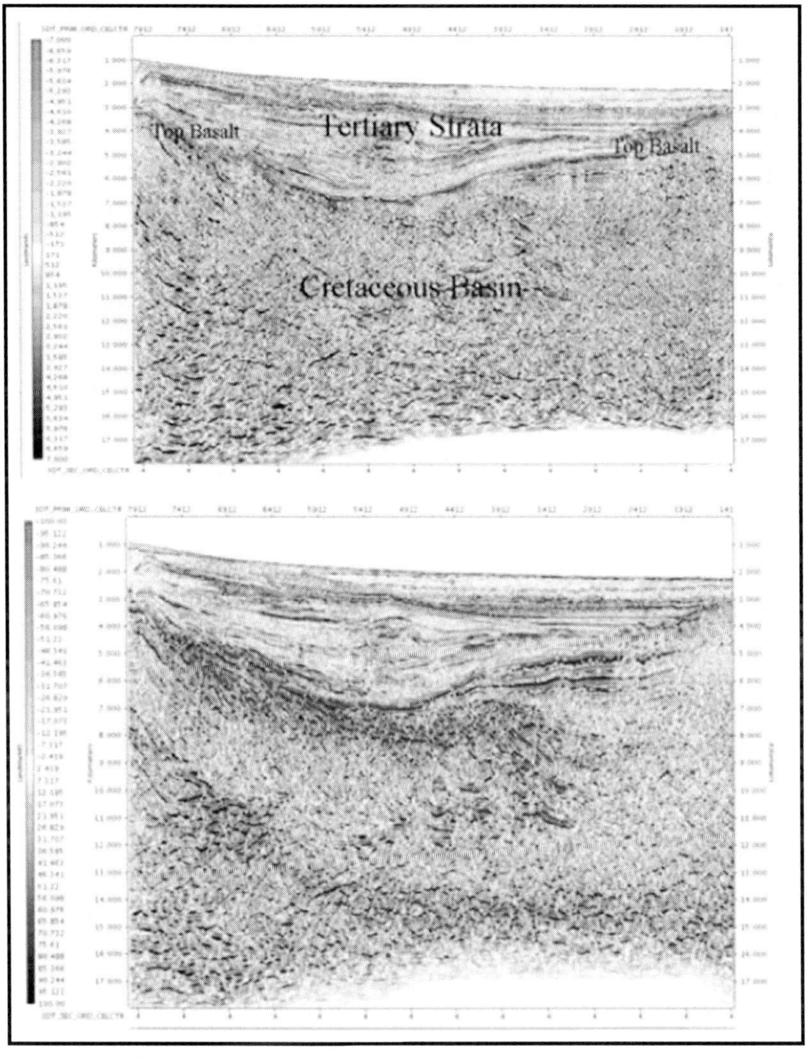

FIGURE 7.5 Comparison of (A) PSTM time converted to depth section and (B) PSDM section *(adapted with permission from Fainstein et al., 2009 © AAPG conference)*. Note the marked improved in the resolution of both Tertiary as well as Mesozoic strata below the basalts. (For color version of this figure, the reader is referred to the online version of this chapter.)

situated on volcanic passive margins and encounter the complex geological situations argued for and against petroleum generation, migration, and entrapment (Mishra et al., 2008). However, MMT in conjunction with high-resolution seismic data may determine the basalt thickness and the sub-basalt Mesozoic sediments below them. Advanced seismic data acquisition techniques, and Controlled Source Electromagnetic (CSEM) experiments together may solve such issues as known from the industry experience in the Konkan-Kerala basin.

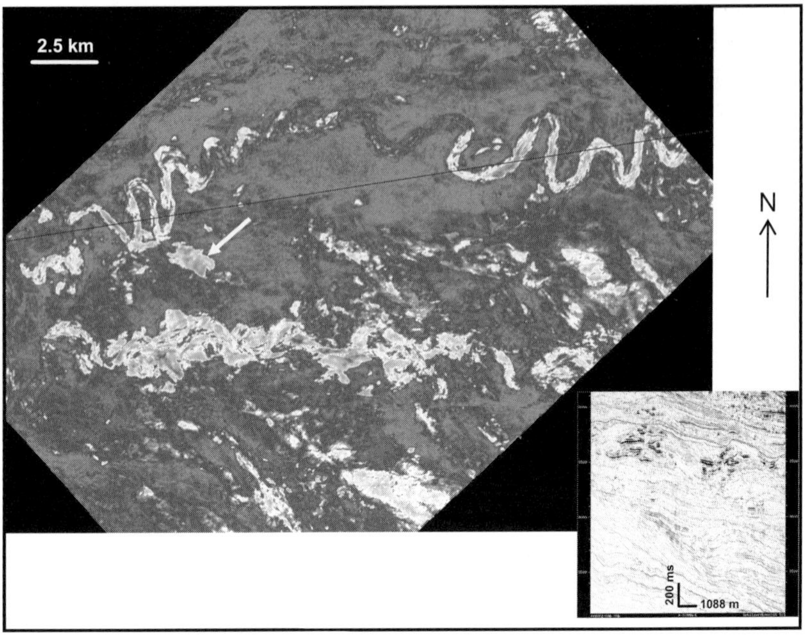

FIGURE 7.6 Results of seismic data analysis in the KG offshore showing sinuous channels and splays depicted through stratal amplitude slice of the horizon marked in blue in the seismic section (shown in the inset) *(adapted with permission from Bastia, 2007 © Technology Publications, Dehradun)*. (For interpretation of the references to color in this figure legend, the reader is referred to the online version of this chapter.)

CSEM is a tool that may extract resistivity which relates to porosity, permeability, cementation, clay content, fractures, and fluid content, all of which are parameters required for modeling and seismic inversion. The complexity of several areas poses a constant challenge that is tackled by combining diverse geophysical synergies. Marine EM and Magnetotelluric (MMT) methods have great potential for the understanding of the West Coast of India affected by the Deccan basalts (Gorain, 2012). Accurate velocity models from the reflection tomographic techniques (Chatterjee et al., 2012) and the 4-C seismic data are useful in pore pressure prediction studies.

7.2.2 Issues Related to Logging and Petrophysics

Once the initial identification of prospective area is established from detailed seismic data, the logging and petrophysical data will be acquired through exploratory drilling which play a significant role in the hydrocarbon exploration strategy. The method has evolved from simple wireline logging to LWD (logging while drilling), and in this regard, the contributions of Schlumberger in terms of the development of modern technology as well as its application for solving complex geological problems are noteworthy. In earlier times, the well data as well as core data used to have applications in facies analysis and for the estimation of paleocurrent direction (Hardas, 1993); however, with the fast growing technology, log analysis acquired significance practically in all aspects of reservoir characterization. While the lithostratigraphy and facies evaluation help to a great extent in refining the existing geological models, the petrophysical interpretation further reduces the uncertainties of models. Further, the borehole imaging is useful for fracture characterization. Petrophysical and formation fluid data are also important in rock physics analysis.

As discussed in previous chapters, most of petroleum production in India originates from the Tertiary formations. Apart from few onshore areas, the future focus of hydrocarbon exploration in the country essentially will depend on the deepwater plays as well as the Mesozoic sediments underlying the Cretaceous basaltic rocks. Most of the deepwater prospects are combination traps of structure/stratigraphy and their total volumes should be estimated by drilling. The most important deepwater reservoir discovered in the KG Basin, the Dhirubhai gas field discovery in 2002 is seen structurally through mapping and stratigraphically through attribute analyses.

The conventional logging became an important tool in the exploration programs of almost all major petroliferous basins in India. The importance of logging data for accurate knowledge of the subsurface can be realized from the large number of wells drilled in two of the most important producing basins of India such as the Cambay basin and the Mumbai High region. For example, the high-resolution borehole imaging of drilled wells in the Mumbai High region has helped to characterize the fracture system in the limestone reservoir. We present here an interesting case study, wherein the high-resolution logs integrated with petrophysical and seismic data helped to resolve some major challenging issues with regard to the identification and characterization of thin beds in the deepwater eastern offshore of India. In view of many oil/gas discoveries along the eastern offshore and the potential of thin beds as a major prospect zones, some background information related to them and various methods of their evaluation are important. A detailed description of the theory as well as workflow methodology followed for the thin-bed studies in the eastern offshore is available in various publications (Bastia, 2007; Bastia et al., 2007; Saxena et al., 2006; Tyagi et al., 2008a,b, 2009a–c; Deepak et al., 2012). Some excerpts from these publications are briefly presented here.

7.2.2.1 Thin Beds as a Challenge

From the petrophysical point of view, thin beds can be defined as those beds that are thinner than the vertical resolution of logging devices (Saxena et al., 2006). Since each logging tool has its own definition of thin beds starting from 2 to 4 in. sonic, deep resistivity to 0.5 in. in borehole resistivity imager and the maximum vertical resolution of conventional logging tool is ~10 in., the bed having thickness less than 10 in. can be designated as thin beds for conventional logs. In thin-bedded reservoir, the conventional logs average out the petrophysical property of sand and shale layer due to lack of vertical resolution. So, the conventional log analysis in such cases leads to underestimation of hydrocarbon saturation and hydrocarbon pore thickness. Therefore, in the case of deepwater channel over bank system, proper characterization of thin beds is always a very challenging task (Tyagi et al., 2008b). As the thin beds contain nearly 30–40% of the in-place resources worldwide (Tyagi et al., 2008a), it is extremely important to find out their true potential with minimum uncertainty. Further, in thin sand–shale sequences, the hydrocarbon potential depends on their thickness, mineralogy, sediment compaction, and resistivity of pays. Due to the presence of conductive clays in between the resistive hydrocarbon sands, electric current makes short circuiting through shales (Worthington, 1997; Saxena et al., 2006) leading to low resistivity (Fig 7.7).

Figure 7.7 further shows that the conventional logs average the petrophysical property of sand and shale layers against the highlighted laminar interval, as a result, the identification of thin beds using conventional logs is very difficult. However, the resistivity image and resistivity anisotropy logs can easily identify the laminated reservoirs.

7.2.2.1.1 Identification of Thin Beds

Though the best way to identify the presence of thin beds is directly from the conventional core data, as it is not cost effective, we need to evolve strategies to find indirect evidences to confirm their presence. Some of the indirect methods such as borehole image, NMR, dipmeter, multicomponent resistivity logs, etc. would be helpful to identify and characterize them. Of these, the bore hole image and multi component resistivity logs are the best to identify and characterize the laminated reservoir. This can be realized in Fig. 7.8,

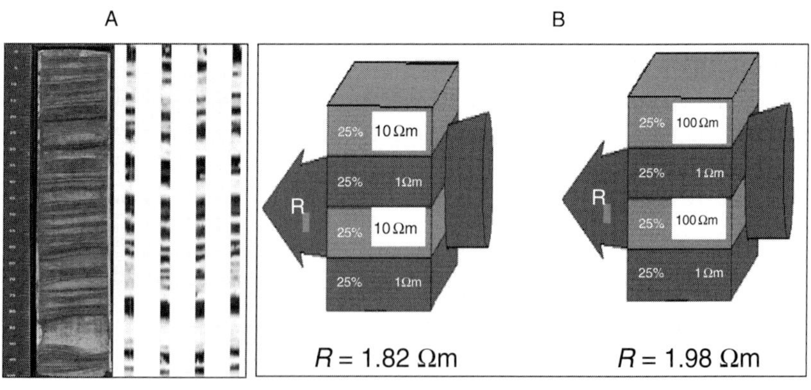

FIGURE 7.7 (A) Core and Image Log over a thin-bedded interval in a well, East Coast of India (B) demonstrating the lack of sensitivity to the characteristics of resistive reservoir layers in the case of thin-bed sequences *(adapted with permission from Saxena et al., 2006 © AP Energy Business Publications PTE Ltd)*. (For color version of this figure, the reader is referred to the online version of this chapter.)

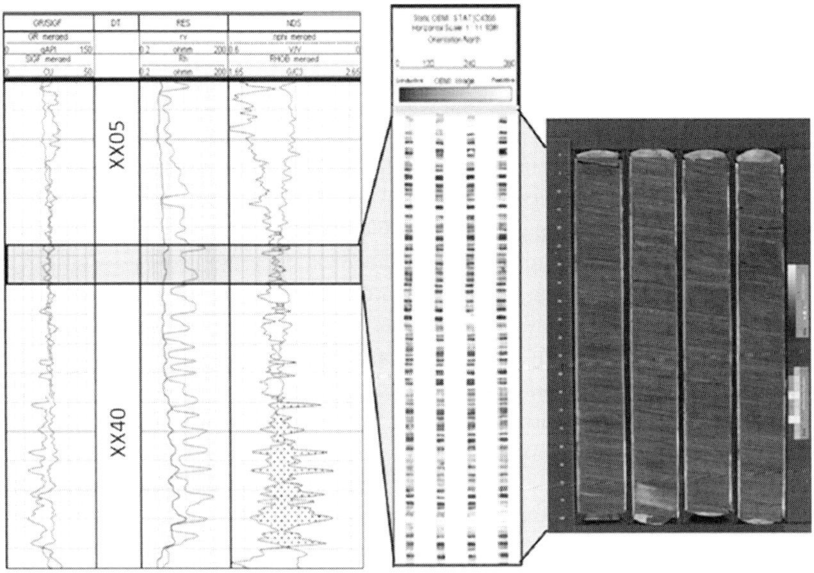

FIGURE 7.8 Signature of conventional logs, vertical resistivity, and image logs against thin beds *(adapted with permission from Deepak et al., 2012 © Society of Petroleum Engineers).* (For color version of this figure, the reader is referred to the online version of this chapter.)

which shows fine sand and shale lamination in the conventional core (after Deepak et al., 2012). While the image log shows the same and the vertical resistivity (red color curve in track 3) also reads high compared to conventional resistivity in this zone, it is not detected in the gamma ray and sigma log (in track 1), conventional resistivity (blue color curve in track 3), and neutron density (in track 4) logs.

7.2.2.1.2 Petrophysical Analysis

As mentioned earlier, when thin laminae of sand and shale are intersected by a wellbore, the high conductivity of the shale leads to electrical current shortage, and as a result, the hydrocarbon-bearing sand layers being more resistive may not be detected. Two different approaches are adopted by all major petroleum companies to resolve this issue as well as to verify the effect of the thin beds for reservoir characterization (e.g., Tyagi et al., 2009). These are (i) thin-bed analysis using the Image tools and (ii) resistivity anisotropy approach using 3D resistivity tools. The first method assumes each single sand lamination as a separate reservoir unit and uses the resistivity borehole image log to improve the vertical resolution of conventional logs in order to obtain the true petrophysical property of sand and shale lamina. Whereas the second method adopts the volumetric approach and uses the resistivity anisotropy data by considering the bunch of sand and shale lamination as a reservoir unit. As the data processing concept and work flow is entirely different for these two methods, it is important to understand them separately in more detail. The methods are briefly presented below (adapted from Tyagi et al., 2008a, 2009a,b).

7.2.2.1.2.1 Analysis Using Resistivity Image Tools For the purpose of identifying thin beds, shallow resistivity tools with high vertical resolution are useful. Generally, such tools are used for measuring the dip, orientation of sedimentation, and in stratigraphic studies; however, they are proved to be quite effective when we integrate them with normal logs for the thin bed analysis.

This approach originally developed and implemented by Alterman and his co-workers is nothing but a process of sharpening the logs to the finest resolution with any of the log data in order to bring the true layer property. The work flow is briefly described here (Fig. 7.9). First, the earth-model parameters in both reservoir and nonreservoir beds are selected from the core data and logs along with all available data from thicker intervals. Then, using high-resolution shallow resistivity logs obtained from Formation Micro Imager, Oil Base Mud Imager, or some other microresistivity device, the Scaled Synthetic Resistivity Curve (SRES) is created which is useful for generating a lithofacies model comprising of five facies such as sand, shale, silt, wet, and tight. After the input

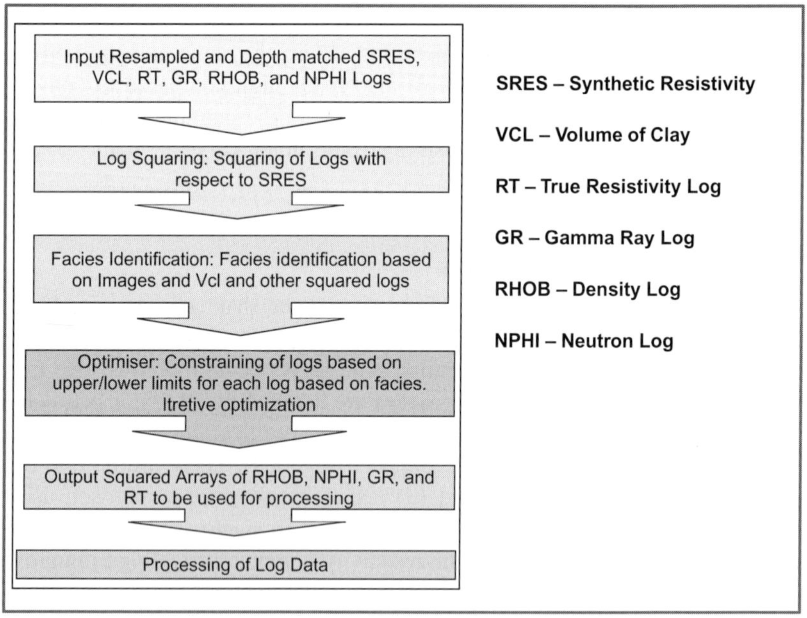

FIGURE 7.9 Flow diagram for the Processing of Logs using Image Logs *(adapted with permission from Tyagi et al., 2008a © AP Energy Business Publications PTE Ltd).* (For color version of this figure, the reader is referred to the online version of this chapter.)

logs are squared, the SRES is used to generate an initial set of modeled square logs for all the input logs. A new set of values for each curve belonging to each facies are then assigned, but the facies transitions are determined by the squared SRES curve. Then, the depth-matched model curves are processed in a constrained optimizer, which iteratively calculates the difference between the squared model convolution and the original log by adjusting the square model curve and recalculating the convolution/log difference. The lower and upper limits for each facies are used for this adjustment. The macrofacies or volume of clay (VCL) will also constrain the optimization. The finalization step takes the optimized squared logs and reoptimizes with no macrofacies constraints. One can also restrict the optimization process for a certain facies. For better matching of the synthetic/convolved logs in that facies, the optimization process can be restricted for that facies, which help clean up the model and assures that the convolved square logs match the original depth-matched logs.

Limiting the above process to five lithofacies, and squaring and convolutions limiting to deep resistivity (RT), density (RHOB), neutron (NPHI), and gamma ray (GR) logs, the processing is run using the core-derived parameters and by calibrating the model at another well for validation. The processed output using the above method in the case of deepwater KG Basin (after Tyagi et al., 2008a) is depicted in Fig. 7.10. This method works very well if the bed thickness is less than the vertical resolution of the resistivity image log (approximately 1 in.). Below this resolution, the method has limitation in that it leads to either underestimation or overestimation (Fig. 7.11) of net pay (Tyagi et al., 2009c). In addition, since the output of this method is of very high resolution, it cannot be directly used in geocellular modeling. In order to address these problems, resistivity anisotropy method was developed.

7.2.2.1.2.2 Resistivity Anisotropy Approach There could be some sand beds which are still thinner than the resolution of the image logs. The dominance of low resistivity layers, in laminated shaly sand formations, pose problems in the log interpretation in conventional resistivity measurements. As majority of the response is due to the shales, the sand layers consisting of hydrocarbon will have little effect on the overall measurement. The conventional resistivity in a laminated shaly sand formation for the sand resistivity (R_{sd}) of 20 Ωm and the shale resistivity (R_{sh}) of 1 Ωm (Fig. 7.12A) shows that as the laminar shale increases the resistivity decreases dramatically (Tyagi et al., 2009a). This leads to large uncertainties in the apparent sand resistivity even for relatively small volumes of laminar shale because the conventional resistivity measurements are made horizontally into the formation. As the current takes the path of least resistance through the more conductive layers of the formation, the greater contrast between the layers poses more problem.

This problem has been resolved after the new-generation 3D induction tools developed by Baker Hughes which measure both horizontal and vertical resistivities of the layered formation. The theoretical formulations for the

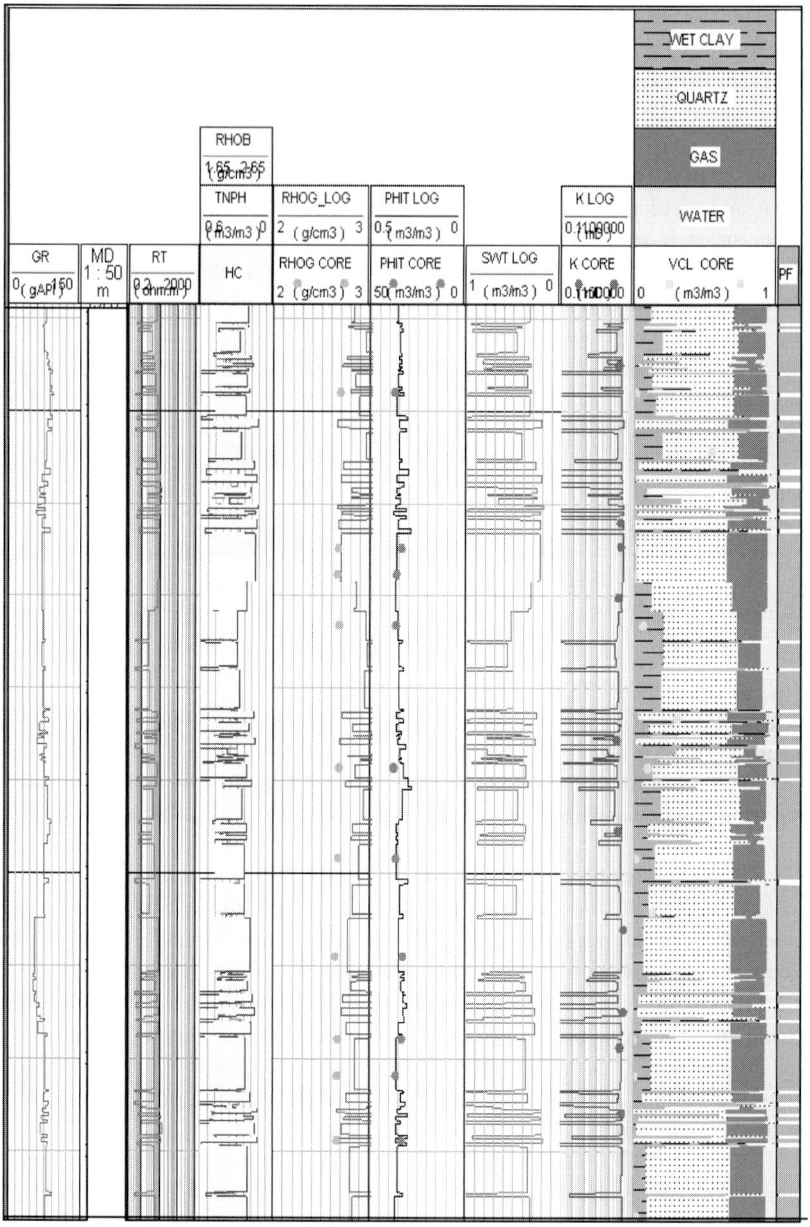

FIGURE 7.10 Sample of processed volumetric data using the Image and Normal logs after sharpening *(adapted with permission from Tyagi et al., 2008a © AP Energy Business Publications PTE Ltd)*. (For color version of this figure, the reader is referred to the online version of this chapter.)

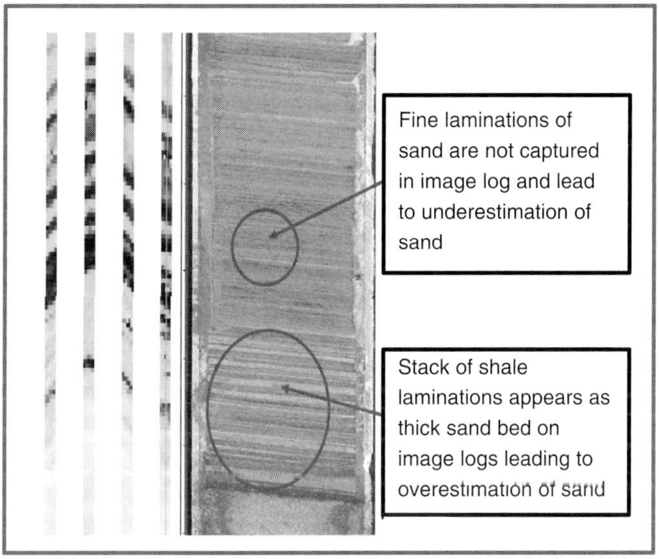

FIGURE 7.11 Issue of underestimation and overestimation of net pay by high-resolution image log *(adapted with permission from Tyagi et al., 2009c © SPWLA)*. (For color version of this figure, the reader is referred to the online version of this chapter.)

resistivity anisotropy have been derived and implemented by Gupta et al. (1998) and Mollison et al. (1999). Making vertical resistivity measurement is advantageous because the current is forced through all layers of the formation and is therefore affected by both components. Figure 7.12B shows the changes in the conventional horizontal resistivity with the vertical measurement with reference to increase in the laminar shale volume. Here, the sand layers have resistivity of 20 Ωm and the shale layers have resistivity of 1 Ωm.

The tri-axial array induction tool measures the formation resistivity both perpendicular (Rv) and parallel (Rh) to the direction of the shale–sand layers. Using Rh and Rv as inputs in a bimodal sand–shale laminar model (Thomas and Stiber, 1975) shown in Fig. 7.12C, we can obtain the two unknowns: R_{shale} and R_{sand} as given below (after Mollison et al., 2000):

- The horizontal resistivity data (Rh) measures the horizontal shale resistivity (Rh_{shale}) and the sand resistivity (R_{sand}) in parallel:

$$\left(\frac{1}{\text{Rh}}\right) = \left(\frac{V_{\text{shale}}}{\text{Rh}_{\text{shale}}}\right) + \left(\frac{V_{\text{sand}}}{R_{\text{sand}}}\right) \tag{7.1}$$

- The vertical resistivity data (Rv) measures the vertical shale resistivity (Rv_{shale}) and sand resistivity (R_{sand}) in series:

$$\text{Rv} = (V_{\text{shale}} \cdot \text{Rv}_{\text{shale}}) + (V_{\text{sand}} \cdot R_{\text{sand}}) \tag{7.2}$$

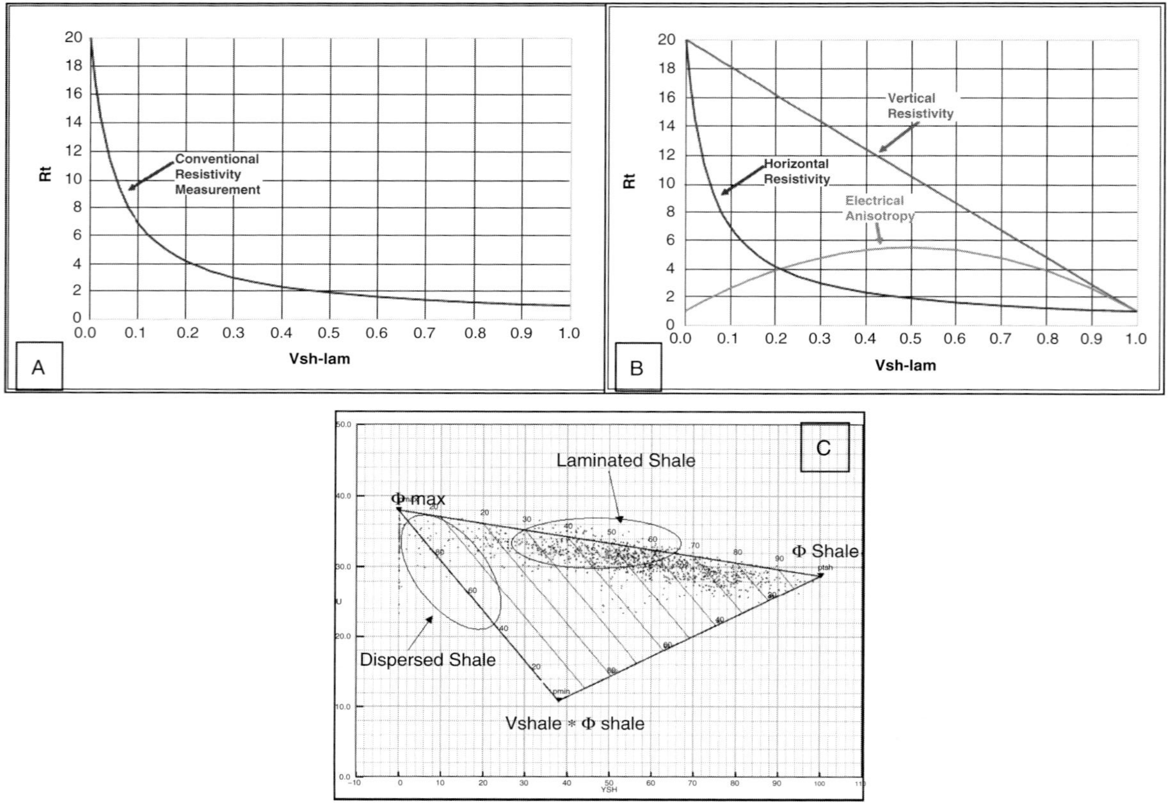

FIGURE 7.12 (A) Theoretical response of a horizontal resistivity measurement in a flat, laminated shaly sand formation, where Rsd = 20 Ωm and Rsh = 1 Ωm. (B) Theoretical responses of horizontal and vertical resistivity measurements in a flat, laminated shaly sand formation, where Rsd = 20 Ωm and Rsh = 1 Ωm. (C) Thomas–Stiber cross plot indicating that most of the shale is laminar *(adapted with permission from Tyagi et al., 2009a © Indian Association of Sedimentologists)*. (For color version of this figure, the reader is referred to the online version of this chapter.)

- The sum of the fractional volumes of the two constituents (i.e., sand and shale are the only two components of the model) is equal to one:

$$V_{sand} + V_{shale} = 1 \tag{7.3}$$

The above system of three equations contains five variables: R_{sand}, Rh_{shale}, Rv_{shale}, V_{shale}, and V_{sand}. To solve this system of equations, the lambda (Rv_{shale}/Rh_{shale}) which is shale micro-anisotropy coefficient will be assumed to be constant throughout the evaluated interval and Rh_{shale} is allowed to vary. Once this coefficient is selected, the number of parameters is reduced to four. The fractional volume of the laminated shale (fV_{shale}) can be computed from standard log interpretation. Water saturation can then be computed using a bimodal approach used for resistivity evaluation (splitting the formation into two constituents: one the sand reservoir with structural shale and the other shale/silt laminae). Given below is the general workflow (after Tyagi et al., 2009b) followed for the determination of Rv, Rh, formation dip, and azimuth through inversion:

1. Estimation of reservoir true resistivity (R_{sand}) and shale true resistivity (Rt_{shale}) from Rv and Rh.
2. Estimation of laminated/dispersed shale fraction and sand porosity estimation using the Thomas and Stiber (1975) method.
3. Calculation of water saturation (S_w) using reservoir true resistivity (R_{sand}) and sand porosity as inputs.
4. Normalization of porosity with respect to total volume.
5. Estimation of gas saturation using sand porosity and R_{sand}.

7.2.2.1.3 Example/Case Study

In order to understand the hydrocarbon potential of deepwater channel-levee complex system along the eastern offshore, a large number of wells have been drilled spanning the area between Cauvery offshore in the south to Mahanadi-NEC basin in the north. The wells have confirmed the presence of different channels in the drilled intervals and show shifting of channel axis with time. The wells have encountered thick (channel) and thin (levee) sands stacked within the hydrocarbon intervals. We present here an important case study of a deepwater well of the eastern offshore in which the resistivity anisotropy approach has been found to be very successful for the characterization of reservoirs in laminated sand shale sequences (after Tyagi et al., 2009c). The procedure followed by them along with final result is summarized below:

The well was logged using all conventional anisotropy, image and formation tester logs. Core sample is collected from the entire reservoir section. Results of the resistivity anisotropy processing are shown in Fig. 7.13. The analysis involves two reservoir formations, the upper sand1 and the lower sand2. The log data correlation study indicates that both the reservoirs are gas bearing, while sand1 is laminated, sand2 is a thick reservoir.

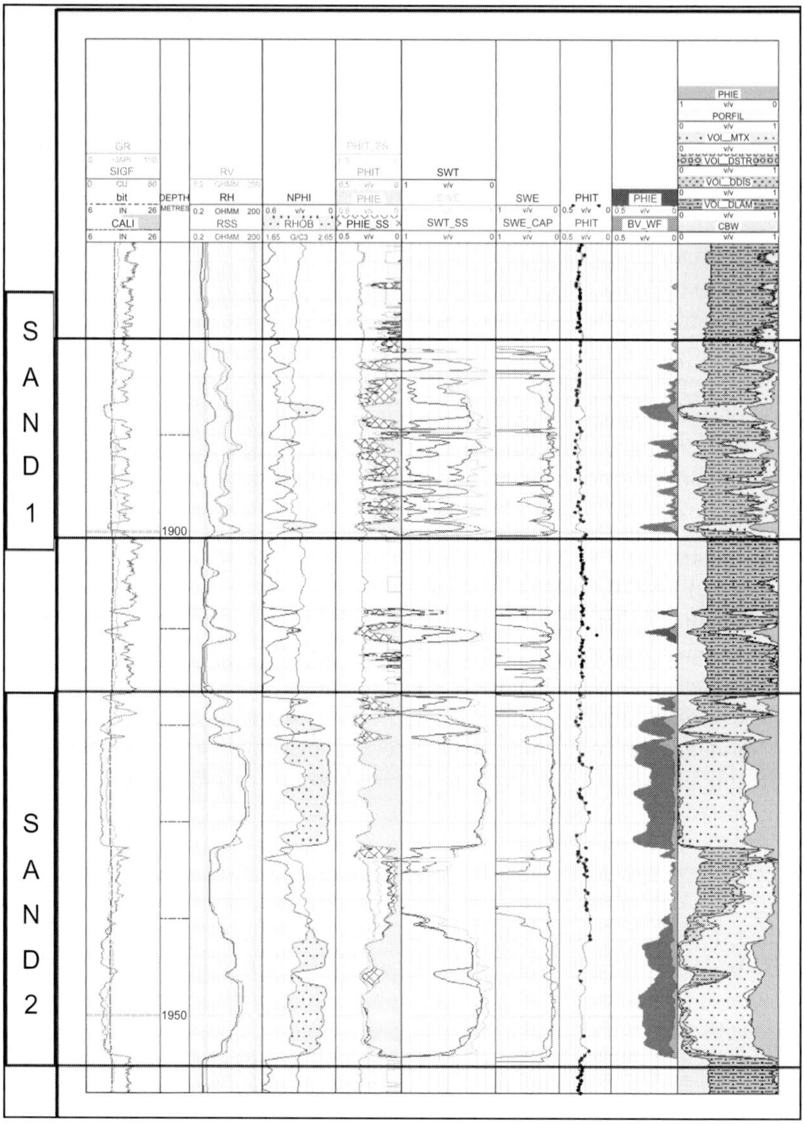

FIGURE 7.13 Results of resistivity anisotropy processing and comparison with core data *(adapted with permission from Tyagi et al., 2009c © SPWLA)* for a deepwater well in the eastern offshore. (For color version of this figure, the reader is referred to the online version of this chapter.)

The shale volume is computed from lithology log sigma. The porosity is obtained from density and neutron logs which shows good match with the core porosity. The shale distribution and laminar sand porosity have been computed using Thomas Steiber cross plot. The laminar sand resistivity is computed from horizontal and vertical resistivity (using Mollison equations). Finally, the water saturation is estimated using Archie equation by taking laminar sand porosity and laminar sand resistivity as inputs. The estimated average water saturation is around 10–20% in sand1 and around 5–10% in sand2. A comparison of the water saturation obtained from the resistivity anisotropy method and the saturation from the capillary pressure curves shows that both are in good agreement.

7.3. DRILLING HAZARDS IN THE INDIAN OFFSHORE—CHALLENGES AND MITIGATION

The tremendous pace at which the offshore exploration activity took place after the initiation of NELP in India has lead to the delineation of several new prospects as well as development of offshore oil/gas fields both in the shallow and deepwater areas. As a result, the offshore areas of India witnessed immense E&P activity. In the offshore areas, particularly in the deeper waters, drilling and other related activities are very challenging and risky, as the safety, suitability, and stability of drilling rigs and other subsea installations should be taken very seriously (Bastia et al., 2010d). This gave rise to many challenging tasks to the E&P industry in India while choosing an appropriate drilling location to ensure safety during the drilling operation. All available technologies have to be effectively utilized for successful mitigation of hazards. On the basis of various drilling challenges in the Indian offshore, and the application of latest technologies, two most important aspects that pose severe threat to the drilling activity, such as the pore pressure prediction and the shallow hazards, are dealt in this section.

7.3.1 Pore Pressure Prediction Studies

We provide here a brief insight about the methodology of pore pressure prediction and present an important case study pertaining to the KG offshore basin which is a major oil/gas-producing basin in the eastern offshore of India.

7.3.1.1 Importance of Pore Pressure Analysis

The pore pressure prediction is very important for successful completion of the well at any location and helps to a great extent in understanding the pressures in nonreservoirs as well as reservoir rocks. During the exploration phase, the pore pressure analysis is useful to study the effectiveness of seals and to rank the prospects (Sayers et al., 2002). With the tremendous pace of technological advancements, the subsurface velocity information and various geophysical attributes obtained from seismic reflection data are routinely integrated with the high-resolution petrophysical and drilling data along with the knowledge of geological processes for the purpose of carrying out accurate pore pressure prediction analysis and to generate 3D pore pressure models. Such an analysis is extremely useful and essential for appropriate selection of mud weight, optimization of casing program, and for taking precautionary measures to mitigate the drilling hazards. As pore pressure-related events have a huge economic impact, all out efforts are needed to avoid incidence while drilling through successful pore pressure estimation. It is well known that drilling rate improves as the mud weight approaches the actual pore pressure gradient, so prediction of pore prediction will not only improve the drilling rate but also lowers the overall cost of drilling. Further, presence of gas can induce abnormally high pressure at very shallow depths, and this kind of shallow gas presents significant risk while drilling. Pore pressure analysis from seismic often helps to identify such zones and reduce the drilling risks.

7.3.1.2 Background Theory

When the pressure of a fluid column at any given depth equals the hydrostatic pressure, it is considered as normal pressure. However, abnormal pressure conditions may sometimes prevail at depth if the pressure of the fluid column is either less or more than the hydrostatic pressure. Pore pressure is the pressure of the fluid in the pore space of the rock, and as mentioned above, when it exceeds the hydrostatic pressure, overpressure situation occurs. From mid-twentieth century, several investigators have been carrying out studies to understand the causative processes and mechanisms of generation of overpressure in sedimentary basins. Swarbrick et al. (2002) came up with a comprehensive solution linking the generating processes of overpressure with the stress and fluid expansion suffered by a compressible rock. Similarly, Bowers (2002) suggested few other mechanisms that account for the abnormal pressure generation in sediments, and according to him, the mechanisms resulting abnormal pressure in sediments can be categorized into two types. These are (i) compaction disequilibrium (Type I) and (ii) mechanisms other than compaction disequilibrium (Type II).

The compaction disequilibrium (Type I) is widely regarded as the primary mechanism of generation of overpressure in sedimentary basins and accounts for most of the pore pressure-related problems worldwide. When the basins undergo compaction, normally equilibrium is maintained between the rates of sediment deposition and subsidence. However, if a layer of low permeability (clay) prevents the escape of pore fluids at rates sufficient to keep up the rate of increase in vertical stress, the pore fluid begins to carry a large part of the load and hence becomes overpressured (Hubbert and Rubey, 1959). In such a situation, the strata retaining more fluid in pore spaces act as an overpressured zone over geologic time. Under the mechanisms other than compaction disequilibrium (Type II), several secondary pressure-generating mechanisms can be listed (Bowers, 2002) and these are related to (i) fluid expansion and (ii) tectonic stress. According to Bowers (2002), all Type II mechanisms can be referred to as unloading (fluid expansion), which means that expansion and migration of trapped pore fluid occur after the compaction phase is over.

7.3.1.3 *Methods of Pore Pressure Estimation*

The general basis for all pore pressure estimation methods is that pore pressure affects the compaction-dependent shale properties such as porosity, density, sonic velocity, and resistivity. The methods of pore pressure estimation can be broadly divided into (i) direct methods and (ii) effective stress methods. The direct methods relate the amount a pore pressure-dependent parameter (resistivity, sonic velocity, etc.) diverges from its normal trend line to the pore pressure gradient at the depth. There are basically two direct methods: the cross plots pioneered by Hottman and Johnson (1965) and overlays first proposed by Pennebaker (1968). On the other hand, the effective stress analysis is mostly based on the effective stress principle of Terzaghi (1943), which states that compaction of geologic materials is controlled by the difference between total confining pressure and the pore fluid pressure. Several effective stress methods (vertical, horizontal, and velocity-effective stress) are available in the literature for the estimation of pore pressure (e.g., Bowers, 1995, 2001; Eaton, 1972; Foster and Whalen, 1968; Ham, 1966; Miller et al., 2002). A detailed description of all the above mentioned methods is out of scope of the present book, hence not provided here.

7.3.1.4 *Pore Pressure Work Flow Methodology*

The pore pressure estimation is carried out by the industry in two broad phases such as the predrill and postdrill. Before drilling a well, predrill pore pressure estimation is carried out in order to help the well designing and casing plan. During and after the completion of the drilling phase, a real-time postdrill pore pressure analysis is made to ascertain the formation pressure at the location. Predrill pressure estimation heavily relies on velocities obtained from the seismic methods which provide changes in interval velocities with depth. The input seismic velocities are mostly derived for imaging/stacking processes. In the absence of any reliable offset well information, accurate predrill prediction is very challenging. Rock physics-based pore pressure modeling using such velocity dataset has inherent drawbacks, and hence, the accuracy of velocity model is significantly important. In addition, these velocity data do not have sufficient vertical resolution to identify and perform pressure analysis. Hottman and Johnson (1965) first showed that log-derived shale properties can be used to calculate the formation pressure by equivalent depth method. With the advent of LWD tools, it became possible to calculate the formation pressures in real time from resistivity logs while drilling. Wireline sonic logs, formation tester data, and the Vertical Seismic Profiling (VSP) data help to calculate the accurate formation pressures after completion of the well. Postdrill analysis of well is important because it gives the calibration points to model regional pore pressure profiles. Lessons learnt from the offset well data are a key input parameter for estimation of predrill pore pressure.

7.3.1.5 Case Studies from the Indian Offshore

Abnormal pressures occur over a wide range of geological conditions in the Indian offshore sedimentary basins and pose serious challenges. The possible cause of abnormal pressuring needs to be very well understood in order to overcome this challenge. According to Law and Spencer (1998), compaction disequilibrium is the dominant cause of abnormal pressure in younger (Tertiary) deltaic sequences, while hydrocarbon generation, aquathermal expansion, and tectonics contribute to the abnormal pressure in older (pre-Tertiary) rocks. Rapid sedimentation rate, thickly deposited sediments, high temperature, clay diagenesis, and deeply buried mobile strata favor abnormal pressures in both the shallow and deeper parts of the basin. The Krishna–Godavari offshore is a typical area of abnormal pressure occurrence along the east coast of India which is characterized by the presence of thick sediments, high sedimentation rate, and deeply buried mobile shale strata favoring the shale tectonics. This is manifested in the form of large extensional growth faults in the shelf and upper slope regions, and mud diapirs and toe thrusts in the deeper parts of the basin (Choudhuri, 2011). The basin has a proven thermogenic petroleum system in Tertiary and Mesozoic, and biogenic petroleum system in the Mio-Pliocene.

With high rate of sediment transport and deposition, overpressure zones occur in the low permeable Mio-Pliocene and Cretaceous shales during burial as a result of undercompaction. So, pore prediction analysis is of utmost importance for safe and fast drilling as well as to meet the desired reservoir depths. For this purpose, Chatterjee et al. (2012) carried out an exercise to demonstrate the effectiveness of advanced velocity analysis for improved predrill pore pressure prediction for a deepwater well located at a water depth of ~600 m in a High Temperature–High Pressure (HTHP) regime (Fig. 7.14). According to these authors, the well was drilled to explore synrift sediments belonging to Mesozoic age to a total depth of more than 4 km with two offset wells A and B (Fig. 7.14) already drilled in the block with similar targets. In the wells A and B, mild overpressure was observed in Mio-Pliocene sediments, whereas both reservoir and nonreservoir rocks belonging to Cretaceous encountered considerable overpressure. According to Sayers et al. (2006), the velocity obtained from seismic data has limitations for pore pressure prediction as the sampling interval for velocity picking is usually very coarse and may not invoke the geological events. However, Chatterjee et al. (2012) effectively integrated the advanced seismic velocity information at the well location for predrill pore pressure prediction analysis with the experience gained at offset wells as a guide. The methodology is follows:

For well A, predrill pore pressure was predicted using dense prestack time migrated (hand picked and stacked velocity) velocity data, while the postdrill pore pressure was estimated using petrophysical log data (resistivity, density, sonic with the VSP data). Considerable variation in the pore pressure is

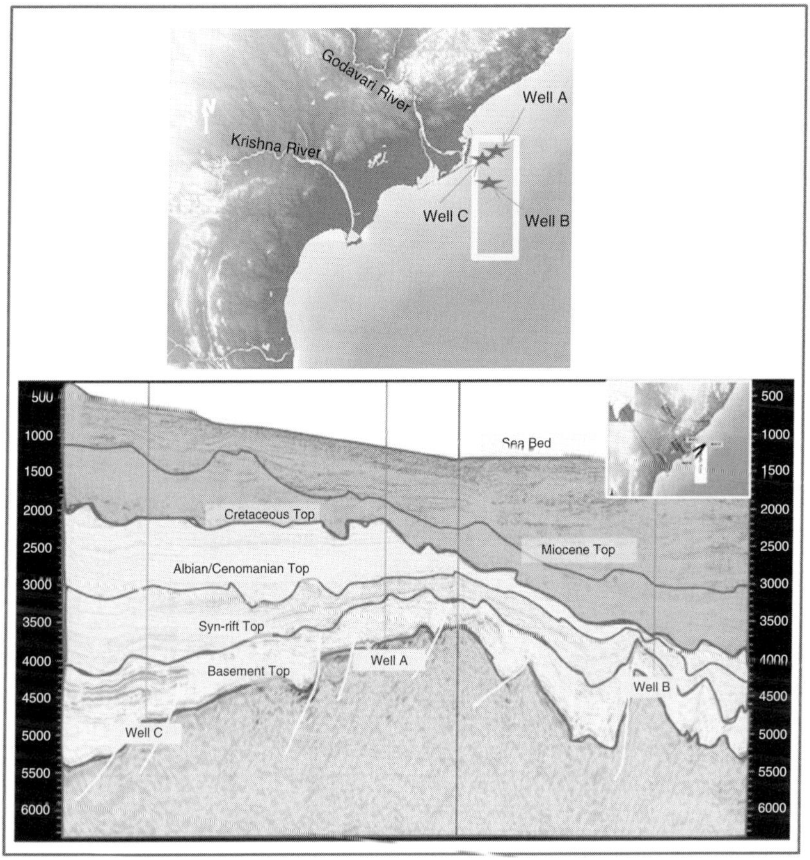

FIGURE 7.14 Seismogeological section showing various stratigraphic units and the rifted basement high in the region of drilled wells in KG offshore *(adapted with permission from Chatterjee et al., 2012 © Society of Petroleum Engineers)*. (For color version of this figure, the reader is referred to the online version of this chapter.)

noticed at the overpressure zone between predrill and postdrill plots (Fig. 7.15). Thus, the well A information was used as a guide to predict pore pressure for well B prior to drilling, but during drilling, a water influx was observed in the pressure regression zone (synrift). Postdrill pore pressure analysis supported the presence of an overpressure zone in the synrift reservoir (Fig. 7.16). In order to overcome this problem, in well C, high-quality velocity data were used (Inversion and AVO) for predrill pore pressure prediction taking well A and B observations as a guideline. The pre and postdrill pore pressure curves of well C are shown in Fig. 7.17. It was observed that there was minor variation in the pre and postdrill curves both in top of overpressure as well as in the synrift region (Chatterjee et al., 2012).

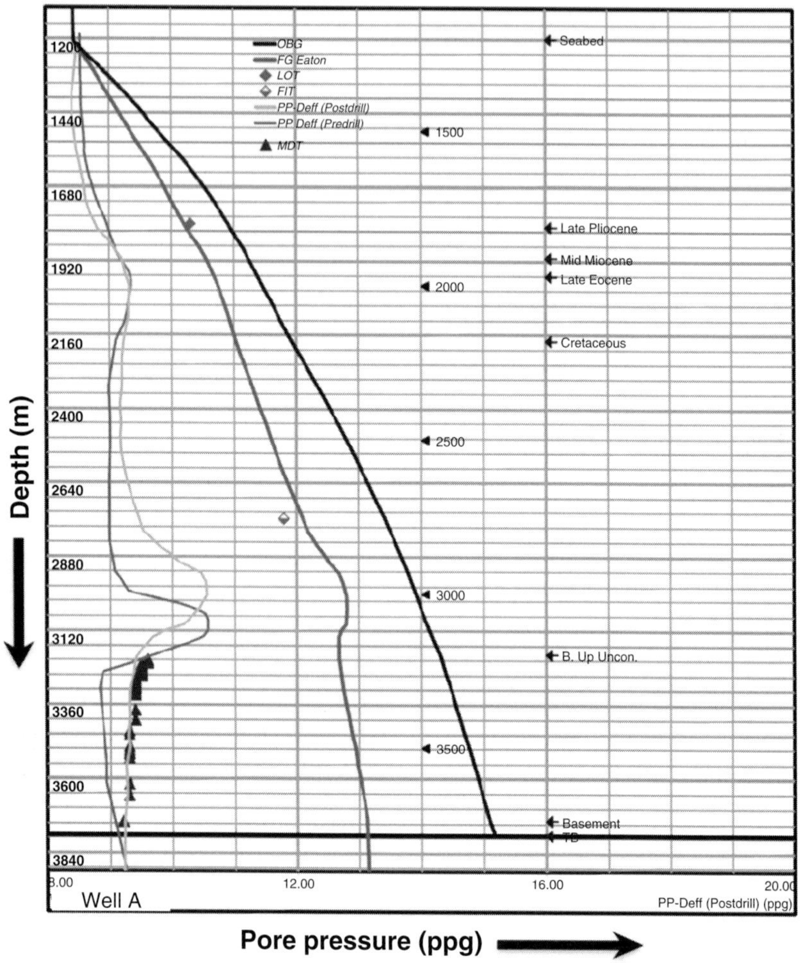

FIGURE 7.15 Pore pressure plot of well A in KG offshore *(adapted with permission from Chatterjee et al., 2012 © Society of Petroleum Engineers)*. (For color version of this figure, the reader is referred to the online version of this chapter.)

7.3.2 Shallow Hazards in the Indian Offshore Areas

Apart from the pore pressure problems related to the drilling activity described above, another major challenging task in the offshore areas is the identification of suitable locations for any kind of operation on the seafloor. Safety, suitability, and stability of drilling rigs as well as other seabed installations are of paramount importance in any offshore exploration activity and should be taken very seriously. There are many locations in the Indian offshore which can give rise to potential geohazards and these need to be identified, understood, and analyzed before venturing into deep marine operations. Geohazards can be simply told as the seabed features or near-surface events that

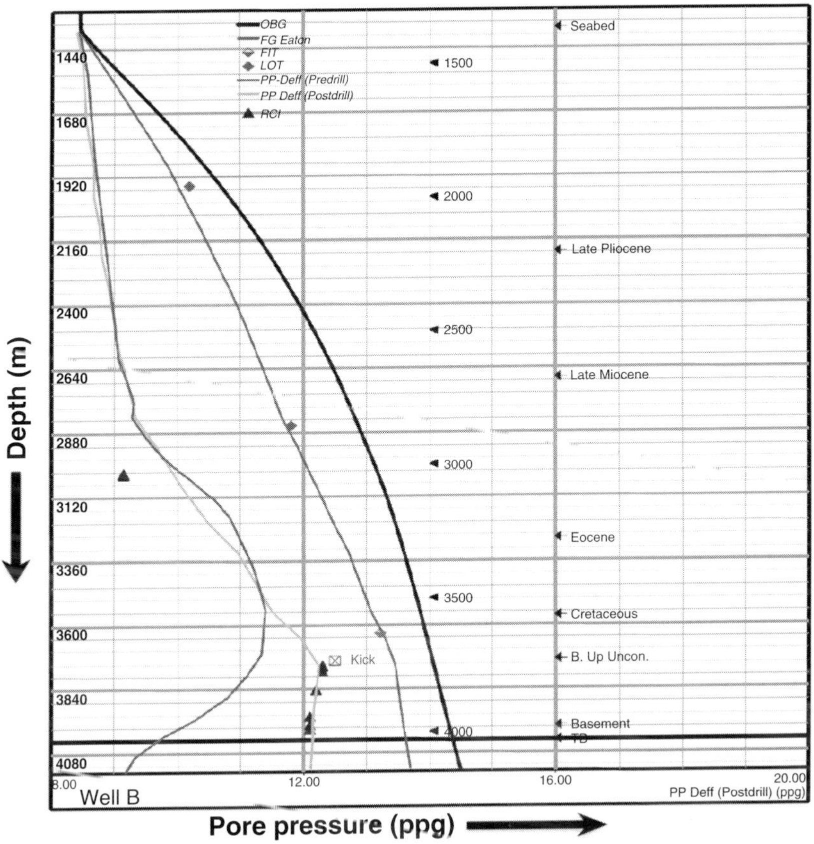

FIGURE 7.16 Pore pressure plot of well B in KG offshore *(adapted with permission from Chatterjee et al., 2012 © Society of Petroleum Engineers)*. (For color version of this figure, the reader is referred to the online version of this chapter.)

create problems for drilling or seabed installations (Bonnell and Mullee, 2000; Yahaya-Joe et al., 2000). High-resolution multibeam bathymetry data as well as side scan sonar data are used to identify various sea bottom features. In addition, large number of 2D seismic profiles or the 3D seismic data (wherever available) can be used for this purpose. From the 3D seismic data, a number of seismic attribute slices are used to describe the features. Time slices are also used to demarcate lateral extent of the corresponding features. This kind of assessment is done on a routine basis to avoid problems while drilling, such as a gas blowout or subsidence of sediments in the well, causing equipment loss, in the worst case, loss of the drilling rig or human lives (Heggland, 2004). Depending on the regional geology, tectonics, and the sediment dispersal pattern, the characteristic geomorphic features related to geohazards vary from place to place. We present here various shallow hazards in the eastern offshore areas of India with the main objective of presenting their characteristics as well as their distribution (after Bastia et al., 2010d).

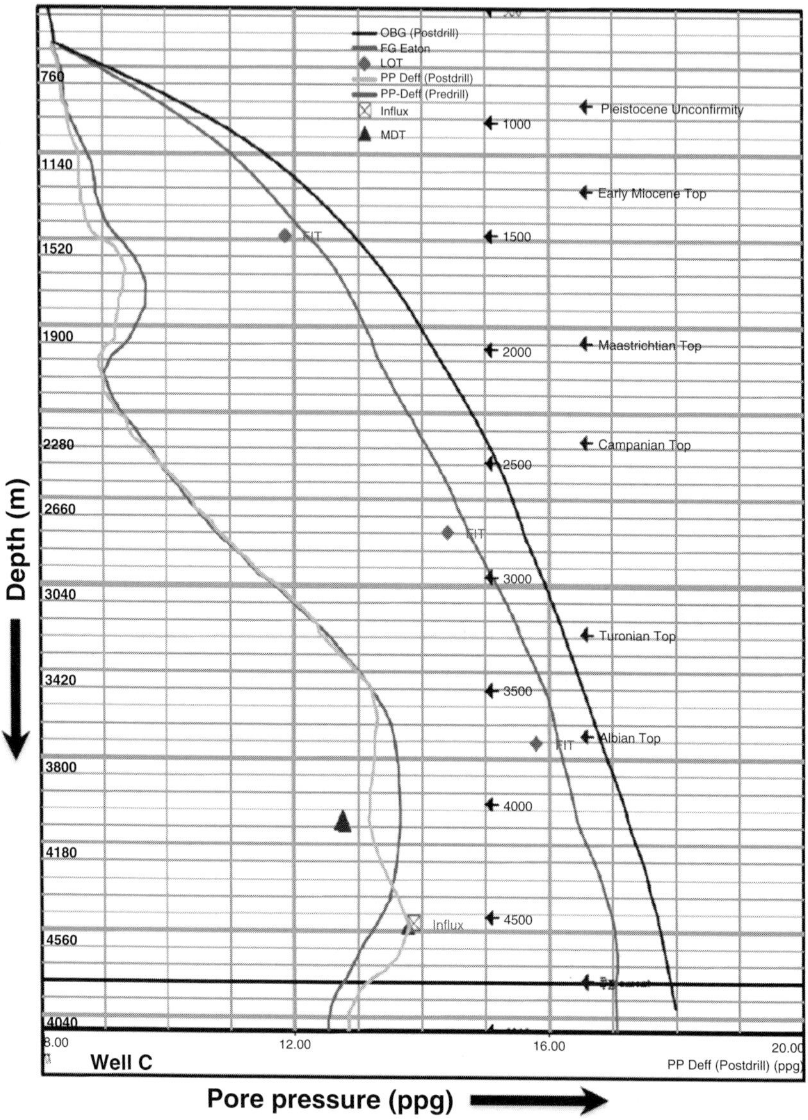

FIGURE 7.17 Pore pressure plot of well C in KG offshore *(adapted with permission from Chatterjee et al., 2012 © Society of Petroleum Engineers)*. (For color version of this figure, the reader is referred to the online version of this chapter.)

7.3.2.1 Shelf-Slope Morphology Along the Margin

The topography shaded relief map of the Indian offshore (Fig. 7.18) suggests that the eastern offshore is characterized by a relatively narrow continental shelf and a relatively wide area comprising the continental slope and rise, whereas the western offshore is characterized by presence of a wide shelf with

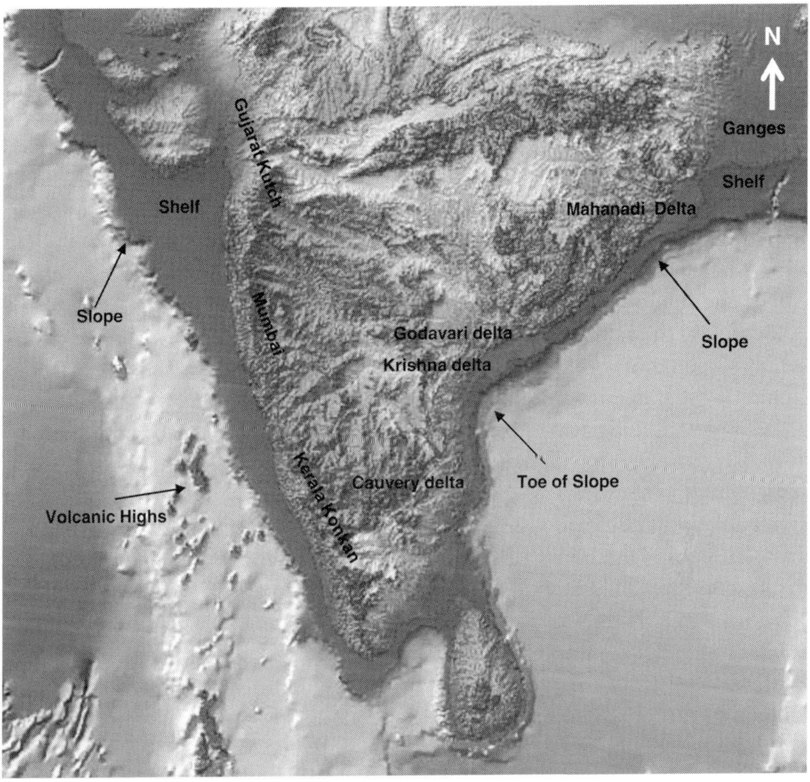

FIGURE 7.18 Topography shaded relief map of the Indian shield and its contiguous offshore areas. (For color version of this figure, the reader is referred to the online version of this chapter.)

width increasing from south to north. The western offshore is prominent for the presence of giant Mumbai High oil field in the north, where shelf is the widest in the entire margin. The drainage pattern of major east flowing river system of the Peninsular India brings huge amount of sediments into the eastern offshore and contributes to the morphological variations along the margin. In the western offshore, except Narmada and Tapti, such major river drainage system is characteristically absent. Further, the deep offshore areas of the western margin are characterized by several isolated topographic features of the Laccadive ridge complex.

7.3.2.2 Various Geohazards and Their Characteristics

7.3.2.2.1 Slope Instability and Slope Canyons

Slope instability is the most serious offshore threat on both local as well as regional scale. The shelf-slope break and the toe of the slope are considered to be important boundary markers between different depositional regimes, and these can be readily recognized from the bathymetry map or slope map of offshore areas. The steepness of slope causes more erosion as well as bypass of sediments and therefore shows failure features. On the other hand, the toe of slope is characterized by smoother seafloor and shows depositional elements. Attention must be paid to such features as they may pose risk to subsurface installations or drilling equipments. Along the eastern offshore, the steepest slope is noticed in southern part in the Cauvery–Palar basin and the steepness decreases northward and becomes minimum in the NEC-Mahanadi basin. Average gradient along the eastern offshore is about 0.5° in the shallow water area having depth less than 60 m, 1.5° to 2° between 100 and 200 m water depth, and increases further to 2.5° to 3° on the continental slope (Bastia et al., 2010d). Regional bathymetry map shows offshore Gujarat is characterized by the presence of well-developed canyons.

Submarine canyons are the conduits for high sediment flows and strong turbidity current, and their stability is mainly related to the seabed slope angles and the canyon walls which modify in time due to dynamic processes taking place in those areas. These are areas to be cautious for any seabed installations as large mass of sediments flow through these canyons resulting strong currents. This is evident from the events like the 1929 Grand Banks and the 1971 Nice/Var canyon-related submarine cable breakages (Day, 2002). A number of deep canyons identified along the eastern offshore (Bastia et al., 2010d) from the multibeam bathymetry and 3D seismic data are presented in Fig. 7.19. Steep walls of the canyons show scar marks related to canyon wall failure.

7.3.2.2.2 Submarine Channels

In the shelf-slope area as well as the deep offshore, a number of straight to highly sinuous submarine channels have been identified (Fig. 7.20). The turbidity current activity along these channels would be hazardous for any kind of subsea engineering activities and therefore are important.

7.3.2.2.3 Mass Transport Complexes/Mud Flow

The deposits or features related to the submarine mass movement processes like slides, slumps, and flows with plastic behavior can be referred to as Mass Transport Complexes (MTCs), and these are generally represented in the seismic sections by chaotic reflectors and transparent zones. The surface morphology of these deposits controls the sediment pathways until it is completely covered (Dykstra, 2005; Dykstra and Kneller, 2002). Kneller and Dykstra (2003) observed that large–scale MTCs can redirect the sediment pathways over hundreds to thousands of square kilometers of area on the seafloor. Therefore, identifying such areas is extremely important because they are hazard prone and any installation along their path can be affected. The seafloor map prepared from the 3D seismic data shows the presence of

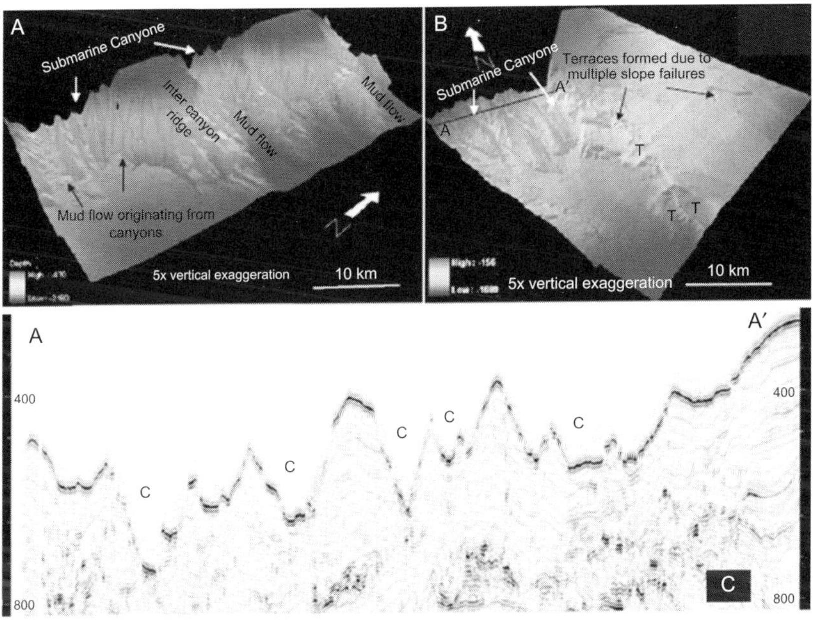

FIGURE 7.19 Slope canyon systems along the eastern offshore. (A) 3D visualization of bathymetry from KG offshore. (B) Submarine canyons and slope failure scarps in a part of KG basin. AA′ is a section across the canyons shown in (C) (*adapted from Bastia et al 2010d*). (For color version of this figure, the reader is referred to the online version of this chapter.)

an active MTC in the Krishna–Godavari basin (Fig. 7.21), and the semblance attribute extracted from this data demonstrates various features associated with mass movements. Many such locations could be identified in the offshore areas only through detailed studies.

7.3.2.2.4 Sediment Waves

From the 3D seismic data, conspicuous wavy-shaped seismic reflectors are identified in the deepwater eastern offshore area and these represent the sediment waves. These are commonly observed on submarine channel levees (McHugh and Ryan, 2000; Normark et al., 2002; Wynn and Stow, 2002) and formed in response to spillover and flow stripping from channels of fine-grained turbidity flows. These can be related to active submarine flows, therefore should be identified and characterized for hazard analysis. Sediment waves have been identified (Fig. 7.22) from a number of locations in the eastern offshore from seismic sections, where they are either associated with the channel-levee complexes or with mass transport complex areas (Bastia et al., 2010d). Active sediment movement and remobilization are noticed and could be hazardous to any sea bottom installation due to their significant height and length.

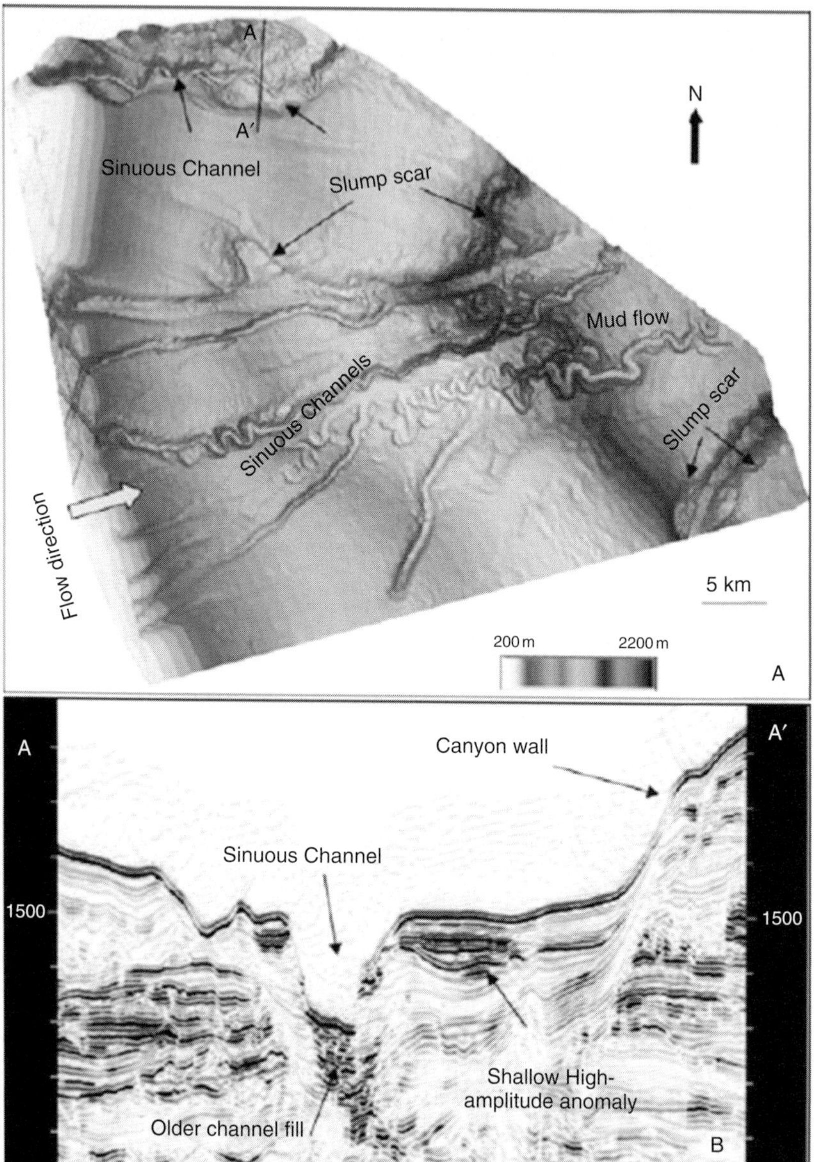

FIGURE 7.20 (A) 3D visualization of bathymetry in a part of Cauvery offshore area depicting sinuous channels and straight channels. (B) Section across a sinuous channel showing high-amplitude seismic reflectors which could be hazardous in case they correspond to shallow gas charged sands (*adapted from Bastia et al., 2010d*). (For color version of this figure, the reader is referred to the online version of this chapter.)

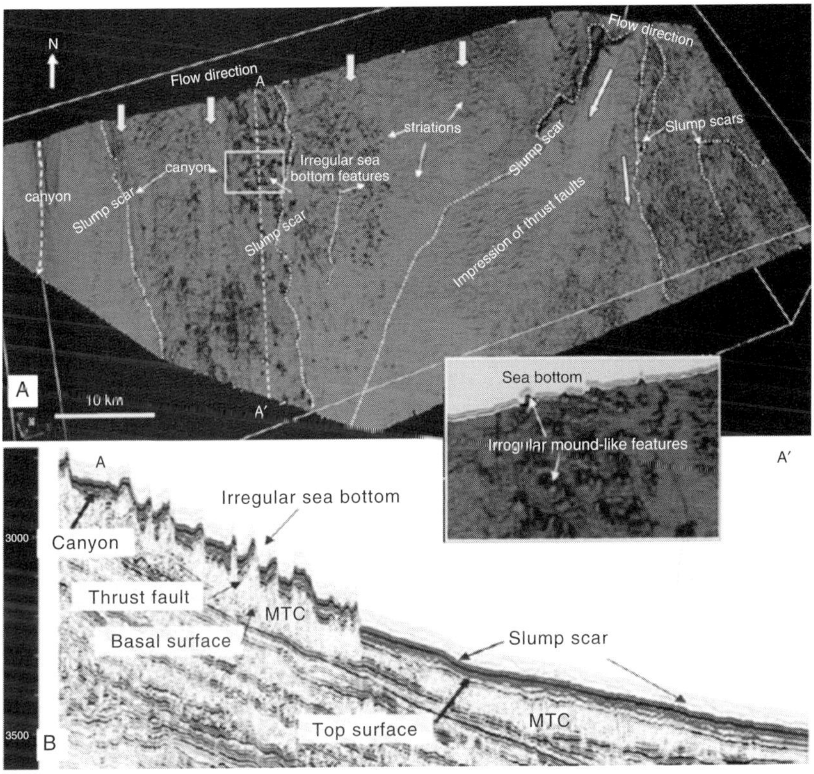

FIGURE 7.21 (A) Semblance attribute overlaid on a seabed relief used to describe the geomorphic features in the eastern offshore. Presence of slump scars and striation marks are characteristic features of a mass transport movement. Block arrows represent different inlets for sedimentation. The inset shows a zoomed feature of the sea mounds. (B) Seismic section showing irregular water bottom formed due to toe thrust related to mass transport movement. MTC, mass transport complex (*adapted from Bastia et al., 2010d*). (For color version of this figure, the reader is referred to the online version of this chapter.)

7.3.2.2.5 Shallow Faults

Shallow faults are smaller in size formed near the water bottom and are not associated with any tectonic movement or seismicity. They produce steeper scarps on the seafloor and can trigger slumps; in certain situations, these act as failure surfaces along which a large mass of younger sediments can move down. An interpreted depth section from offshore Godavari basin shows the presence of shallow faults (Fig. 7.23). These could be interpreted as syn-sedimentary listric faults associated with gravitational sediment loading (Bastia et al., 2010d). Identifying such areas is important as any drilling or heavy subsea installation in this zone can lead to mass failure.

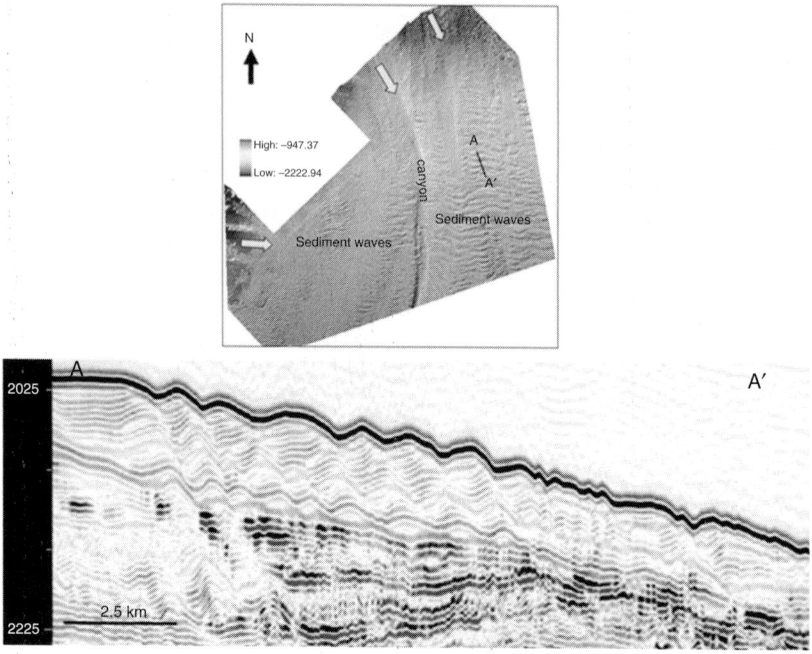

FIGURE 7.22 (A) Multibeam bathymetry map from deepwater Mahanadi offshore showing the presence of sediment waves and a well-developed submarine canyon running from north to south. (B) Seismic section (TWT) showing seismic signature of the sediment waves. As can be seen from the section, each wave is having a length of around 1 km and height of 30–50 m which describes the high energy associated with the formation (*adapted from Bastia et al., 2010d*). (For color version of this figure, the reader is referred to the online version of this chapter.)

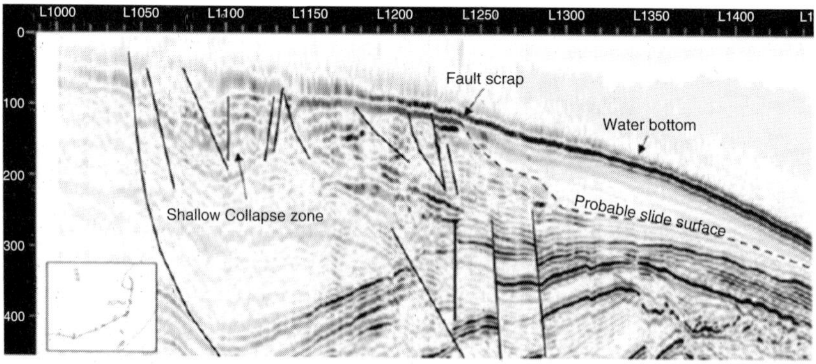

FIGURE 7.23 A depth section from shallow water region of Godavari offshore showing interpreted shallow faults. These faults can act as conduit for gas seep or planes for sliding and can create hazard to drilling operations or any subsea installation (*adapted from Bastia et al., 2010d*). (For color version of this figure, the reader is referred to the online version of this chapter.)

7.3.2.2.6 Gas Hydrates

The gas hydrate regions are zones of potential slope failure (Kvenvolden, 1999) in the margins and therefore can pose threat to many seabed operations. The primary parameters that affect the hydrate stability zone in marine environments are the water depth, the seafloor temperature, and the geothermal gradient (Sain et al., 2011). These are generally stable under low temperatures and high pressures (Kvenvolden and Barnard, 1983). The hydrate stability zone maps for the Indian offshore have been prepared by several researchers (Rao et al., 1998; Sain et al., 2011; Sain and Gupta, 2012). The presence of Bottom-Simulating Reflector (BSR) sequence which is the main marker for the presence of gas hydrates has been observed in the seismic sections in several areas of the Indian offshore (Karisiddiah and Subba Raju, 2002; Mathur et al., 2007; Ramana et al., 2009; Sain et al., 2011; Sain and Gupta, 2012); Satyavani et al., 2008; Subrahmanyam et al., 1998) and helps in identifying potential zones of gas hydrate occurrence. Few representative seismic sections depicting the presence of BSRs in the KG offshore and the Mahanadi offshore areas are shown in Fig. 7.24. Detailed analysis made using the seismic data and shallow core revealed the presence of huge gas hydrate zone in the KG offshore (Lee and Collett, 2009). The gas hydrate stability thickness map of the Indian offshore areas is shown in Fig. 7.25 (after Sain and Gupta, 2012), and this map provides a first-order information of spatial and depth domains of probable gas hydrate occurrence. Further, from the hazard point of view, the region of gas hydrates could significantly reduce the submarine slope stability (Nixon and Grozic, 2000), and sometimes may turn out to be areas of potential geohazard particularly when shallow gas is present under overpressure conditions.

7.3.2.2.7 Mud Volcanoes

Submarine mud volcanic activity may affect drilling operations, rig installations, and pipeline routings. The presence of mud volcanoes is reported from both passive as well as active margin setting (Milkov, 2000). In the KG offshore area, the presence of one such mud volcano region appearing as a mound-like feature has been mapped from the 3D seismic as well as multibeam bathymetry data. On a seismic section (Fig. 7.26), it represents as a vertical zone with transparent seismic signature and looks like a pillar-type structure (Bastia et al., 2010d). The amplitudes within the pillar have been wiped out with a mound sitting above it. The reflection-free zone below the center of the mud volcano could represent the feeder dyke (Graue, 2000; Stewart and Davies, 2006). Such features may affect the stability of bore holes drilled for oil and gas production as the mud is associated with poorly constrained pore pressure and fracture strength (Ebrom et al., 2004).

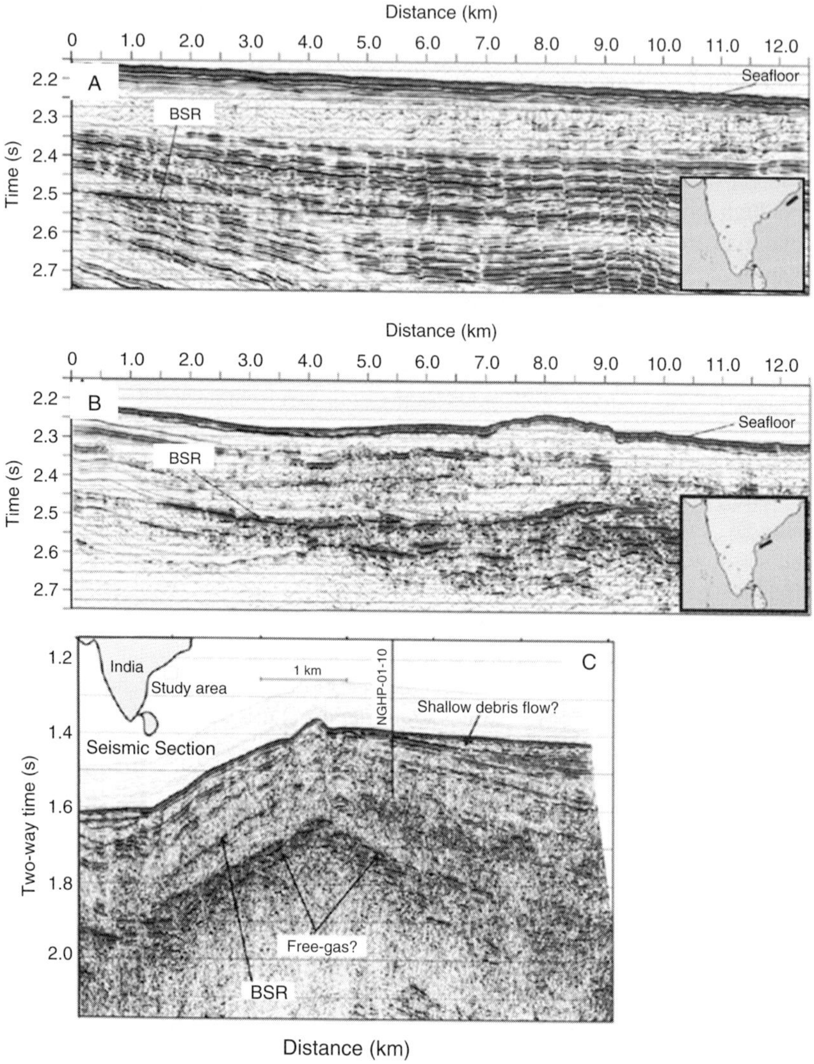

FIGURE 7.24 Seismic sections showing the presence of bottom simulating reflector (BSR) (A) in Mahanadi offshore and (B) in KG offshore. (C) Presence of gas hydrate in 120-m-thick fractured shale in KG offshore (*adapted from Sain and Gupta, 2012*). (For color version of this figure, the reader is referred to the online version of this chapter.)

7.3.2.2.8 Pockmarks

The concave, crater-like depressions on the seabed associated with the release of gas or fluids from the subsurface can be termed as pockmarks (Hovland and Judd, 1988; King and MacLean, 1970). These are generally formed in soft, fine-grained seabed sediments and represent an important global

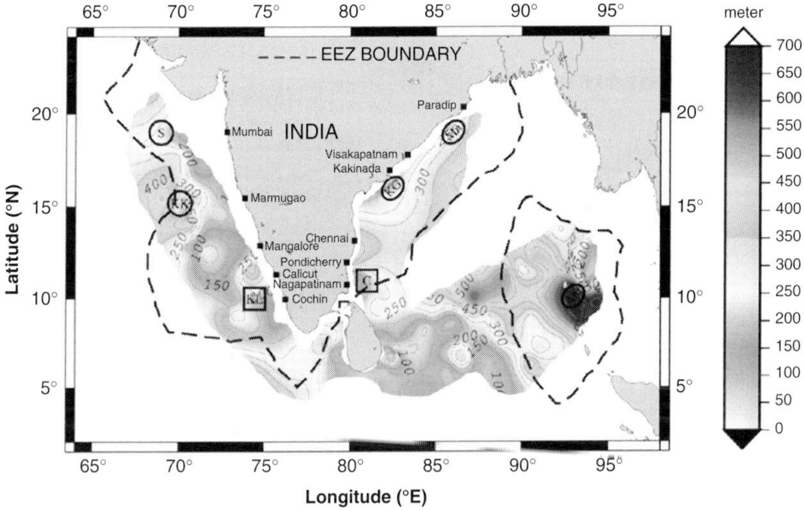

FIGURE 7.25 Gas hydrate stability thickness map for the Indian offshore area (*adapted from Sain and Gupta, 2012*) showing prospective zones of gas hydrate occurrence (shown by circle, oval, and square symbols). Mn, Mahanadi; Am, Andaman; KG, Krishna–Godavari; C, Cauvery; KL, Kerala-Laccadive; KK, Kerala-Konkan; S, Saurashtra. (For color version of this figure, the reader is referred to the online version of this chapter.)

mechanism for degassing deeply buried sediments thereby contributing significantly methane into the atmosphere (Judd, 2005; Kvenvolden and Rogers, 2005). The high backscatter values from the calcareous material over these features help to identify them in the multibeam bathymetry. Significant presence of pockmarks at number of locations in the eastern as well as western offshore is noticed in high-resolution multibeam bathymetry dataset and few such areas are presented here (Figs. 7.27). Along the western offshore, starting from the Kerala–Konkan in the south to the Kutch offshore in north, pockmarks of various size and shape have been noticed. Presence of such dense cluster of pockmarks could be dangerous if they are active and hence detailed studies are required.

7.3.2.2.9 Gas Chimney

The upward migration of gas from the deeper parts gives rise to gas chimneys and causes problems during drilling (Heggland, 2004). The gas is transported upward along with the migrating pore water and forms an inverted cone shape characterized by low amplitude. Bastia et al. (2010d) identified one such area characterizing the presence of gas chimney in the offshore KG basin. The high-amplitude anomalies identified near to the gas chimney indicate probable gas charge along the fault. For this reason,

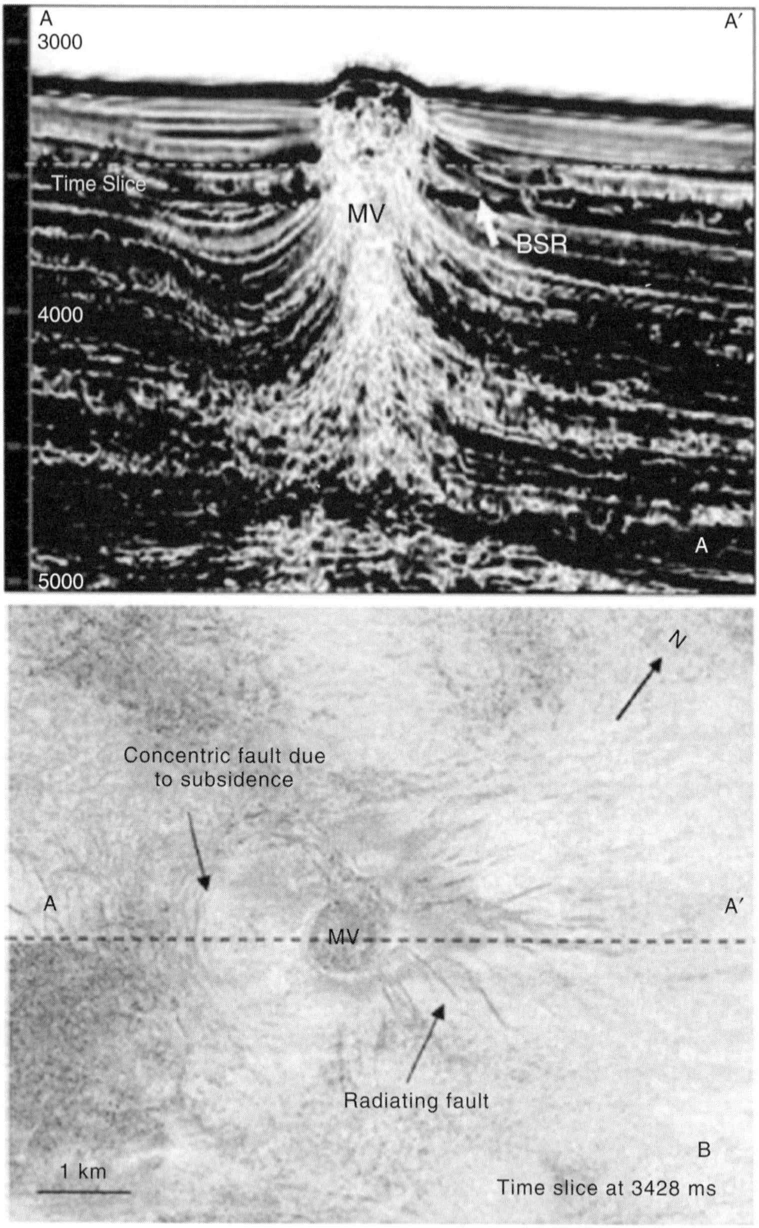

FIGURE 7.26 (A) Reflection strength extracted along a seismic section in the KG offshore showing that white-colored plume-shaped feature corresponds to loss of amplitude zone which is identified as the mud volcano. At around 3500 ms, a seismic reflector cross-cutting other reflectors is interpreted to be a BSR. (B) Time slice of coherency attribute showing the mud volcano and related features. In the time slice, the circular feature depicts the mud volcano (*adapted from Bastia et al., 2010d*). (For color version of this figure, the reader is referred to the online version of this chapter.)

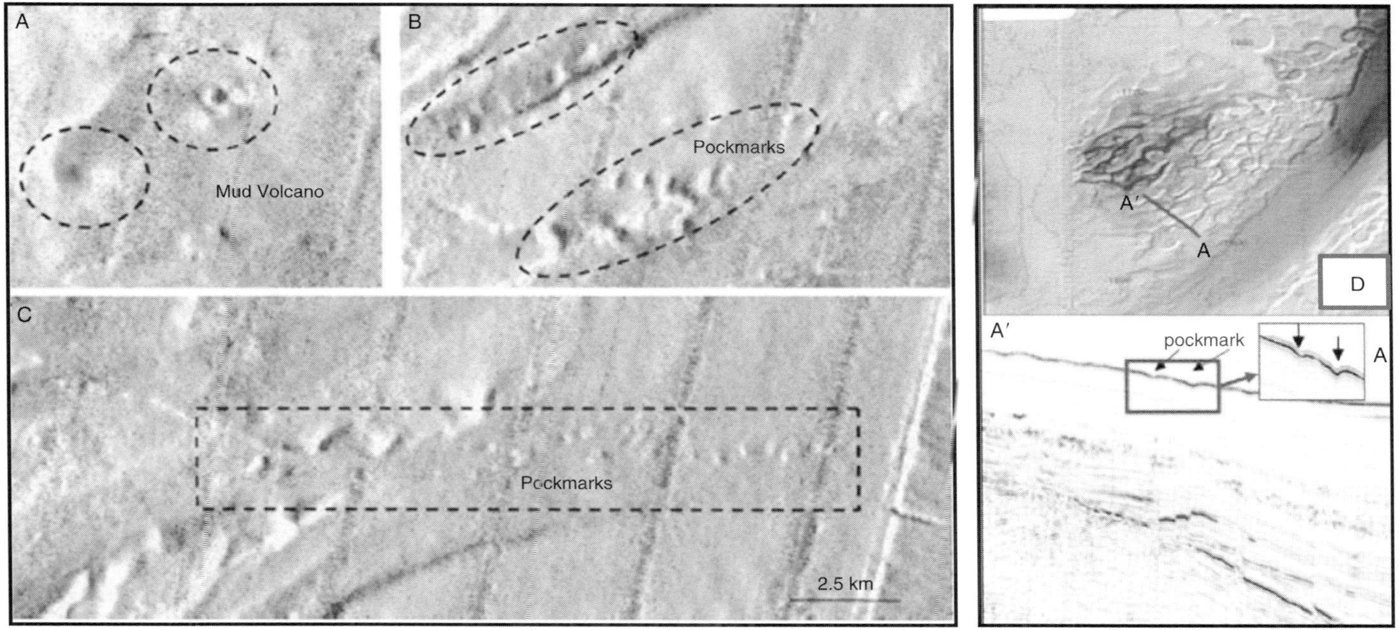

FIGURE 7.27 Backscatter image overlaid on multibeam bathymetry data from deepwater area of Krishna basin shows identified mud volcanoes and pockmarks in the area. (A) Two topographic positive mounds interpreted as mud volcanoes are circled. Small depressions on sea bottom (in B and C) are interpreted as pock marks (*adapted from Bastia et al., 2010d*). In (D), the multibeam bathymetry in KK offshore is shown which reveals the presence of a large mound and the section below shows its association with a fault (*adapted from Mahapatra et al., 2008*). (For color version of this figure, the reader is referred to the online version of this chapter.)

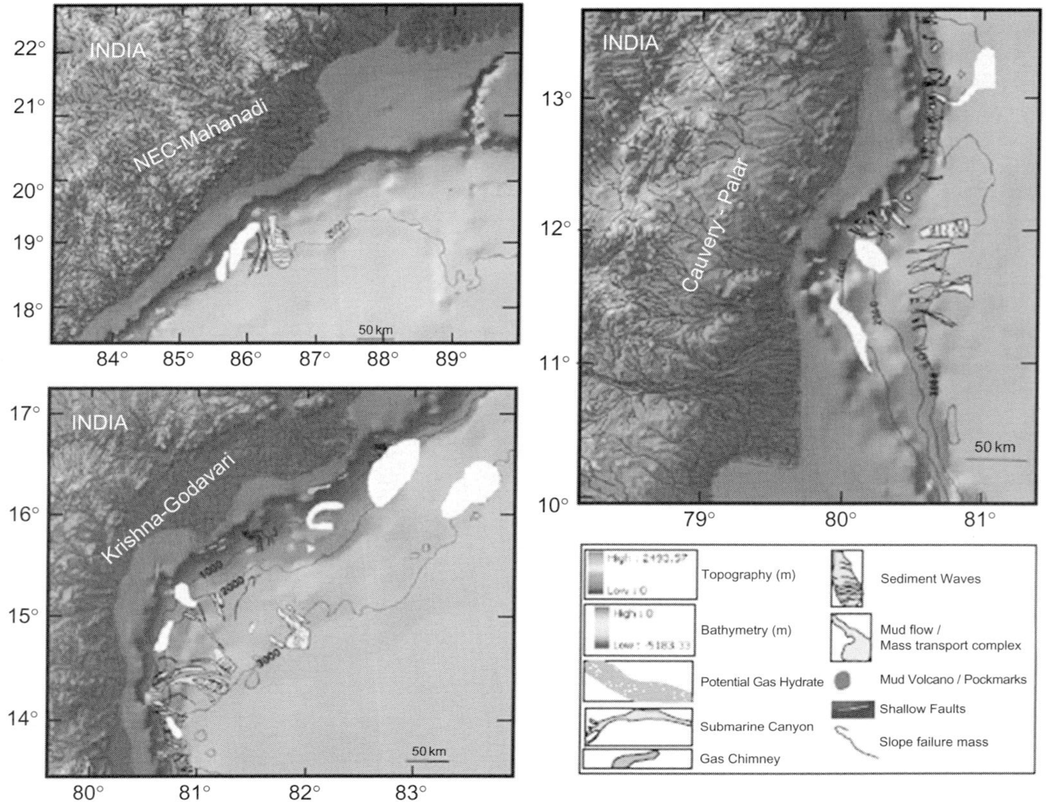

FIGURE 7.28 Map summarizing the identified marine geohazards along the eastern offshore areas (*adapted from Bastia et al., 2010d*). (For color version of this figure, the reader is referred to the online version of this chapter.)

amplitude maps are prepared to identify the anomalous areas representative of shallow gas accumulations as well as faults. As these are areas of active gas seepage, one should be cautious for undertaking drilling activity.

A summary of various geohazards identified in the eastern offshore areas is presented in Fig. 7.28. Shallow core analysis would be more useful for a detailed understanding of these features. Further, the above results do not rule out the presence of any other type of geohazards and more data will be required to properly characterize them.

Chapter 8

Summary

The tectonic framework of the present-day continental margin of India and the wide variations in morphology, structure, sedimentation, volcanic activity, and so on observed along its various segments basically reflect that these segments have evolved through a variety of tectonomagmatic processes and varied breakup history (passive/active rifting, shearing) during their evolution. It is very well established from the geological and geophysical investigations of the margins as well as the contiguous oceanic areas that the breakup processes between India and its conjugate eastern Gondwanaland continents are essentially controlled by the Precambrian structural grain that includes the cratons and mobile belts. Early breakup history of the eastern Gondwanaland continents is extremely important for understanding the complete basin evolution process of the individual margin segments and can be achieved through carefully derived reconstruction models of the Indian Ocean guided by the geological and geophysical data of the margins. Unfortunately, though well-constrained plate reconstruction models of the Indian Ocean are available until the Cretaceous, the role of India in the Gondwanaland dispersal remained very uncertain in the Mesozoic. This is partly due to the lack of clearly discernible Mesozoic magnetic anomaly identifications along the Indian continental margins as well as its conjugate segments. Further, building up of a tighter reconstruction becomes very difficult due to lack of full understanding on the nature of crust in certain critical areas of the continental margins as well as large aseismic ridges/plateaus in the Indian Ocean. The large scale magmatic activity and plume tracks/volcanic traces at the margins obliterate the signatures of crust that had formed by normal seafloor spreading processes. For example, in the case of Indian continental margins, complete understanding of rift–drift history and clear identification of Continent Ocean Boundary could not be clearly established due to factors such as the multiphase rifting episodes along the east as well as west coast, presence of thick volcanic traps above the sediments/Precambrian basement in the western offshore, controversies with regard to nature of crust below Laccadive Ridge in the west and 85° E Ridge in the east, strike-slip motion along the southeast coast and presence of thick sediments in the northeast coast, the style of rifting between India and Madagascar, extent of Rajmahal and Deccan volcanism, and

Developments in Petroleum Science, Vol. 59. DOI: 10.1016/B978-0-444-53604-4.00008-9 the post breakup volcanism in the Mahanadi offshore. Though recent geoscientific investigations could partially resolve some of the above mentioned issues, a concerted effort in terms of acquiring state-of-art deep seismic reflection as well as seismic refraction data definitely is need of the hour along with deep drilling in certain critical areas. Further, these issues also have direct relevance to proper evaluation of petroleum system along the Indian continental margins.

8.1. PETROLEUM SYSTEMS IN THE INDIAN OFFSHORE BASINS

The varied tectonic history of the east and west coasts, the India–Asia collision process that led to the development of Indus Fan in the west and the Bengal Fan in the east, and the drainage pattern of major east flowing river system during Tertiary all together influenced the manner in which major petroliferous basins in the onshore and offshore realms have developed. Based on the evolution, structural style, depositional history, and major tectonic trends, the sedimentary basins extending from the land to the deepwater areas of the Indian continental margins have been grouped into the Kutch–Saurashtra, Mumbai and the Konkan–Kerala offshore basins along the western margin, and the Cauvery–Palar, Krishna–Godavari, and Mahanadi-NEC (Bengal) onshore–offshore basins along the eastern margin of India. Large variations in the sedimentation history and depositional style along the margins gave rise to the diversity in basins and significantly influenced the development of offshore reservoirs. The structural and petroleum setup of these basins is very useful and compiled in Table 8.1. While the reservoirs in the western offshore are generally carbonates, the clastics are the main pay zones in the eastern offshore.

In the context of Indian continental margins, the earlier exploration efforts have primarily focused close to the coast until when the deepwater exploration came into focus during early 2000. Subsequently, acquisition of new high-resolution geoscientific data connects the east coast margin beyond the 85° E ridge and west coast margin beyond the Laccadive Ridge and Laxmi basin area. These data provide meaningful insight to the possible petroleum system over the entire unexplored stretch.

8.1.1 Eastern Offshore

The Cauvery basin is characterized by nearly 8000-m-thick sediments ranging from the Jurassic to recent with play types having stratigraphic traps (pinchouts, wedgeouts, lenticular sand) and other structural traps. The Krishna–Godavari basin is another significant offshore oil-producing basin of India, apart from the Mumbai High and is the site for several new oil and gas discoveries particularly in the deepwater areas. The estimated sedimentary section is $\sim$7000 m ranging from Proterozoic to Recent with play

TABLE 8.1 Summary of Type, Source, Reservoir, Entrapment, and Play Distribution of Indian Offshore Basins

Basin	Sub-Basin	Margin Type	Evolution	Petroleum System				Future Trend
				Principal Source	Principal Reservoir	Principal Entrapment	Play Type	
A. East Coast of India								
Bengal		Passive	Multiphase rift with pronounced major unconformity	Postrift Tertiary **TH** and **BI**	Clastics and carbonates	Stratigraphic	DW channel-levee complex, wedgeout	85° East ridge carbonates
Mahanandi		Passive	Multiphase rift with pronounced major unconformity	Postrift Tertiary **TH** and **BI**	Clastics and carbonates	Stratigraphic	DW channel-levee complex, wedgeout	85° East ridge carbonates
Krishna–Godavari	Godavari	Passive	Multiphase rift with pronounced major unconformity	Syn-rift Mesozoic **TH**, postrift Tertiary **TH** and **BI**	Clastics	Structural and stratigraphic	DW channel-levee complex, wedge out, synrift fault closure	Postrift Cretaceous
	Krishna	Passive	Multiphase rift with pronounced major unconformity	Syn-rift Mesozoic **TH**, postrift Tertiary **TH** and **BI**	Clastics	Structural and stratigraphic	DW channel-levee complex, wedgeout, synrift fault closure	Tertiary fold-thrust belt
Cauvery–Palar	Penar	Passive	Multiphase Rift	Syn-rift and postrift Mesozoic **TH**	Clastics	Structural	Synrift fault closure, wedgeout	Postrift Cretaceous
	Palar	Passive	Strike slip dominated rift	Syn-rift and postrift Mesozoic **TH**	Clastics	Structural	Synrift fault closure, wedgeout	Postrift Cretaceous
	Cauvery	Passive	Multiphase rift with pronounced major unconformity	Syn-rift and postrift Mesozoic **TH**	Clastics	Structural	Synrift fault closure, wedgeout	Postrift Cretaceous
Andaman		Active	Back-arc spreading	Tertiary **TH** and **BI**(?)	Clastics	Stratistructural	Accretionaryprism, fault closure	Fore-arc trend
B. West Coast of India								
Kutch		Intracratonic	Multiphase rift	Syn-rift Mesozoic **TH**	Clastics and carbonates	Stratistructural	Synrift fault closure	Tertiary carbonates
Saurashtra		Passive	Multiphase rift	Postrift Tertiary **TH**	Clastics and carbonates	Stratistructural	Carbonate buildups	Sub-basalt trend
Mumbai		Passive	Strike-slip dominated rift	Postrift Tertiary **TH**	Carbonates	Structural	Carbonate buildups, anticlinal domes	Satellite structure flanking major domes
Kerala–Konkan	Konkan	Passive	Strike-slip dominated rift	Syn-rift Mesozoic **TH**(?)	Clastics(?)	Structural	Sub-basalt clastics(?)	Sub-basalt rift trend, anticlinal domes
	Kerala	Passive	Strike-slip dominated rift	Syn-rift Mesozoic **TH**(?)	Clastics(?)	Structural	Sub-basalt clastics (?) anticlinal domes	Sub-basalt rift trend, anticlinal domes

TH, thermogenic; **BI**, biogenic; **DW**, deepwater.

types such as structural highs/fault closures, unconformity-related traps, wedgeouts, and channel-levee systems. The Mahanadi-NEC basin lies along the east coast north of Visakhapatnam off the Orissa and West Bengal coast and covers the deepwater prospective areas of 85° E Ridge and the Bengal Fan. The sedimentary section in the Mahanadi offshore is of the order of 5000 m ranging from early Cretaceous to Recent with major play types such as anticlinal/fault-related, wedgeouts, carbonate buildups, and turbidite submarine fans. In the NEC region, the sedimentary section is estimated to be more than 9000 m with play types as canyon cut/fill, channel-levee complexes, and turbidite submarine fans similar to the Mahanadi offshore.

The eastern offshore basins exhibit a variety of petroleum systems and play types (Table 8.1). The Bengal–Mahanadi basins to the north have a huge pile of Tertiary sediments derived primarily from the Himalayan rivers. The sedimentary package decreases toward Cauvery–Palar in the south which dominantly comprises Mesozoic sediments. The Krishna–Godavari basins in the center have considerable thickness both in the Tertiary and Mesozoic depositional units. The multiple rifting episodes are mainly responsible for the source–reservoir distribution. Strike-slip movements have been noticed sporadically over the rift which has also controlled the depositional systems. Deepwater channel-levee complex and wedgeout along with syn-rift fault closures form the major play types.

8.1.1.1 Reservoir Potential of the Deepwater Depositional System of the Bengal Fan

The channel-levee complexes are important elements in the deepwater depositional systems and have been the focus of hydrocarbon exploration in many submarine fan systems of world oceans. We discuss here the possible depositional model (Fig. 8.1) for the Bengal Fan (Bastia et al., 2010c). During the precollision (pre-Eocene) times, major influx of sediments is from the west, and thermogenic model best explains the hydrocarbon generation during this time. The presence of 85° E ridge system gave rise to restricted mini basins which increased preservation potential for source rock due to restricted circulation. Wedges against the 85° E ridge might act as good entrapment locations for pre-Eocene petroleum system. Though input from the west continues also in the postcollision period, the postcollision Bengal Fan is believed to contain good reservoir because of huge sediment supply and the possible location should be channel-levee complexes. The basin floor fan/frontal splay characterized by high-amplitude reflection packets are also expected to have good reservoir potential if the connectivity and amalgamation of sand deposit could be maintained. Further, the exploration target also depends on the location of sand deposit with respect to the fan system. While the frontal splay/basin floor is more important in the distal part, the channel-levee deposits play an important role in present-day upper and middle fan. It is relevant to note that the

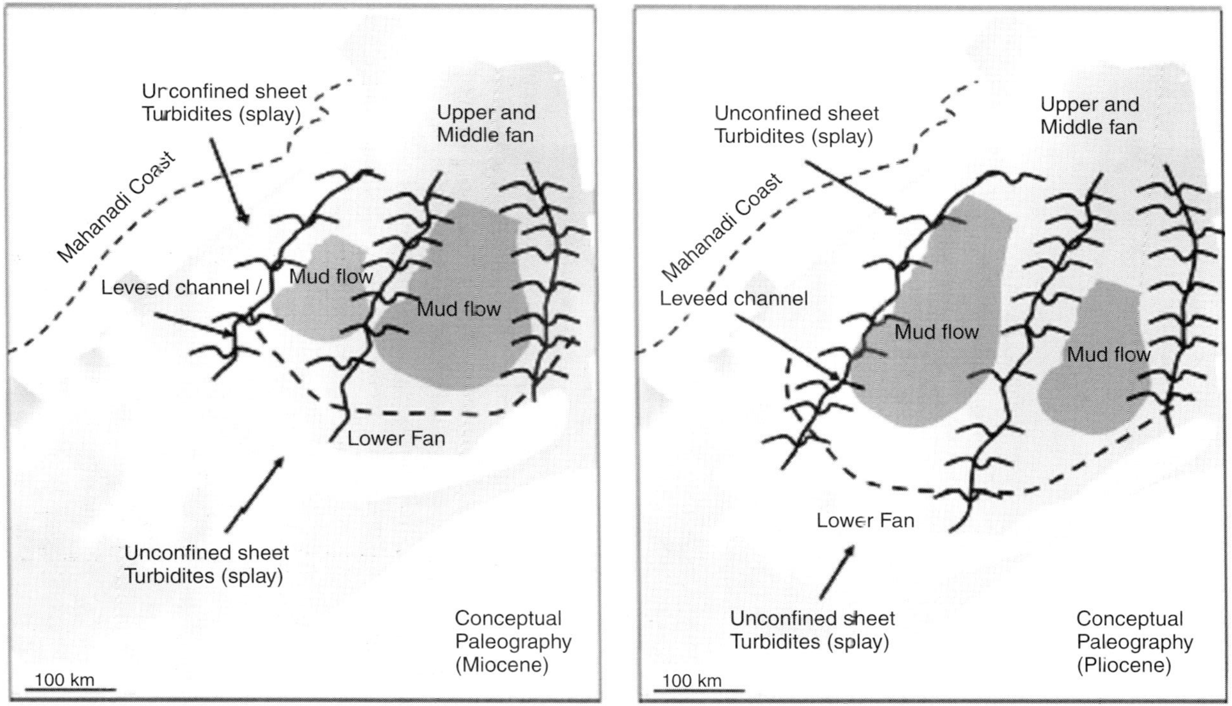

FIGURE 8.1 Conceptual paleogeography in the Miocene and Pliocene times indicating seaward shifting of facies during progradation (adapted from Bastia et al., 2010c). The extent of mud flows and expected splay deposits are shown. (For color version of this figure, the reader is referred to the online version of this chapter.)

shallower Mio-Pliocene petroleum system is more likely to have mixed thermogenic and biogenic source, unlike the deeper Cretaceous–Paleocene section. In the younger section, we can expect pure stratigraphic entrapment condition. For the mud-rich system like the Bengal Fan, vertical seal is not a problem; however, for updip seal, the truncation of reservoir facies by younger mud flows is favorable (Bastia et al., 2010c).

8.1.2 Andaman Basin

In contrast to the above basins, the Andaman basin to the far-east is an active margin. The Andaman and Nicobar Islands, lying along an Island Arc system, encompass an outer-arc accretionary wedge as well as the fore-arc, volcanic arc, and the back-arc basins. The estimated sedimentary section in both the fore-arc and back-arc areas exceeds 8000 m. In the fore-arc, the main speculative play types consist of stratigraphic pinchout, turbidite reservoirs. The deeper lithology consists of shallow water-rimmed platform carbonates. In the back-arc basin, there are major inverted anticline structures forming large four-way dip closures and three-way fault closures. The accretionary prisms and fault closures in the back-arc and fore-arc trends are the key petroleum systems. The deepwater back-arc basin in the Andaman offshore is a major frontier province.

8.1.3 Western Offshore Basins

The Kutch–Saurashtra basin comprises a total sedimentary section of ~7000 m ranging from Mesozoic to Recent with the carbonate buildups/platforms, pinchouts/wedgeouts, and transpressional rollovers as play types. The Mumbai offshore basin is characterized by the Tertiary sedimentary section of nearly 7000 m ranging from Paleocene to Recent with play types such as platform and fractured carbonates, reefal buildups, anticlinal/fault-related closures, wedgeouts against basement highs, and tilted fault blocks. The Konkan–Kerala basin is the largest frontier deepwater basin. The sedimentary section is estimated to be ~5500 m ranging from Cretaceous to Recent with basalt cover all over and is expected to be underlain by thick Mesozoic prospective section. The major speculative play types are structural highs, carbonate buildups, turbidites, and distal fans. This basin attracted the attention of geoscientists and operators in India, as the hydrocarbon shows and geochemical studies indicate potential hydrocarbon reservoirs beneath the basalt. The Mesozoic sediments underlying the Cretaceous basaltic rocks are the targets. As there are fewer wells with conventional well logs run which partially penetrated the basalt, the focus on petrophysical log evaluation in basalts is essential in the context of exploration for sub-basalt reservoirs. Therefore, it can be concluded that the western offshore basins in general are characterized by predominantly carbonate buildup associated with large structural features (Table 8.1). While the Mumbai offshore basin is very well explored and is a prolific petroleum-producing basin, the Konkan–Kerala offshore in the south and Kutch–Saurashtra offshore in the north still remain as a frontier province, particularly with reference to the Mesozoic sub-basalt plays, as the sub-basalt clastics have the potential to open a new play type in this corridor.

References

REFERENCES

Acharyya, S.K., 2000. Comment on "Crustal structure based on gravity-magnetic modelling constrained from seismic studies under Lambert Rift, Antarctica and Godavari and Mahanadi rifts, India and their interrelation ship" by D.C. Mishra. Earth Planet. Sci. Lett. 179, 595–598.

Acharyya, S.K., 2007. Collisional emplacement history of the Naga-Andaman ophiolites and the position of the eastern Indian suture. J. Asian Earth Sci. 29, 229–242.

Acharyya, S.K., Ray, K.K., Roy, D.K., 1989. Tectono-stratigraphy and emplacement history of the ophiolite assemblage from the Naga Hills and Andaman Island Arc, India. J. Geol. Soc. India 33 (1), 4–18.

Agarwal, A.K., Weaver, J.T., 1989. Regional electromagnetic induction around the Indian Peninsula and Sri Lanka: a three-dimensional numerical model study using the thin sheet approximation. Phys. Earth Planet. Inter. 54, 320–331.

Agarwal, P.K., Pandey, O.P., Negi, J.G., 1992. Madagascar: a continental fragment of the paleo super Dharwar craton of India. Geology 20, 543–546.

Ajay, K.K., Chaubey, A.K., Krishna, K.S., Rao, D.G., Sar, D., 2010. Seaward dipping reflectors along the SW continental margin of India: evidence for volcanic passive margin. J. Earth Syst. Sci. 119 (6), 803–813.

Alam, M., Curray, J.R., Chowdhury, M.L.R., Gani, M.R., 2003. An overview of the sedimentary geology of the Bengal basin in relation to the regional tectonic framework and basin fill history. Sediment. Geol. 155, 179–208.

Allegre, C.J., Birch, J.L., Campus, F., Courtillot, V., 1999. Age of the Deccan Traps using ^{187}Re–^{189}Os systematics. Earth Planet. Sci. Lett. 170, 197–204.

Allen, P.A., Allen, J.R., 1991. Basin Analysis: Principal and Applications. Blackwell Science Publications, London.

Amir, A., 1992. Channel-levee on the Indus Deep-sea Fan. Ph.D. Thesis. University of Wales, UK, 241 pp.

Amir, A., Kenyon, N.H., Cramp, A., Kidd, R.B., 1996. Morphology of channel-levee system on the Indus Deep-Sea Fan, Arabian Sea. Pak. J. Hydrocarbon Res. 8, 43–53.

Anand, S.P., Rajaram, M., Majumdar, T.J., Bhattacharyya, R., 2009. Structure and tectonics of 85° E ridge from analysis of geopotential data. Tectonophysics 478, 100–110.

Andersen, O., Knudsen, P., 2008. The DNSC08 Global gravity and bathymetry. In: European Geophysical Union meeting, 14–18 April, Vienna, Austria.

Arthurton, R.S., Farah, A., et al., 1982. The Late Cretaceous-Cenozoic history of western Baluchistan Pakistan-the northern margin of the Makran subduction complex. Geol. Soc. of Lond., 10, 373–385.

Auden, J.B., 1949. Dykes in western India. A discussion of their relationships with the Deccan Traps. Trans. Natl. Inst. Sci. India 3, 123–157.

Ayub, A., 1992. Channel-levee systems on the Indus Deep-Sea Fan. Ph.D. Thesis. University of Wales, Cardiff.

Babenko, K.M., Panayev, V.A., Svistunov, Y.I., Shelzinger, A.Y., 1981. Tectonics of the eastern part of the Arabian Sea, according to seismic data. Geotectonics 16, 139–147.

Baburao, V., Atchuta Rao, D., Sankaranarayana, P.V., Ratnam, C., 1982. Aeromagnetic survey over parts of Mahanadi basin and the adjoining offshore region, Orissa. Geophys. Res. Bull. 20, 219–226.

Backman, J., Duncan, R.A., et al., 1988. College Station, TX (Ocean Drilling Program) Proc. ODP, Init. Repts. 115 p.

Baillie, P., Barber, P.M., Deighton, I., Gilleran, P.A., Jinadasa, W.A., Shaw, R.D., 2004. Petroleum systems of the deepwater Mannar basin, offshore Sri Lanka, Proceedings, Deepwater and Frontier Exploration in Asia & Australasia Symposium. Dec. 2004, DFE04-PO-015, 1-13.

Baksi, A.K., 2001. The Rajahmundry Traps, Andhra Pradesh: evaluation of their petrogenesis relative to the Deccan Traps. Proc. Indian Acad. Sci., vol. 110. 397–407.

Baksi, A.K., Barmann, T.R., Paul, D.K., Farrar, E., 1987. Widespread early Cretaceous flood basalt volcanism in eastern India: geochemical data from Rajmahal-Bengal-Sylhet Traps. Chem. Geol. 63, 133–141.

Balakrishna, T.S., Sarma, D.S., 1981. Tectonics of west coast of India. Bull. Oil Nat. Gas Comm. India 18 (2), 25–26.

Balan, K.C., Dhar, P.C., Banerjee, B., Chari, M.V.N., Aggarwal, A., Thomas, N.J., 1993. Quantitative genetic modelling of Cauvery Basin. In: Proceedings of the 2nd Seminar on Petroliferous Basins of India, vol. 1. Indian Petroleum Publishers, Dehradun, India, pp. 127–160.

Banerjee, B., Shaw, R.K., 2001. Marine magnetic surveys around Barren Island volcano, Andaman Sea. Curr. Sci. 81, 817–818.

Banerjee, B., Sengupta, B.J., Banerjee, P.K., 1995. Signals of Barremian (116 Ma) or younger oceanic crust beneath the Bay of Bengal along 140°N latitude between 810°E and 930°E. Mar. Geol. 128, 17–23.

Banerji, R.K., 1981. Cretaceous—Eocene sedimentation, tectonism and biofacies in the Bengal Basin, India. Palaeogeogr. Palaeoclimatol. Palaeoecol. 34, 57–85.

Barron, E.J., 1987. Cretaceous plate tectonic reconstructions. Palaeogeogr. Palaeoclimatol. Palaeoecol. 59, 3–29.

Barton, P.J., Owen, T.R.E., Whiif, R.S., 1990. The deep structure of the cast Oman continental margin: preliminary results and interpretation. Technophysics 173, 319–331.

Basavaraju, M.H., Govindan, A., 1997. First record of Permian palynofossils in subsurface sediments of Cauvery basin, India. J. Geol. Soc. India 50, 571–576.

Bassias, Y., 1992. Petrological and geochemical investigations of rocks form the Davie Fracture Zone (Mozambique Channel) and some tectonic implications. J. Afr. Earth Sci. 15, 321–339.

Bastia, R., 2006. An overview of Indian sedimentary basins with special focus on emerging east coast deepwater frontiers. Leading Edge 25, 818–829.

Bastia, R., 2007. Geologic Setting and Petroleum Systems of India's East Coast Offshore Basins Concept and Applications. In: Technology Publications, Dehradun, India, p. 204.

Bastia, R., 2010. Expanding exploration frontiers of India. 8th Biennial International conference and Exposition on Petroleum Geophysics, SPG Hyderabad P-419, 1–5.

Bastia, R., Nayak, P., 2006. Tectonostratigraphy and depositional patterns in Krishna Offshore Basin. Bay of Bengal, The Leading Edge 25, 839–845.

Bastia, R., Tyagi, A.K., Saxena, K., Klimentos, T., Altman, R., Alderman, S., Bahuguna, S., 2007. Evaluation of low resistivity pay deepwater turbidites using constrained thin bed analysis. In: SPE Annual Technical Conference and Exhibition Held at Anaheim, California, 11–14 November 2007 SPE 110752, 1–14.

Bastia, R., Radhakrishna, M., Srinivas, T., Nayak, S., Nathaniel, D.M., Biswal, T.K., 2010a. Structural and tectonic interpretation of geophysical data along the Eastern Continental

Margin of India with special reference to the deep water petroliferous basins. J. Asian Earth Sci. 39, 608–619.

Bastia, R., Radhakrishna, M., Suman, D., Kale, A., Cateneanu, O., 2010b. Delineation of 85°E ridge and its structure in Mahanadi offshore basin, Eastern Continental Margin of India (ECMI) from seismic reflection imaging. Mar. Petrol. Geol. 27, 1841–1848.

Bastia, Rabi, Suman, Das, Radhakrishna, M., 2010c. Pre- and post-collisional depositional history in the upper and middle Bengal fan and evaluation of deepwater reservoir potential along the northeast Continental Margin of India. Marine and Petroleum Geology 27, 2051–2061.

Bastia, R., Radhakrishna, M., Nayak, S., 2010d. Identification and Characterization of marine geohazards in the deepwater eastern offshore of India: constraints from multibeam bathymetry, side scan sonar and 3-D high resolution seismic data. Natural Hazards 57, 107–120.

Bastia, R., Reeves, C.V., Rao, D.P., D'Silva, K., Radhakrishna, M., 2010e. Paleogeographic reconstruction of East Gondwana and evolution of the Indian Continental Margin. DCS-DST Newsletter, August, 1–8.

Basu, D.N., Banerjee, A., Tamhane, D.M., 1982. Facies distribution and petroleum geology of the Bombay offshore basin, India. J. Petrol. Geol. 5 (1), 51–75.

Behera, L., Sain, K., Reddy, P.R., 2004. Evidence of underplating from seismic and gravity studies in the Mahanadi delta of eastern India and its tectonic significance. J. Geophys. Res. 109, B12311. doi:10.1029/2003JB002764.

Bell, D.W., 2002. Velocity estimation for pore-pressure prediction. In: Huffman, A.R., Bowers, G. L. (Eds.), Pressure Regimes in Sedimentary Basins and Their Prediction, AAPG Memoir, vol. 76. 177–215.

Ben-Avraham, Z., Bunce, E.T., 1977. Geophysical study of the Chagos-Laccadive Ridge, Indian Ocean. J. Geophys. Res. 82, 1295–1305.

Ben-Avraham, Z., Hartnady, C.J.H., le Roex, A.P., 1995. Neotectonic activity on continental fragments in the Southwest Indian Ocean. Agulhas Plateau and Mozambique Ridge. J. Geophys. Res. 100, 6199–6211.

Berger, P., Cemy, A., Hook, P., et al., 1983. Well Evaluation Conference India. Schlumberger, 263 pp.

Bergh, H.W., 1977. Mesozoic sea floor off Dronning Maud Land, Antarctica. Nature 269, 686–687.

Bergh, H.W., Norton, I.O., 1976. Prince Edward fracture zone and the evolution of the Mozambique Basin. J. Geophys. Res. 81, 5221–5239.

Besse, J., Courtillot, V., 1988. Paleogeographic maps of the continents bordering the Indian Ocean since the early Jurassic. J. Geophys. Res. 93, 11791–11808.

Bhattacharya, G.C., Chaubey, A.K., 2003. Western Indian ocean—a glimpse of the Tectonic Scenario. In: Sen Gupta, R., Desa, E. (Eds.), The Indian Ocean—A Perspective. Oxford-IBH, New Delhi, India, pp. 691–729.

Bhattacharya, G.C., Subrahmanyam, V., 1986. Extension of the Narmada-Son lineament on the continental margin off Saurashtra, western India as obtained from magnetic measurements. Mar. Geophys. Res. 8, 329–344.

Bhattacharya, A., Reddy, C.S.S., Srivastav, S.K., 1993. Remote sensing for active volcano monitoring in barren Island, India. In: Ninth Thematic Conference on Geologic Remote sensing, Pasadena, California 993–1003.

Bhattacharya, G.C., Chaubey, A.K., Murty, G.P.S., Srinivas, K., Sarma, K.V.L.N.S., Subrahmanyam, V., Krishna, K.S., 1994. Evidence for seafloor spreading in the Laxmi Basin, northeastern Arabian Sea. Earth Planet Sci. Lett. 125, 211–220.

Bird, D.E., 2001. Shear margins: continent—ocean transform and fracture zone boundaries. Leading Edge 20 (2), 150–159.

Bisen, M., Sriram, K., Gupte, S., 2010. Structure and tectonics of deep-water Kutch-Saurashtra area, Western India. 8th Biennial International Conference & Exposition on Petroleum Geophysics, Hyderabad 2010, 244–251.

Biswal, T.K., Sinha, S., 2004. Fold-Thrust-Belt structure of the Eastern Ghats Mobile Belt: a line of correlation between India and Antarctica in the Gondwana. Gondwana Res. 7, 43–56.

Biswas, S.K., 1974. Landscape of Kutch—a morphotectonic analysis. Ind. J. Earth Sci. 1, 177–190.

Biswas, S.K., 1980. Structure of Kutch-Kathiawar Region, W. India. In: Proceedings of the 3rd Indian Geological Congress, Pune 255–272.

Biswas, S.K., 1981. Basin framework, palaeo environment and depositional history of the Mesozoic sediments of Kutch basin, Western India. Q. J. Geol. Min. Met. Soc. India 53 (1/2), 56–85.

Biswas, S.K., 1982. Rift basins in the western margin of India and their hydrocarbon prospects. Bull. Am. Assoc. Petrol. Geol. 66, 1497–1513.

Biswas, S.K., 1987. Regional tectonic framework, structure and evolution of the western marginal basins of India. Tectonophysics 135, 307–327.

Biswas, S.K., 1991. Exploration in Bombay Offshore Basin - An overview. Petroliferous Basins India 15–25.

Biswas, S.K., 1993. Major Neotectonic events during the Quaternary in Krishna-Godavari. Curr. Sci. 64, 797–803.

Biswas, S.K., 2005. A review of structure and tectonics of Kutch basin, western India, with special reference to earthquakes. Curr. Sci. 88 (10), 1592–1600.

Biswas, S.K., Deshpande, S.V., 1983. Geology and hydrocarbon prospects of Kutch, Saurashtra, and Narmada basins. Pet. Asia J. 6, 111–126.

Biswas, S.K., Singh, N.K., 1988. Western continental margin of India and hydrocarbon potential of deep sea basins. In: 7th Offshore South Asia Conference, Singapore170–181.

Biswas, S.K., Bhasin, A.L., Jokhan, R., 1993. Classification of Indian sedimentary basins in the framework of plate tectonics. In: Biswas, S.K., Alak, D., Garg, P., Pandey, J., Maithani, A., Thomas, N.J. (Eds.), Proceedings of the 2nd Seminar on Petroliferous Basins of India, vol. 1. Indian Petroleum Publishers, Dehradun, India, pp. 1–46.

Black, L.P., Harley, S.L., Sun, S.S., McCulloch, M.T., 1987. The Rayner Complex of east Antarctica: complex isotopic systematics within a Proterozoic mobile belt. J. Metamor. Geol. 5, 1–26.

Blanford, W.T., 1867. Traps and intertrappean beds of western and central India. Mem. Geol. Surv. India 6, 155–156.

Blanford, W.T., Blanford, H.F., Theobald, W., 1856. On the geological structure and relations of the Talcheer Coal-field in the district of Cuttack. Geol. Surv. India Memoir 1, 1–98.

Bénard, F., Callot, J.-P., Vially, R., Schmitz, J., Roest, W., Patriat, M., Loubrieu, B., The Extra-Plac Team, 2010. The Kerguelen Plateau: records from a long-living/composite microcontinent. Mar. Petrol. Geol. 27, 633–649.

Bonatti, E., Hamlyn, R., 1978. Mantle uplifted block in the western Indian Ocean. Science 201, 249–251.

Bonnell, S.A., Mullee, J.E., 2000. Assessment and evaluation of deepwater, continental slope geohazards. Offshore Technology Conference, Houston. Paper no. OTC 12070, pp. 1–13.

Borissova, I., Coffin, M.F., Charvis, P., Operto, S., 2003. Structure and development of a microcontinent: Elan Bank in the southern Indian Ocean. Geochem. Geophys. Geosyst. 4 (9), 16. doi:10.1029/2003GC000535.

Bosence, D.W.J., 1997. Mesozoic rift basins of Yemen. Mar. Pet. Geol. 14, 611–616.
Bott, M.H.P., 1980. Mechanisms of subsidence at passive continental margins. In: Bally, A., Bender, P., McGetchin, T., Walcott, R. (Eds.), Dynamics of Plate Interiors. American Geophysical Union/Geological Society of America, Washington, DC/Boulder, Co, pp. 27–35.
Bott, M.H.P., 1982. Origin of the lithosphere tension causing basin formation. Philos. Trans. R. Soc. Lond. Ser. A 305, 319–324.
Bott, M.H.P., Kusznir, N.J., 1979. Stress distributions associated with compensated plateau uplift structures with application to the continental splitting mechanism. Geophys. J. R. Astron. Soc. 56, 451–459.
Bouma, A.H., 2000. Coarse-grained and fine-grained turbidite systems as end member models: applicability and dangers. Mar. Petrol. Geol. 17, 137–143.
Bouma, A.H., Stelting, C.E., Coleman, J.M., 1985. Mississippi Fan: internal structure and depositional processes. Geo-Mar. Lett. 3, 147–153.
Bowers, G.L., 1995. Pore pressure estimation from velocity data: accounting for overpressure mechanism besides undercompaction. Soc. Petrol. Eng. Drill. Complet. 4 (10), 89–95.
Bowers, G., 2001. Determining an appropriate pore pressure estimation strategy. In: Offshore Technology Conference, Houston, Texas, April 30–May 3, 2001 OTC 13042.
Bowers, G.L., 2002. Detecting high overpressure. Leading Edge 21 (2), 174–177.
Bowin, C., 1973. Origin of the Ninety East Ridge from studies near the equator. J. Geophys. Res. 78, 6029–6043.
Bradley, D.C., 2008. Passive margins through earth history. Earth Sci. Rev. 91, 1–26.
Brown, R.W., Rust, D.J., Summerfield, M.A., Gleadow, A.J., Dewitt, M.C.J., 1990. An Early Cretaceous phase of accelerated erosion on the south-western margin of Africa: evidence from apatite fission track analysis and the offshore sedimentary record. Nucl. Tracks Rad. Meas. 17, 339–350.
Brune, J.N., Priestley, K., 1989. Anomalous crustal structure beneath the Bay of Bengal and passive oceanic sedimentary basins. Proc. Indian Acad. Sci. 98, 25–30.
Brune, J.N., Singh, D.D., 1986. Continent-like crustal thickness beneath the Bay of Bengal sediments. Bull. Seismol. Soc. Am. 76, 191–204.
Brune, J.N., Curray, J., Dorman, L., Raitt, R., 1992. A proposed super-thick sedimentary basin, Bay of Bengal. Geophys. Res. Lett. 19, 565–568.
Buck, W.R., 1986. Small-scale convection induced by passive rifting: the cause for uplift of rift shoulders. Earth Planet. Sci. Lett. 77, 362–372.
Burchfiel, B.C., 1993. Tectonic evolution of the Tibetan Plateau and adjacent regions. Geol. Soc. Am. Abstr. Program. 25 (6), A-39.
Burgath, K., Von Rad, U., Van der Linden, W., Block, M., Khan, A.A., Roeser, H.A., Weiss, W., 2002. Basalt and Peridotite recovered from Murray Ridge: are they of supra-subduction origin? In: Clift, P.D., Kroon, D., Gaedicke, C., Craig, J. (Eds.), The Tectonic and Climatic Evolution of the Arabian Sea Region 195, Geol. Soc. Lond. Special Publications, pp. 117–135.
Burke, K., 1977. Aulacogenes and continental breakup. Ann. Rev. Earth Planet. Sci. 5, 371–396.
Busby-Spera, C., 1985. A sand-rich submarine fan in the lower Mesozoic Mineral King Caldera Complex, Sierra Nevada, California. J. Sedimen. Res. 55, 376–391.
Butler, R., 1995. When did India hit Asia? Nature 373, 20–21.
Campanile, D., Nambiar, C.G., Bishop, P., Widdowson, M., Brown, R., 2007. Sedimentation record I the Konkan-Kerala basin: implications for the evolution of the Western Ghats and the Western Indian Passive margin. Basin Res. doi:10.1111/j.1365-2117.2007.00341.x.

Catuneanu, O., 2005. Principles of Sequence Stratigraphy. Elsevier, Amsterdam, 375 pp.

Chand, S., Subrahmanyam, C., 2003. Rifting between India and Madagascar—mechanism and Isostasy. Earth Planet. Sci. Lett. 210, 317–332.

Chand, S., Radhakrishna, M., Subrahmanyam, C., 2001. India-East Antarctica margins: rift shear tectonic setting inferred from gravity and bathymetry data. Earth Planet Sci. Lett. 185, 225–236.

Chandra, K., Philip, P.C., Sridharan, P., Chopra, V.S., Saha, P.K., 1991. Petroleum sourtce-rock potentials of the Cretaceous transgressive-regressive sedimentary sequences of the Cauvery Basin. In: Fan, P., Li., Yu, X., Huang, Y. (Ed.). Proceedings of the 2nd Conference on Petroleum Geochemistry and Exploration in the Afro-Asian Region. J. Southeast Asian Earth Sci. 5 (1–4), 367–371.

Chandra, K., Raju, D.S.N., Mishra, P.K., 1993. Sea level changes, anoxic conditions, organic matter enrichment and petroleum source rock potential of the Cretaceous sequences of the Cauvery Basin, India. Source rocks in a Sequence Stratigraphic Framework. American Association of Petroleum Geologists Studies in Geology 37, 131–146 (AAPG, Tulsa, OK, USA).

Chandra, K., Raju, D.S.N., Bhandari, A., Mishra, C.S., 2005. Petroleum systems in the Indian sedimentary basins: stratigraphic and geochemical perspectives. In: Raju, D.S.N., Peters, J., Shanker, R., Kumar, G. (Eds.), An Overview of Litho-, Bio-, Chrono-, Sequence Stratigraphy and Sea Level Changes of Indian Sedimentary Basins. Association of Petroleum Geologists, Dehradun, India, pp. 186–203.

Chandrasekhar, D.V., Mishra, D.C., 2002. Some geodynamic aspects of Kutch basin and seismicity: an insight from gravity studies. Curr. Sci. 83, 492–498.

Chandrasekhar, D.V., Mishra, D.C., Poornachandra Rao, G.V.S., Mallikharjuna Rao, J., 2002. Gravity and magnetic signatures of volcanic plugs related to Deccan volcanism in Saurashtra, India and their physical and geochemical properties. Earth Planet. Sci. Lett. 201, 277–292.

Chandrasekharam, D., 1985. Structure and evolution of the western continental margin of India deduced from gravity, seismic, geomagnetic and geochronological studies. Phys. Earth Planet. Int. 41, 186–198.

Chandrasekharam, D., 2003. Deccan flood basalts. In: Mahadevan, T.N., Arora, B.R., Gupta, K.R. (Eds.), Indian Continental Lithosphere: Emerging Research Trends, Geol. Soc. India Memoir, vol. 53. 197–214.

Chandrasekharam, D., Santo, A.P., Capaccioni, B., Vaselli, O., Alam, M.A., Manetti, P., Tassi, F., 2009. Volcanological and petrological evolution of Barren Island (Andaman Sea, Indian Ocean). J. Asian Earth Sci. 35, 469–487.

Chari, M.V.N., Sahu, J.N., Benerjee, B., Zutshi, P.L., Chandra, K., 1995. Evolution of the Cauvery Basin, India from subsidence modeling. Mar. Pet. Geol. 12, 667–675.

Chatterjee, P.K., 1967. Geology of the main Islands of the Andaman area. In: Proc. Symp. Upper Mantle Project, Hyderabad, India 348–362.

Chatterjee, A., Mondal, S., Basu, P., Patel, B.K., 2012. Pore Pressure Prediction Using Seismic Velocities for Deepwater High Temperature- High Pressure Well in Offshore Krishna Godavari Basin, India, SPE Oil and Gas India Conference and Exhibition held in Mumbai, India, 28–30 March 2012. SPE 153764, 1–20.

Chaubey, A.K., Ramana, M.V., Sarma, K.V.L.N.S., Krishna, K.S., Murty, G.P.S., Subrahmanyam, V., Mittal, G.S., Drolia, R.K., 1991. Marine geophysical studies over the 85° E ridge, Bay of Bengal. In: First International Seminar & Exhibition on Exploration Geophysics in 1990s, held at Hyderabad. Extended Abstracts 2, India, 508–515, 25–30 November.

Chaubey, A.K., Bhattacharya, G.C., Murthy, G.P.S., Desa, M., 1993. Spreading history of the Arabian Sea: some new constraints. Mar. Geol. 112, 343–352.

Chaubey, A.K., Bhattacharya, G.C., Gopala Rao, D., 1995. Sea floor spreading magnetic anomalies in the southeastern Arabian Sea. Mar. Geol. 128, 105–114.

Chaubey, A.K., Bhattacharya, G.C., Murty, G.P.S., Srinivas, K., Ramprasad, T., Gopala Rao, D., 1998. Early Tertiary seafloor spreading magnetic anomalies and paleo-propagators in the northern Arabian Sea. Earth Planet. Sci. Lett. 154, 41–52.

Chaubey, A.K., Gopala Rao, D., Srinivas, K., Ramprasad, T., Ramana, M.V., Subrahmanyam, V., 2002a. Analyses of multichannel seismic reflection, gravity and magnetic data along a regional profile across the central-western continental margin of India. Mar. Geol. 182, 303–323.

Chaubey, A.K., Dyment, J., Bhattacharya, G.C., Royer, J.Y., Srinivas, K., Yatheesh, V., 2002b. Paleogene magnetic isochrons and paleo-propagators in the Arabian and Eastern Somali basins, northwest Indian Ocean. In: Clift, P.D., Kroon, D., Gaedicke, C., Craig, J. (Eds.), The Tectonic and Climatic Evolution of the Arabian Sea Region. Geol. Soc. London Spec. Publ. 195, 71–85.

Chauhan, O.S., Almeida, F., 1993. Influences of Holocene sea level, regional tectonics, and fluvial, gravity and slope currents induced sedimentation on the regional geomorphology of the continental slope off northwestern India. Mar. Geol. 112, 313–328.

Chopra, S., Huffman, A., 2006. Velocity determination for pore-pressure prediction. Leading Edge 25 (12), 1502–1515.

Choudhuri, M., 2011. Deep water fold-thrust belts in Krishna-Godavari basin, east coast of India (abs). In: 7th International Conference on Asian Marine Geology, 11–14 October 2011, Goa, pp. 137.

Choudhuri, M., Guha, D., Dutta, A., Sinha, S., Sinha, N., 2006. Spatio-temporal variations and kinematics of shale mobility in the Krishna-Godavari basin, India. In: AAPG Hedberg Conference, Port of Spain, Trinidad and Tobago, 5–7 June.

Chaudhuri, A., Rao, M.V., Dobriyal, J.P., Saha, G.C., Chidambaram, L., Mehta, A.K., Ramana, L.V., Murthy, K.S., 2009. Prospectivity of Cauvery basin in deep syn-rift sequences, SE India. AAPG Annual Convention and Exhibition, Denver, Colorado, USA, 7–10 June 2009.

Clift, P.D., Gaedicke, C., 2002. Acclerated mass flux to the Arabian Sea during the Middle-Late Miocene. Geology 30, 207–210.

Clift, P.D., Shimizu, N., Layne, G., Gaedicke, C., Schlüter, H.U., Clark, M., Amjad, S., 2001. Development of the Indus Fan and its significance for the erosional history of the western Himalaya and Karakoram. Geol. Soc. Am. Bull. 113, 1039–1051.

Clift, P.D., Gaedicke, C., Edwards, R., et al., 2002. The stratigraphic evolution of the Indus Fan and the history of sedimentation in the Arabian Sea. Mar. Geophys. Res. 23, 223–245.

Cloetingh, S., Wortel, J.R., 1985. Regional stress field of the Indian plate. Geophys. Res. Lett. 12, 77–80.

Cochran, J.R., 1983. Effects of finite extension times on the development of sedimentary basins. Earth Planet. Sci. Lett. 66, 289–302.

Cochran, J.R., 1990. Himalayan lift, sea level, and the record of Bengal Fan sedimentation at the ODP Leg 116 sites. In: Cochran, J.R., Stow, D.A. (Eds.), Proceedings of the Oceanic Drilling Program, Scientific Results, Leg 116. Oceanic Drilling Program, College Station, Texas 397–414.

Coffin, M.F., 1992. Subsidence of the Kerguelen Plateau: the Atlantis concept. In: Wise Jr., S.W., Schlich, R. et al., (Eds.), Proc. ODP, Sci. Results 945–949.

Coffin, M.F., Eldholm, O., 1994. Large igneous provinces: crustal structure, dimensions, and external consequences. Rev. Geophys. 32, 1–36.

Coffin, M.F., Rabinowitz, P.D., 1987. Reconstruction of Madagascar and Africa: evidence from the Davies Fracture Zone and Western Somali Basin. J. Geophys. Res. 92, 9385–9406.

Coffin, M.F., Rabinowitz, P.D., 1992. The Mesozoic East African and Madagascar conjugate continental margins: stratigraphy and tectonics. In: Watkins, J.S., Zhiqiang, F., McMillen, K.J. (Eds.), Geology and Geophysics of Continental Margins, Am. Assoc. Petrol. Geol. Mem., vol. 53. 207–246.

Coffin, M.F., et al., 2000. Proceedings of the Ocean Drilling Program, Initial Reports, vol. 183. Ocean Drilling Program, College Station, TX.

Coffin, M., Pringle, M.S., Duncan, R.A., Gladczenko, T.P., Storey, M., Müller, R.D., Gahagan, L.A., 2002. Kerguelen Hotspot magma output since 130 Ma. J. Petrol. 43, 1121–1139.

Coleman, J.M., Prior, D.B., Lindsay, J.F., 1983. Deltaic influences on shelf-edge instability processes. In: Stanely, D.J., Moore, G.T. (Eds.), The Shelfbreak: Critical Interface on Continental Margins, 121–137 SEPM Special Publication No. 33, Tulsa.

Collier, J.S., Sansom, V., Ishizuka, O., Taylor, R.N., Minshull, T.A., Whitmarsh, R.B., 2008. Age of Seychelles–India break-up. Earth Planet. Sci. Lett. 272, 264–277.

Colombo, D., De Stefano, S., 2007. Geophysical modeling via simultaneous joint inversion of seismic, gravity and electromagnetic data: application to pre-stack depth imaging. Leading Edge 26 (3), 326–331.

Courtillot, V., Besse, J., Vandamme, D., Montigny, R., Jaeger, J.J., Cappetta, H., 1986. Deccan flood basalts at the Cretaceous/Tertiary boundary? Earth Planet. Sci. Lett. 80, 361–374.

Courtillot, V., Feraud, G., Maluski, H., Vandamme, D., Moreau, M.G., Besse, J., 1988. Deccan flood basalts and the cretaceous/tertiary boundary. Nature 333, 843–846.

Courtillot, V., Jaupart, C., Manighetti, I., Tapponnier, P., Besse, J., 1999. On causal links between flood basalts and continental breakup. Earth Planet. Sci. Lett. 166 (3–4), 177–195. doi:10.1016/S0012-821X(98) 00282-9.

Cox, K.G., 1988. Inaugural address. In: Subbarao, K.V. (Ed.), Deccan Flood Basalts, Geol. Soc. India Mem. 10, 15–22.

Cox, K.G., 1989. The role of mantle plumes in the development of continental drainage patterns. Nature 342, 873–877.

Curray, J.R., 1991. Possible greenschist metamorphism at the base of a 22-km sedimentary section, Bay of Bengal. Geology 19, 1097–1100.

Curray, J.R., 1994. Sediment volume and mass beneath the Bay of Bengal. Earth Planet. Sci. Lett. 125, 371–383.

Curray, J.R., 2005. Tectonics and history of the Andaman Sea region. J. Asian Earth Sci. 25, 187–232.

Curray, J.R., Moore, D.G., 1971. Growth of the Bengal deep sea fan and denudation in the Himalayas. Geol. Soc. Am. Bull. 82, 563–573.

Curray, J.R., Moore, G.D., 1974. Sedimentary and tectonic processes in the Bengal Deep Sea Fan and Geosyncline. In: Burke, C.A., Drake, C.L. (Eds.), The Geology of Continental Margin. Springer-Verlag, New York, pp. 617–627.

Curray, J.R., Munasinghe, T., 1989. Timing of Intraplate deformation, northeastern Indian Ocean. Earth Planet. Sci. Lett. 94 (1–2), 71–77.

Curray, J.R., Munasinghe, T., 1991. Origin of the Rajmahal Traps and the 85°E Ridge: preliminary reconstructions of the trace of the Crozet hotspot. Geology 19, 1237–1240.

Curray, J.R., Moore, D.G., Lawver, L.A., Emmel, F.J., Raitt, R.W., Henry, M., Kieckheffer, R., 1979. Tectonics of the Andaman Sea and Burma. In: Watkins, J.S., Montaderi, L., Dickenson, P.W. (Eds.), Geological and Geophysical Investigations of Continental Margins. American Association of Petroleum Geologists, Tulsa, pp. 189–198 Memoir 29.

Curray, J.R., Emmel, F.J., Moore, D.G., Raitt, R.W., 1982. Structure, tectonics and geological history of the northeastern Indian Ocean. In: Nairn, A.E.M., Stehli, F.G. (Eds.), The Ocean Basins and Margins 6. The Indian Ocean. Plenum Press, New York, pp. 399–450.

Curray, J.R., Emmel, F.J., Moore, D.G., 2003. The Bengal fan: morphology, geometry, stratigraphy, history and processes. Mar. Petrol. Geol. 19, 1191–1223.

Damuth, J.E., Flood, R.D., 1985. Amazon Fan, Atlantic Ocean. In: Bouma, A.H., Normark, W.R., Barnes, N.E. (Eds.), Submarine Fans and Related Turbidite Systems: Frontiers in Sedimentary Geology. Springer, New York, pp. 97–106.

Dangwal, V., Sengupta, S., Desai, A.G., 2008. Speculated petroleum systems in Deep Offshore Mahanadi Basin in Bay of Bengal, India. In: International Conference & Exposition on Petroleum Geophysics, Hyderabad, pp. 1e7.

Das, S.R., Ray, A.K., 1976. Fracture pattern, hot springs and lineaments of the west coast. Geol. Surv. India, 125 Year News Letter, pp. 1–6.

Das sharma, S., Chakrabharti, M.K., Bisht, S.S., 1993. Shallow marine area, Bengal Basin—the site of an ancient fan. In: Biswas, S.K. et al., (Ed.), Proceedings of the 2nd Seminar on Petroliferous Basins of India, vol. 1. 61–82.

Dasgupta, S., Mukhopadhyay, M., 1993. Seismicity and plate deformation below the Andaman arc, northeastern Indian Ocean. Tectonophysics 225, 529–542.

Dasgupta, S., Mukhopadhyay, M., 1997. Aseismicity of the Andaman subduction zone and recent volcanism. J. Geol. Soc. India 49, 513–521.

Dasgupta, S., Mukhopadhyay, M., Bhattacharya, A., Jana, T.K., 2003. The geometry of the Burmese–Andaman subducting lithosphere. J. Seismol. 7, 155–174.

Dasgupta, S., Mukhopadhyay, B., Bhattacharya, A., 2007. Seismicity pattern in north Sumatra–Great Nicobar region: in search for precursor for the 26 December 2004 earthquake. J. Earth Syst. Sci. 116, 215–223.

Datta, A.K., Bedi, T.S., 1968. Faunal aspect and the evolution of the Cauvery basin. Mem. Geol. Soc. Ind. 2, 168–177.

Davis, E.E., Lister, C.R.B., 1974. Fundamental of ridge crest topography. Earth Planet. Sci. Lett. 21, 405–413.

Davison, I., 1997. Wide and narrow margins of the Brazilian South Atlantic. J. Geol. Soc. Lond. 154, 471–476.

Day, K., 2002. Deepwater Canyon slope stability. Offshore site investigation and geotechnics 'Diversity and Sustainability'. In: Proceedings of an International Conference, 26–28 November 2002, London.

Deepak, V., Tyagi, A.K., Singh, A., Vidya, M., Pandey, V., Saxena, K., 2012. A New Petrophysical Interpretation Approach To Characterize The Thinly Laminated Reservoir Using Conventional Tools, SPE Oil and Gas India Conference and Exhibition, 28–30 March 2012, Mumbai, India, MS-152912.

Dehadrai, P.V., 1970. An outline of the stratigraphy and structure of the sedimentary basins in India. In: U.N. ECAFE Mineral Resources Development Series No. 36, pp. 93–94.

DeMets, C., Gordon, R.G., Argus, D.F., Stein, S., 1990. Current plate motions. Geophys. J. Int. 101, 425–478.

DeMets, C., Gordon, R.G., Argus, D.F., Stein, S., 1994a. Effect of recent revisions to the geomagnetic reversal timescale. Geophys. Res. Lett. 21, 2191–2194.

DeMets, C., Gordon, R.G., Vogt, P., 1994b. Location of the Africa–Australia–India triple junction and motion between the Australian and Indian plates: results from an aeromagnetic investigation of the Central Indian and Carlsberg ridges. Geophys. J. Int. 119, 893–930.

DeMets, C., Gordon, R.G., Argus, D.F., 2010. Geologically current plate motions. Geophys. J. Int. 181, 1–80.

Deptuck, M., Steffens, G.S., Barton, M., Pirmez, C., 2003. Architecture of upper channel-belts on the Niger Delta slope and in the Arabian Sea. Mar. Petrol. Geol. 20, 649–676.

Desa, M., Ramana, M.V., Ramprasad, T., 2006. Seafloor spreading magnetic anomalies south off Sri Lanka. Mar. Geol. 229, 227–240.

Deshmukh, S.S., 1982. Volcanological and petrological appraisal on Deccan basalts. Science Lecture Series, vol. 1. Gondwana Geological Society, India, pp. 30–56.

Dessai, A.G., Bertrand, H., 1995. The 'Panvel Flexure' along the western Indian continental margin: an extensional fault structure related to Deccan magmatism. Tectonophysics 241, 165–178.

Detrick Jr., R.S., Purdy, G.M., 1980. The crustal structure of the Kane fracture zone from seismic refraction studies. J. Geophys. Res. 85, 3759–3777.

Devey, C.W., Lightfoot, P.C., 1986. Volcanological and tectonic context of stratigraphy and structure in the Western Deccan Traps. Bull. Volcanol. 43, 125–126.

Dewey, J.F., Burke, K., 1974. Hotspots and continental break-up: implications for collisional orogeny. Geology 2, 57–60.

Dewey, J.F., Shackleton, R.M., Chang, C., Sun, Y., 1988. The tectonic evolution of the Tibetan Plateau. Philos. Trans. R. Soc. Lond. 327, 379–413.

Dewey, J.F., Cande, S., Pitman III, W.C., 1989. Tectonic evolution of the India/Eurasia collision zone. Eclogae Geol. Helv. 82, 717–734.

DGH, 2010. Petroleum exploration & production activities, India annual report 2008–2009. Directorate General of Hydrocarbons Publication, p. 124.

DGH, 2011. Hydrocarbon exploration & production activities, India annual report 2009–2010. DGH Publication.

Dhar, P.C., Bhattacharya, S.K., 1993. Status of exploration in the Cambay Basin. In: Biswas, S.K. (Ed.), Proceedings of the Second Seminar on Petroliferous Basins of India, vol. 2. Indian Petroleum Publication, Dehradun, pp. 1–33.

Dick, H.J.B., Lin, J., Schouten, H., 2003. An ultraslow-spreading class of Ocean Ridge. Nature 426, 405–412.

Dickinson, W.R., 1974. Plate tectonics and sedimentation. In: Dickinson, W.R. (Ed.), Tectonics and Sedimentations Soc. Pet. Eng., Spec. Publ. 22, 1–27.

Dickinson, W.R., 1976. Plate tectonic evolution of sedimentary basins: American Association of Petroleum Geologists. Continuing Education Course Notes Series 1.

Dirghangi, R.S., Kothari, V., Waraich, R.S., Baruah, R.M., Lal, N.K., Zutshi, P.L., 2000. 3rd Conference and Exposition on Petroleum Geophysics, New Delhi, India, pp. 278–283.

Dixit, M.M., Satyavani, N., Sarkar, D., Khare, P., Reddy, P.R., 2000. Velocity inversion in the Lodhika well area, Saurashtra peninsula, western India. First Break 18, 499–504.

Drake, C.L., Girdler, R.W., 1964. A geophysical study of the Red Sea. Geophys. J. R. Astron. Soc. 8, 473–495.

Droz, L., Bellaiche, G., 1991. Seismic facies and geologic evolution of the central portion of the Indus Fan. In: Weimer, P., Link, M.H. (Eds.), Seismic Facies and Sedimentary Processes of Submarine Fans and Turbidite Systems. Springer Verlag, Heidelberg, pp. 383–402.

Droz, L., Mougenot, D., 1987. Mozambique upper fan origin of depositional units. AAPG Bull. 71 (11), 1355–1365.

Dudley, D.R., Claypool, G.E., 1981. Generation, accumulation, and resource potential of biogenic gas. AAPG Bull. 65 (1), 5–25 January 1981.

Duncan, R.A., 1978. Geochronology of basalts from the Ninetyeast Ridge and continental dispersion in the Eastern Indian Ocean. J. Volcanol. Geotherm. Res. 4, 283–305.

Duncan, R.A., 1990. The volcanic record of the Reunion hot spot. Proc. Ocean. Drill. Prog. Results 5, 3–10.

Duncan, R.A., Pyle, D.G., 1988. Rapid eruption of the Deccan flood basalts at the Cretaceous/Tertiary boundary. Nature 333, 841–843.

Duncan, R.A., Storey, M., 1992. The life cycle of Indian Ocean hotspots. In: Duncan, R.A. (Ed.), Synthesis of Results from Scientific Drilling in the Indian Ocean. Geophys. Monogr. Ser., vol. 70. AGU, Washington, DC, pp. 91–104.

Duncan, R.A., Rea, D.K., Kidd, R.B., Rad, U.V., Weissel, J.K., 1992. Synthesis of results from Scientific Drilling in the Indian Ocean. Geophysical Monograph 70. American Geophysical Union 225–244.

Dykstra, M., 2005. Dynamics of sediment mass-transport from the shelf to the deep sea. Ph.D. Thesis. University of California, Santa Barbara, 152 pp.

Dykstra, M., Kneller, B., 2002. Internal architecture, external geometry, and cyclicity of mass transport complexes in the Gulf of Mexico from 3-D seismic data. In: AGU Annual Conference, San Francisco, USA.

Dyment, J., 1998. Evolution of the Carlsberg Ridge between 60 and 45 Ma: ridge propagation, spreading asymmetry, and the Deccan-Reunion hotspot. J. Geophys. Res. 103B, 24067–24084.

Eaton, B.A., 1972. Graphical method predicts geopressure worldwide. World Oil 182, 51–56.

Ebrom, D., Heppard, P., Thomsen, L., Mueller, M., Harrold, T., Phillip, L., Watson, P., 2004. Effective stress and minimum velocity trends (abs.). Society of Exploration Geophysicists Technical Program Expanded Abstracts, pp. 1615–1618.

Edwards, R.A., Minshull, T.A., White, R.S., 2000. Extension across the Indian–Arabian plate boundary: the Murray Ridge. Geophys. J. Int. 142, 461–477.

Eguchi, T., Uyeda, S., Maki, T., 1979. Seismotectonics and tectonic history of the Andaman sea. Tectonophysics 57, 35–51.

Elliott, G.M., Berndt, C., Parson, L.M., 2009. The SW African volcanic rifted margin and the initiation of the Walvis Ridge, South Atlantic. Mar. Geophys. Res. 30, 207–214.

Ellouz-Zimmermann, N., Lallemant, S.J., Castilla, R., Mouchot, N., Leturmy, P., Battani, A., Buret, C., Cherel, L., Desaubliaux, G., Deville, E., Ferrand, J., Lügcke, A., Mahieux, G., Mascle, G., Mühr, P., Pierson-Wickmann, A.-C., Robion, P., Schmitz, J., Danish, M., Hasany, S., Shahzad, A., Tabreez, A., 2007. Offshore frontal part of the Makran accretionary prism: the Chamak Survey (Pakistan). In: Brun, J.P., Oncken, O., Weissert, H., Dullo, C. (Eds.), Frontiers in Earth Sciences, Thrust Belts and Fore Land Basins, 351–366, Springer Publishers, Chapter 18.

Embleton, B.J.J., McElhinney, M.W., 1975. The palaeo position of Madagascar: palaeomagnetic evidence from the Isalo group. Earth Planet. Sci. Lett. 27, 329–341.

Emmel, F.J., Curray, J.R., 1985. Bengal Fan, Indian Ocean. In: Bouma, A.H., Normark, W.R., Barnes, N.E. (Eds.), Submarine Fans and Related Turbidite Systems. Springer, New York, pp. 107–112.

Eremenko, N.A., Negi, B.S., 1968. Tectonic Map of India, 1:20,00,000 Scale. ONGC, Dehradun, India.

Eremenko, N.A., Sastri, V.V., 1977. On the petroleum geology of Andaman Islands. Bull. ONGC 14, 1–13.

Eremenko, N.A., Negi, B.S., Kasianov, M.V., et al., 1969. Tectonic map of India. Bull. Oil Nat. Gas Comm. India 6, 1–11.

Erlank, A.J., Reid, D.C., 1974. Geochemistry, mineralogy and petrology of basalts, Leg 25 DSDP. In: Schlick, R., Simpson, E.S.W. et al., (Eds.), Initial Reports of Deep Sea Drilling Project, Leg 25. U.S. Govt. Printing Office, Washington, DC.

Fainstein, R., Kalra, R., 2010. Regional seismic imaging link deepwater reservoirs offshore South/Southeast Asia. SEG Workshop, Kuala Lumpur.

Fainstein, R., Kalra, R., Prasad, G.K., Chandrashekhar, S., Rao, C.V., 2012. Modern Sub-Basalt Seismic Imaging-Deepwater Realm Offshore Southwest India, 9th Biennial International conference and Exposition on Petroleum Geophysics. SPG Hyderabad P-241, 1–7.

Fainstein, R., Traylen, S.F., Vasilyev, P., Stempel, A., 2009. Time/depth imaging offshore India, SEG Houston. International Exposition and Annual Meeting 2009, 1039–1043.

Falvey, D.A., 1974. The development of continental margins in plate tectonic theory. APEA J. 14, 95–106.

Fedorov, L.V., Ravich, M.G., Hofmann, J., 1982. Geologic comparison of southeastern Peninsular India and Sri Lanka with a part of East Antarctica (Enderby Land, MacRobertson Land, and Princess Elizabeth Land). In: Craddock, C. (Ed.), Antarctic Geoscience. University of Wisconsin Press, Madison, pp. 73–78.

Fertl, W.H., 1976. Abnormal Formation Pressures. Elsevier, Amsterdam, 382 pp.

Fisher, R.L., Sclater, J.G., McKenzie, D.P., 1971. Evolution of the Central Indian Ridge, Western Indian Ocean. Geol. Soc. Am. Bull. 82, 553–562.

Fisher, R.L., Bunce, E.T., Cernock, P.J., Clegg, D.C., Cronan, D.S., Diamani, V., Dmitriev, L.V., Kinsman, D.J.J., Roth, P.H., Thiede, J., Vincent, E., 1974. Initial Reports of the Deep Sea Drilling Project, vol. 24. U.S. Government Printing Office, Washington, DC, 1–1183.

Fleitout, L., Dalloubeix, C., Moriceau, C., 1989. Small-wavelength geoid and topography anomalies in the South-Atlantic Ocean: a clue to new hotspot tracks and lithospheric deformation. Geophys. Res. Lett. 16, 637–640.

Flood, R.D., Piper, D.J.W., 1997. Amazon Fan sedimentation: the relationship to equatorial climate change, continental denudation, and sea-level fluctuations. In: Flood, R.D., Piper, D.J.W., Klaus, A., Peterson, L.C. (Eds.), Proceedings of the Ocean Drilling Program, Scientific Results, Leg 155. Ocean Drilling Program, College Station, TX, pp. 653–675.

Flood, R.D., Manley, P.L., Kowsmann, R.O., Appi, C.J., Primez, C., 1991. Seismic facies and late quaternary growth of Amazon submarine fan. In: Weimer, P., Link, M.H. (Eds.), Seismic Facies and Sedimentary Processes of Submarine Fans and Turbidite Fans and Turbidite Systems. Springer, New York, pp. 415–433.

Flood, R.D., Piper, D.J.W., Klaus, A., 1995. Proceedings Ocean drilling Program, Initial Reports. Ocean Drilling Program, College Station, TX.

Forsyth, D., Uyeda, S., 1975. On the relative importance of the driving forces of plate motion. Geophys. J. R. Astron. Soc. 43, 163–200.

Fournier, M., Patriat, P., Leroy, S., 2001. Reappraisal of the Arabia–India–Somalia triple junction kinematics. Earth Planet. Sci. Lett. 189, 103–114.

Fowler, C.M.R., 1990. Solid earth: an introduction to global geophysics. Cambridge University Press, Cambridge, U.K., p. 472.

Fowler, S., Mckenzie, D.P., 1989. Gravity studies of the Rockall and Exmouth plateau using SEASAT altimetry. Basin Res. 1, 27–34.

Fowler, S.R., White, R.S., Spence, G.D., Westbrook, G.K., 1989. The Hatton Bank continental margin. II. Deep structure from two-ship expanding spread seismic profiles. Geophys. J. 96, 295–309.

Francis, T.J.G., Raitt, R.W., 1967. Seismic refraction measurements in the southern Indian Ocean. Geophys. Res. 72, 3015–3041.

Francis, T.J.G., Shor, G.G., 1966. Seismic refraction measurements in the northwest Indian Ocean. J. Geophys. Res. 71, 427–449.

Franke, D., Schnabel, M., Ladage, S., Tappin, D.R., Neben, S., Djajadihardja, Y.S., Müller, C., Kopp, H., Gaedicke, C., 2008. The great Sumatra-Andaman earthquakes—imaging the boundary between the ruptures of the great 2004 and 2005 earthquakes. Earth Planet. Sci. Lett. 269, 118–130. doi:10.1016/j.epsl.2008.01.047.

Frey, F.A., Coffin, M.F., Wallace, P.J., Quilty, P.G., 1996. Leg 183 synthesis: Kerguelen plateau-Broken ridge-a large igneous province. In: Proceedings of the Ocean Drilling Program. Scientific Results, vol. 183. 1–101.

Fruehn, J., White, R.S., Minshull, T.A., 1997. Integral deformation and compaction of the Makran accretionary wedge. Terra Nova 9 (3), 101–104.

Fullerton, L.G., Sager, W.W., Handschumacher, D.W., 1989. Late Jurassic-Early Cretaceous evolution of the eastern Indian Ocean adjacent to Northwest Australia. J. Geophys. Res. 94, 2937–2953.

Fuloria, R.C., 1993. Geology and hydrocarbon prospects of Mahanadi Basin, India. In: Biswas, S. K. et al., (Ed.), Proceedings of the 2nd Seminar on Petroliferous Basins of India, vol. 1. Indian Petroleum Publishers, Dehradun, India, pp. 355–369.

Fuloria, R.C., Pandey, R.N., Bharali, B.R., Mishra, J.K., 1992. Stratigraphy, structure and tectonics of Mahanadi offshore Basin. In: Recent Geoscientific Studies in the Bay of Bengal and the Andaman Sea. J. Geol. Soc. India 29, 255–265 (special publication).

Gaedicke, C., Schlüter, H.U., Roeser, H.A., Prexl, A., Schreckenberger, B., Meyer, H., Reichert, C., Clift, P., Amjad, S., 2002. Origin of the northern Indus Fan and Murray Ridge, Northern Arabian Sea: interpretation from seismic and magnetic imaging. Tectonophysics 355, 127–143.

Gaina, C., Mueller, R.D., Brown, B., Ishihara, T., 2003. Microcontinent formation around Australia. Geol. Soc. Aust. 22, 399–410 Special publication.

Gaina, C., Müller, R.D., Brown, B., Ishihara, T., Ivanov, S., 2007. Breakup and early seafloor spreading between India and Antarctica. Geophys. J. Int. 170, 151–169.

Gallagher, K., Brown, R., 1997. The onshore record of passive margin evolution. J. Geol. Soc. Lond. 54, 451–457.

Gallagher, K., Brown, R., Johnson, C., 1998. Fission track analysis and its applications to geological problems. Annu. Rev. Earth Planet. Sci. 26, 519–572.

Gallardo, L.A., Meju, M.A., 2007. Joint 2D cross-gradient imaging of magnetotelluric and seismic travel-time data for structural and lithological classification. Geophys. J. Int. 169, 1261–1272. doi:10.1111/j.1365-246X. 2007.03366.x.

Gansser, A., 1964. Geology of the Himalayas. Interscience, London, 289 pp.

Gansser, A., 1981. The geodynamic history of the Himalaya. In: Gupta, H.K., Delany, F.M. (Eds.), Zagros, Hindu Kush, Himalayan Geodynamic Evolution. Geodyn. Ser., vol. 3. AGU, Washington, DC, pp. 111–121.

Gaur, V.K., Rai, S.S., Reddy, K.S.N., Rao, C.V.R., Ramesh, D.S., Nagesh, D.S., Suryaprakasam, K., 1989. Geotomography: deep structural study of Deccan volcanic province. Annual report of NGRI 1988–89, 30–31.

Gersonde, R., Spieß, V., Flores, J.A., Hagen, R.A., Kuhn, G., 1998. The sediments of Gunnerus Ridge and Kainan Maru Seamount (Indian sector of Southern Ocean). Deep Sea Res. I 45, 1515–1540.

Ghosal, D., 2008. Evaluation of crustal structure beneath the central part of the Bay of Bengal from surface wave dispersion studies. In: 7th International Conference and Exposition on Petroleum Geophysics. Society of Petroleum Geophysicists, Hyderabad, pp. 250–257.

Ghosh, B.N., Zutshi, P.L., 1989. Indian west coast shelf break tectonic features. Geol. Surv. Ind. Spl. Pub. 24, 309–318.

Gilchrist, A.R., Summerfield, M.A., 1990. Differential denudation and flexural isostasy in the formation of rifted margin upwarps. Nature 346, 739–742.

Gilchrist, A.R., Summerfield, M.A., 1991. Denudation, isostasy and landscape evolution. Earth Surf. Processes Landforms 16, 555–562.

Gilchrist, A.R., Kooi, H., Beaumont, C., 1994. The post-Gondwana geomorphic evolution of southwestern Africa: implications for the controls on landscape development from observations and numerical experiments. J. Geophys. Res. 99, 12211–12228.

Gohl, K., 2008. Antarctica's continent-ocean transitions: consequences for tectonic reconstructions, and the 10th ISAES Editorial Team. In: Cooper, A.K., Barrett, P.J., Stagg, H.,

Storey, B., Stump, E., Wise, W. (Eds.), Antarctica: A Keystone in a Changing World. Proceedings of the 10th International Symposium on Antarctic Earth Sciences. The National Academies Press, Washington, DC. doi:10.3133/of2007-1047.

Govindan, A., 1982. Imprints of Global 'Cretaceous anoxic events' in the East Coast Basins of India and their Implications. Oil and Natural Gas Commission Bulletin 19, 257–270.

Golynsky, A.V., Alyavdin, S.V., Masolov, V.N., Tscherinov, A.S., Volnukhin, V.S., 2002. The composite magnetic anomaly map of the East Antarctica. Tectonophysics 347, 109–120.

Gombos, A.M., Powell, W.G., Norton, I.O., 1995. The tectonic evolution of western India and its impacts on hydrocarbon occurrences: an overview. Sediment. Geol. 96, 119–129.

Gopala Rao, D., Bhattacharaya, G.C., Subba Raju, L.V., Ramana, M.V., Subrahmanyam, V., Raju, K.A.K., Ram Prasad, T., Chaubey, A.K., 1987. Regional marine geophysical studies of the south western continental margin of India. In: Rao, T.S.S., Natarajan, R., Desai, B.N., Swamy, G.N., Bhat, S.R. (Eds.), Contributions in Marine Sciences. NIO, Goa, pp. 427–437.

Gopala Rao, D., Krishna, K.S., Sar, D., 1997. Crustal evolution and sedimentation history of the Bay of Bengal since the Cretaceous. J. Geophys. Res. 102, 17747–17768.

Gopala Rao, D., Paropkari, A.L., Krishna, K.S., Chaubey, A.K., Ajay, K.K., Kodagali, V.N., 2010. Bathymetric highs in the mid-slope region of the western continental margin of India—structure and mode of origin. Mar. Geol. 276 (1–4), 58–70.

Gordon, R.G., DeMets, C., 1989. Present-day motion along the Owen Fracture Zone and Dalrymple Trough in the Arabian Sea. J. Geophys. Res. 94, 5560–5570.

Gordon, R.G., DeMets, C., Argus, D.F., 1990. Kinematic constraints on distributed lithospheric deformation in the equatorial Indian Ocean from present motion between the Australian and Indian plates. Tectonics 9, 409–422.

Gorain, S., 2012. Mesozoic Prospectivity of Kerala Konkan Offshore Basin. 9th Biennial International conference and Exposition on Petroleum Geophysics. SPG Hyderabad P-117, 1–6.

Goslin, J., Recq, M., Schlich, R., 1981. Structure profonde du plateau de Madagascar: relations avec le plateau de Crozet. Tectonophysics 76, 75–97.

Govindan, A., 1984. Stratigraphy and sedimentation of East Godavari subbasin, Andhra Pradesh. In: Bhandari, L.L., Venkatachala, B.S., Mitra, P., Kumar, R.K., Srivatava, D.C., Nanjunda Swamy, S. (Eds.), Petroliferous Basins of India. Petrol. Asia J. 7 (1), 132–146. Himachal Times Group Publication, Dehradun, India.

Govindan, A., 1993. Cretaceous anoxic events, sea level changes and microfauna in Cauvery Basin, India. In: Biswas, S.K. et al., (Ed.), Proceedings of the 2nd Seminar on Petroliferous Basin of India, vol. 1. Indian Petroleum Publishers, Dehradun, India, pp. 161–176.

Govindan, A., Ravindran, C.N., 1996. Cretaceous biostratigraphy and sedimentation history of Cauvery Basin, India. In: XV Indian Colloquium Micropal. Strat., Dehradun19–31.

Graue, K., 2000. Mud volcanoes in deep water Nigeria. Mar. Petrol. Geol. 17, 959–974.

Green, A.G., 1972. Seafloor spreading in the Mozambique channel. Nat. Phys. Sci. 236, 19–21.

Gunnell, Y., Fleitout, L., 1998. Shoulder uplift of the Western Ghats Passive margin, India: a denudational model. Earth surf. Processes Landforms 23, 391–404.

Gunnell, Y., Fleitout, L., 2000. Morphotectonic evolution of the Western Ghats, India. In: Summerfield, M.A. (Ed.), Geomorphology and Global Tectonics. John Wiley & Sons, Chichester; New York, pp. 321–338.

Gunnell, Y., Radhakrishna, B.P., 2001. Sahyadri, the great escarpment of the Indian subcontinent. Mem. Geol. Soc. India 47, 717.

Gupta, S.K., 2006. Basin architecture and petroleum system of Krishna Godavari basin, east coast of India. Leading Edge 25, 830–837.

Gupta, R.P., Husain, R., Maurya, S.N., Lal, N.K., 2001. Gondwana sediments in Godavari Offshore—implications for tectonostratigraphic evolution and hydrocarbon prospectivity of east coast of India. Gondwana Res. 4, 625.

Guzman-Speziale, M., Ni, J.F., 1996. Seismicity and active tectonics of the Western Sunda Arc. In: Yin, A., Mark Harrison, T. (Eds.), The Tectonic Evolution of Asia, 63–84.

Ham, H.H., 1966. A method of estimating formation pressures from Gulf coast Well logs: Gulf Coast. Assoc. Geol. Soc. Trans. 16, 185–197.

Hamblin, K.W., Christiansen, E.H., 2003. Earths Dynamic Systems. Prentice Hall, 10th Edition, pp. 816.

Hamilton, W., 1979. Tectonics of the Indonesian Region. U.S.G.S. Prof. Paper 1078, 345 pp.

Haq, B.U., Hardenbol, J., Vail, P.R., 1987. Chronology of fluctuating sea levels since the Triassic. Science 235, 1156–1167.

Hardas, M.G., 1993. Depositional pattern of Tatipaka-Pasarlapudi sands. In: Proceedings of the 2nd Seminar on Petroliferous Basins of India, vol. 1. 255–290.

Harinarayana, T., 1999. Combination of EM and DC measurements for upper crustal studies. Surveys Geophys. 20 (3–4), 257–278.

Hayes, D.E., Frakes, L.A., et al., 1975. Init. Repts. DSDP 28, pp. 1017.

Hübscher, C., Spiess, V., Breitzke, M., Weber, M.E., 1997. The youngest channel-levee system of the Bengal Fan: results from digital sediment echosounder data. Mar. Geol. 141, 125–145.

Heezen, B.C., Tharp, M., 1964. Physiographic Diagram of the Indian Ocean, the Red Sea, the South China Sea, the Sulu Sea, and the Celebes Sea. Geological Society of America, New York.

Heezen, B.C., Tharp, Marie, 1966. Physiography of the Indian Ocean. Philos. Trans. R. Soc. Lond. A259, 137.

Heggland, R., 2004. Hydrocarbon migration and accumulation above salt domes—risking of prospects by the use of gas chimneys. Presented at the 24th Annual Gulf Coast Section, SEPM.

Heirtzler, J.R., Burroughs, R.H., 1971. Madagascar's palaeoposition: new data from the Mozambique Channel. Science 174, 488–490.

Hofmann, C., Féraud, G., Courtillot, V., 2000. 40Ar/39Ar dating of mineral separates and whole rocks from the Western Ghats lava pile: further constraints on duration and age of the Deccan Traps. Earth Planet. Sci. Lett. 180, 13–27.

Holbrook, W.S., Reiter, E.C., Purdy, G.M., Sawyer, D., Stoffa, P.L., Austin, J.A., Oh, J., Makris, J. J., 1994. Deep structure of the U.S. Atlantic continental margin, offshore South Carolina, from coincident ocean bottom and multichannel seismic data. J. Geophys. Res. 99, 9155–9178.

Holmes, Watkins, D.K., 1992. Synthesis of Results from Scientific Drilling in the Indian Ocean.

Holt, W.E., Stern, T.A., 1991. Sediment loading on the western margin of the New Zealand Continent: implications for the strength of young continental lithosphere. Earth Planet. Sci. Lett. 107, 523–538.

Hopper, J.R., Funck, T., Tucholke, B.E., Louden, K.E., Holbrook, W.S., Larsen, H.C., 2006. A deep seismic investigation of the Flemish Cap margin: implications for the origin of deep reflectivity and evidence for asymmetric breakup between Newfoundland and Iberia. Geophys. J. Int. 164, 501–515.

Horsfield, S.J., Whitmarsh, R.B., White, R.S., Sibuet, J.C., 1993. Crustal structure of the Goban Spur rifted continental margin NE Atlantic. Geophys. J. Int. 119, 1–19.

Hottman, C.E., Johnson, R.K., 1965. Formation pressure from log-derived shale properties. Paper (SPE1110). Presented at 40th Annual SPE Meeting, Denver.

Hovland, M., Judd, A.G., 1988. Seabed Pockmarks and Seepages, Impact on Geology, Biology and the Marine Environment. Graham and Trotman, London, p 293.

Hubbert, M.K., Rubey, W.W., 1959. Role of fluid pressure in the mechanics of overthrust faulting. I: mechanics of fluid filled porous solids and its applications to overthrust faulting. Geol. Soc. Am. Bull. 70, 115–166.

Hubscher, C., Spieb, V., Breitzke, M., Weber, M.E., 1997. The youngest channel-levee system of the Bengal Fan: results from digital sediment echosounder data. Mar. Geol. 141, 125–145.

Hurst, A., Fraser, A.J., Fraser, S.I., Hadler-Jacobsen, F., 2005. Deep-water clastic reservoirs: a leading global plays in terms of reserve replacement and technical challenges. In: Dore, A.G., Vining, B. (Eds.), Petroleum Geology: North-West Europe and Global Perspective. Proceedings of the 6th Petroleum Geology Conference. The Geological Society, London, pp. 1111–1120.

Husain, R., Mitra, T., Gupta, R.P., Trivedi, K.B., Ghosh, A., Baghel, S.P.S., 2000. Tectono-stratigraphy of the East coast of India vis-a-vis development of cretaceous petroleum systems. In: Petroleum Geochemistry & Exploration in the Afri-Asian Region, 5th Intl. Conf & Exbn. 5–27 November.

Ingle, S., Weis, D., Frey, F.A., 2002. Indian continental crust recovered from Elan Bank, Kerguelen Plateau (ODP Leg 183, Site 1137). J. Petrol. 43, 1241–1257.

Isacks, B., Oliver, J., Sykes, L.R., 1968. Seismology and the new global tectonics. J. Geophys. Res. 73, 5855–5899.

Jagannathan, C.R., Ratnam, C., Baishya, N.C., Dasgupta, U., 1983. Geology of the offshore Mahanadi Basin. In: Bhandari, L.L. et al., (Ed.), Petroliferous Basins of India. KDM Institute of Petroleum Exploration, Dehradun, India, pp. 101–104.

Janardhan, A.S., 1999. Southern granulite terrain, south of the Palghat-Cauvery shear zone: implications for India-Madagascar connection. Gondwana Res. 2, 463–469.

Jenkins, S.I., 1992. Mesozoic to recent sedimentary facies development of the Seychelles marginal basin. In: Plummet, Ph.S. (Ed.), First Indian Ocean Petroleum Seminar. Proc. UN Sem., Seychelles 261–272 1990.

Johnson, B.D., Powell, C.M., Veevers, J.J., 1976. Spreading history of the Eastern Indian Ocean and greater India's flight from Antarctica and Australia. Geol. Soc. Am. Bull. 87, 1560–1566.

Johnson, B.D., Powell, C.M., Veevers, J.J., 1980. Early spreading history of the Indian Ocean between India and Australia. Earth Planet. Sci. Lett. 47, 131–143.

Jokat, W., Boebel, T., König, M., Meyer, U., 2003. Timing and geometry of early Gondwana breakup. J. Geophys. Res. 108 (B9), 2428. doi:10.1029/2002JB001802.

Judd, A., 2005. Gas emissions from mud volcanoes. In: Mud Volcanoes, Geodynamics and Seismicity. Proceedings of the NATO Advanced Research Workshop on Mud Volcanism, Geodynamics and Seismicity. Baku, Azerbaijan, 20–22 May 2003. Springer, The Netherlands.

Kahle, H.-G., Naini, B.R., Talwani, M., Eldholm, O., 1981. Marine geophysical study of the Comorin Ridge, North Central Indian Basin. J. Geophys. Res. 86, 3807–3814.

Kaila, K.L., 1988. Mapping the thickness of Deccan trap flows in India from DSS studies and inferences about a hidden Mesozoic basin in the Narmada-Tapti region. In: Subba Rao, K.V. (Ed.), Deccan Flood Basalts, Mem. Geol. Soc. Ind., vol. 10. 91–116.

Kaila, K.L., Tewari, H.C., 1985. Structural trends in the Cuddapah basin from deep seismic soundings (DSS) and their tectonic implications. Tectonophysics 115, 69–86.

Kaila, K.L., Chowdhury, R.K., Reddy, P.R., Krishna, V.G., Narain, Hari, Subbotin, S.I., Sollogub, V., Chekunov, A.V., Kharetchko, G.E., Lazarenko, M.A., Ilchenko, T.V., 1979. Crustal structure along the Kavali-Udipi profile in the Indian Peninsular Shield from Deep Seismic Sounding. J. Geol. Soc. India 20, 307–333.

Kaila, K.L., Reddy, P.R., Dixit, M.M., 1981a. Deep crustal structure at Koyna, Maharashtra, indicated by deep seismic soundings. J. Geol. Soc. India 22, 1–16.

Kaila, K.L., Murty, P.R.K., Rao, V.K., Kharetcheko, G.E., 1981b. Crustal structure from deep seismic sounding along the Koyna II (Kelsi-Loni) profile in the Deccan Trap area, India. Tectonophys 73, 365–384.

Kaila, K.L., Krishna, V.G., Mall, D.M., 1981c. Crustal structure along Mehmadabad-Billimora profile in the Cambay basin, India from deep seismic soundings. Tectonophysics 76, 99–130.

Kaila, K.L., Murthy, P.R.K., Mall, D.M., Dixit, M.M., Sarkar, D., 1987. A deep seismic sounding along Hirapur-Mandla profile, Central India. Geophys. J. R. Astro. Soc. 89, 399–404.

Kaila, K.L., Tewari, H.C., Mall, D.M., 1987. Crustal Structure and Delineation of Gondwana Sediments in the Mahanadi Delta Area, India, from Deep Seismic Soundings. J. Geol. Soc. India 29, 293–308.

Kaila, K.L., Tewari, H.C., Sarma, P.L.N., 1988. Crustal structure from deep seismic sounding studies along Navibandar-Amreli profiles in Saurashtra, India. In: Deccan Volcanism and Related Basalt Provinces in Other Parts of the World, Mem. Geol. Soc. Ind., vol. 3. 218–232.

Kaila, K.L., Murthy, P.R.K., Rao, V.K., Venkateswarlu, N., 1990a. Deep seismic sounding in the Godavari graben and Godavari (coastal) basin, India. Tectonophysics 173, 307–317.

Kaila, K.L., Tewari, H.C., Krishna, V.G., Dixit, M.M., Sarkar, D., Reddy, M.S., 1990b. Deep seismic sounding studies in the north Cambay and Sanchor basins. India. Geophys. J. Int. 103, 621–637.

Kaila, K.L., Reddy, P.R., Mall, D.M., Venkateswarlu, N., Krishna, V.G., Prasad, A.S.S.S.R.S., Rao, I.B.P., 1992. Crustal structure of the West Bengal basin from deep seismic sounding investigations. Geophys. J. Int. 111, 45–66.

Kalyanasundaram, R., Vijayalakshmi, K.G., 1991. Paleogeography of Cauvery basin (ONGC Report-unpublished).

Kamesh Raju, K.A., Rama Prasad, T., Rao, P.S., Ramalingeswara Rao, B., Varghese, J., 2004. New insights into the tectonic evolution of the Andaman basin, northeast Indian Ocean. Earth Planet. Sci. Lett. 221, 145–162.

Karisiddiah, S.M., Subba Raju, L.V., 2002. Scenario of Gas charged sediments and Gas hydrates in the western continental margin of India. J. Geophys. 23, 33–41.

Karner, G.D., Watts, A.B., 1982. On isostasy at Atlantic-type continental margins. J. Geophys. Res. 87, 2923–2948.

Katz, M.B., Premoli, C., 1979. India and Madagascar in Gondwanaland based on matching Precambrian lineaments. Nature 279, 312–315.

Kearey, P., Klepeis, K.A., Vine, F.J., 2009. Global Tectonics. Wiley-Blackwell 496.

Kenneth, D., Nathaniel, D.M., Bastia, R., Verma, R., Srinivas, T., Sreejesh, V.S., 2008. Southwestern Continental Margin of India: A Poly-Rift Basin, 7th Biennial International conference and exposition on petroleum Geophysics. Hyderabad 2008, P-404, 1-5.

Kent, R., 1991. Lithospheric uplift in Eastern Gondwana: evidence for a long-lived mantle plume system. Geology 19, 19–23.

Kent, R.W., Storey, M., Saunders, A.D., 1992. Large igneous provinces: sites of plume impact or plume incubation. Geology 20, 891–894.

Kent, W., Saunders, A.D., Kempton, P.D., Ghose, N.C., 1997. Rajmahal Basalts, eastern India: mantle sources and melt distribution at a volvanic rifted margin. In: Mahoney, J.J., Coffin, M.F. (Eds.), Large Igneous Provinces: Continental, Oceanic and Planetary Flood Volcanism, Geophys. Monogr. Am. Geophys. Union, vol. 100. 145–182.

Kent, R.W., Pringle, M.S., Muller, R.D., Saunders, A.D., Ghose, N.G., 2002. 40Ar/39Ar geochronology of the Rajmahal Basalts, India, and their relationship to the Kerguelen Plateau. J. Pet. Technol. 43, 1141–1153.

Kenyon, N.H., Amir, A., Cramp, A., 1995. Geometry of the younger sediment bodies of the Indus Fan. In: Pickering, K.T. et al., (Ed.), Atlas of Deep Water Environments: Architectural Style in Turbidite Systems. Chapman and Hall, London, pp. 89–93.

Khain, V.Y., 1992. The role of rifting in the evolution of the Earth's crust. Tectonophysics 215, 1–7.

Khain, V.E., Polyakova, I.D., 2008. Large and giant hydrocarbon accumulations in the transitional continent–ocean zone. Geotectonics 42 (3), 163–175.

Khan, P.K., Chakraborty, P.P., 2005. Two-phase opening of Andaman Sea: a new seismotectonic insight. Earth Planet. Sci. Lett. 229, 259–271.

Kieckhefer, R.M., Shor, G.G., Curray, J.R., SugiarTa, W., Hehuwat, F., 1980. Seismic refraction studies of the Sunda Trench and forearc basin. J. Geophys. Res. 85, 863–890.

Kieckhefer, R.M., Moore, G.F., Emmel, F.J., 1981. Crustal structure of the Sunda fore arc region west of central Sumatra from gravity data. J. Geophys. Res. 86, 7003–7012.

King, W., 1881. The geology of Pranhita-Godavari Valley. Geol. Surv. India Mem. 18, 195–263.

King, L.C., 1967. The Morphology of the Earth, second ed. Oliver & Boyd, Edinburgh.

King, L.H., MacLean, B., 1970. Pockmarks on the Scotian Shelf. Geol. Soc. Am. Bull. 81, 3141–3148.

Klootwijk, C.T., Gee, J.S., Peirce, J.W., Smith, G.M., McFadden, P.L., 1992. An early India-Asia contact: paleomagnetic constraints from Ninetyeast Ridge, ODP Leg 121. Geology 20, 395–398.

Kneller, B., Dykstra, M., 2003. External geometry and internal architecture of mass-transport complexes, and the creation of ponded turbidities on the slope. In: Hodgson, D.M., Flint, S.S., Garfield, T. (Eds.), Submarine Slope Systems: Processes, Products, Prediction. Geological Society of London, Liverpool, p. 57.

König, 2005. Processing of shipborne magnetometer data and revision of the timing and geometry of the Mesozoic break-up of Gondwana Ph.D. Thesis. Alfred Wegener Institute for Polar and Marine Research, Bremerhaven.

Kolla, V., 2007. A review of sinuous channel avulsion patterns in some major deep-sea fans and factors controlling them. Mar. Pet. Geol. 24, 450–469.

Kolla, V., Coumes, F., 1987. Morphology, internal structure, seismic stratigraphy and sedimentation of the Indus Fan. Am. Assoc. Petrol. Geol. Bull. 71, 650–677.

Kolla, V., Coumes, F., 1990. Extension of structural and tectonic transform from the Indian subcontinent into the eastern Arabian Sea. Mar. Petrol. Geol. 7, 188–196.

Kolla, V., Kostecki, J.A., Robinson, F., Biscaye, P.E., 1981. Distributions and origins of clay minerals and quartz in surface sediments of the Arabian Sea. J. Sediment. Petrol. 51, 563–569.

Kolla, V., Bourges, P., Urruty, J.M., Safa, P., 2001a. Evolution of deepwater Tertiary sinuous channels offshore, Angola (West Africa) and implications to reservoir architecture. AAPG Bull. 85, 1373–1405.

Kolla, V., Bourges, P., Urruty, J.M., Safa, P., 2001b. Evolution of deepwater Tertiary sinuous channels offshore Angola (west Africa) and implications for reservoir architecture. Am. Assoc. Pet. Geol. Bull. 85 (8), 1373–1405.

Kooi, H., Beaumont, C., 1994. Escarpment evolution on high-elevation rifted margins; Insights derived from a surface-processes model that combines diffusion, advection and reaction. J. Geoph. Res. 99, 12191–12209.

Kooi, H., Cloetingh, S., Burrus, J., 1992. Lithospheric necking and regional isostasy at extensional basins 1. Subsidence and gravity modeling with an application to the Gulf of Lions margin (SE France). J. Geophys. Res. 97, 17553–17571.

Krishna, K.S., 2003. Structure and evolution of the Afanasy Nikitin seamount, buried hills and 85E ridge in the northeastern Indian Ocean. Earth and Planetary Science Letters 209, 379–394.

Krishna, V.G., Kaila, K.L., Reddy, P.R., 1991. Low velocity layers in the subcrustal lithosphere beneath the Deccan Traps region of western India. Phys. Earth Planet. Inter. 67, 288–302.

Krishna, K.S., Gopala Rao, D., Ramana, M.V., Subrahmanyam, V., Sarma, K.V.L.N.S., Pilipenko, A.I., Shcherbakov, V.S., Radhakrishna Murthy, I.V., 1995. Tectonic model for the evolution of oceanic crust in the northeastern Indian Ocean from the Late Cretaceous to the early Tertiary. J. Geophys. Res. 100, 20,011–20,024.

Krishna, K.S., Gopala Rao, D., Subba Raju, L.V., Chaubey, A.K., Shcherbakov, V.S., Pilipenko, A.I., Radhakrishna Murthy, I.V., 1999. Paleocene on-axis hotspot volcanism along the Ninetyeast Ridge: an interaction between the Kerguelen hotspot and the Wharton spreading center. Proc. Indian Acad. Sci. Earth Planet. Sci. 108, 255–267.

Krishna, K.S., Neprochnov, Y.P., Gopala Rao, D., Grinko, B.N., 2001. Crustal structure and tectonics of the Ninetyeast Ridge from seismic and gravity studies. Tectonics 20, 416–433.

Krishna, K.S., Gopala Rao, D., Sar, D., 2006. Nature of the crust in the Laxmi Basin (14°–20° N), western continental margin of India. Tectonics 25, TC1006. doi:10.1029/2004 TC001747.

Krishna, K.S., Michael, Laju, Bhattacharyya, R., Majumdar, T.J., 2009. Geoid and gravity anomaly data of conjugate regions of Bay of Bengal and Enderby Basin—new constraints on breakup and early spreading history between India and Antarctica. J. Geophys. Res. 114, B03102. doi:10.1029/2008JB005808.

Krishnan, M.S., 1953. The structural and tectonic history of India. Mem. Geol. Surv. India 81, 2–12.

Kudrass, H.R., Michels, K.H., Wiedicke, M., Suckow, A., 1998. Cyclones and tides as feeders of a submarine canyon off Bangladesh. Geology 26 (8), 715–718.

Kuehl, S.A., Hariu, T.M., Moore, W.S., 1989. Shelf sedimentation off the Ganges–Brahmaputra river system: evidence for sediment bypassing to the Bengal Fan. Geology 17, 1132–1135.

Kukowski, N., Schillhorn, T., Flueh, E.R., Huhn, K., 2000. Newly identified strike-slip plate boundary in the northeastern Arabian Sea. Geology 28, 355–358.

Kumar, S., 1981. Geodynamics of Burma and Andaman–Nicobar region on the basis of tectonic stresses and regional. Tectonophysics 79, 75–85.

Kumar, S.P., 1983. Geology and hydrocarbon prospects of Krishna-Godavari and Cauvery basins. In: Bhandari, L.L. (Ed.), Petroliferous Basins of India. Pet. Asia J., vol. 6. Himachal Times Group Publication, Dehradun, India, pp. 57–65.

Kumar, Anil, Pande, K., Venkatesan, T.R., Bhaskara Rao, Y.J., 2001. The Karnataka Late Cretaceous dykes as products of the Marion hot spot at the Madagascar–India break-up event: evidence from ^{40}Ar–^{39}Ar geochronology and geochemistry. Geophys. Res. Lett. 28, 2715–2718.

Kumar, Anil, Dayal, A.M., Padmakumari, V.M., 2003. Kimberlite from Rajmahal magnetic province: Sr-Nd-Pb isotopic evidence for Kerguelen plume derived magma. Geophys. Res. Lett. 30, 9–11.

Kumar, D., Bastia, R., Guha, D., 2004. Prospect hunting below Deccan basalt: imaging challenges and solutions. First Break 22, 35–39.

Kvenvolden, K.A., 1999. Potential effects of gas hydrate on human welfare. Proc. Natl. Acad. Sci. U.S.A. 96, 3420–3426.

Kvenvolden, K.A., Barnard, L.A., 1983. Hydrate of natural gas in continental margins. In: Watkins, J.S., Drake, C.L. (Eds.), Studies in Continental Margin Geology, Am. Assoc. Pet. Geol. Mem., vol. 34. 31–640.

Kvenvolden, K.A., Rogers, B.W., 2005. Gaia's breath—global methane exhalations. Mar. Pet. Geol. 22 (4), 579–590.

Lal, N.K., Siawal, A., Kaul, A.K., 2009. Evolution of east coast of India—a plate tectonic reconstruction. J. Geol. Soc. India 73, 249–260.

Larson, R.L., 1975. Late Jurassic sea floor spreading in the eastern Indian Ocean. Geology 3, 69–71.

Larson, R.L., 1977. Early Cretaceous breakup of Gondwanaland off western Australia. Geology 5, 57–60.

Laughton, A.S., 1975. General bathymetric chart of the oceans (Gebco), Sheet 5-05. Can. Hydrogr. Serv., Ottawa, ON, Canada.

Laughton, A.S., Whitmarsh, R.B., Jones, M.T., 1970. The evolution of the Gulf of Aden. Philos. Trans. R. Soc. Lond. A. 267, 227–266.

Law, B.E., Spencer, C.W., 1998. Abnormal pressures in hydrocarbon environments. Am. Assoc. Pet. Geol. Mem. 70, 1–11.

Lawver, L.A., Royer, J.Y., Sandwell, D.T., Scotese, C.R., 1991. Crustal development: Gondwana break-up. In: Thomson, M.R.A., Crame, J.A., Thomson, J.W. (Eds.), Geological Evolution of Antarctica. Cambridge University Press, Cambridge, pp. 533–539.

Lawver, L.A., Gahagan, L.M., Dalziel, I.W.D., 1999. A tight fit—early Mesozoic Gondwana, a plate reconstruction perspective. Memoirs of National Institute of Polar Research, 53, pp. 214–229 (special issue).

Lay, T., Kanamori, H., Ammon, C.J., Nettles, M., Ward, S.N., Aster, R.C., Beck, S.L., Bilek, S.L., Brudzinski, M.R., Butler, R., DeShon, H.R., Ekstrom, G., Satake, K., Sipkin, S., 2005. The great Sumatra–Andaman Earthquake of 26 December 2004. Science 308, 1127–1133.

Le Pichon, X., Sibuet, J.C., 1981. Passive margins: a model of formation. J. Geophys. Res. 86, 3708–3720.

Lee, M.W., Collett, T.S., 2009. Gas hydrate saturations estimated from fractured reservoir at Site NGHP-01-10, Krishna–Godavari Basin, India. J. Geophys. Res. 114, B07102.

Lee, T.T., Lawver, L.A., 1995. Cenozoic plate reconstruction of Southeast Asia. Tectonophysics 251, 85–138.

Leroy, S., Gente, P., Fournier, M., d'Acremont, E., Patriat, P., Beslier, M.-O., Bellahsen, N., Maia, M., Blais, A., Perrot, J., Al-Kathiri, A., Merkouriev, S., Fleury, J.-M., Ruellan, P.-Y., Lepvrier, C., Huchon, P., 2004. From rifting to spreading in the eastern Gulf of Aden: a geophysical survey of a young oceanic basin from margin to margin. Terra Nova 16, 185–192.

Levell, B., Argent, J., Dore, A.G., Fraser, S., 2010. Passive margins: overview. In: Vining, B.A., Pickering, S.C. (Eds.), Petroleum Geology: From Mature Basins to New Frontiers—Proceedings of the 7th Petroleum Geology Conference. Geological Society, London, pp. 823–830. doi:10.1144/0070823.

Lisker, F., Fachmann, S., 2001. Phanerozoic history of the Mahanadi region, India. J. Geophys. Res. B 106, 22027–22050.

Liu, C.S., Sandwell, D.T., Curray, J.R., 1982. The negative gravity field over the 85°E ridge. J. Geophys. Res. 87, 7673–7686.

Liu, C.S., Curray, J.R., McDonald, J.M., 1983. New constraints on the tectonic evolution of the Eastern Indian Ocean. Earth Planet. Sci. Lett. 65, 331–342.

Lopez, M., 2001. Architecture and depositional pattern of the Quarternary deep-sea fan of the Amazon. Mar. Pet. Geol. 18, 479–486.

Luyendyk, B.P., Rennick, W., 1977. Tectonic history of aseismic ridges in the eastern Indian Ocean. Geol. Soc. Am. Bull. 88, 1347–1356.

Magoon, L.B., Dow, W.G., 1994. The petroleum system. In: Magoon, L.B., Dow, W.G. (Eds.), The Petroleum System-From Source to Trap, AAPG Memoir, vol. 60. 3–24.

Mahalik, N.K., 1995. Geomorphology. In: Mohanty, B.K., Mahalik, N.K., Mishra, R.N. (Eds.), Geology and Mineral Resources of Orissa. SGAT, Bhubaneswar, pp. 1–7.

Mahoney, J.J., Coffin, M.F., 1997. Large Igneous Provinces: Continental, Oceanic, and Planetary Flood Volcanism. In: Geophys. Monogr., vol. 100. American Geophysical Union, Washington, DC.

Mahoney, J.J., Macdougall, J.D., Lugmair, G.W., Gopalan, K., 1983. Kerguelen hotspot source for Raimahal Traps and Ninetyeast Ridge? Nature 303, 385–389.

Maia, M., Diament, M., Recq, M., 1990. Isostatic response of the lithosphere beneath the mozambique ridge (SW Indian Ocean) and geodynamic implications. Geophys. J. Int. 100, 337–348.

Maksoud, J., 2007. Indian exploration moves into deepwater frontiers, NELP VI brings in record bids. Pennwell Offshore Magazine. pp. 38–42.

Mall, D.M., Rao, V.K., Reddy, P.R., 1999. Deep sub-crustal features in the Bengal basin: Seismic signatures for plume activity. Geophys. Res. Lett. 26, 2545–2548.

Mallik, J., Mathew, G., Greiling, R.O., 2009. Magnetic fabric analysis of fault propagation folds in the Eastern Kachchh, Western India. Tectonophysics 473, 428–445.

Malod, J.A., Droz, L., Mustafa Kemal, B., Patriat, P., 1997. Early spreading and continent to oceanic basement transition beneath the Indus deep sea fan: northeastern Arabian Sea. Mar. Geol. 141, 221–235.

Manglik, A., Verma, S.K., 1998. Delineation of sediments below flood basalts by joint inversion of seismic and magnetotelluric data. Geophys. Res. Lett. 25, 4015–4018.

Manighetti, I., Tapponnier, P., Courtillot, V., Gruszow, S., Gillot, P.Y., 1997. Propagation of rifting along the Arabia-Somalia plate boundary: the Gulfs of Aden and Tadjoura. J. Geophys. Res. 102, 2682–2710.

Mann, P., Gahagan, L., Gordon, M., 2001. Tectonic setting of the world's giant oil fields. World Oil 222, 78–79.

Mareschal, M., Vasseur, V.G., Srivastava, B.J., Singh, R.N., 1987. Induction models of Southern India and effect of offshore Geology. Phys. Earth Planet Inter. 45, 137–148.

Markl, R.G., 1974. Evidence for the breakup of eastern Gondwanaland by the early Cretaceous. Nature 251 (5472), 196–200.

Markl, R.G., 1978. Further evidence for the early cretaceous breakup of Gondwanaland off southwestern Australia. Mar. Geol. 26, 41–48.

Mathew, G., Singhvi, A.K., Karanth, R.V., 2006. Luminescence chronometry and geomorphic evidence of active fold growth along the Kachchh Mainland Fault (KMF), Kachchh, India: seismotectonic implications. Tectonophysics 422, 71–87.

Mathur, R.B., Nair, K.R., 1993. Exploration of Bombay offshore basin. In: Proceedings of the 2nd Seminar on Petroleum Basins of India, vol. 2. KDMIPE and ONGC, Indian Petroleum Publishers, Dehradun, India, pp. 365–396.

Mathur, M.C., Sinharay, S., Kamath, V., Singh, R.J., 2007. Gas hydrate resource potential of Indian continental margin—a case study from eastern margin of India. In: Proceedings of the National Seminar on Modern Trends in Geophysical Sciences and Techniques, ISM, Dhanbad157–166.

Maurin, T., Rangin, C., 2009. Impact of the 90 degrees E ridge at the Indo-Burmese subduction zone imaged from deep seismic reflection data. Mar. Geol. 266, 143–155.

McElhinney, M.W., Embleton, B.J., Daly, L., Pozzi, J.P., 1976. Palaeomagnetic evidence for the location of Madagascar in Gondwana land. Geology 4, 455–457.

McElhinny, M.W., Embleton, B.J.J., Daly, L., Pozzi, J.-P., 1976. Paleomagnetic evidence for the location of Madagascar in Gondwanaland. Geology 4, 355–358.

McHargue, T.R., Webb, J.E., 1986. Internal geometry, seismic facies, and petroleum potential of canyons and inner fan channels of the Indus Submarine Fan. AAPG Bull. 70, 161–180.

Mchugh, C.M.G., Ryan, W.B.F., 2000. Sedimentary features associated with channel overbank flow: examples from the Monterey Fan. Mar. Geol. 163, 199–215.

McKee, E.H., 1974. K-Ar ages of deep-sea basalts, Indian Ocean, Leg 25, Deep Sea Drilling Project. In: Simpson, E.S.W., Schlich, R. et al., (Eds.), Initial Reports of the Deep Sea Drilling Project, vol. 25. U.S. Government Printing Office, Washington, DC, pp. 553–554.

McKenzie, D.P., 1978. Some remarks on the development of sedimentary basins. Earth Planet. Sci. Lett. 40, 25–32.

McKenzie, D.P., 1984. A possible mechanism for epeirogenic uplift. Nature 307, 616–618.

McKenzie, D.P., Bickle, M.J., 1989. The volume and composition of melt generated by extension of the Lithosphere. J. Pet. 29, 625–679.

McKenzie, D.P., Scalter, J.G., 1971. The evolution of the Indian Ocean since the late Cretaceous. Geophys. J. R. Astron. Soc. 25, 437–528.

Mercuriev, S., Patriat, P., Sochevanova, N., 1996. Evolution de la dorsale de Carlsberg: evidence pour une phase d'expansion tres lente entre 40 et 25Ma (A18 a A7). Oceanol. Acta 19 (1), 1–13.

Merh, S.S., 1995. Geology of Gujarat. Memoir of Geological Society of India, pp. 222.

Michael, L., Krishna, K.S., 2011. Dating of the 85° E Ridge (northeastern Indian Ocean) using marine magnetic anomalies. Curr. Sci. 100 (9), 1314–1322.

Michels, K.H., Kudrass, H.R., Hubscer, C., Suckow, A., Wiedicke, M., 1998. The submarine delta of the Ganges-Brahmaputra: cyclone-dominated sedimentation patterns. Mar. Geol. 149, 133–154.

Miles, P.R., Munschy, M., Segoufin, J., 1998. Structure and early evolution of the Arabian Sea and East Somali basin. Geophys. J. Int. 134, 876–888.

Milkov, A.V., 2000. Worldwide distribution of submarine mud volcanoes and associated gas hydrates. Mar. Geol. 167, 29–42.

Miller, T.W., Luk, C.H., Olgaard, D.L., 2002. The interrelationship between overpressure mechanisms and in situ stress. AAPG Memoir 76, pp. 13–20.

Milliman, J.D., Meade, R.H., 1983. World-wide delivery of river sediment to the ocean. J. Geol. 91, 1–21.

Minshull, T.A., Lane, C.I., Collier, J.S., Whitmarsh, R.B., 2008. Rifting and magmatism in the northeastern Arabian Sea. Nat. Geosci. 1 (7), 463–467. doi:10.1038/ngeo228.

Mishra, K.S., 2001. The 'Panvel Flexure': part of a large structure in Western India. Geol. Soc. India Mem. 47, 225–232.

Mishra, D.C., Babu Rao, V., Laxman, G., Rao, M.B.S.V., Venkatarayudu, M., 1987. Three dimensional structural model of Cuddapah basin and adjacent eastern part from geophysical structures. Memoir Geol. Soc. India 6, 313–330.

Mishra, D.C., Gupta, S.B., Venkatarayudu, M., 1989. Godavari rift and its extension towards the east coast of India. Earth Planet. Sci. Lett. 94, 344–352.

Mishra, D.C., Sekhar, D.V.C., Raju, D.C.V., Kumar, V.V., 1999. Crustal structure based on gravity–magnetic modelling constrained from seismic studies under Lambert Rift, Antarctica and Godavari and Mahanadi rifts, India and their interrelationship. Earth Planet. Sci. Lett. 172, 287–300.

Mishra, D.C., Bijendra Singh Gupta, S.B., Prabhakar Rao, M.R.K., Singh, A.P., Chandrasekhar, D.V., Hodlur, G.K., Rao, M.B.S.V., Tiwari, V.M., Laxman, G., Venkata Raju, D.Ch., Kumar, V.P., Rajesh, R.S., Babu Rao, V., Chetty, T.R.K., 2001. Major lineaments and gravity–magnetic trends in Saurashtra, India. Curr. Sci. 80 (8), 1059–1067.

Mishra, C.K., Nathaniel, D.M., Bastia, R., Muralikrishna, A., Verma, R., Srinivas, T., Megha, S., 2008. An insight into Sub Basalt seismic imaging: Examples from offshore Kerala Konkan

Basin, 7th Biennial International Conference and Exposition on Petroleum Geophysics, SPG Hyderabad 2008 1–5 P-401.

Mitchell, A.H.G., 1985. Collision-related fore-arc and back-arc evolution of the northern Sunda Arc. Tectonophysics 116, 323–334.

Mitra, D.S., Agarwal, R.P., 1991. Geomorphology and petroleum prospects of Cauvery basin, Tamilnadu, based on interpretation of Indian Remote Sensing Satellite (IRS) data. J. Indian Soc. Rem. Sens. 19 (4), 263–268.

Müller, R.D., Royer, J.-Y., Lawver, L.A., 1993. Revised plate motions relative to the hotspots from combined Atlantic and Indian Ocean hotspot tracks. Geology 21, 275–278.

Müller, R.D., Gaina, C., Tikku, A., Mihut, D., Cande, S.C., Stock, J.M., 2000. Mesozoic/Cenozoic tectonic events around Australia. In: Richards, M.A., Gordon, R., van der Hilst, R.D. (Eds.), The History and Dynamics of Plate Motion. Geophys. Monogr. Ser., vol. 121. AGU, Washington, DC, pp. 161–188.

Mohan, M., 1985. Geohistory analysis of Bombay high region. Mar. Pet. Geol. 2, 350–360.

Mohapatra, P., Sivarama Krishnan, T.E., Rajagopalan, S., Sethi, J.R., 1996. Producing Indian Basins and their reserve potential. Bull. Oil Nat. Gas Corp. Ltd. Vol. 33. (1), 45–76.

Mohinuddin, S.K., Satyanarayana, K., Rao, G.N., 1993. Cretaceous sedimentation in the subsurface of Krishna Godavari basin. J. Geol. Soc. India 41 (6), 533–539.

Moore, G.F., Curray, J.R., 1980. Structure of the Sunda trench lower slope off Sumatra from multichannel seismic reflection data. Mar. Geophys. Res. 4, 319–340.

Moore, D.G., Curray, J.R., Raitt, R.W., Emmel, F.J., 1974. Stratigraphic-seismic section correlations and implications to Bengal Fan history. In: von der Borch, C.C., Sclater, J.C., et al. (Eds.), Initial Reports of the Deep Sea Drilling Project, Leg 22, pp. 403–412.

Morgan, W.J., 1981. Hot spot tracks and the opening of the Atlantic and Indian Ocean. In: Emiliani, E. (Ed.), The Sea, vol. 7. Wiley Intersciences, New York, pp. 443–475.

Morgan, J.V., Barton, P.J., White, R.S., 1989. The Hatton Bank continental margin—III. Structure from wide-angle OBS and multichannel seismic refraction profiles. Geophys. J. Int. 98, 367–384.

Mountain, G.S., Prell, W.L., 1989. Geophysical reconnaissance survey for ODP Leg 117 in the northwest Indian Ocean. Proc. Ocean Drill. Program Initial Rep. 117, 51–64.

Mountain, G.S., Prell, W.L., 1990. A multiphase plate tectonic history of the southeast continental margin of Oman. In: Robertson, A.H.F. et al., (Ed.), The Geology and Tectonics of the Oman Region. Geological Society, London, pp. 725–743 Special Publication 49.

Mukhopadhyay, M., 1984. Seismotectonics of subduction and back arc rifting under the Andaman Sea. Tectonophysics 108, 229–239.

Mukhopadhyay, M., 1988. Gravity anomalies and deep structure of the Andaman arc. Mar. Geophys. Res. 90, 197–210.

Mukhopadhyay, M., Krishna, M.R., 1991. Gravity field and deep structure of the Bengal Fan and its surrounding continental margins, northeast Indian Ocean. Tectonophysics 186, 365–386.

Mukhopadhyay, M., Krishna, M.R., 1992. Geophysical evidences for crustal transition under the eastern continental margin of India. Geol. Surv. Ind. Spec. Publ. 29, 87–113.

Mukhopadhyay, M., Krishna, M.R., 1995. Gravity anomalies and deep structure of the Ninetyeast Ridge north of the equator, eastern Indian Ocean—a hot spot trace model. Mar. Geophys. Res. 17, 201–216.

Murthy, N.G.K., Ahmad, M., 1970. Gondwana rocks of Chinglepet district, Tamil Nadu, India (abs.). In: International Symposium on Stratigraphy and Mineral Resources of Gondwana System, Aligarh, India. Aligarh Muslim University, Aligarh, India, pp 28.

Murthy, I.V.R., Babu, S.B., 2006. Structure of Charnockitic basement in a part of the Krishna–Godavari basin, Andhra Pradesh. J. Earth Syst. Sci. 115 (4), 387–393.

Murthy, K.V.S., Singh, R.P., Nagar, V.K., 1973. Hydrocarbon Prospects of Mahanadi Basin. Unpublished report of O.N.G. Commission of India, 30 pp.

Murthy, K.S.R., Rao, T.C.S., Malleswara Rao, M.M., 1987. Delineation of structural lineaments from marine magnetic anomalies of Lawson's Bay (Vishakhapatnam), East Coast of India. Ind. J. Mar. Sci. 16, 19–21.

Murthy, K.S.R., Rao, T.C.S., Subrahmanyam, A.S., Rao, M.M.M., Lakshminarayana, S., 1993. Structural lineaments from the magnetic anomaly map of the eastern continental margin of India (ECMI) and NW Bengal Fan. Mar. Geol. 114, 171–183.

Murthy, K.S.R., Subrahmanyam, A.S., Lakshminarayana, S., Chandrasekhar, D.V., Rao, T.C.S., 1995. Some geo-dynamic aspects of the Krishna-Godavari basin, east coast of India. Cont. Shelf Res. 15 (7), 779–785.

Murthy, K.S.R., Subrahmanyam, V., Subrahmanyam, A.S., Murty, G.P.S., Sarma, K.V.L.N.S., 2010. Land-ocean tectonics (LOTs) and the associated seismic hazard over the Eastern Continental Margin of India (ECMI). Nat. Hazards 55, 167–175. doi:10.1007/s11069-010-9523-8.

Murty, K.V.S., Ramakrishna, M., 1980. Structure and tectonics of Godavari-Krishna coastal sedimentary basins. Bull. Oil Nat. Gas Corp. 17, 147–158.

Murty, G.P.S., Subrahmanyam, A.S., Murthy, K.S.R., Sarma, K.V.L.N.S., 2002. Evidence of fault reactivation off Puducherry coast from marine geophysical data. Curr. Sci. 83, 1446–1449.

Mutter, J.C., 1993. Margins declassified. Nature 364, 393–394.

Naini, B.R., Kolla, V., 1982. Acoustic character and thickness of sediments of the Indus fan and the continental margin of western India. Mar. Geol. 47, 181–195.

Naini, B.R., Leyden, R., 1973. Ganges cone: a wide-angle seismic reflection and refraction study. J. Geophys. Res. 78, 8711–8720.

Naini, B.R., Talwani, M., 1982. Structural framework and the evolutionary history of the continental margin of Western India. In: Watkins, J.S., Drake, C.L. (Eds.), Studies in Continental Margin Geology. American Association of Petroleum Geologists, pp. 67–191.

Nair, K.M., Singh, N.K., Jokhan, R., Gavarshetty, C.P., Muraleekrishanan, B., 1992. Stratigraphy and sedimentation of Bombay offshore basin. J. Geol. Soc. India 40, 415–442.

Narain, H., Kaila, K.L., Verma, R.K., 1968. Continental margins of India. Can. J. Earth Sci. 5, 1051–1065.

Nathaniel, D.M., D'Silva, K., Muralikrishna, A., Chandra Rao, D., 2011. Synergy of concepts and technology redefines the WCMI and enhances petroleum exploration opportunities. In: 7th International Conference on Asian Marine Geology, 11–14 October 2011.

National Geophysical Research Institute, 1978. Gravity Map of India, Scale 1:5 Million, Hyderabad, India.

Nayak, G.K., Rama Rao, Ch, 2002. Structural configuration of Mahanadi offshore basin, India: an aeromagnetic study. Mar. Geol. Res. 23, 471–479.

Nayak, S., Das, S., Bastia, R., Bhabesh Kar, 2011. Petroleum System Related to 85° E Ridge: a Conceptual Study. Extended abstract presented at GEO-India, Greater Noida, New Delhi, India, January 12–14.

Nayak, S., Das, S., Bastia, R., Kar, B., 2012. Lava Delta below 85° E Ridge, Mahanadi offshore basin: Identification, Characterization and Implication on hydrocarbon Prospectivity, 9th Biennial International conference and Exposition on Petroleum Geophysics, SPG Hyderabad, 1-5.

Nelson, C.H., Nilsen, T.H., 1984. Modern and ancient deep-sea fan sedimentation. Soc. Econ. Paleontol. Mineral. Short Course 14, 404 pp.

Nicolaysen, K., Bowring, S., Frey, F., Weis, D., Ingle, S., Pringle, M., Coffin, M.F., 2001. Provenance of Proterozoic garnet-biotite gneiss recovered from Elan Bank, Kerguelen Plateau, southern Indian Ocean. Geology 29, 235–238.

Nielsen, T.F.D., Brooks, C.K., 1981. The E. Greenland rifted continental margin: an examination of the coastal flexure. J. Geol. Soc. Lond. 138, 559–568.

Nixon, M.F., Grozic, J.L.H., 2000. A simple model for submarine slope stability analysis with gas hydrates. Norwegian J. Geol. 86, 309–316.

Nogi, Y., Seama, N., Isezaki, N., Fukuda, Y., 1996. Magnetic anomaly lineations and fracture zones deduced from vector magnetic anomalies in the West Enderby Basin. In: Storey, B.C., King, E.C., Livermore, R.A. (Eds.), Weddell Sea Tectonics and Gondwana Breakup, 265–273 Microcontinents around Australia 409, Geological of London Society Special Publication 108.

Nogi, Y., Nishi, K., Seama, N., Fukuda, Y., 2004. An interpretation of the seafloor spreading history of the west Enderby basin between initial breakup of Gondwana and anomaly C34. Mar. Geophys. Res. 25, 219–229.

Normark, W.R., 1970. Growth patterns of deep sea fans. Am. Assoc. Pet. Geol. Bull. 54, 2170–2195.

Normark, W.R., Piper, D.J.W., Posamentier, H., Pirmez, C., Migeon, S., 2002. Variability in form and growth of sediment waves on turbidite channel levees. Mar. Geol. 192, 23–58 S0025–3227(02)00548-0.

Norton, I.O., Sclater, J.G., 1979. A model for the evolution of the Indian Ocean and the breakup of Gondwanaland. J. Geophys. Res. 84, 6803–6830.

Ollier, C.D., Pain, C.F., 1997. Equating the basal unconformity with the palaeoplain: a model for passive margins. Geomorphology 19, 1–15.

Ollier, C.D., Powar, K.B., 1985. The Western Ghats and the morphotectonics of Peninsular India. Z. Geomorph. N.F. 54, 57–69.

Olsen, K.H., 1995. Continental Rifts: Evolution, Structure, Tectonics. Elsevier, Amsterdam pp. 466.

Olsen, K.H., Morgan, P., 1995. Introduction: progress in understanding continental rifts. In: Olsen, K.H. (Ed.), Continental Rifts: Evolution, Structure, Tectonics. Elsevier, Amsterdam, pp. 3–21.

ONGC, 2006. Upstream India: 50 Golden Years of Oil and Natural Gas Corporation. ONGC Publication, p. 278.

Pal, T., Chakraborty, P.P., Gupta, T.D., Singh, C.D., 2003. Geodynamic evolution of the outer-arc-forearc belt in the Andaman Islands, the central part of the Burma-Java subduction complex. Geol. Mag. 140, 289–307.

Pande, K., 2002. Age and duration of the Deccan Traps, India: a review of radiometric and paleomagnetic constraints. Proc. Indian Acad. Sci 111, 115–123.

Pandey, J., Dave, A., 1998. Stratigraphy of Indian petroliferous basins. In: XVI Indian Colloquium on Micropaleontology and Stratigraphy, Goa, p. 248.

Pandey, D., MacGregor, L., Sinha, M., Singh, S., 2008. Feasibility of using the magnetotelluric method for sub-basalt imaging at Kachchh, India. Appl. Geophys. 5 (1), 74–82.

Pandey, D., Singh, S., Sinha, M., MacGregor, L., 2009. Structural imaging of Mesozoic sediments of Kachchh, India, and their hydrocarbon prospects. Mar. Pet. Geol. 26, 1043–1050.

Pateria, M.L., Rangaraju, M.K., Raiverman, V., 1992. A note on the structure and stratigraphy of Bay of Bengal sediments. In: Recent Geoscientific Studies in the Bay of Bengal and the Andaman Sea, Spec. Publ. Ser. Geol. Surv. India, vol. 29. 21–23.

Patriat, Ph., Achache, J., 1984. India-Eurasia collision chronology has implications for crustal shortening and driving mechanism of plates. Nature 311, 615–621.

Patriot, J.P., 1980. Existence d'anomalies mésozoiques dans le bassin de Somalie, implications pour les rélations Afrique-Antarctique-Madagascar. C.R. Acad. Sci. Paris, Sér. B2 91, 85–88.

Peirce, J.W., 1978. The northward motion of India since the Late Cretaceous. Geophys. J. R. Astron. Soc. 5 (2), 277–311.

Peirce, J.W., et al., 1989. Proceedings of the Ocean Drilling Program, Initial Reports. Ocean Drill. Program, College Station, TX, vol. 121, 1000 pp.

Peirce, J.W., Weissel, J.K., et al., 1989. Proc. ODP, Init Repts. 121, 1000 pp.

Pennebaker, E.S., 1968. Seismic data indicate depth, magnitude of abnormal pressure. World Oil 166, 73–78.

Peter, G., Weeks, L.A., Burns, R.E., 1966. A reconnaissance geophysical survey in the Andaman Sea and across the Andaman-Nicobar island arc. J. Geophys. Res. 71, 495–509.

Petroy, D.E., Wiens, D.A., 1989. Historical seismicity and implications for diffuse plate convergence in the northeast Indian Ocean. J. Geophys. Res. 94, 12301–12319.

Kearey, P., Klepeis, K.A., Vine, F.J., 2009. Global Tectonics. Wiley-Blackwell, Oxford 496.

Pickering, K.T., Hiscott, R.N., Hein, F.J., 1989. Deep Marine Environments: Clastic Sedimentation and Tectonics. Unwin Hyman, London, 416 p.

Pilipenko, A.I., 1996. Fracture zones of the Ninetyeast Ridge area, Indian Ocean. Geotectonics 30, 441–451 (Engl. Transl.).

Pirmez, C., Flood, R.D., 1995. Morphology and structure of Amazon channel. In: Flood, R.D., Piper, D.J.W., Klaus, A. (Eds.), Proceedings of the Ocean Drilling Program, Initial Report. Ocean Drilling Program, College Station, TX, pp. 23–45.

Powell, C., McA, Conaghan, P.J., 1973. Plate tectonics and the Himalayas. Earth Planet. Sci. Lett. 20, 1–12.

Powell, C., McA, Roots, S.R., Veevers, J.J., 1988. Pre-breakup continental extension in East Gondwanaland and the early opening of the eastern Indian Ocean. Tectonophysics 155, 261–283.

Prabhakar, K.N., Zutshi, P.L., 1993. Evolution of southern part of Indian east coast basins. J. Geol. Soc. India 41, 215–230.

Prell, W.L., Niitsuma, N., et al., 1989. Proceedings of the Ocean Drilling Program Initial Reports, vol. 117. (College Station, TX, Ocean Drilling Program).

Prins, M.A., Stuut, J.-B.W., et al., 2000. Grain-size distribution of siliciclastics in the Arabian Basin: a proxy for changes in climate and sea level. In: Poster Abstr., BSRG, Durham, 15–16 December 1995.

Prince, R.A., Forsyth, D.W., 1988. Horizontal extent of anomalously thin crust near the Vema Fracture Zone from the three-dimensional analysis of gravity anomalies. J. Geophys. Res. 93, 8051–8063.

Prins, M.A., Postma, G., Cleveringa, J., Cramp, A., Kenyon, N.H., 2000. Controls on terrigenous sediment supply to the Arabian Sea during the late Quaternary: the Indus Fan. Mar. Geol. 169, 327–349.

Qayyum, M., Lawrence, R.D., Niem, A.R., 1997. Discovery of the palaeo-Indus delta-fan complex. Geol. Soc. Lond. J. 154, 753–756.

Qayyum, M., Niem, A.R., Lawrence, R.D., 2001. Detrital modes and provenance of the Paleogene Khojak Formation in Pakistan:Implications for early Himalayan orogeny and unroofing. Geol. Soc. Am. Bull. 113, 320–332.

Qureshy, M.N., 1981. Gravity anomalies, isostacy and crust–mantle relations in Deccan traps and contiguous regions, India. In: Subba Rao, K.V., Sukheswala, R.N. (Eds.), Deccan Volcanism, Mem. Geol. Soc. Ind., vol. 3. 184–197.

Radhakrishna, B.P., 1989. Suspect tectono-stratigraphic terrane elements in the Indian subcontinent. J. Geol. Soc. India 34 (1), 1–24.

Radhakrishna, B.P., 1993. Neogene uplift and geomorphic evolution of the Indian Peninsula. Curr. Sci. 64, 787–792.

Radhakrishna, M., Sanu, T.D., 2002. Shallow seismicity, stress distribution and crustal deformation pattern in the Andaman-West Sunda arc and Andaman Sea, Northeastern Indian Ocean. J. Seismol. 6, 25–41.

Radhakrishna, B.P., Vasudev, V.N., 1977. The early Precambrian of the Southern Indian Shield. J. Geol. Soc. India 18, 523–541.

Radhakrishna, T., Dalimeyer, R.D., Joseph, M., 1994. Paleomagnetism and 36Ar/40Ar Vs 39Ar/40Ar isotope correlation ages of dyke swarms in central Kerala, India: tectonic implications. Earth Planet. Sci. Lett. 21, 213–226.

Radhakrishna, M., Chand, S., Subrahmanyam, C., 2000. Gravity anomalies, sediment loading and lithospheric flexure associated with the Krishna-Godavari basin, eastern continental margin of India. Earth Planet. Sci. Lett. 175, 223–232.

Radhakrishna, M., Verma, R.K., Arts, K.P., 2002. Lithospheric structure below the eastern Arabian Sea and adjoining West Coast of India based on integrated analysis of gravity and seismic data. Mar. Geophys. Res. 23, 25–42.

Radhakrishna, M., Lasitha, S., Mukhopaghyay, M., 2008. Seismicity gravity anomalies and lithospheric structure of Andaman arc, NE Indian Ocean. Tectonophysics 460, 248–262.

Radhakrishna, M., Subrahmanyam, C., Twinkle, D., 2010. Thin oceanic crust below Bay of Bengal inferred from 3-D gravity interpretation. Tectonophysics 493, 93–105.

Radhakrishna, M., Rao, G.S., Nayak, Satyabrata, Bastia, R., Twinkle, D., 2012. Early Cretaceous fracture zones in the Bay of Bengal and their tectonic implications: constraints from multichannel seismic reflection and potential field data. Tectonophysics 522–523, 187–197.

Radha Krishan, Joppen, T., Schmidt, R., Rama, P., Sinha, A.K., Prabhakar, V., Mayo, S., 2012. Tectonics, structural style and petroleum system modeling to understand generation and migration of hydrocarbon in Kerala Konkan Area, Western Offshore, India, Society of Petroleum Geophysicists. In: 9th Biennial International Conference and Exposition in Petroleum Geophysics, P-032, 1–7.

Rai, K., Rawat, N.S., Verma R.P., Pardeep Kumar, 2004. High Resolution Reservoir Characterization of Bassein Limestone in Neelam Field of Mumbai Offshore Basin using FMS Image Data, 5th Conference & Exposition on Petroleum Geophysics, Hyderabad-2004, India 124–134.

Rajaram, M., Anand, S.P., Erram, V.C., 2000. Magnetic studies over Krishna-Godavari basin in Eastern Continental margin of India. Gondwana Res. 3, 385–393.

Rajaram, M., Anand, S.P., Balakrishna, T.S., 2006. Composite magnetic anomaly map of India and its contiguous regions. J. Geol. Soc. India 68, 569–576.

Raju, A.T.R., 1979. Basin analysis and petroleum exploration with some examples from Indian sedimentary basins. J. Geol. Soc. India 20, 49–60.

Raju, D.S.N., Ramesh, P., 2005. Biochronostratigraphy and paleoecology of Raghavapuram shale equivalents in subsurface of Krishna-Godavari basin, India—an updated. In: Raju, D.S.N., Peters, J., Shankar, R., Kumar, G. (Eds.), An Overview of Litho-Bio-Chrono-Sequence Stratigraphy and Sea Level Changes of Indian Sedimentary Basins, Assoc. Pet. Geol. Spec. Publ. No. 1, 80–83.

Raju, D.S.N., Jaiprakash, B.C., Ravindran, C.N., Kalyansundar, R., Ramesh, P., 1994. The magnitude of hiatuses and sea level changes across K/T boundary in Cauvery and Krishna-Godavari basins. J. Geol. Soc. India 44, 301–315.

Raju, D.S.N., Ramesh, P., Mohan, S.K.G., Uppal, S., 2005. Sequence and bio-chronostratigraphic subdivisions of the Cretaceous and Cenozoic of India with notes on pre-Cretaceous events: an overview. In: Raju, D.S.N., Peters, J., Shankar, R., Kumar, G. (Eds.), Assoc. Pet. Geol. Spec. Publ. No. 1, 97–103.

Ramakrishnan, M., Vaidyanathan, R., 2008. Geology of India Vol. I & II, Geological Society of India, Bangalore.

Ramana, M.V., Nair, R.R., Sarma, K.V.L.N.S., Ramprasad, T., Krishna, K.S., Subrahmanyam, V., D'Cruz, M., Subrahmanyam, C., Paul, J., Subrahmanyam, A.S., Chandra Sekhar, D.V., 1994. Mesozoic anomalies in the Bay of Bengal. Earth Planet. Sci. Lett. 121, 469–475.

Ramana, M.V., Subrahmanyam, V., Chaubey, A.K., Ramprasad, T., Sarma, K.V.L.N.S., Krishna, K.S., Desa, M., Murty, G.P.S., 1997. Structure and origin of the 85°E ridge. J. Geophys. Res. 102, 17995–18012.

Ramana, M.V., Krishna, K.S., Ramprasad, T., Desa, M., Subrahmanyam, V., Sarma, K.V.L.N.S., 2003. Structure and tectonic evolution of the northeastern Indian Ocean. In: SenGupta, R., Desa, E. (Eds.), The Indian Ocean: A Perspective, vol. 2. Oxford & IBH, New Delhi, India, pp. 731–816.

Ramana, M.V., Ramprasad, T., Paropkari, A.L., Borole, D.V., Rao, B.R., Karisiddaiah, S.M., Desa, M., Kocherla, M., Joao, H.M., Lokabharati, P., Gonsalves, M.J., Pattan, J.N., Khadge, N.H., Babu, C.P., Sathe, A.V., Kumar, P., Sethi, A.K., 2009. Multidisciplinary investigations exploring indicators of gas hydrates occurrence in the Krishna-Godavari Basin offshore, east coast of India. GeoMar. Lett. 29, 25–38.

Ramani, K.K.V., Naidu, B.D., Giridhar, M., 2000. Reassessment of hydrocarbon potential of Cauvery basin, India—a quantitative genetic model approach. In: Petroleum Geochemistry & Exploration in the Afro-Asian Region, 5th International Conference & Exhibition, 25–27 November 2000, New Delhi.

Ramasamy, S.M., Joyce, E.B., Bishop, I., 2001. Tectonically induced environment problems on and off Pondicherry coast, Tamil Nadu—a vision through remote sensing. In: 22nd Asian Conference on Remote Sensing, Singapore, 5–9 November.

Ramam, P.K., Murthy, V.N., 1997. Geology of Andhra Pradesh. Geological Society of India, Bangalore, pp. 98.

Rangaraju, M.K., Aggarwal, A., Prabhakar, K.N., 1993. Tectono-stratigraphy, structural styles, evolutionary model and hydrocarbon prospects of Cauvery and Palar Basins, India. In: Biswas, S.K., Alak, D., Garg, P., Pandey, J., Maithani, A., Thomas, N.J. (Eds.), Proceedings of the 2nd Seminar on Petroliferous Basins of India, vol. 1. Indian Petroleum Publishers, Dehradun, India, pp. 331–354.

Rao, T.C.S., 1967. Seismic refraction measurements in the Gulf of Cambay and Kutch. In: Proc. Symp. Upper Mantle Project, Geophys. Res. Board, NGRI 342–347.

Rao, D.G., 1984. Marine magnetic anomalies off Ratnagiri, western continental shelf of India. Mar. Geol. 81, 103–110.

Rao, G.N., 2001. Sedimentation, stratigraphy and petroleum potential of Krishna-Godavari basin, east coast of India. AAPG Bull. 85 (9), 1623–1643.

Rao, T.C.S., Rao, V.B., 1985. Geophysical studies over the continental margins of the East coast of India. Mar. Geol. 67, 151–161.

Rao, P.R., Srivastava, D.C., 1984. Regional seismic facies analysis of Western Offshore India. Bull. ONGC 21, 83–96.

Rao, R.P., Talukdar, S.N., 1980. Petroleum geology of Bombay High field, India. In: Halbouty, M.T. (Ed.), Giant Oil and Gas Fields of the Decade 1968–1978, Am. Assoc. Pet. Geol. Mem., vol. 30. 487–506.

Rao, Y.H., Reddy, S.I., Khanna, R., Rao, T.G., Thakur, N.K., Subrahmanyam, C., 1998. Potential distribution of methane hydrates along the Indian continental margins. Curr. Sci. 74, 466–468.

Raval, U., Veeraswamy, K., 2003a. India-Madagascar separation: breakup along a pre-existing mobile belt and chipping of the craton. Gondwana Res. 6, 467–485.

Raval, U., Veeraswamy, K., 2003b. Modification of geological and geophysical regimes due to interaction of mantle plume with Indian lithosphere. J. Virtual Expl. 12, 117–143.

Ravikumar, M., Rao, N.P., Chalam, S.V., 1996. A seismotectonic study of the Burma and Andaman arc regions using centroid moment tensor data. Tectonophysics 253, 155–165.

Ravindran, C.N., Kundu, R.N., Dasgupta, A., Ramah, M., Bhatia, B., Nalini, R., Mishra, C.S., Mehrotra, K.L., 2012. Cretaceous sedimentation history and hydrocarbon potential of Kutch offshore. Bull. ONGC 47 (1), 16–26.

Reddy, S.I., Roychowdhury, K., Drolia, R.K., Ashalatha, B., Mittal, G.S., Subrahmanyam, C., Singh, R.N., 1988. On the structure of the western continental margin of Mangalore coast, India. J. Assoc. Expl. Geophys. 9, 181–189.

Reddy, P.R., Chandrakala, K., Srdhar, A.R., 2000. Crustal velocity structure of the Dharwar craton, India. J. Geol. Soc. Ind. 55 (4), 381–386.

Reddy, P.R., Sarkar, D., Sain, K., Mooney, W.D., 2001. Report on collaborative scientific study at USGS, Menlo Park, USA, 19 p (30 October–31 December 2001).

Reddy, P.R., Venkateswarulu, N., Prasad, A.S.S.S.R.S., Koteswar Rao, P., 2002. Basement structure below the coastal belt of Krishna-Godavari Basin: correlation between seismic structure and well information. Gondwana Res. 5, 513–518.

Reeves, C.V., 2003. East Africa and Western India: passive margins from the evolution of a complex ocean. In: AAPG International Meeting, Barcelona, September 2003. Poster in five panels.

Reeves, C.V., 2009. Re-examining the evidence from plate-tectonics for the initiation of Africa's passive margins. Geological Society of Houston/Petroleum Exploration Society of Great Britain. London, September 9–10.

Reeves, C.V., de Wit, M.J., 2000. Making ends meet in Gondwana: retracing the transforms of the Indian Ocean and reconnecting continental shear zones. Terra Nova 12 (6), 272–280.

Reeves, C.V., Sahu, B.K., de Wit, M.J., 2002. A re-examination of the paleo-position of Africa's eastern neighbours in Gondwana. J. Afr. Earth Sci. 34, 101–108.

Reeves, C.V., de Wit, M.J., Sahu, B.K., 2004. Tight reassembly of Gondwana exposes Phanerozoic shears in Africa as global tectonic players. Gondwana Res. 7, 7–19.

Richards, M.A., Duncan, R.A., Courtillot, V.E., 1989. Flood basalts and hot spot tracks: plume heads and tails. Science 246, 103–107.

Richards, M., Bowman, M., Reading, H., 1998. Submarine-fan systems I: characterization and stratigraphic prediction. Mar. Pet. Geol. 15, 689–717.

Richardson, A.R., Solomon, R.M., Sleep, S.C., 1979. Tectonic stress in the plates. Rev. Geophys. Space Phys. 17, 981–1019.

Robb, M.S., Taylor, B., Goodliffe, A.M., 2005. Re-examination of the magnetic lineations of the Gascoyne and Cuvier Abyssal Plains, off NW Australia. Geophys. J. Int. 163, 42–55.

Roberts, G., Rutherford, K., O'Brien, C., 2008. Observations on the petroleum potential of deep offshore west coast India from newly reprocessed 2D seismic data. First Break 26, 77–86 (September).

Rodger, M., Watts, A.B., Greenroyd, C.J., Peirce, C., Hobbs, R.W., 2006. Evidence for unusually thin oceanic crust and strong mantle beneath the Amazon Fan. Geology 34, 1081–1084.

Rodolfo, K.S., 1969. Bathymetry and marine geology of the Andaman basin and its tectonic implications for SE Asia. Am. Assoc. Pet. Geol. Bull. 52, 2422–2437.

Roeser, H.A., Fritsch, J., Hinz, K., 1996. The development of the crust off Dronning Maud Land, East Antarctica. In: Storey, B.C., King, E.C., Livermore, R.A. (Eds.), Weddell Sea Tectonics and Gondwana Break-Up, vol. 108. Geological Society, London, pp. 243–264 Special Publication.

Roots, W.D., Veevers, J.J., Clowes, D.F., 1979. Lithospheric model with thick oceanic crust at the continental boundary: a mechanism for shallow spreading ridges in young oceans. Earth Planet. Sci. Lett. 43, 417–433.

Rotstein, Y., Munschy, M., Bernard, A., 2001. The Kerguelen Province revisited: additional constraints on the early development of the South Indian Ocean. Mar. Geophy. Res. 22, 81–100.

Rowley, D.B., 1996. Age of initiation of collision between India and Asia: a review of stratigraphic data. Earth Planet. Sci. Lett. 145, 1–13.

Rowley, D.B., Sahagian, D.L., 1986. Depth-dependent stretching: a different approach. Geology 14, 32–35.

Roy, T.K., 1983. Geology and hydrocarbon prospects of Andaman–Nicobar basin. Pet. Asia J. 37–50.

Roy, S.K., 1992. Accretionary prism in Andaman fore arc. Geological Survey of India, Special Publication 29, 273–278.

Roy, S.K., Das Sharma, S., 1993. Evolution of Andaman forearc basin and its hydrocarbon potential. In: Biswas, S.K. (Ed.), Petroliferous Basins of India. Indian Petroleum Publishers, Dehradun, India, pp. 407–428.

Roy, A., Pal, T., Bhattacharya, A., Ghosh, B., 2008. Emplacement of lithotectonic units of accretionary prism and forearc setting in subduction complex—a geodynamic approach in Andaman Islands. Int. Assoc. Gondwana Res. Conf. Ser. 5, 230–231.

Royden, L., Keen, C.E., 1980. Rifting process and thermal evolution of the continental margin of eastern Canada determined from subsidence curves. Earth Planet. Sci. Lett. 51, 343–361.

Royer, J.-Y., Chang, T., 1991. Evidence for relative motions between the Indian and Australian Plates during the last 20 m.y. from plate tectonic reconstructions: implications for the deformation of the Indo-Australian Plate. J. Geophys. Res. 96 (B11), 11779–11802.

Royer, J.-Y., Coffin, M.F., 1992. Jurassic to Recent plate tectonic reconstructions in the Kerguelen Plateau region. Proc. Ocean Drill. Program Sci. Results 120, 917–928.

Royer, J.-Y., Sandwell, D.T., 1989. Evolution of the eastern Indian ocean since the late cretaceous: constraints from GEOSAT altimetry. J. Geophys. Res. 94, 13755–13782.

Royer, J.-Y., Sclater, J.G., Sandwell, D.T., 1989. A preliminary tectonic fabric chart of the Indian Ocean. Proc. Ind. Acad. Sci. (Earth Planet. Sci.) 98, 7–24.

Royer, J.-Y., Peirce, J.W., Weissel, J.K., 1991. Tectonic constraints on hotspot formation of the Ninetyeast Ridge. In: Peirce, J., Weissel, J. et al., (Eds.), Proceedings of the Ocean Drilling Program, Scientific Results, vol. 121. Ocean Drilling Program, College Station, TX.

Royer, J.Y., Chaubey, A.K., Dyment, J., Bhattacharya, G.C., Srinivas, K., Yatheesh, V., Ramprasad, T., 2002. Paleogene plate tectonic evolution of the Arabian and Eastern Somali basins. In: Clift, P.D., Croon, D., Gaedicke, C., Craig, J. (Eds.), The Tectonic and Climatic Evolution of the Arabian Sea Region. Geological Society, London, pp. 7–23 Special Publications, vol. 195.

Rust, R.J., Summerfield, M.A., 1990. Isopach and borehole data as indicators of rifted margin evolution in southwestern Africa. Mar. Pet. Geol. 7, 277–287.

Sager, W.W., Paul, C.F., Krishna, K.S., Pringle, M., Eisin, A.E., Frey, F.A., Gopala Rao, D., Levchenko, O., 2010. Large fault fabric of the Ninetyeast Ridge implies near-spreading ridge formation. Geophys. Res. Lett. 37, L17304. doi:10.1029/2010GL044347.

Sahni, A., Kumar, V., 1974. Paleogene paleobiogeography of the Indian subcontinent. Palaeogeogr. Palaeoclimatol. Palaeoecol. 15, 209–226.

Sahu, J.N., 2005. Krishna Godavari offshore basin emerging to be a world class petroleum province. DEW J. 28–34 July Issue.

Sahu, J.N., 2008. Hydrocarbon Potential and Exploration Strategy of Cauvery Basin, India. Technology Publications, Dehradun, pp. 314.

Sahu, J.N., 2010. Hydrocarbon Exploration Opportunities in Krishna Godavari Basin, India. Technology Publications, Dehradun, pp. 190.

Sahu, J.N., Zutshi, P.L., Shukla, S.D., 1995. Seismic sequence analysis of Tanjore subbasin, Cauvery basin. India Pro. Petrotech-95, 9–12 Jan, New Delhi, Technology Trend in Petroleum Industry, vol.1, pp. 257–262.

Sain, K., Zelt, C.A., Reddy, P.R., 2002. Imaging of subvolcanic Mesozoics using traveltime inversion of wide-angle seismic data in the Saurashtra peninsula of India. Geophys. J. Int. 150, 820–826.

Sain, K., Gupta, H.K., 2012. Gas hydrates in India: potential and development. Gondwana Res. 22, 645–657.

Sain, K., Rajesh, V., Satyavani, N., Subbarao, K.V., Subrahmanyam, C., 2011. Gas hydrates stability thickness map along the Indian continental margin. Mar. Pet. Geol. 28, 1779–1786.

Saki, T., Tamura, Y., Tokuhashi, S., Kodato, T., Mizukoshi, I., Amano, H., 1987. Preliminary report of geological and geophysical surveys of Queen Maud Land, east Antarctica. Proceedings of the NIPR Symposium Antarctic Geoscience, vol. 1. 23–40.

Sandwell, D.T., 1992. Antarctic marine gravity field from high-density satellite altimetry. Geophys. J. Int. 109, 437–448.

Sandwell, D.T., Smith, W.H.F., 1997. Marine Gravity anomaly from GEOSAT and ERS-1 Satellite Altimetry. J. Geophys. Res. 102, 10039–10054.

Sar, D., Maheshwari, M.K., Rangarajan, S., Bahuguna, C.S., 2009. Eighty five degree east ridge and its hydrocarbon potential. Geohorizons July, 15–23.

Sarkar, A., Datta, A.K., Poddar, B.C., Bhattacharya, B.K., Kollapuri, V.K., Sanwal, R., 1996. Geochronological studies of Mesozoic igneous rocks from eastern India. J. Southeast Asian Earth Sci. 13, 77–81.

Sarkar, D., Chandrakala, K., Padmavathi Devi, P., Sridhar, A.R., Sain, K., Reddy, P.R., 2001. Crustal velocity structure of western Dharwar craton, South India. J. Geodyn. 31, 227–241.

Sarma, K.V.L.N.S., Subrahmanyam, V., Subrahmanyam, A.S., Murty, G.P.S., Murthy, K.S.R., 2009. Significance of Gundlakamma river (Krishna Basin) fault over Eastern Continental Margin of India—a qualitative appraisa. Curr. Sci. 98, 1438–1439.

Sastri, V.V., Sinha, R.N., Singh, G., Murti, K.V.S., 1973. Stratigraphy and Tectonics of sedimentary basins on east coast of Peninsular India. AAPG Bull. 57, 655–678.

Sastri, V.V., Raju, A.T.R., Sinha, R.N., Venkatachala, B.S., 1974. Evolution of the Mesozoic sedimentary basins on the east coast of India. APEA J. 14, 29–41.

Sastri, V.V., Venkatachala, B.S., Narayanan, V., 1981. The evolution of the east coast of India. Palaeogeogr. Palaeoclimatol. Palaeoecol. 36, 23–54.

Satpal, O.P., Singh, D., Chaterjee, S.M., Sawai, S., 2006. Integrated interpretation for sub-basalt imaging in Saurashtra Basin, India. Lead Edge Mag. 882–885 July Issue.

Satyavani, N., Sain, K., Lall, M., Kumar, B.J.P., 2008. Seismic attribute study for gas hydrates in the Andaman offshore, India. Mar. Geophys. Res. 29, 167–175.

Savostin, L.A., Kerusov, I.N., (Eds.), 2002. Mumbai deep offshore region and Konkan Kerala deep offshore region. Reports by LARGE (Laboratory of Regional Geodynamics) for the DGH.

Saxena, K., Tyagi, A., Klimentos, T., Morriss, C., Mathew, A., 2006. Evaluating Thin Bedded Reservoirs with the Rt Scanner. Petromin, Kuala Lumpur.

Sayers, C.M., Johnson, G.M., Schlumberger, Denyer, G., 2002. Pore pressure prediction from Seismic tomography. Offshore Technology Conference. doi: 10.2118/59122-MS.

Schlich, R., 1975. Structure et age de l'ocean Indien occidental, Mem. Hors. Ser. Soc. Geol. Fr, 6, 103 pp.

Schlich, R., 1982. The Indian Ocean: aseismic ridges, spreading centers and ocean basins. In: Nairn, A.E., Stehli, F.G. (Eds.), The Ocean Basins and Margins, vol. 6. Plenum, New York, pp. 51–147.

Schlich, R., Wise, S.W., et al. 1989. Proc.ODP, Init. Repts, 120. Ocean Drilling Program, College Station, TX.

Schwenk, T., Spieß, V., 2009. Architecture and stratigraphy of the Bengal Fan as response to tectonic and climate revealed from high-resolution seismic data. In: Kneller, B.C., Martinsen, O. J., McCaffrey, B. (Eds.), External Controls on Deep-Water Depositional Systems, 107–132 SEPM Special Publication 92.

Schwenk, T., Spieß, V., Hübscher, C., Breitzke, M., 2003. Frequent channel avulsions within the active channel–levee system of the middle Bengal Fan—an exceptional channel–levee development derived from parasound and hydrosweep data. DeepSea Res. II 50 (5), 1023–1045.

Schwenk, T., Spieß, V., Breitzke, M., Hübscher, C., 2005. The architecture and evolution of the Middle Bengal Fan in vicinity of the active channel–levee system imaged by high-resolution seismic data. Mar. Pet. Geol. 22, 637–656.

Sclater, J.G., Fisher, R.L., 1974. Evolution of the east central Indian Ocean, with emphasis on the tectonic setting of the Ninetyeast Ridge. Geol. Soc. Am. Bull. 85, 683–702.

Sclater, J.G., Ritter, U., Hilton, W., Meinke, L., Fisher, R.L., 1971. Charts of residual magnetic field plotted along track for the Indian Ocean. Scripps Institution of Oceanography, La Jolla, CA, 40 pp (Ref. 71–77).

Sclater, J.G., Bowin, C., Hey, R., Hoskins, H., Peirce, J., Phillips, J., Tapscott, C., 1976a. The Bouvet triple junction. J. Geophys. Res. 81, 1857–1869.

Sclater, J.G., Luyendyk, B.P., Meinke, L., 1976b. Magnetic lineations in the southern part of the Central Indian Basin. Geol. Soc. Am. Bull. 87, 371–378.

Sclater, J.G., Fisher, R.L., Patriat, P., Tapscott, C., Parsons, B., 1981. Eocene to recent development of the Southwest Indian Ridge, a consequence of the evolution of the Indian Ocean Triple Junction. Geophys. J. R. Astron. Soc. 64, 587–604.

Scrutton, R.A., 1978. Davie fracture zone and the movement of Madagascar. Earth Planet. Sci. Lett. 39, 84–88.

Scrutton, R.A., Heptonstall, W.B., Peacock, J.H., 1981. Constraints on the motion of Madagascar with respect to Africa. Mar. Geol. 43, 1–20.

Segoufin, J., 1978. Anomalies magnetiques mesozoiques dans le bassin de Mozambique (Mesozoic magnetic anomalies in the Mozambique Basin). Comptes Rendus Hebdomadaires des Seances de l'Academie des Sciences 287, 109–112.

Segoufin, J., Patriot, P., 1980. Existence d'anomalies mésozoiques dans le bassin de Somalie, implications pour les rélations Afrique-Antarctique-Madagascar. C.R. Acad. Sci. Paris Sér. B291, 85–88.

Sengor, A.M.C., Burke, K., 1978. Relative timing of rifting and volcanism on Earth and its tectonic implications. Geophys. Res. Lett. 5, 419–421.

Shanmugam, G., 2003. Deep-marine tidal bottom currents and their reworked sands in modern and ancient submarine canyons. Mar. Pet. Geol. 20, 471–491.

Shanmugam, G., 2006. Deep-Water Processes and Facies Models: Implications for Sandstone Petroleum Reservoirs. Elsevier, Amsterdam, 476 p.

Shanmugam, G., Moiola, R.J., 1988. Submarine fans: characteristics, models, classification, and reservoir potential. Earth Sci. Rev. 24, 383–428.

Sheena, V.D., Radhakishna, M., Subrahmanyam, C., 2007. Estimates of effective elastic thickness along the Southwest Continental Margin of India using Coherence analysis of gravity and bathymetry data—geodynamic implications. J. Geol. Soc. India 70, 475–487.

Sheth, H.C., 1998. A reappraisal of the coastal Panvel flexure, Deccan Traps, as a listric-fault-controlled reverse drag structure. Tectonophysics 294, 143–149. doi:10.1016/S0040-1951(98)00148-6.

Sheth, H.C., 2005. From Deccan to Réunion: no trace of a mantle plume. In: Foulger, G.R., Natland, J.H., Presnall, D.C., Anderson, D.L. (Eds.), Plates, Plumes, and Paradigms, 477–501 Geological Society of America, Special Paper, 388, chapter 29.

Sheth, H.C., Ray, J.S., Bhutani, R., Kumar, A., Awasthi, N., 2010. The latest (2008–09) eruption of Barren Island volcano, and some thoughts on its hazards, logistics and geotourism aspects. Curr. Sci. 98, 620–626.

Shirley, K., 2002 (October). Global Depths Have Great Potential. AAPG Explorer 23, pp. 16–17.

Shurr, G.W., Ridgley, J.L., 2002. Unconventional shallow biogenic gas systems. Am. Assoc. Pet. Geol. Bull. 86, 1939–1969.

Simkin, T., Siebert, L., 1994. Volcanoes of the World. Geoscience, Tucson, 249 pp.

Simpson, E.S.W., et al., 1974. Site 249. In: Initial Reports of the Deep Sea Drilling Project, vol. 25. U.S. Government Printing Office, Washington, DC, pp. 287–346.

Simpson, E.S.W., Sclater, J.G., Parsons, B., Norton, I.O., Meinke, L., 1979. Mesozoic magnetic lineations in the Mozambique Basin. Earth Planet. Sci. Lett. 43, 260–264.

Singh, A.P., 1999. The deep crustal accretion beneath the Laxmi ridge in the northeastern Arabian Sea: the plume model again. J. Geodynamics 27, 609–622.

Singh, B., Arora, K., 2008. Geophysical exploration for petroleum in the subtrappean Mesozoic sedimentary formations of India. Mem. Geol. Soc. India 68, 237–258.

Singh, N.K., Lal, N.K., 1993. Geology and petroleum prospects of Konkan-Kerala Basin. In: Proceedings of the 2nd Seminar on Petroleum Basins of India, vol. 2. KDMIPE and ONGC, India Petroleum Publishers, Dehradun, India, pp. 461–469.

Singh, D., Alat, C.A., Singh, R.N., Gupta, V.P., 1997. Source rock characteristics and hydrocarbon generating potential of Mesozoic sediments in Lodhika Area, Saurashtra Basin, Gujarat, India. In: Proceedings of the 2nd International Petroleum Conference and Exploration Petrotech-97, New Delhi, India 205–220.

Singh, S., Jain, A.K., Barby, M.E., 2007. SHRIMP U-Pb ~1860 Ma anorogenic magmatic signatures from NW Himalaya: implications for palaeoproterozoic assembly. In: Palaeoproterozoic Supercontinent and Global Evolution. Geological Society, London.

Singh, S.C., Carton, H., Chauhan, A., Androvandi, S., Davaille, A., Dyment, J., 2008. Evidence for the formation of an extremely thin crust due to Plume–Ridge interaction in the Indian Ocean (Ab). In: American Geophysical Union, Fall Meeting T53G.015.

Sinha, M.C., Louden, K.E., Parsons, B., 1981. The crustal structure of the Madagascar Ridge. Geophys. J. R. Astron. Soc. 66, 351–377.

Smith, A.G., Hallam, A., 1970. The fit of the southern continents. Nature 225, 139–144.

Smith, W.H.F., Sandwell, D.T., 1997. Global sea floor topography from satellite altimetry and ship depth soundings. Science 277, 1956–1962.

Solli, K., Kuvaas, B., Kristoffersen, Y.K., Leitchenkov, G., Guseva, J., Gandjukhin, V., 2007. Seismic morphology and distribution of inferred glaciomarine deposits along the East Antarctic continental margin, 20°E–60°E. Mar. Geol. 237, 207–223.

Soman, K., 1997. Geology of Kerala, Geological Society of India, pp.280.

Somen, M., Verma, R., Kenneth, D., Banerjee, S., Bastia, R., Nathaniel, D.M., 2011. Sub-basalt hydrocarbon prospectivity in Kerala-Konkan offshore basin, India : A basin modeling approach, 2nd South Asian Geoscience conference and exhibition. GEO India 2011, New Delhi, pp. 1–5, 12-14th January 2011.

Spohn, T., Schubert, G., 1982. Modes of mantle convection and the Removal of heat from the Earth's interior. J. Geophys. Res. 87 (Nb6), 4682–4696.

Sreejith, K.M., Krishna, K.S., Bansal, A.R., 2008. Structure and isostatic compensation of the Comorin Ridge, north Central Indian Ocean. Geophys. J. Int. 175, 729–741.

Sreejith, K.M., Radhakrishna, M., Krishna, K.S., Majumdar, T.J., 2011. Development of the negative gravity anomaly of the 85°E Ridge, northeastern Indian Ocean—a process oriented modeling approach. J. Earth Syst. Sci. 120, 605–615.

Sriram, K., Gupte, S.S., Kothari, V., Bisen, M., Waraich, R.S., 2006. Structure and evolution of Saurashtra Arch in Kutch-Saurashtra Deepwater Area, Western India. 6th International Conference & Exposition on Petroleum Geophysics, Kolkata 21–25.

Stagg, H.M.J., Colwel, J.B., Direen, N.G., O'Brien, P.E., Bernardel, G., Borissova, I., Brown, B.J., Ishihara, T., 2004. Geology of the continental margin of Enderby and Mac. Robertson Lands, East Antarctica: insights form a regional data set. Mar. Geophys. Res. 25, 183–218.

Stagg, H.M.J., Colwell, J.B., Direen, N.G., O'Brien, P.E., Brown, B.J., Bernardel, G., Borissova, I., Carson, L., Close, D.B., 2005. Geological framework of the continental margin in the region of the Australian Antarctic Territory. Geoscience Australia Record 2004/25, 356 pp (CD-ROM).

Steckler, M.S., Akhter, S.H., Seeber, L., 2008. Collision of the Ganges–Brahmaputra Delta with the Burma Arc: implications for earthquake hazard. Earth Planet. Sci. Lett. 273, 367–378.

Stewart, S.A., Davies, R.J., 2006. Structure and emplacement of mud volcano systems in the South Caspian Basin. AAPG Bull. 90, 71–786.

Storey, B.C., 1995. The role of mantle plumes in continental breakup: case histories from Gondwanaland. Nature 377, 301–308.

Storey, B.C., Alabaster, T., Hole, M.J., Pankhurst, R.J., Wever, H.E., 1992. Role of subduction-plate boundary forces during the initial stages of Gondwana break-up: evidence from the proto-Pacific margin of Gondwana. In: Storey, B.C., Alabaster, T., Pankhurst, R.J. (Eds.), Magmatism and the Causes of Continental Break-up. Geological Society, London, Special Publications, 68, pp. 149–163.

Storey, M., Mahoney, J.J., Saunders, A.D., Duncan, R.A., Kelley, S.P., Coffin, M.F., 1995. Timing of hot spot-related volcanism and the breakup of Madagascar and India. Science 267, 852–855.

Stow, D.A.V., et al., 1985. Sedimentary, tectonic, and sea-level controls. In: Bouma, A.H. et al., (Ed.), Submarine Fans and Related Turbidite Systems. Springer-Verlag, New York, pp. 15–22.

Stow, D.A.V., Amano, K., Balson, P.S., et al., 1990. Sediment facies and processes on the distal Bengal fan. In: Cochran, J.R., Stow, D.A.V. (Eds.), Proceedings of the Ocean Drilling Program Leg 116, Scientific Results. Ocean Drilling Program, College Station, TX, pp. 377–396.

Subba Rao, K.V., Sukheswala, R.N., 1981. Deccan volcanism and related basalt provinces in other parts of the world. Geological Society of India, 474 pp (Memoir, vol. 3).

Subrahmanya, K.R., 1998. Tectono-magmatic evolution of the West Coast of India. Gondwana Res. 1 (3/4), 319–327.

Subrahmanyam, C., Chand, S., 2006. Evolution of the passive continental margins of India—a geophysical appraisal. Gondwana Res. 10, 167–178.

Subrahmanyam, C., Singh, R.N., 1992. Geotectonics of the Bay of Bengal. Indian J. Pet. Geol. 1, 161–180.

Subrahmanyam, V., Ramana, M.V., Gopala Rao, D., 1993. Reactivation of Precambrian faults on the southwestern continental margin of India: evidence from gravity anomalies. Tectonophysics 219, 327–339.

Subrahmanyam, V., Gangadhara Rao, M., Subba Raju, L.V., 1993a. Subsurface Pre-Cambrian ridge on the continental shelf off western India between Coondapoor and Kasaragod. Mar. Geol. 112, 329–341.

Subrahmanyam, V., Ramana, M.V., Gopala Rao, D., 1993b. Reactivation of Precambrian faults on the southwestern continental margin of India : evidence from gravity anomalies. Tectonophysics 219, 327–339.

Subrahmanyam, V., Gopala Rao, D., Ramana, M.V., Krishna, K.S., Murty, G.P.S., Rao, M.G., 1995. Structure and tectonics of the southwestern continental margin of India. Tectonophysics 249, 267–282.

Subrahmanyam, C., Reddi, S.I., Thakur, N.K., Rao, T.G., Sain, K., 1998. Gas-hydrates: a synoptic view. J. Geol. Soc. India. 52 (5), 497–512.

Subrahmanyam, C., Thakur, N.K., Gangadhara Rao, T., Ramana, M.V., Subrahmanyam, V., 1999. Tectonics of the Bay of Bengal, northeastern Indian Ocean: new insights from satellite derived gravity and shipborne geophysical data. Earth Planet. Sci. Lett. 171, 237–251.

Subrahmanyam, V., Subrahmanyam, A.S., Murthy, K.S.R., Murty, G.P.S., Sarma, K.V.L.N.S., Suneetha Rani, P., Anuradha, A., 2006. Precambrian mega lineaments across the Indian sub-continent—preliminary evidence from offshore magnetic data. Curr. Sci. 90, 578–580.

Subrahmanyam, C., Gireesh, R., Chand, S., Kamesh Raju, K.A., Gopala Rao, D., 2008a. Geophysical characteristics of the Ninety East Ridge–Andaman Island arc/trench convergent zone. Earth Planet. Sci. Lett. 266, 29–45. doi:10.1016/j.epsl. 2007.10.016.

Subrahmanyam, V., Subrahmanyam, A.S., Murty, G.P.S., Murthy, K.S.R., 2008b. Morphology and tectonics of Mahanadi Basin, northeastern continental margin of India from geophysical studies. Mar. Geol. 253, 63–72.

Subrahmanyam, A.S., Subrahmanyam, V., Sarma, K.V.L.N.S., Murty, G.P.S., Murthy, K.S.R., 2010. Land–ocean tectonic signatures of the Krishna–Godavari Basin, a qualitative evaluation. Curr. Sci. 98 (4), 485–487.

Sukheshwala, R.N., Poldervaart, A., 1958. Deccan basalts of the Bombay area, India. Bull. Geol. Soc. Am. 69, 1475–1495.

Swarbrick, R.E., Osborne, M.J., Yardley, G.S., 2002. Comparison of overpressure magnitude resulting from the main generating mechanisms. In: Huffman, A.R., Bowers, G.L. (Eds.), Pressure Regimes in Sedimentary Basins and Their Prediction, AAPG Memoir, vol. 76. 1–12.

Talukdar, S.N., 1982. Geology and hydrocarbon prospects of east coast basins of India and their relationships to evolution of the Bay of Bengal. Unpublished paper distributed at Offshore South East Asia 82 Conference, Singapore.

Talwani, M., Reif, C., 1998. Laxmi ridge—a continental sliver in the Arabian Sea. Mar. Geophys. Res. 20, 259–271.

Tarling, D.H., Kent, P.E., 1976. The Madagascar controversy still lives. Nature 261, 304–305.

Terzaghi, K., 1943. Theoretical Soil Mechanics. John Wiley and Sons, Inc., New York.

Thakur, N.K., Mahashabde, M.V., Arora, B.R., Singh, B.P., Srivastava, B.J., Prasad, S.N., 1986. Geomagnetic variation anomalies in Peninsular India. Geophys. J. R. Astron. Soc. 86, 839–854.

Thakur, S.S., Arasu, R.T., Subrahmanyam, V.S.R., Srivastava, A.K., Murthy, A.V.S., 1999. Basin configuration in Konkan deep waters, Western Indian offshore. J. Geol. Soc. India 53, 79–88.

Thakur, S., Tyagi, A.K., Anish, T.K., Saxena, K., 2010. An Integrated Petrophysical Study for deep water turbidite reservoir to improve Hydrocarbon Saturation, 8th Biennial International conference and Exposition on Petroleum Geophysics, SPG Hyderabad, 1–7 P-419.

Thomas, N.J., Sharma, V.N., 1993. Thermal evolution of source rocks in Cauvery Basin. In: Biswas, S.K. (Ed.), Proceedings of the 2nd Seminar on Petroliferous Basins of India, vol. 1. 245–254.

Thomas, E.C., Stiber, S.J., 1975. The distribution of shale in sandstone and its effects upon porosity. In: 16th Annual Logging Symposium, SPWLA, 4–7 June.

Tingay, M.R.P., Hillis, R.R., Swarbrick, R.E., Morley, C.K., Damit, A.R., 2009. Origin of overpressure and pore-pressure prediction in the Baram province, Brunei. AAPG Bull. 93 (1), 51–74.

Todal, A., Eldholm, O., 1998. Continental margin off western India and Deccan Large Igneous Province. Mar. Geophys. Res. 20, 273–291.

Trehu, A.M., Ballard, A., Dorman, L.M., Gettrust, J.F., Klitgord, K.D., Schreider, A., 1989. Structure of the lower crust beneath the Carolina trough, US Atlantic continental margin, J. Geophys. Res. 94B, 10585–10600.

Tyagi, A., Bastia, R., Das, M., 2008a. Evaluating Hydrocarbon Potential of Thin-Bedded Reservoirs. Petromin, pp. 26–35.

Tyagi, A.K., Bastia, R., Das, M., 2008b. Identification and Evaluation of the thin bedded reservoir potential in the East Coast Deep Water Basins of India. 7th Biennial International Conference and Exposition on Petroleum Geophysics, SPG Hyderabad 2008 1–7 P-317.

Tyagi, A., Bastia, R., Das, M., 2009a. Potential evaluation of thin bedded sand shale sequences in deep water settings. J. Indian Assoc. Sedimentol. 28 (1), 9–16.

Tyagi, A.K., Bastia, R., Das, S., 2009b. Petrophysical analysis of laminated sand shale sequence in deepwater setting. AAPG Conference, RIO, Brazil November 1–6.

Tyagi, A.K., Guha, R., Deepak, V., Saxena, K., 2009c. Challenges in the reservoir characterization of a laminated sand shale sequence. Second SPLWA Conference, Mumbai, November 1–8.

Uchupi, E., Swift, S.A., Ross, D.A., 2002. Tectonic geomorphology of the Gulf of Oman Basin. In: Clift, P.D., Kroon, D., Gaedicke, C., Craig, J. (Eds.), The Tectonic and Climatic Evolution of the Arabian Sea Region 195, Geol. Soc. Lond. Special Publications, pp. 37–69.

Uddin, A., Lundberg, N., 1998. Cenozoic history of the Himalayan–Bengal system: sand composition in the Bengal basin, Bangladesh. Geological Society of America Bulletin 110, 497–511.

Uyeda, S., Kanamori, H., 1979. Back arc opening and the mode of subduction. J. Geophys. Res. 84, 1049–1061.

Vail, P.R., Mitchum Jr., R.M., Todd, R.G., Widmier, J.M., Thompson, S.I., Sangree, J.B., Bubb, J.N., Hatlelid, W.G., 1977. Seismic stratigraphy and global changes of sea level. In: Seismic Stratigraphy-Applications to Hydrocarbon Exploration. American Association of Petroleum Geologists, Tulsa, OK, pp. 49–212.

Valdiya, K.S., 1984. Evolution of the Himalaya. Tectonophysics 105, 229–248.

Valdiya, K.S., 2010. The Making of India-Geodynamics Evolution. Macmillan India Limited, New Delhi, pp. 816.

Vandamme, D., Courtillot, V., Besse, J., Montigny, R., 1991. Palaeomagnetism and age determination of the Deccan traps (India) results of Nagpur–Bombay traverse and review of earlier work. Rev. Geophys. 29, 259–290.

Vandenbroucke, M., Behar, F., Rudkiewicz, J.L., 1999. Kinetic modelling of petroleum formation and cracking: implications from the high pressure/high temperature Elgin Field (UK, North Sea). Org. Geochem. 30, 1105–1125.

Veevers, J.J., 2009. Palinspastic (pre-rift and -drift) fit of India and conjugate Antarctica and geological connections across the suture. Gondwana Res. 16, 90–108.

Venkatachala, B.S., Rawat, M.S., 1972. Palynology of the Tertiary sediments in the Cauvery Basin Palaeocene-Eocene palynoflora from the sub-surface. In: Ghosh, A.K. et al., (Ed.), Seminar on Palaeopalynology and Indian Stratigraphy 292–335.

Venkatarengan, R., 1987. Depositional systems and thrust areas of exploration. Bull. OGC 24 (1), 56–69.

Venkatarengan, R., Ray, D., 1993. Geology and petroleum systems, Krishna-Godavari basins. In: Biswas, S.K., Alak, D., Garg, P., Pandey, J., Maithani, A., Thomas, N.J. (Eds.), Proceedings of the 2nd Seminar on Petroliferous Basins of India, vol. 1. Indian Petroleum Publishers, Dehradun, India, pp. 331–354.

Venkatarengan, R., Prabhakar, K.N., Singh, D.N., Awasthi, A.K., Misra, P.K., Reddy, P.K., Roy, S.K., 1993. Lithostratigraphy of Indian petroliferous basin. Document VII: Cauvery basin, ONGC document (Unpublished).

Venkatesan, T.R., Pande, K., Gopalan, K., 1993. Did Deccan volcanism pre-date the Cretaceous transition. Earth. Planet. Sci. Lett. 119, 181–190.

Verma, R.K., Mukhopadhyay, M., 1977. An analysis of the gravity field in northeastern India. Tectonophysics 42, 283–317.

Verma, R.K., Rao, S.C., Satyanarayana, Y., 1993. Gravity field and evolution of Krishna-Godavari and Cauvery basins of India. Indian J. Pet. Geol. 2 (1), 39–52.

Vijaya Rao, V., Sain, K., Reddy, P.R., Mooney, W.D., 2006. Crustal structure and tectonics of the northern part of the Southern Granulite Terrane. India. Earth Planet. Sci. Lett. 251, 90–103.

Von der Borch, C.C, Sclater, J.G, et al., 1978. Synthesis of Deep-Sea Drilling Results in the Indian Ocean. Mar. Geol 26, 1–175.

Von der Borch, C.C., 1978. Synthesis of Deep-Sea Drilling Results in the Indian Ocean. Mar. Geol 26, 1–175.

Von der Borch, C., Sclater, J.G, et al., 1974. Init. Repts. DSDP 22, 890 pp.

Von der Borch, C.C., Sclater, J.C., et al., 1974a. Initial Reports of the Deep Sea Drilling Project, Leg 22. US Government Printing Office, Washington, DC, 890 pp.

Von der Borch, C.C., Sclater, J.G., Shipboard Scientific Party, 1974b. Site 218. Init. Rep. Deep Sea Drill. Proj. 22, 325–348.

Von Rad, U., Tahir, M., 1997. Late Quaternary sedimentation on the outer Indus shelf and slope (Pakistan): evidence from high-resolution seismic data and coring. Mar. Geol. 138, 193–236.

Von Rad, U., Exon, N.F., Boyd, R., Haq, B.U., 1992. Mesozoic paleoenvironment of the rifted margin off NW Australia (ODP Legs 122/123). In: Duncan, R.A. et al., (Ed.), Synthesis of Results from Scientific Drilling in the Indian Ocean. Geophys. Monogr. Ser., vol. 70. AGU, Washington, DC, pp. 157–184. doi:10.1029/GM070p0157.

Watts, A.B., 1988. Gravity anomalies, crustal structure and flexure of the lithosphere at the Baltimore Canyon Trough. Earth Planet. Sci. Lett. 89, 221–238.

Watts, A.B., 2001. Isostasy and Flexure of the Lithosphere. Cambridge University Press, Cambridge, U.K., 472 pp.

Watts, A.B., Cox, K.G., 1989. The Deccan Traps: an interpretation in terms of progressive lithospheric flexure in response to a migrating load. Earth Planet. Sci. Lett. 93, 85–97.

Watts, A.B., Fairhead, J.D., 1999. A process-oriented approach to modeling the gravity signature of continental margins. Leading Edge 18, 258–263.

Watts, A.B., Marr, C., 1995. Gravity anomalies and the thermal and mechanical structure of rifted continental margins. In: Banda, E. (Ed.), Rifted Ocean-Continent Boundaries. Kluwer Academic Publishers, Norwell, MA, pp. 65–94.

Watts, A.B., Torné, M., 1992. Subsidence history, crustal structure and thermal evolution of the Valencia Trough: a young extensional basin in the western Mediterranean. J. Geophys. Res. 97 (20), 20 021-20,041.

Weber, M.E., Wiedicke-Hombach, M., Kudrass, H.R., Erlenkeuser, H., 1997. Active growth of the Bengal Fan during sea-level rise and highstand. Geology 25 (4), 315–318.

Weedon, G.P., McCave, I.N., 1991. Mud turbidites from the Oligocene and Miocene Indus Fan at sites 722 and 731 on the Owen Ridge. In: Prell, W.L., Niitsuma, N., Emeis, K. et al., (Eds.), Proceedings of the Ocean Drilling Program, Scientific Results, vol. 116. Ocean Drilling Program, College Station, TX, pp. 215–220.

Weeks, L.A., Harbison, R.A., Peter, G., 1967. Island arc system in Andaman sea. Am. Assoc. Pet. Geol. Bull. 51, 1803–1815.

Weis, D., Frey, F.A., 1991. Isotope geochemistry of Ninetyeast Ridge basement basalts: Sr, Nd, and Pb evidence for involvement of the Kerguelen hot spot. In: Weissel, J., Peirce, J., Taylor, E., Alt, J. et al., (Eds.), Proceedings of the Ocean Drilling Program, Scientific Results, vol. 121. Ocean Drilling Program, College Station, TX, pp. 591–610.

Weis, D., White, W.M., Frey, F.A., Duncan, B., Dehn, J., Fisk, M., Ludden, J., Saunders, A., Storey, M., 1992. The influence of mantle plumes in generation of Indian Ocean crust. In: Synthesis of Results from the Scientific Drilling in the Indian Ocean. Geophys. Monogr., vol. 70. AGU, Washington, DC, pp. 57–89.

Weis, D., Ingle, S., Damasceno, D., Frey, F.A., Nicolaysen, K., Barling, J., 2001. Origin of continental components in Indian Ocean basalts: evidence from Elan Bank (Kerguelen Plateau, ODP Leg 183, Site 1137). Geology 29, 147–150.

Weissel, J.K., Karner, G.D., 1989. Flexural uplift of rift flanks due to mechanical unloading of the lithosphere during extension. J. Geophys. Res. 94, 13919–13950.

Weissel, J., Peirce, J., Taylor, E., Alt, J., et al., 1991. Proc. ODP, Sci. Results 121, 990 pp.

White, R.S., 1982. Deformation of the Makran accretionary sediment prism in the Gulf of Oman (North-West Indian Ocean). In: Leggett, J.K. (Ed.), Trench-Forearc Geology: Sedimentation and Tectonics on Modern and Ancient Active plate Margins, vol. 10. 357–372 Geological Society Special Publication 10.

White, R.S., 1992. Magmatism during and after continental break up. In: Storey, B.C., Alabaster, T., Pankshurst, R.J. (Eds.), Magmatism and the Causes of Continental Break Up, Geol. Soc. Lond. (Special Publication), vol. 68. 1–16.

White, N., McKenzie, D., 1988. Formation of the "steer's head" geometry of sedimentary basins by differential stretching of the crust and mantle. Geology 16, 250–253.

White, R.S., McKenzie, D., 1989. Magmatism at rift zones: the generation of volcanic continental margins and flood basalts. J. Geophys. Res. 94, 7685–7729.

Whitmarsh, R.B., 1974. Some aspects of plate tectonics in the Arabian Sea. Init. Rep. DSDP 23, 527–535.

Whitmarsh, R.B., 1979. The Owen Basin off the southeast margin of Arabia and the evolution of the Owen Fracture Zone. Geophys. J. R. Astron. Soc. 58, 441–470.

Wiens, D.A., DeMets, C., Gordon, R.G., Stein, S., Argus, D., Engeln, J.F., Lundgren, P., Quible, D., Stein, C., Weinstein, S., Woods, D.F., 1985. A diffuse plate boundary model for Indian Ocean tectonics. Geophys. Res. Lett. 12, 429–432. doi:10.1029/GL012i007p00429.

Windley, B.F., Razafiniparany, A., Razakamanana, T., Ackermand, D., 1994. Tectonic framework of the Precambrian of Madagascar and its Gondwana connections. Geol. Rundschau 83, 642–659.

Worthington, P.F., 1997. Recognition and Development of Low-Resistivity Pay. In: SPE Asia Pacific Oil and Gas Conference and Exhibition 14–16 April 1997, Kuala Lumpur, Malaysia, MS-38035, 15 pp.

Wynn, R.B., Stow, D.A.V., 2002. Classification and characterisation of deep-water sediment waves. In: Wynn, R.B., Stow, D.A.V. (Eds.), Recognition and Interpretation of Deep-Water Sediment Waves: Implications for Palaeoceanography, Hydrocarbon Exploration and Flow Process Interpretation, Mar. Geol. (Special Issue), vol. 192. 7–22.

Yahaya-Joe, U., Okikiade, F., Ghosh, D.P., 2000. Geohazard Analysis in Deepwater Nigeria. Offshore Technology Conference, Houston. Paper no. OTC 11924, pp. 1–8.

Yatheesh, V., Bhattacharya, G.C., Mahender, K., 2006. The terrace like feature in the mid-continental slope region off Trivandrum and a plausible model for India Madagascar juxtaposition in immediate pre-drift scenario. Gondwana Res. 10, 179–185.

Yatheesh, V., Bhattacharya, G.C., Dyment, J., 2009. Early oceanic opening off Western India–Pakistan margin: the Gop Basin revisited. Earth Planet. Sci. Lett. 284 (2009), 399–408.

Yoshida, M., Funaki, M., Vintage, P.W., 1992. Proterozoic to Mesozoic East Gondwana. The Juxtaposition of India, Srilanka and Antarctica. Tectonics 11, 381–391.

Yoshida, M., Rajesh, H.M., Santosh, M., 1999. Juxtaposition of India and Madagascar: a perspective. Gondwana Res. 2, 449–462.

Index

Note: Page numbers followed by "*f*" indicate figures, and "*t*" indicate tables.

D

F

G

H

I

K

L

M

N

O

S

T

U

V

W